Monographs in Electrochemistry

Series Editor

Fritz Scholz, University of Greifswald, Greifswald, Germany

Surprisingly, a large number of important topics in electrochemistry are not covered by up-to-date monographs and series on the market, some topics are even not covered at all. The series "Monographs in Electrochemistry" fills this gap by publishing in-depth monographs written by experienced and distinguished electrochemists, covering both theory and applications. The focus is set on existing as well as emerging methods for researchers, engineers, and practitioners active in the many and often interdisciplinary fields, where electrochemistry plays a key role. These fields range – among others – from analytical and environmental sciences to sensors, materials sciences and biochemical research.

More information about this series at http://www.springer.com/series/7386

László Péter

Electrochemical Methods of Nanostructure Preparation

Springer

László Péter
Wigner Research Centre for Physics
Budapest, Hungary

ISSN 1865-1836 ISSN 1865-1844 (electronic)
Monographs in Electrochemistry
ISBN 978-3-030-69116-5 ISBN 978-3-030-69117-2 (eBook)
https://doi.org/10.1007/978-3-030-69117-2

This Springer imprint is published by the registered company Springer Nature Switzerland AG
The registered company address is: Gewerbestrasse 11, 6330 Cham, Switzerland

Preface of Series Editor

Current chemical and electrochemical literature abounds in titles referring to nano, and so one may ask, why another book dealing with nano structures? I hope that the readers of this monograph will agree when I claim that this book is something else, that it is really written to support the scientific community for a long time. Although the innovation rate in the area of nanostructures is very high both with respect to synthesis and characterization, there will remain a solid body of knowledge, which is presented in this monograph. It is a big advantage of this monograph that it is "from one mould", and not a collection of reviews from many different authors. The author, László Péter from the Wigner Research Centre for Physics in Budapest, Hungary, is a distinguished electrochemist with long experience in electrochemical plating and nano-structuring of surfaces. He is also a highly gifted lecturer and teacher, which allowed him to give this book a remarkably didactic character. This monograph attests to his encyclopaedic knowledge of the literature, which he is able to overview and to disseminate among a wide scientific audience, not only to specialists, but also to the uninitiated, not only to electrochemists, but also to any scientists interested in nano structures.

Greifswald, Germany
June 2020

Fritz Scholz

Contents

About the Author

László Péter is a scientific advisor at the Wigner Research Centre for Physics (Budapest, Hungary). He graduated as a teacher of physics and chemistry at the Eötvös University of Budapest in 1992 and obtained the Ph.D. degree at the same university in 1995. After spending 2 years in the USA as a postdoctoral fellow and 1 year in Japan as a visitor scientist, he returned to his home country and started working in the predecessor of the Wigner Research Centre. His main interest is electrochemistry; in particular, the formation of solid phases in electrochemical processes and the physical properties of electrodeposited materials. Beside electrochemistry, he deals with various fields of experimental physical chemistry and research aspects of industrial problems. His publication list includes more than 100 research papers and 2 book chapters. He is the founding secretary of the conference series called International Workshops on Electrodeposited Nanostructures (EDNANO). In 2013, he became the doctor of the Hungarian Academy of Sciences. From 2020, he is one of the topical editors of *Journal of Solid State Electrochemistry* (Springer).

About the Series Editor

Fritz Scholz is Professor at the University of Greifswald, Germany. Following studies of chemistry at Humboldt University, Berlin, he obtained a Dr. rer. nat. and a Dr. sc. nat. (habilitation) from that University. In 1987 and 1989 he worked with Alan Bond in Australia. His main interest is in electrochemistry and electroanalysis. He has published more than 300 scientific papers, and he is editor and co-author of the book "Electroanalytical Methods" (Springer, 2002, 2005, 2010, and Russian Edition: BINOM, 2006), coauthor of the book "Electrochemistry of Immobilized Particles and Droplets" (Springer 2005), co-editor of the "Electrochemical Dictionary" (Springer, 2008; 2nd ed. 2012), and co-editor of volumes 7a and 7b of the "Encyclopedia of Electrochemistry" (Wiley-VCH 2006). In 1997 he has founded the *Journal of Solid State Electrochemistry* (Springer) and serves as Editor-in-Chief since that time. In 2014 he has founded the journal *ChemTexts–The Textbook Journal* (Springer). He is the editor of the series "Monographs in Electrochemistry" (Springer) in which modern topics of electrochemistry are presented. Scholz introduced the technique 'Voltammetry of Immobilized Microparticles' for studying the electrochemistry of solid compounds and materials, he introduced three-phase electrodes to determine the Gibbs energies of ion transfer between immiscible liquids, and currently he is studying the interaction of free oxygen radicals with metal surfaces, as well as the interaction of liposomes with the surface of mercury electrodes in order to assess membrane properties. He is also interested in the history of science and co-editor of the English translation of Wilhelm Ostwald's autobiography (Springer

2017), and editor of the book “Electrochemistry in a Divided World. Innovations in Eastern Europe in the 20th Century” (Springer 2015).

Abbreviations

3DOM	Three-dimensionally ordered macroporous
AAO	Anodic aluminum oxide
AES	Auger electron spectroscopy
AFM	Atomic force microscopy
ALD	Atomic layer deposition
AMR	Anisotropic magnetoresistance
CMA	Compositionally modulated alloy
CMMC	Compositionally modulated multilayer coating
CMWP	Convolutional multiple whole profile fitting
CNT	Carbon nanotube
CVD	Chemical vapour deposition
D3PIE	Dynamic three-phase interline electropolymerization
DCT	Diblock copolymer template
DHBT	Dynamic hydrogen bubble template
DLS	Dynamic light scattering
DLVO (theory)	Derjaguin–Landau–Wervey–Overbeek (theory)
DMSO	Dimethyl sulfoxide
DO	Directionally oriented
DSC	Differential scanning calorimetry
DWCNT	Double-wall carbon nanotube
EBSD	Electron backscattered diffraction
ECP	Electrochemical printing
EC-ALD	Electrochemical atomic layer deposition
EC-FPN	Electrochemical fountain pen nanofabrication
EDS or EDX	Energy-dispersive X-ray spectrometry
EELS	Electron energy loss spectroscopy
EIS	Electrochemical impedance spectroscopy
ESED	Electrochemical step-edge decoration
EXAFS	Extended X-ray Absorption Fine Structure
FM	Ferromagnetic
FONGUE	Filamentary one-dimensional nanocrystal growth in an ultra-dilute electrolyte

GISAXS	Grazing incidence small-angle X-ray scattering
GIXRD	Grazing incidence X-ray diffraction
GMR	Giant magnetoresistance
HOPG	Highly oriented pyrolytic graphite
IR	Infrared
ITIES	Interface between two immiscible electrolyte solutions
ITO	Indium-tin-oxide
LECD	Localized electrochemical deposition
LEED	Low-energy electron diffraction
LN	Less noble
MCE	Magnetic composite electroplating
MFM	Magnetic force microscopy
MHD	Magnetohydrodynamic
ML	Monolayer (Chap. 4) *or* Multilayer (Chap. 5)[1]
MMC	Metal matrix composite
MN	More noble
MW	Microwave
MWCNT	Multiwall carbon nanotube
NM	Non-magnetic
PAA	Porous anodic alumina
PAS	Positron annihilation spectroscopy
PEO	Poly(ethylene oxide)
PMMA	Poly(methylmetacrylate)
PS	Polystyrene
PTFE	Poly(tetrafluorethylene)
PTRF-XAFS	Polarization-dependent total-reflection fluorescence X-ray absorption fine structure
QCM	Quartz crystal miocrobalance
RBS	Rutherford backscattering spectroscopy
RDE	Rotating disc electrode
RF	Radiofrequency
RRDE	Rotating ring-disc electrode
SAM	Self-assembled monolayer
SCC	Stress corrosion cracking
SEBALD	Selective electrodesorption-based atomic layer deposition
SECM	Scanning electrochemical microscopy
SEM	Scanning electron microscopy
SERS	Surface-enhanced Raman spectroscopy
SILAR	Successive ionic layer adsorption and reaction
SLRR	Surface-limited redox replacement
SNMS	Secondary neutral mass spectrometry
SPION	Superparamagnetic iron oxide nanoparticle

[1]The dual notation could not be resolved because the same abbreviation was strongly embedded into the literature of various fields.

SPL	Scanning probe lithography
SPM	Superparamagnetic
SQUID	Superconducting quantum interference device
STM	Scanning tunneling microscopy
SWCNT	Single-wall carbon nanotube
TBN	Tip-based nanomanufacturing
TEM	Transmission electron microscopy
THF	Tetrahydrofurane
ToF-SIMS	Time-of-flight secondary ion mass spectrometry
TUME	Tunneling ultramicroelectrode
UIC	Uniform injection cell
ULSI	Ultra-large-scale integration
UPD	Underpotential deposition
UPS	Ultraviolet photoelectron spectroscopy
UV	Ultraviolet
WDS	Wavelength-dispersive X-ray spectroscopy
XAFS	X-ray absorption fine structure
XANES	X-ray absorption near-edge structure
XPS	X-ray photoelectron spectroscopy
XRD	X-ray diffractometry
XRF	X-ray fluorescence spectroscopy
bcc	Body-centered cubic
fcc	Face-centered cubic
hcp	Hexagonal close-packed
nc	Nanocrystalline

Remarks on the Notation of the Alloys, Composites and Multilayers

A–B	Binary alloy of A and B with an undefined composition
A_xB_y	Binary alloy of A and B in the mole ratio indicated with x and y
A(B)	Composite in which the matrix of the material A contains inclusions (or precipitation) of the material B
A/B	Multilayered material with layers of A and B

Part I
Background

Chapter 1
Introduction

1.1 The "Nano" Era

It is really scarce that the start of a period of science can be kept track of so well as the turn towards nanoscale and atomic-level manipulation of materials. The launching talk of nanoscience is attributed to the famous physicist Richard Feynman, already a Nobel Prize laureate at that time, who gave voice to a really novel view in December 1959 at the California Institute of Technology in Pasadena, California, USA [1]. His speach is an excellent example of the blend of two things: the far-reaching vision of a great mind and the limitations of one's thought by the experience. While Feynman outlined the way of miniaturization and information storage at the small scale on the basis of lithography and information reading by electron microscopy, he also mentioned the atomic-scale manipulation of material as a tool to be created.

What Feynman meant as solid-state nanoscience (and what does not necessarily refers to completely independent chemical entities) was well established in the previous century during the establishment of colloid chemistry. As it was stated in a very early and carefully written paper of Thomas Graham [2], the "colloidal condition of matter" was distinguished from "crystalloids", mostly based on their transport and crystallization properties. Later, Wolfgang Ostwald (1907) introduced the term of "colloidal state" instead of colloidal matter, hence distinguishing the dispersity of the material from its chemical nature. The debate on the nature of colloidal systems (solution or suspension) was clearly decided by the discovery of the ultramicroscope by Richard Zsigmondy, which emerged him to among the Nobel laureates (1925). The ultramicroscope first used the scattered light for the analysis of a system and indeed "shed light" on the heterogeneous nature of colloidal solutions. Even though colloid science could slowly establish itself as an independent branch of chemistry, it has long been dealing with solution-based dispersed systems only, and its merge with modern solid-state nanoscience is still going on nowadays. The hard process of the general acceptance of colloid science was nicely summarized in 1955 by Hauser [3], a prominent researcher of the field, giving also the tribute to many peers.

L. Péter, *Electrochemical Methods of Nanostructure Preparation*,
Monographs in Electrochemistry, https://doi.org/10.1007/978-3-030-69117-2_1

Electroplating has long developed in parallel to and independently of either colloid chemistry or any other means of microscopic manipulation of materials. The first successful electroplating was carried out by Luigi Brugnatelli in 1805, only five years after Volta's publication of his electrical pile. Since Brugnatelli's work was denied to be published by the French Academy of Sciences, it took nearly three and a half decades while electroplating was rediscovered in parallel by Henry and George Elkington (in Britain) and by Moritz Hermann von Jacobi (in Russia), the latter becoming the author of the first electrodeposition monograph [4]. While electrodeposition was first used for decorative purposes, the second half of the nineteenth century gave rise to the establishment of the plating industry and electrorefining procedure of various metals (Wohlwill process).

Various concepts of electrochemical science and galvanotechnology could be seen already in Jacobi's work [4]. By the middle of the twentieth century, electrodeposition could be much embedded into the knowledge of chemical thermodynamics and kinetics. Monographs published at that time either for electrodeposition [5] or galvanotechnology [6] indicate the completion of the theoretical background. The theories concerning crystal growth were also applied to electrocrystallization by this time, mostly based on the efforts of Stranski and Kaishev, also improved later by Budevski and Milchev.

The first detailed monograph on electrodeposition of alloys was published only in 1963 [7]. Brenner's book was used for decades as the primary reference for the scientific literature on electrodeposition, and some key categories outlined therein are used in an essentially unchanged form still nowadays (i.e, codeposition modes of metals). Many important works published later [8–11] enriched the scientific and technological literature with bunch of new data, but they primarily served an industry that focused on the production of large-scale and non-structured coatings. It happened only in the twenty-first century that monographs on nanoscale electrochemistry [12–21] and electrochemistry-based nanofabrication [22–24] appeared, including fields also much beyond electrodeposition.

The list below offers a short overview on some major important milestones of the development of micro- and nanoscience with relevance in electrochemistry, although by far not all are considered as a part of pure electrochemistry.

1982 Scanning Tunneling Microscopy [25]
1985 Electron tunneling measurement through single-atom junctions [26]
1986 STM for solid–liquid interfaces [27]
1986 Atomic Force Microscopy [28]
1988 STM as applied for electrodes [29–34]
1989 Scanning Electrochemical Microscopy [35]
1990 Nanoelectrodes of 10 Angstrom [36]
1995 Electrolysis in porous oxide nano-templates [37, 38]
2001 Measurement of single-molecule conductivity [39]
2001 *In-situ* video-STM of electrode surfaces [40]
2003 *In-situ* TEM for electrochemical systems [41].

Although the vision of the atomic-scale manipulation of materials, including the molecular-level drive of the chemical reactions [1] is very appealing, the big majority of the electrochemical processes involving nanomaterials is based on classical electrochemical techniques and do not require nanoscale tools at all. This is why the understanding of both the formation of nanostructures and their electrochemical background should go along with each other.

As it was emphasized also by Feynman [1], nanostructuring is by far not only about miniaturization. Manufacturing a device in a miniature form can save material and energy, but nanostructuring often leads us to a field where the counterparts of the effects occurring cannot be found in the conventional macroscopic world. The increase of the efficiency of a nanostructured catalyst can be explained simply by the reduced transfer time of the intermediate from one active center to another, and the process is easily elucidated by a downscaling. However, it is rather unusual that the charging of a metal nanoparticle is accompanied by a step in the electrical potential, although this behavior is a direct consequence of the change of the density of the surface charge. Here, the quantized nature of the charge is of importance, while the description of the *potential—charge density* relationship can be fully classical. The mechanical behavior of a defect-free single crystal cannot be described by the same manner as the dislocation slipping and elongation is treated in the mechanics of macroscopic objects. Similarly, the magnetic behavior of a single-domain particle is different from a macroscopic multidomain magnet. The electrical resistivity of a metallic structure also changes as the characteristic length in the nanostructure becomes comparable to the mean free path of the electrons (quantum confinement). Magnetic objects in the close vicinity of each other may also exhibit a magnetic coupling interaction that is not known in any sense in the macroscopic world—this gives rise to the special magnetoresistance behavior of nanometric magnetic/non-magnetic multilayers. A nanostructured metal surface may interact with the light in a manner that has no parallel phenomenon on a plain metal surface and what can be explained by the relevant surface plasmon effects.

The above examples clearly show that "nanostructuring", if treated in the right way, always must have the thoughtful background. It is not enough to see that in the field of electrochemical publications the ratio of nanostructure-related papers grew from 20 to 50% in the decade between 2005 and 2015. Should this make nanostructuring a fashionable field, the works with high-quality and long-lasting impact will be those which can validate the necessity of nanostructuring without any autotelism.

1.2 The Concept of This Book

The family of nanostructures prepared by means of either a method fully relying on electrochemistry or an electrochemistry-assisted preparation procedure is very wide. Not only do these nanomaterials differ a lot in the fine details of the preparation methods and basic class of materials (like metal or ceramics), but the composition, the characteristic feature size and the possible application goals are all very diverse. Therefore, it is very difficult to find a systematic approach that is capable of giving home to all these materials with the least possible compromise. It is to be admitted that several categorization systems might be appropriate for the complete description of electrochemical nanomaterial preparation, depending on how the largest units of the scheme are selected. While each system can be logical in some means, the categorization always reflects to some extent the author's taste. At the same time, it is not sure that each single description system can serve the easy understanding of the audience. However, one of the prime points of view of the author of this book was to offer a didactically smooth approach to the field that can be understood also by beginners without compromising the pursuit that state-of-the-art information should be disseminated also for the advanced readers. This book structure definitely aims at eliminating the obvious disadvantage of edited books where the selection of the chapter topics is always somewhat arbitrary and the depth of the discussion may also vary. Concerning the fields of specification, this book is primarily meant to those who approach nanostructure preparation from the field of electrochemistry and wish to understand both the electrochemical background and the mechanisms behind the electrochemical manufacturing processes. Theories behind the processes are often mentioned qualitatively only. The missing quantitative information can be found in the relevant literature for which a rich citation background is offered. It was the pronounced intention of the author to avoid the mistake of many textbooks that write about a topic without pointing to the original works where the information comes from, hence attributing the wisdom of the community of researchers solely to the author.

The view-points detailed above led to the conclusion that the morphological aspects should be followed, as already emphasized in the choice of the topic. The shape of the entity or the key feature of the nanostructure and that of the material used for its preparation will serve as the major guideline. This gave the title of the major parts. Part I comprises this Introduction (Chap. 1), the description of the electrochemical background (Chap. 2) and the review of the investigation methods that are important in the study of some nanostructure classes (Chap. 3). Parts II and III cover the methods that make use of electrochemical processes working in the "classical" potential regime of electrochemistry; i.e., in practically all these processes, the electrode potential can be considered as the driving force in accord with the activation theory of the charge transfer. These parts include processes where the application of a non-structured (Part II) or textured (Part III) electrode can be used for the preparation of nanostructures, respectively. Within these parts of the book, the chapters are organized in accord with both the principles of the formation of the nanostructures as well as character of the nanomaterials to be obtained. Part IV summarizes high-voltage methods where the charge transfer or the current flow cannot be elucidated on the basis of the activation theory. The chapters in this part offer an overview on processes where the ion motion has to be activated in electrically non-conducting materials (like in the formation of anodic alumina during anodization) and the preparation of carbon nanostructures with high-voltage discharge methods. Although the basic understanding of these processes is not as deep as that of the classical electrochemical reactions, they cannot miss from a comprehensive overview. Very importantly, the product of the high-voltage preparation methods is very often the starting materials of other nanostructure synthesis methods discussed in the earlier parts of this book.

The approach of this book does not take into account the actual technological relevance of the systems discussed. As nanotechnology develops, methods and systems that are not in the focus of the presently available applications may gain more attention. Therefore, a detailed picture on the great variety of the field is believed to be a merit that the reader may benefit from. Whenever it is possible, the methods listed in this book are referred to with the names commonly found in the literature, which is hoped to help the reader to link the content of this book to new works to be published later and using the same terminology.

It is the author's hope that this book will be useful for students and researchers at various fields. Since there is a fast-increasing gap between the role of electrochemical methods in nanostructure preparation and the content of the courses offered in this field, the book can be an equally important resource for teachers composing their own curriculum on electrodeposited nanostructures as well as for their students. Also, it is hoped that researchers dealing with one or another field of nanotechnology can use it as a primary source of information. During the topic selection of this book, it was kept in mind that researchers dealing with electrodeposited nanostructures may have a very versatile education background that includes chemistry, chemical engineering, physics, materials science and even electrical engineering. Therefore, a little theoretical background of electrochemistry was blended with the materials science aspects of the related fields. Perhaps specialists of some of the fields discussed can have a deeper insight into their specific topics; however, a comprehensive book never can be a reference work that contains all information and each single resource about a field. The information content was meant to range to a level that assists the reader to start seeking further readings.

The lifetime of a book, especially nowadays, cannot be more than about two decades. Over such a period, science itself as well as our view on both nature and scientific methods develop so much that the contemplation of the reference works has to be renewed, even in the case when neither the basis of a field nor the factual information on how things work can change fundamentally. Beside the paradigm changes, the toolset of science and the accuracy improvement of the characterization instruments can also rationalize the needs of redesigning the knowledge dissemination pattern of a field. Being aware of all these limitations, it is hoped very much that this book can contribute to the formation of the contemplation of the growing generation whose members will play a role in the modernization of both the field itself and its education background.

Acknowledgements The author thanks the series editor, Prof. Fritz Scholz for the invitation to write this book as well as the tight supervision of the manuscript, the improvement of which was greatly due to the careful editorial guide.

The author is to acknowledge Dr. Imre Bakonyi for the two-decade-long cooperation and the immense personal help offered during this period. The author is particularly grateful to Dr. Bakonyi for the introduction to various fields of solid-state nanoscience and for the ceaseless instigation for accurate and comprehensive work.

This work could not be born without the supportive work environment of the author, including the Hungarian Academy of Sciences as a whole and in particular, the Wigner Research Centre for Physics (by the time of the publication of this book, a part of the Eötvös Loránd Research Network). The service of the librarians of the latter institute was invaluable for collecting the literature background of the present book.

And last but definitely not least, the author must express his deepest acknowledgement to his family. Being patient and tolerant with a family member who works as a scientist is hard in general, but it is particularly challenging when this scientist is engaged in a work that lasts years. Thank you for the immense lenience it required.

References

1. Feynman RP (1992) There's plenty of room at the bottom. Eng Sci 23(5):22–36
2. Graham T (1861) Phyl Trans R Soc Lond 183–224
3. Hauser EA (1955) J Chem Educ 32:2–9
4. Jacobi MH (1840) Die Galvanoplastik. Eggers et Cie, Sankt Petersburg
5. Fischer H (1954) Elektrolytische Abscheidung und Elektrokristallization von Metallen. Springer, Berlin
6. Pfanhauser F (1949) Galvanotechnik, 9th edn. Akademische Verlagsgesellschaft Geest & Portig K.K, Leipzig
7. Brenner A (1963) Electrodeposition of Alloys. Principles and Practice, Academic Press, New York and London
8. Safranek WH (1974) The properties of electrodeposited metals and Alloys. A Handbook. American Elsevier Publishing Company Inc., New York
9. Dini JW (1993) Electrodeposition. The Materials Science of Coatings and Substrates, Noyes Publications, Park Ridge, NJ, US
10. Kanani N (2004) Electroplating — Basic Principles, Processes and Practice. Elsevier, Oxford, UK
11. Schlesinger M, Paunovich M (eds) (2010) Modern electroplating. John Wiley and Sons, Hoboken, NJ, USA
12. Hodes G (ed) (2001) Electrochemistry of nanomaterials. Wiley-VCH, Weinheim
13. Watanabe T (2004) Nano-plating. Microstructure control theory of plated film and data base of plated film microstructure. Elsevier, Amsterdam
14. Lin Y, Nalwa HS (eds) (2009) Handbook of electrochemical nanotechnology. American Scientific Publishers, Valencia, CA, USA
15. Schmuki P, Virtanen S (eds) (2009) Electrochemistry at the nanoscale. Springer Science+Business Media, New York
16. Osaka T, Datta M, Shacham-Diamand Y (eds) (2010) Electrochemical nanotechnologies. Springer, New York
17. Wei Di (ed) (2012) Electrochemical nanofabrication. Pan Stanford Publishing, Singapore
18. Aliofkhazraei M (ed) (2014) Modern electrochemical methods in nano, surface and corrosion science. InTech, Rijeka, Croatia
19. Mirkin MV, Amenia S (2015) Nanoelectrochemistry. CRC Press, Boca Raton
20. Aliofkhazraei M, Makhlouf ASH (eds) (2016) Handbook of nanoelectrochemistry. Springer International Publishing, Cham, Switzerland
21. Nasirpouri F (2017) Electrodeposition of nanostructured materials. Springer International Publishing, Cham, Switzerland
22. Wuthrich R (2000) Micromachining using electrochemical discharge phenomenon. Elsevier, Amsterdam

23. Aliofkhazraei M, Rouhaghdam AS (2010) Fabrication of nanostructures by plasma electrolysis. Wiley-VCH, Weinheim, Germany
24. Bhattacharyya B (ed) (2015) Electrochemical micromachining for nanofabrication. MEMS and nanotechnology, William Andrew, India
25. Binnig G, Rohrer H, Gerber Ch, Weibel E (1982) Phys Rev Lett 49:57–61
26. Moreland J, Ekin JW (1985) J Appl Phys 58:3888–3895
27. Sonnenfeld R, Hansma PK (1986) Science 232:211–213
28. Binnig G, Quate CF, Gerber C (1986) Phys Rev Lett 56:930–933
29. Wiechers J, Twomey T, Kolb DM, Behm RJ (1988) J Electroanal Chem 248:451–460
30. Otsuka I, Iwasaki T (1988) J Microsc–Oxford 152:289–297
31. Sonnenfeld R, Schardt BC (1988) Appl Phys Lett 49:1172–1174
32. Fan FRF, Bard AJ (1988) Anal Chem 60:751–758
33. Green MP, Richter M, Xing X, Scherson D, Hanson KJ, Ross PN, Carr R, Lindau I (1988) J Microsc–Oxford 152:823–829
34. Hottenhuis MHJ, Mickers MAH, Gerritsen JW, Van Der Eerden JP (1988) Surf Sci 206:259–278
35. Bard AJ, Fan FRF, Kwak J, Lev O (1989) Anal Chem 61:132–138
36. Penner RM, Heben MJ, Longin TL, Lewis SL (1990) Science 250:1118–1121
37. Martin CR (1994) Science 266:1961–1966
38. Matsuda H, Fukuda K (1995) Science 268:1466–1468
39. Cui XD, Primak A, Zarate X, Tomfohr J, Sankey OF, Moore AL, Moore TA, Gust D, Harris G, Lindsay SM (2001) Science 294:571–574
40. Magnussen OM, Zitzler L, Gleich B, Vogt MR, Behm RJ (2001) Electrochim Acta 46:3725–3733
41. Williamson MJ, Tromp RM, Vereecken PM, Hull R, Ross FM (2003) Nature Mater 2:532–536

Chapter 2
Electrochemistry and Electrodeposition

2.1 The Goal of This Overview

It may sound trivial that a book published as a volume of an electrochemistry book series does not need an introduction dealing with electrochemistry. The author really wishes it could be true. However, in spite of the large number of presently available electrochemistry books, there are only very few that deal with the important aspects of the metal deposition process [1–6], the dissolution of metals [7] and miscellaneous materials science aspects of electrochemistry [8, 9]. These works mostly focus on general aspects of the topic of choice and not all can be readily used as a primary source concerning electrochemical nanostructure formation. For the above reasons, a short but essential summary is offered here to the reader for sparing much time in the further literature search as well as clarifying the terms used hereinafter.

The terms used throughout this book are meant to be in accord with the IUPAC recommendations as they are defined in the Electrochemical Dictionary [10], a comprehensive source of definitions and historical descriptions.

2.2 Charge Transfer in Heterogeneous Electrochemical Systems

Electrochemistry deals with chemical processes where charged species are involved. We can speak about *electrochemical reactions* where charge transfer takes place between chemical species. If the charge transfer occurs between species that are a part of different phases, we encounter *heterogeneous electrochemistry* (also often termed as *interfacial electrochemistry*, but the latter term is difficult to use if there is no well-defined single interface or if multiple interfaces occur in a heterogeneous system). The phases involved in a heterogeneous electrochemical reaction may be similar or different concerning the charge carriers. If two non-miscible liquids, each containing ionic solutes (i.e., *electrolyte*s) and hence exhibiting ionic conductance,

L. Péter, *Electrochemical Methods of Nanostructure Preparation*,
Monographs in Electrochemistry, https://doi.org/10.1007/978-3-030-69117-2_2

contact each other, an interface occurs where electrochemical reactions may take place. When one of the conducting phases in contact with each other is an electron conductor and the other is an ionic conductor, an *electrode* is formed. The concept of electrode always involves more than one conducting phases with different charge carrier mechanisms. A detailed analysis of the scientific approach to the electrode definition can be found elsewhere [11]. What is important for us is that the region where the transition from a particular conduction mechanism to another one takes place has the properties that give rise to the functionality of the electrode, should this transition zone be a single well-defined interface or a thick layer of gradually varying composition. The charge (Q) passing through an electrode is related to the amount of material (n, measured in mol) undergoing an electrochemical reaction, the proportionality factor being the Faraday constant F (96,485 C mol^{-1}), and the number of electrons taking part in the electrode reaction, z. If a single electrode reaction takes place, we get the simple form of Faraday's law:

$$Q = zFn \tag{2.1}$$

Carrying out a derivation with respect of the time and dividing both sides of the equation with the electrode surface area, the relationship of the surface-normalized reaction rate (v) and the current density (j) is obtained:

$$j = zFv \tag{2.2}$$

2.3 Electrodes and Electrochemical Cells

If two electrodes share an ionic conducting phase (or, these ionically conducting phases with possibly different composition at least contact each other without a change in the conduction mechanism), an *electrochemical cell* is made. Since the study of an electrode always involves the possibility of charge transfer between the electron-conducting and ion-conducting phases and charge accumulation within any of these phases is impossible, the study of one particular electrode requires another electrode, too; i.e., electrochemical cells are required for any measurement.

The potential difference between two electrodes of a cell is the so-called *cell potential difference* (E_{cell}), regardless of the nature of the electrodes. In the description of an electrochemical cell, three features of the cell are inherently connected with each other: (i) the cell diagram, (ii) the cell reaction and (iii) the cell potential difference. In accord with the convention accepted nowadays, the cell reaction should be written in a way so that the flow of the positive charge within the ion-conducting phase points from the electrode shown in the left-hand side of the cell diagram to the right one. This also means that cell reaction has to be written so that an oxidation process takes place at the electrode at left. The cell potential difference is obtained as the potential of the electrode shown at the right minus the potential of the electrode shown at left.

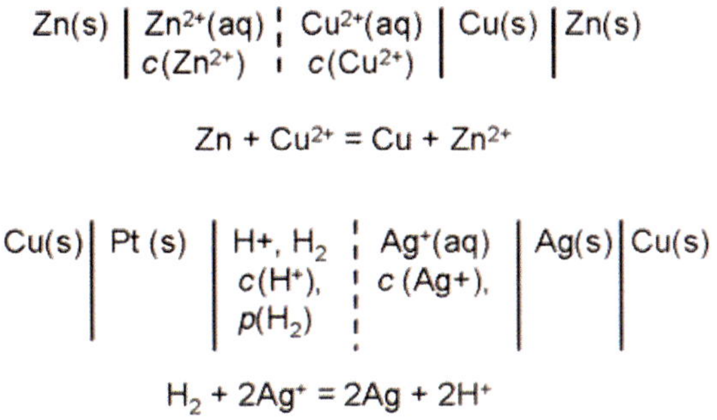

Fig. 2.1 Cell diagram of two electrochemical cells and the corresponding equation of the cell reaction: the so-called Daniell-cell (top) and another cell with a hydrogen reference electrode (bottom). (The standard hydrogen electrode serves as the zero point of the potential scale in aqueous electrochemistry)

If $E_{\text{cell}} > 0$, the direction of the spontaneous reaction coincides with the formulation of the cell reaction (for $E_{\text{cell}} < 0$, the direction of the spontaneous cell reaction is the opposite). The cell diagram and the cell reaction of two different electrochemical cells are indicated in Fig. 2.1. It has to be noted that the measurement of the potential difference is always carried out between metals of identical quality, which is also indicated in the cell diagram.

The measurement of the electrode potentials is based on the standardization of one of the electrodes in a cell that is henceforth named as the *reference electrode*. The convention is that a standard hydrogen electrode should be used as a general reference of electrode potentials, although various other reference electrode systems are common in the practice. As a reference, the standard hydrogen electrode is at the left side of the cell diagram, and the process taking place there is oxidation ($H_2 \rightleftarrows 2H^+ + 2e$), from which it follows that the reaction of interest proceeding at the other electrode must always be written as a reduction.

The above described reference standardization process is similar to that in any branch of physics where *conservative fields* are involved, should the zero level refer to the basic point of potential scale in an electrostatics or in a gravitational field. Such a choice of reference point is always indispensable so that one can speak about the potential of a single point or the potential of a single electrode, even though the measurement must be based on a comparison within a complete electrochemical cell.

The reference to the standard hydrogen electrode makes it possible to evaluate the potential of an equilibrium electrode as a function of the most important chemical variables, which are generally the activity of the reactants involved in the electrode reaction. The result for a reaction formulated as $0 = -ze + \sum \nu_i \mathrm{A}_i$ (A being the symbol of the components and ν is the stoichiometry coefficient of the component, while i runs from 1 to all reactants and products) is the well-known *Nernst equation*:

$$E = E^0 - \frac{RT}{zF} \ln \prod_i \frac{a_i^{\nu_i}}{a_i^0} \tag{2.3}$$

or

$$E = E^0 - \frac{RT}{zF} \sum_i \nu_i \ln \frac{a_i}{a_i^0} \tag{2.4}$$

where ε^0 is the standard electrode potential, T is the absolute temperature, R is the gas constant (8.314 J mol^{-1} K^{-1}), a is the activity of the reactant/product, and a^0 is the standard value of the activity for a particular component. It cannot be emphasized strongly enough that for the validity of the Nernst equation, an appropriate equilibrium must prevail within all participating phases and at all interfaces wherever equilibrium can hold at all.

2.4 Electrode Classification Based on the Electrode Reaction(s)

It is of great importance to see a clear classification of electrodes. An elementary approach to the classification is to enumerate how many different elementary charge transfer reactions may take place at the electrode. If the number of the charge transfer reactions is zero, the electrode can be polarized very easily; i.e., the electrode potential can be set up by passing a very little charge through the electrode. This is the so-called *charging* or *capacitive current*. This corresponds to the charge of less than a monolayer of atoms/ions, and no current flows in the steady-state. These systems are often referred to as *ideally polarizable electrodes*. The processes taking place on such an electrode will be further explained in the discussion of the electrical double layer (Sect. 2.5). Examples for such electrodes are the noble metals or carbon in contact with an electrolyte solution containing non-reactive electrolyte(s) as solute and polarized only within the stability limit of the solvent.

If one single charge transfer process takes place on an electrode, we arrive at the classification system based on the nature of phases involved in the electrode construction. The electrode reaction on an *electrode of the first kind* involves one chemical element and the ions produced from this element where the ions are present in the ionic conducting phase (electrolyte solution or melt). Typical examples are the metals immersed in the solution of the salts of the same metal with fairly high solubility (like silver in silver nitrate solution with the electrode reaction of $Ag^+ + e \rightleftarrows Ag$). Electrodes in which the metal taking part in the reaction is present in an amalgam phase also belong here, just like those in which the metal salt in the solution forms a complex compound. The equilibrium potential of a simple (non-complexed) metal electrode is as follows:

$$E = E^0 + \frac{RT}{zF} \ln \frac{a_{Me^{z+}}}{a^0} \tag{2.5}$$

When the metal ion in the solution forms a complex with a ligand L^{y-}:

$$Me^{z+} + kL^{y-} = [MeL_k]^{(z-ky)+} \tag{2.6}$$

and the complex formation equilibrium can be characterized with a stability constant $K_{ST} \gg 1$:

$$K_{ST} = \frac{a_{MeL_k^{(z-ky)+}}}{a_{Me^{z+}} a_{L^{y-}}^k}, \tag{2.7}$$

the equilibrium potential of the metal electrode can be written by replacing the free metal ion activity in the Nernst equation:

$$E = E^0 - \frac{RT}{zF} \ln \frac{K_{ST}}{\left(a^0\right)^{-k}} + \frac{RT}{zF} \ln \frac{a_{MeL_k^{(z-ky)+}}}{a^0} - \frac{kRT}{zF} \ln \frac{a_{L^{y-}}}{a^0} \tag{2.8}$$

This substitution can be made because of the validity of the so-called "zeroth law" of thermodynamics on the transference of the equilibrium. This means that the electrode potential of a complex metal electrode is calculated as that of a simple metal electrode, even though the electrode reaction mechanism in a simple and a complex metal electrode is clearly different. The sum of the first two terms in Eq. 2.8 is also called as the standard electrode potential of the complex metal electrode. The signs in Eq. 2.8 indicate that the complex formation shifts the metal ion/metal equilibrium to more negative potentials, simply because of the increase of the stability of the metal ion in the solution due to the complex formation. This can be exploited later during the deposition of the alloys where the difference in the deposition potential of the alloy components is of high importance (Sect. 2.12.)

Another example for the electrode of the first kind is the hydrogen electrode where molecular hydrogen and solvated protons take part in the electrode reaction ($H^+ + e \rightleftarrows 1/2H_2$; here, the electron-conducting phase is not a reaction partner but may serve as a catalyst in the reaction). Various gas electrodes, even though many of them have not yet been implemented as equilibrium electrodes, also belong to this class (like the chlorine electrode with the electrode reaction $Cl_2 + 2e \rightleftarrows 2Cl^-$).

The *electrode of the second kind* can be easily derived from a simple electrode that involves the formation of a metal ion from its parent metal. The important difference here is that the metal is immersed into a saturated solution of its weakly soluble metal salt, and the concentration of the anion is often set by adding its compound with a different non-electroactive metal. Well-known examples are the silver/silver chloride electrode (Ag/AgCl) or the calomel electrode (Hg/Hg_2Cl_2), and the chloride ion concentration can be adjusted with an appropriate solution of NaCl or KCl. Such electrodes are often used as reference electrode in experimental cells due to their high stability. The stability of the commercial reference electrodes of the second kind partly originates from the high dispersity of the contacting phases. For a better understanding of the polarizability of the electrodes, see Sect. 2.8 on the electrode kinetics.

In the case of *redoxi electrodes*, the electron-conducting phase serves as an electron exchange surface, and both the reduced and the oxidized form of the redox pair are solvated in the ion conducting phase.

When several elementary electrode reactions take place on an electrode, a *mixed potential* can be measured. The special case when each single electrode reaction can lead to an equilibrium can only be achieved with extreme concentration ratios of the components taking part in various electrode reactions since a 60 mV difference in the electrode potential for a reaction with $z = 1$ requires an order of magnitude change in the activity of a component.

When several electrode reactions take place on one electrode but the net current is zero at the same time, the common case is that each reaction has a dominant direction. This happens during corrosion, electroless metal deposition and cementation processes. Such electrodes can never be in equilibrium, and their electrode potential is set by the kinetic properties of the reactions going on, providing that the total current is zero. A more detailed picture of such processes will be elucidated in Sect. 2.8 where the basic kinetic relationships are explained.

2.5 The Nature of the Electron Conductor/Solution Interface

In an equilibrium bulk solution of an electrolyte, the temporal average of the concentration of any charge carrier is the same as its analytical concentration. However, when we fix a coordinate system to a particular ion, we find that ions of the opposite charge are more abundant in the close vicinity of this ion. The distribution of ions around each other is influenced by electrostatic forces and the thermal fluctuations, and the elucidation of this problem requires the treatment of the Poisson–Boltzmann equation. This field, together with the estimation of the activity coefficients of the electrolytes, is dealt with by the Debye–Hückel theory (not discussed here in detail; see, e.g., Sects. 2.4–2.5 of [12]).

The presence of an interface between two phases of different conductivity mechanism can modify the even concentration distribution of the charged species for various reasons. In the forthcoming part of this chapter, the train of thoughts will refer to an electrode in which the electron-conducting part is an inert metal that itself does not undergo any charge transfer reaction and serves as either a source or a well of electrons. Similarly, the electrolyte dissolved in the solution is taken as non-reactive. The practical consequences of the discussion will be valid for reactive electrodes, too; nevertheless, the understanding of the basic phenomena will be easier by considering a case without any side process.

Shortly after the elaboration of the theory of electrolytic dissolution of ionic compounds in solution, Helmholtz assumed (1879) that the surface of the solid in contact with the electrolyte solution can be charged, and the ions bearing the opposite charge may accumulate at the surface. The charged metal and the ion layer in the solution together make an electrical double layer. It is straightforward that the modification of the charge on the surface of the metal leads to a redistribution of the ions near the metal/solution interface. Such an interface behaves in a similar way as an electrostatic capacitor, and the ionic conductor/electrolyte solution interface always exhibits capacitive properties. Later, Gouy and Chapman completed Helmholtz's theory by assuming a distribution of ions in the solution near a charged surface in accord with the thermal fluctuations, hence modifying the parallel-plate model and introducing the concept of the *diffuse double layer*. This assumption led again to the Poisson–Boltzmann problem, with the difference as compared to the environment of a dissolved ion that the spherical ion distribution turns into a planar one. This consideration could explain that the differential capacitance of the interface is not constant but depends on the surface density of the charge carriers already accumulated therein. However, another limitation came from the fact that the medium was treated as a continuum. The proper constraint was set later by Stern (1924) by defining the closest approach of an ion to the surface as the radius of this ion. The introduction of the distance of closest approach resulted in a maximum of the capacitance of the interface, hence eliminating a weakness of earlier theories.

The picture on a metal/solution interface can be further complicated if specific adsorption of ions at the solid surface is also assumed, which shifts the purely physical view on the interface towards a chemically more realistic contemplation. The goal here is not the exact quantitative treatment of all related phenomena (that can be found elsewhere; see, e.g., Chap. 2 of [3], Chap. 3 of [12] and [13]). Instead, a reliable qualitative picture is offered that is suitable to elucidate the capacitance-related interfacial phenomena. For this purpose, the scheme in Fig. 2.2 is recommended.

A detailed description of the electrochemical double layer involves the following concepts: the inner Helmholtz plane is defined at the centre of the strongly adsorbed non-hydrated ions. The outer Helmholtz plane lies at the centre of hydrated ions at the closest approach to the metal surface while the hydration shell is intact. The diffuse double layer ranges to a distance from the metal surface where the concentration of the cations and anions is equal to their bulk concentration and the influence of the charged surface decays. The characteristic thickness of the diffuse double layer (χ) is inversely proportional to the square root of the electrolyte concentration. It is practically negligible at large electrolyte concentrations (if $c > 0.1$ mol dm^{-3}, $\chi <$ 1 nm) but can range to several hundred nanometers if $c < 10^{-6}$ mol dm^{-3}.

It is a key question in the study of the electrical double layer of electrodes at which potential the charge of the surface becomes zero. This *potential of zero charge* (E_{pzc}) can be established mostly indirectly, either from the minimum of the differential capacitance as a function of the electrode potential, from the maximum of the surface tension of a liquid metal or from sensitive chronocoulometric measurements carried out during the immersion of the metal into the electrolyte solution under potentiostatic control. However, regardless of the exact value of E_{pzc}, in experiments with varying potential the capacitance effects have to be taken into account. A positive change in electrode potential leads to a positive double layer charging current and vice versa.

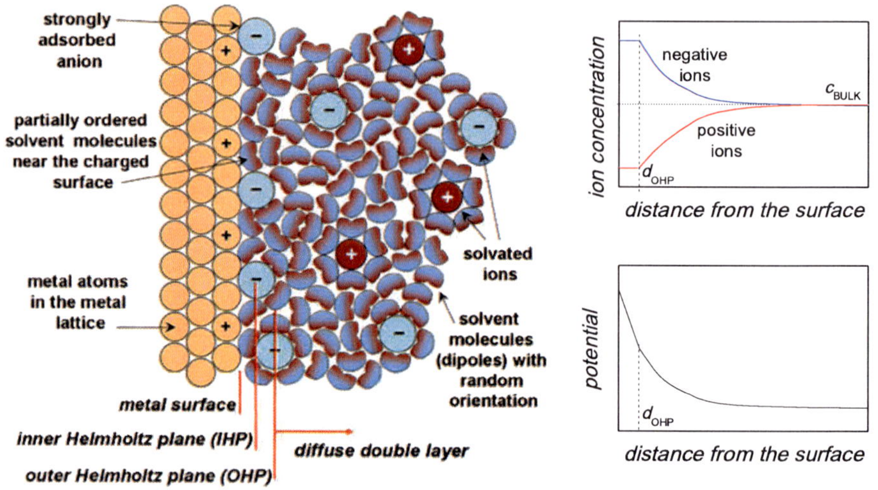

Fig. 2.2 **Left**: Schematic picture of the distribution of strongly adsorbed and solvated ions in the electrolyte solution near a charged metal surface. The orientation of the solvent molecules is partially ordered near the surface and becomes random as the electrical field decays. Observe that the planar projection of the species implies an inherent distortion to the real particle configuration. **Right**: Ion concentrations and electrical potential as a function of the distance from the surface in the absence of strongly adsorbed ions

The double layer phenomenon manifests itself as a transient effect upon the step-wise change in the electrode potential, which is in excess to the Faraday current that is mostly the subject to be studied.

The discussion above refers to the metals as electron conductors. For metals, any change in the charge distribution due to the electrode polarization results in a charge accumulation restricted to the surface, and no change in the charge carrier density within the metal can take place as the interfacial charge is modified. A fundamentally different situation is encountered when the electron conductor in a semiconductor in which the charge carrier density is by orders of magnitude smaller. (Although the nature of the charge carrier in semiconductors may also vary, being electrons or holes for n-doped and p-doped semiconductors, respectively, the nature of the charge carrier distribution as a result of the potential change is the same.) For semiconductors, the analogy of the potential of zero charge is the flat-band potential where an even distribution of the charge carriers prevail. The change in the electrode potential and the charge accumulation on the surface leads to a distortion of the even charge carrier distribution decaying from the surface towards the bulk of the semiconductor. An important difference as compared to the diffuse double layer in the electrolyte solution is that the semiconductor has one dominant charge carrier type whose density is modified locally, while in the electrolyte solution, both positive and negative charge carriers are redistributed due to the surface charging.

2.6 Electrodes in Experimental Cells

When electrochemical cells are used in practice, the number of electrodes is often larger than two, and the role of the electrode may be various. Therefore, it is worthwhile looking over the terms occurring in connection with practical electrochemical cells where the electrode names are related to their functionality, in contrast to the categorization presented earlier concerning the nature of the electrode processes (Sect. 2.4).

The *working electrode* is the one at which the processes taking place are in the focus of interest. It is generally desired that both the potential of this electrode and the current passed are monitored. However, the potential reference and the current lead for the cell are usually not the same. For potential reference, or *reference electrode*, the best choice is to use an equilibrium electrode whose potential is fixed and stable. Alternatively, if the system does not make it possible to use an equilibrium electrode for potential reference, a non-equilibrium electrode with a sufficient stability at the time scale of the experiment can also be used, which is often termed as a quasireference electrode. Regardless of the realization of the potential reference, the stability of the system requires that the reference electrode should not carry any current, which makes it necessary to use a *counter or auxiliary electrode* that can be loaded with the same current as the working electrode. The potential of the counter electrode is seldom monitored but its potential is important because the total cell voltage (i.e., the potential difference between the working and the counter electrode) determines the power needed to operate a cell. Therefore, a counter electrode usually exhibits a large surface area as compared to the working electrode for decreasing the total cell voltage. Another important requirement towards the counter electrode is that its reaction product should not contaminate the working electrode environment. This condition is often fulfilled by constructing cells with several compartments in which the product of the reaction taking place on one electrode cannot reach the other one.

A peculiar feature of electrochemistry is that an electrochemical cell can be operated in various modes. If the total current is controlled, we encounter the galvanostatic control of the cell. In contrast, when the potential of the working electrode is controlled, one arrives at the potentiostatic operation. Both operation modes can be carried out in a more advanced manner, by prescribing sweeps or pulses. However, whichever signal shape is used for the controlled parameter, a current control is equivalent to the regulation of the reaction rate, while the potential control is a tool of the regulation of the driving force of the processes.

Finally, it is important to mention that electrodes may get various names based on the sign of the current passing through them. For the *anode*, the current is positive that is associated with the flux of positively charged species from the electron-conducting phase towards the ionically conducting phase, which is equivalent to that an oxidation reaction takes place on this electrode. The *cathode* is an electrode where reduction takes place and the charge flux is the opposite than at the anode. These categories are not an inherent feature of the electrodes but change as the current direction is reversed. Nevertheless, it is common (though formally incorrect) that some electrode components in batteries are named as "cathode material" or "anode material". These notations always refer to the current direction of the battery operation (discharge), while the roles are exchanged for the charging process.

2.7 Polarization and Electrochemical Devices

The term *polarization* stems historically from the age when semiconductor-based devices were not yet available, and a galvanostat could be built only by using a high-capacity power supply and a sufficient serial shunting resistor. Also in this case, the potential of the electrode of interest could be monitored with respect to a reference electrode (or indicator electrode), but here, the reference electrode was not a part of the regulation circuit (see Fig. 2.3, left). The phenomenon that the potential of the electrode of interest changed as a result of the current passed though this electrode was named as *polarization*. When the electrode of interest is an equilibrium electrode, the deviation from the equilibrium potential is named as the *overvoltage*.

Although the electrode potential has been regarded since the very early age of the kinetic theory as the driving force of the charge transfer processes, from experimental point of view, the change in electrode potential appeared to be the result of the current passing through an electrode. This controversy accompanied the development of the terminology of electrochemistry and gives rise to much misunderstanding even nowadays.

The appearance of electrochemical devices capable of the potential control of electrodes, the *potentiostats*, has not resolved the conflict of the terminology and the experiment. While the conventions retained the notion of polarization in its original

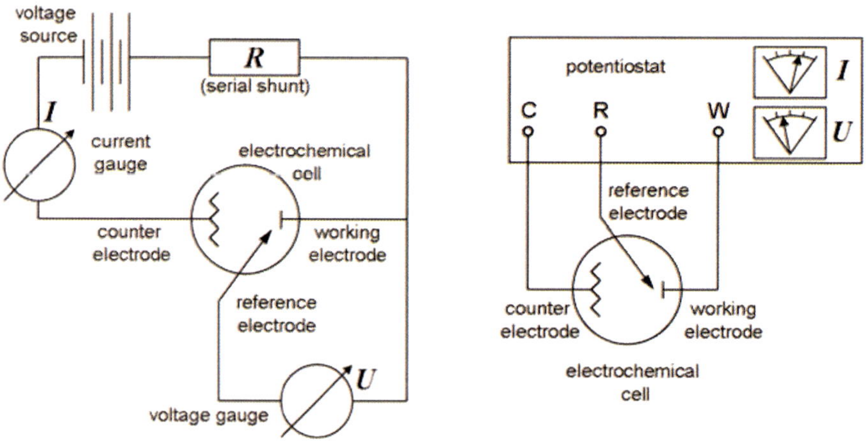

Fig. 2.3 Comparison of the galvanostat (**left**) and the potentiostat (**right**), with emphasis on the role of the reference electrode

meaning, the laboratory practice also renamed the set-up of the potential as polarization, even though this can be carried out with a large number of electrodes without passing a constant current. Especially for beginners, polarization often means something that the potentiostat does, taking advantage of the simplicity of the expression as compared to the "tuning of the electrode potential".

The situation might look even more complicated if we take a look at the basic circuitry and the operation of potentiostats with various internal connection modes. From the view-point of the electrical engineer, the potentiostat operation can be described by saying that the device passes so much charge or current through the system so that the potential of the working electrode achieves the predefined value (Fig. 2.4a). (In this figure, the working electrode is grounded, which is the most common mode of implementation. Nevertheless, the forthcoming discussion is also valid for other electrical connections where either the reference or the counter electrode is grounded.)

The picture on the electrodes in an operating electrochemical cell becomes more complex if a tentative equivalent circuit of the cell is also embedded into the circuit representing the potentiostat. As shown in Fig. 2.4b, each part of the cell can be represented with some circuit element, and some of them can be very complex involving various parts that often cannot be found among passive electrical circuit

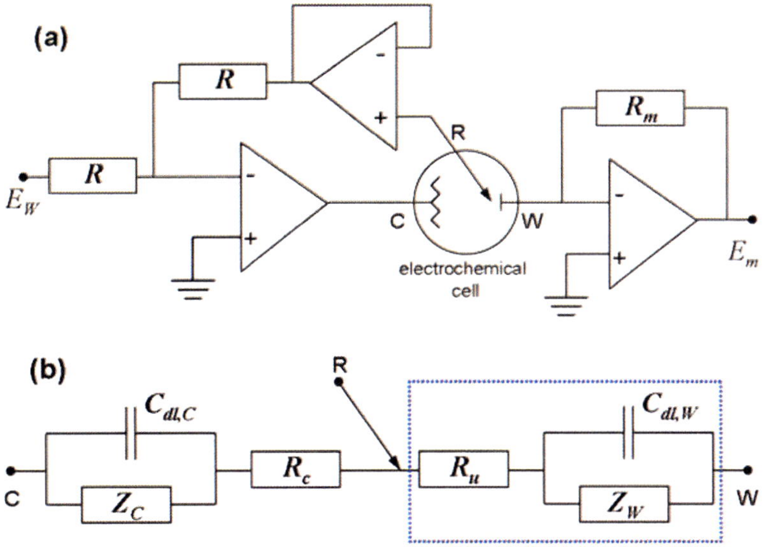

Fig. 2.4 **a** Potentiostatic circuit. E_W: potential of the working electrode, R_m: measuring resistor, E_m: voltage proportional to the current ($I = E_m/R_m$). **b** Schematic representation of the equivalent circuit of a complete electrochemical cell. Z: impedance of the Faraday reaction taking place on the electrode, C_{dl}: double layer capacitance, R_u: uncompensated resistance that gives rise to the ohmic drop, R_c: compensated resistance. Indices W and C and refer to the working and counter electrode, respectively. The dotted square frames the equivalent circuit elements that can be measured as the impedance of the working electrode

elements. The potentiostat ensures that the impedance of the reaction taking place at the counter electrode does not impact the measurement. However, it is to be noted that the physical distance of the working and the reference electrode can never be infinitely small, and the current flowing through the cell always gives rise to a so-called *ohmic drop* which stems from the resistivity of the media between the two electrodes. This results in that the potential-controlled cell operation must be treated with a great care when either the current is relatively large or when the reference position cannot be chosen close enough to the working electrode. In such cases, *in situ* or *ex situ* corrections of the electrode potential may be necessary. Also, literature data reporting extreme working electrode potential values should always be scrutinized for both their origin and meaning.

2.8 Basic Electrode Kinetics

2.8.1 Activation Control

The rate law for electron transfer reactions can be elucidated in the same way as for reactions of any other type; namely, with the activation complex theory. For a simple reaction where the reactants, A and B make an activation complex, the reaction rate can be written as

$$v = kc_A c_B \exp\left(-\Delta G^{\#}/RT\right) \tag{2.9}$$

where $\Delta G^{\#}$ is the activation Gibbs free energy of the reaction. If the reaction involves a charge transfer at a phase boundary, then $\Delta G^{\#}$ is a function of the potential of the phases involved, and it changes with a term $fzF\Delta E$ (where $0 < f < 1$). The reason why not the entire electrical energy, $zF\Delta E$, is considered for the change of the activation energy is that the energy level of the reactants, that of the activation complex and also that of the product is a function of the electrical potential of the relevant phase. This brings an asymmetry factor to the kinetic equations of the electrode reactions, leading to the following forms for an anode reaction formulated as Red = Ox + ze:

$$j_A = k_A c^0_{\mathrm{RED}} zF \exp(\alpha zFE/RT) \tag{2.10}$$

and for its opposite cathode reaction

$$j_C = -k_C c^0_{\mathrm{Ox}} zF \exp(-(1-\alpha)zFE/RT), \tag{2.11}$$

and the sum of the current density corresponding to all partial reactions yields the total current density:

$$j_{\text{TOTAL}} = j_A + j_C = k_A c^0_{\text{RED}} zF \exp(\alpha zFE/RT) \\ - k_C c^0_{Ox} zF \exp(-(1-\alpha)zFE/RT) \quad (2.12)$$

For the above equation, it was applied that $j = zFv$, and that the changes of the activation energy of the anode and cathode reactions as a result of the electrode potential change are not independent. The asymmetry parameter α is often close to 0.5, and the zero in the superscript of the concentrations indicates that their value right at the electrode surface should be considered. This chapter deals with the case where the surface concentration of the reactant can be taken equal to the bulk concentration, and the impact of the transport processes will be discussed later.

The *current density—potential* relationship of the partial reactions define the partial polarization curves of the anode and cathode reactions, whose sum is the polarization curve that can be measured in practice. If the reaction can lead to equilibrium at E^0, then the absolute value of the current density corresponding to either the anodic or cathodic reaction (which are obviously equal) is called the exchange current density, j_{EX}. The polarization curve and the method of the determination of the exchange current density are shown in Fig. 2.5 for constant surface concentration of the reactants.

The so-called Tafel constants (i.e., the slope of the linear sections in the logarithmic plot of the polarization curve, see Fig. 2.4b,) are often referred to in mV/decade unit, which is the inverse of the line slopes shown the graph. The reason of this denotation is that the origin of this analysis stems from the early days of quantitative electrochemistry when only the *E(j)* relationship could be measured (see also the relevant remarks in Sect. 2.7). The Tafel slop can indicate the asymmetry of the activation process if the α parameter deviates from the mean value of 0.5 and the electrode reaction involves a single electron transfer without any kinetic complications. However, when the whole electrode reaction can be formulated with a sequence of several

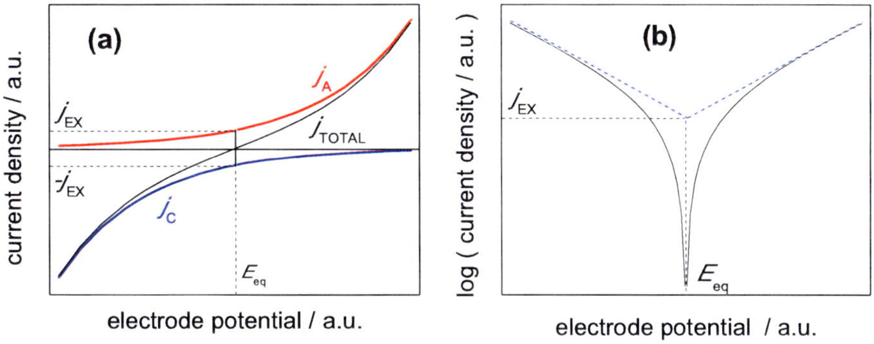

Fig. 2.5 **a** Partial polarization curves of the anodic and cathodic reactions on an electrode where the surface concentration of the reactants can be taken constant (activation control). The equilibrium potential is indicated with E^0. **b** Determination of the exchange current density and the so-called Tafel slopes

consecutive electron transfer steps, the apparent Tafel slope can be obtained by the analysis of the complete kinetic scheme. In this case, the relative hindrance of the consecutive charge transfer processes is a crucial parameter.

2.8.2 Influence of Reactant Transport on the Electrode Processes

Transport of various components in an electrochemical system may take place by three main transport processes: *diffusion*, *migration* and *convection*. If the rate of the transport process is determined by the gradient of the chemical potential of the reactant, the transport is pure diffusional. When the driving force of the transport process is provided by the electric field in the media, we can speak about migration. Finally, if the transport is provided by the displacement of the elemental volume of a fluid media by external mechanical forces, the transport is convective. Whichever means of the transport dominates in a particular system, the transport process must be regarded as a process in series with the electron transfer step.

It is common in the electrochemical laboratory practice that the number of transport processes concerning the reactant is reduced. When a supporting electrolyte is used in basic laboratory experiments and the concentration of the reactant is small, the migration of the reactant(s) can be neglected. However, it should be remembered that this case practically never happens in an industrial plating system where the reactant is a major constituent of the bath. Convective transport can be provided by either gas bubble agitation or mechanical stirring, but well-defined convection can be achieved mostly by the rotation of the electrode (rotating disc or cylinder electrode), for which exact mathematical description is also available. Diffusion is always to be considered. Even for an approximate and qualitative picture on the transport processes prevailing around an electrode, one must deal with at least Fick's first law, simply written in a one-dimensional form as follows:

$$J = \frac{1}{A}\frac{\mathrm{d}n}{\mathrm{d}t} = -D\frac{\mathrm{d}c}{\mathrm{d}x} \tag{2.13}$$

with the usual meaning of the variables (J: flux of the reactant, A: surface area of the electrode, D: diffusion coefficient).

From the point of view of the thickness of the electrolyte solution layer where the reactant concentration is different from its bulk value, one can speak about either a finite or an infinite diffusion layer, which belong to controlled hydrodynamics (stirred solution or rotating electrode) and stagnant solution, respectively. In the next chapter, various examples will be shown for measurements performed under either of these circumstances. Here, the evaluation of the formulae will refer to a well-defined diffusion layer thickness, similarly to the general approach of many textbooks, also for sake of simplicity. First, the electrode will be planar and macroscopic that makes it possible to treat the diffusion field as being planar. In such cases, the 2.13 differential equation is reformulated with the help of finite differences, d_N denoting the thickness of the so-called Nernstian diffusion layer in which all concentration differences decay and in which distance the bulk concentrations are already valid. Although the concept of the Nernstian diffusion layer is an idealization and can be treated as an extrapolated parameter only, it proves to be very useful in the handling of the transport equations, as it can be seen below:

$$\frac{j}{zF} = D\frac{c^* - c_0}{d_N} \tag{2.14}$$

It is straightforward that the surface concentration of the reactant, c_0, is positive or zero; in the latter case, the right-hand side of Eq. 2.14 exhibits a maximum. This is the case of the *diffusion-limited current density* where the surface concentration of the reactant tends to approach zero, which means that any species of the reactant reaching the surface undergoes the electrode reaction immediately. It cannot be emphasized strongly enough that limiting current density can occur also in such conditions when the transport of the reactant is by far not merely diffusional. Therefore, the diffusion limitation has to be treated cautiously. From Eq. 2.14, the diffusion-limited current density is obtained as

$$j_L = zFD\frac{c^*}{d_N}, \tag{2.15}$$

and the combinations of Eqs. 2.14 and 2.15 lead to

$$\frac{c_0}{c^*} = 1 - \frac{j}{j_L} \tag{2.16}$$

This helps to eliminate the surface concentration of the reactant from equations like 2.10. Assuming that only the reduced form of the redox couple is present initially in the solution and one should account only for the oxidation reaction:

$$j = kc^*\left(1 - \frac{j}{j_L}\right)zF\exp(\alpha zFE/RT), \tag{2.17}$$

whose rearrangement yields the *j(E)* relationship desired for the elucidation of the measurements:

$$j = j_L\frac{c^*k\exp(\alpha zFE/RT)}{j_L + c^*k\exp(\alpha zFE/RT)} \tag{2.18}$$

The analysis of Eq. 2.18 shows that once E is moderate and j_L is much larger than the other term in the denominator, one gets back the exponential increase of the current with electrode potential. In the opposite case when the exponential term dominates the denominator, the current density is reduced to j_L. The shape of the polarization curve is shown in Fig. 2.6.

It is a crucial point in the understanding of the role of the transport in an electrode reaction that one needs a realistic image on the concentration profile of the reactant in the neighborhood of the electrode surface. In the right side of Fig. 2.6, the subsequent sections of the polarization curve are matched with the relevant concentration profile of the reactant in the solution. It is evident that the "activation control" is related to a negligible depletion of the reactant in the electrolyte solution in the vicinity of the electrode. However, when the diffusion-limited current density is achieved, the transport rate cannot increase and the current density is unaffected by any further change in the driving force of the electron transfer step.

Electrochemical preparation methods, especially those leading to micro- and nanostructures, are often based in non-steady-state polarization. Therefore, it is of outstanding importance to understand how the concentration profile develops in time after the change in the electrode polarization conditions. Figure 2.7 summarizes such

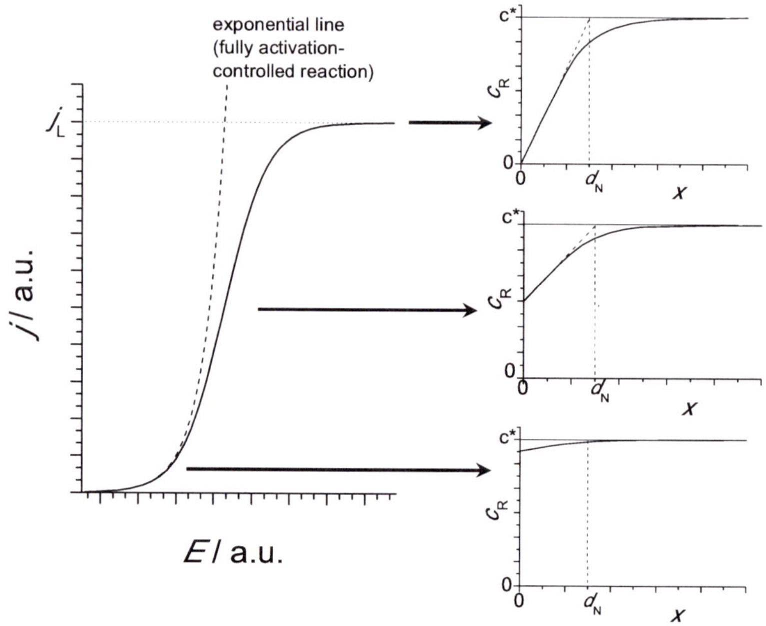

Fig. 2.6 Left: Polarization curve for single-step electrochemical reaction under the influence of mass transport control on a macroscopic electrode (one-dimensional mass transport). The dashed line shows the exponential current increase for the case when the surface concentration of the reactant is equal to the bulk one, regardless of the current density. **Right**: Concentration profile curves of the reactant in a solution under hydrodynamic control for three characteristic sections of the polarization curve. From bottom to top: activation control, mixed activation and mass transport control, full mass transport control (at the diffusion limited current density). Symbols: X: distance from the electrode, c_R: reactant concentration, c^*: bulk concentration of the reactant, d_N: thickness of the Nernstian diffusion layer

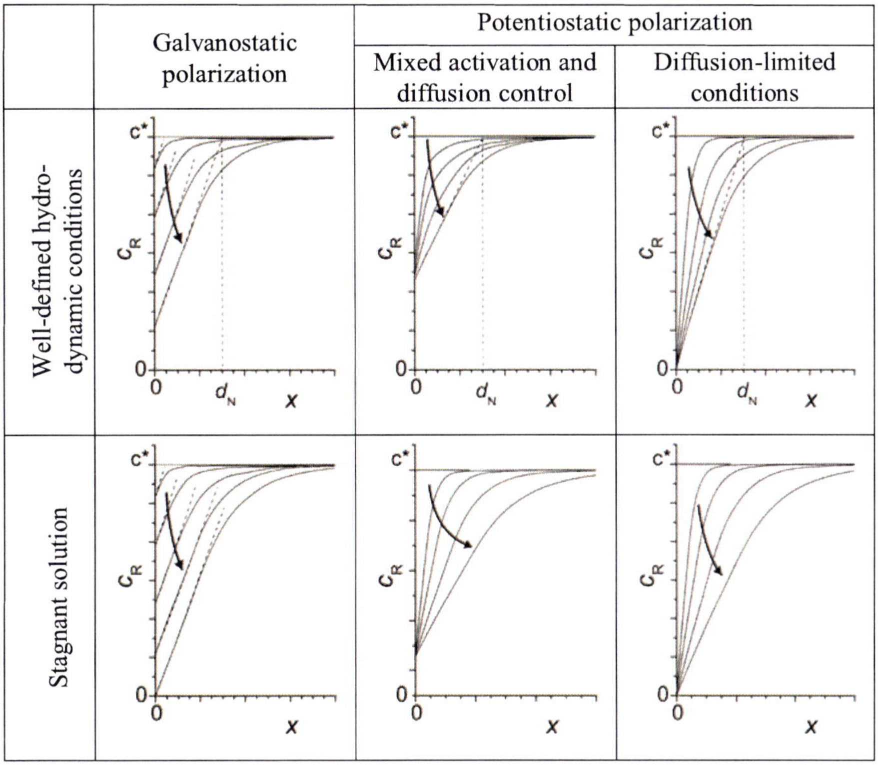

Fig. 2.7 Temporal evolution of the concentration profile of a reactant during electrochemical reaction with various boundary conditions and with different regulated electrical parameters. Symbol meanings are the same as for Fig. 2.5. Thick curved arrows indicate curves that belong to increasing time after the start of the polarization

changes for both galvanostatic and potentiostatic polarization when the current or the potential changes from one value to another one stepwise, and the start of the electrode process is associated with the change in the electrical parameter. A detailed description of the related concept can also be found in the general literature [12].

In Fig. 2.7, the cases of the finite-thickness (Nernstian) and semi-infinite diffusion layers are clearly distinguished, and the response of the non-regulated electrical parameter is also indicated. For understanding of the graphs, it is essential to observe that a fixed current is always associated with the flux of the corresponding reactant (which leads to a constant-slope section in the concentration profile curves near $X = 0$, as shown by the dashed lines next to the curves Therefore, a fixed current in a stagnant solution will sooner or later lead to diffusion-limited conditions, as shown by the last concentration profile curve in the relevant graph of Fig. 2.6. In such cases, the constant current can be maintained only with the onset of another electrode reaction. The variation of the surface concentration of the reactant is given by the Sand equation:

$$c_0(t) = c^* - \frac{2j}{zF}\left(\frac{t}{\pi D}\right)^{-1/2} \tag{2.19}$$

The rearrangement of Eq. 2.19 can yield the time period while galvanostatic conditions can be maintained until the onset of another electrochemical process.

In contrast to the current regulation, the electrode potential sets a driving force, as a result of which the flux of the reactant may vary as the electrode reaction proceeds. The most commonly applied experimental setup is when the electrode is surrounded with a stagnant solution and the electrode potential is instantaneously set to a value where the reaction of interest takes place under diffusion-limited conditions. This results in a current decay in accord with the Cottrell equation:

$$j(t) = zFc\left(\frac{D}{\pi t}\right)^{1/2} \tag{2.20}$$

It is particularly important to remark in a work dealing with micro- and nanostructures that the above analysis is valid for macroscopic and planar electrodes with characteristic lateral size of at least several tenths of millimetres and above, and the side effects are negligible. When the electrode size decreases but the electrode is *recessed* (i.e., the solution which contains the reactant is a similarly narrow column as the electrode itself, like in a nanochannel), the above description is still valid. However, if the electrode size is below about 100 μm and it is *not recessed* (i.e., the solution which the reactants can originate from extends to a much larger distance in both parallel and perpendicular directions relative to the electrode surface), the diffusion field becomes spherical. For solving such problems, the Fick equations have to be rewritten in terms of spherical polar co-ordinates. The solution of the relevant equations is available in the general electrochemistry literature. What is important to remember for a proper qualitative picture is that a micro- or nanoelectrodes in a stagnant solution behaves the same way as a macroscopic electrode in a well-stirred solution in the sense that a well-defined limiting current can be measured for these electrodes. The shape of the polarization curve on a microelectrode is similar to that shown in Fig. 2.6, although the concentration profiles are different. It is important to know about microelectrodes that the limiting current scales with their radius, unlike for macroscopic electrodes on which the current is surface-proportional. The steady-state limiting current on flat disc-shaped a microelectrode can be written as

$$I_L = 4zFDc^*r \tag{2.21}$$

where the meaning of the variables is as usual and the dimensionless numeric factor depends on the electrode shape.

2.8.3 *Basic Voltammetric Experiments for Metal Deposition*

The most commonly applied laboratory method nowadays to obtain a basic picture on the electrochemical reactions is the *cyclic voltammetry*. For such experiments, the solution usually contains only one form of a redox couple. The concept of the method is based on a potential scan at a fixed rate ν from a potential where no reaction can take place to a potential where the electrode reaction of the solute proceeds under diffusion-limited conditions, and then, the sweep is reversed. During the backward scan, the product can be detected, provided that it does not undergo any further reaction and the process leading to its formation is reversible. The cyclic voltammogram obtained for a fully reversible reaction of a solute is shown in Fig. 2.8. It is to be noted that a large variety of the cyclic voltammograms can be obtained as the electrode reaction becomes more complicated (subsequent electrochemical reactions, adsorption of the reactant/intermediate/product, chemical reaction steps proceeding or following the electron transfer step etc.).

As opposed to a reaction scheme where both the reactant and the reaction product are solutes, metal deposition and dissolution lead to different cyclic voltammograms. While the study of the electrode reaction of solute compounds often requires a wide range of sweep rates up to several Volts per second or even to higher rates, metal deposition experiments are seldom carried out with scan rates more than 10 mV s^{-1}. Hence, the occurrence of the peaks is less important, and different other features of the voltammograms are to be observed.

Figure 2.9 shows cyclic voltammograms that can be obtained for solutions of low metal ion concentration (typically, $c < 10$ mM) at a small sweep rate ($\nu = 1\ldots5$ mV s^{-1}). If the metal deposition process is reversible and the deposition takes

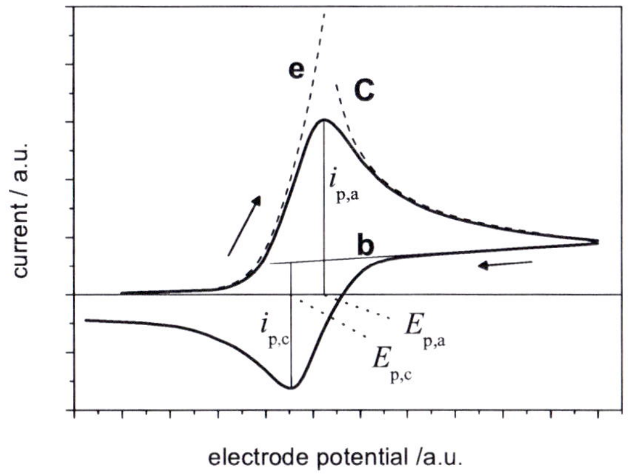

Fig. 2.8 Cyclic voltammogram for a solute that undergoes a simple electrochemical transformation. Arrows indicate the sweep direction. **e** exponential current rise corresponding to a fully activation-controlled reaction; **C**: current decay corresponding to the Cottrell equation; **b**: baseline for the negative-going scan. Indices indicate the anodic (a), cathodic (c) and peak (p) values of the electrode potential and the current

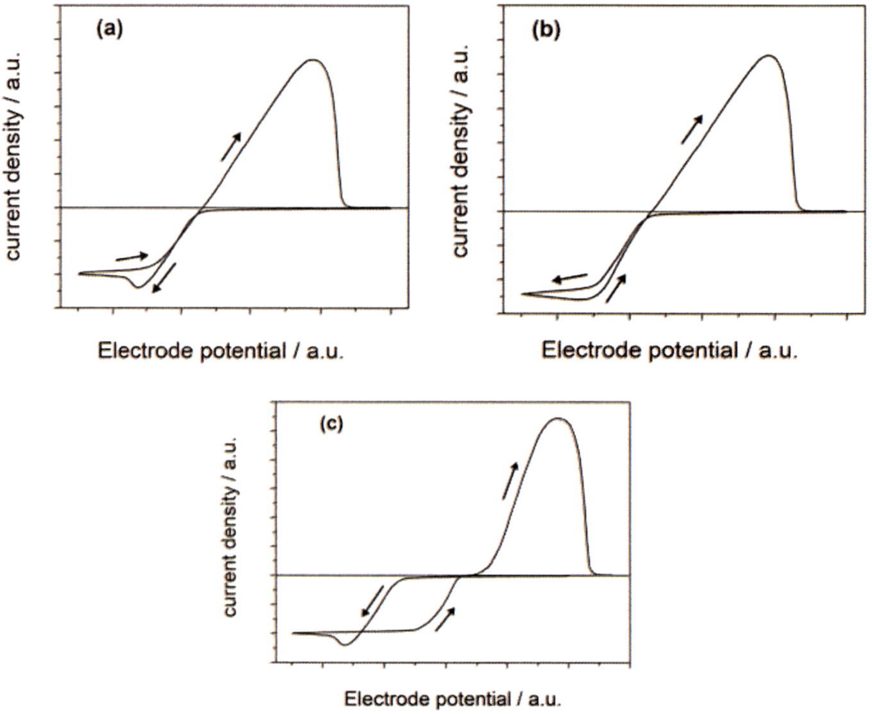

Fig. 2.9 Low-rate cyclic voltammograms for metal deposition from solutions of small metal ion concentration (c_{Me} < 10 mM). **a** Metal with high exchange current density, reversible deposition, smooth deposit. **b** Metal with high exchange current density, reversible deposition, increasing surface area during the deposition. **c** Deposition with a nucleation barrier, metal with small exchange current density

place without a nucleation barrier, the onset potential for the deposition and for the dissolution is essentially the same (Fig. 2.9a, b). Such voltammograms can often be obtained if the exchange current density of the metal electrode is large (Pb, Cu, Ag deposition from non-complexed electrolyte solutions). If the metal thus produced does not develop dendrites even in the diffusion-limited potential regime, the positive-going (backward) segment of the cyclic voltammogram exhibits smaller current densities than the negative-going segment, both because of the solution depletion and the capacitive effects due to the sweep (Fig. 2.9a).

However, if the metal being deposited develops dendrites, the surface area increases quite much simply because of the large amount of the deposit, and a current increase can be seen in the positive-going segment of the voltammogram (Fig. 2.9b). This trend is explained with the fact that the metal ion transport around protrusions and particularly around dendrites is more intensive, and the planar diffusion conditions are no longer valid. Tendencies on the surface roughness change will be further discussed in Sect. 2.11. Figure 2.9c shows the case when the nucleation barrier of the metal being deposited hinders the deposit formation at the equilibrium potential of the metal ion/metal system, and a significant anomalous hysteresis occurs in the voltammogram. In contract to the low-concentration case analysed here, at high metal ion concentration the cathodic current increases exponentially without a limiting current, and the cathodic part of the voltammogram hence becomes nearly featureless.

As the cathodic part of the cyclic voltammogram for a metal deposition is different from the case discussed in connection with Fig. 2.8, there is a deviation in the anodic part, too. For an actively dissolving metal, the dissolution starts with an exponential increase in the current since the source of the reactant, i.e., the metal, is not limited as long as the anion transport can compensate the appearance of the metal cations in the vicinity of the electrode. However, when a thin metal deposit produced in the earlier period of the same experiment is fully consumed, the anodic current drops very suddenly, and the current peak exhibits the shape of the so-called *stripping peak* (all graphs in Fig. 2.9). Further cases of the anodic behaviour of the metals and alloys will be discussed in Sect. 2.13.

2.9 Towards the Electrodeposition of Metals: Crystals and Their Surfaces

2.9.1 Basic Crystallography

Metal surfaces used in the common practice of electrodeposition are seldom perfect; rather, technical work pieces are essentially always disordered because of the presence of crystals of different phases and also of different orientation. Nevertheless, the elucidation of the deposition processes is based on the analysis of single-crystal faces, both in the theoretical approach and in experiments. Metallic elements can form face-centred cubic (fcc), hexagonal close-packed (hcp) and body-centred cubic (bcc) crystals. The structures of a few metals are listed in Table 2.1.

For the sake of simplicity, the characteristic features of crystal faces are often explained by using fcc crystals. This has an important practical reason. Metals commonly used in electrochemical experiments as single crystals (platinum-group metals, gold, silver and copper) all have an fcc structure, and the atomic arrangement of some facets of the fcc and hcp crystals is very similar. Metals having a protecting oxide layer and those spontaneously reacting with the water cannot be studied in

Table 2.1 Crystal structure of a few metallic elements

fcc	hcp	bcc
Au, Pd, Ag, Pt, Ir, Cu, Ni, Pb, Al Co (high temperature)	Cd, Cr, Zn, Ti Co (low temperature)	Nb, Fe, V

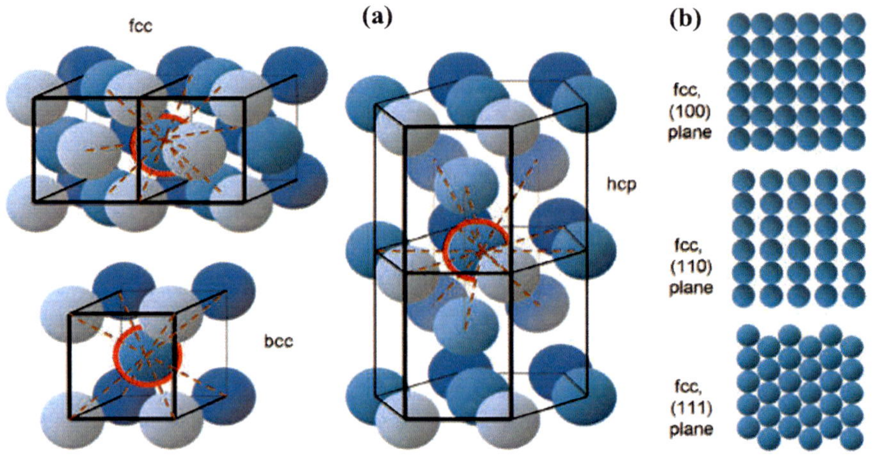

Fig. 2.10 **a** Atomic arrangement in the most important crystal structures. Colours vary with depth and dotted lines indicate the coordination mode of the central atom marked with a solid border. **b** Atomic arrangement on the most important surfaces of an fcc crystal

electrochemical processes in the same way, but it is assumed that the basis of the coating formation on these metals is essentially identical.

Figure 2.10a shows the atomic arrangement of the three basic crystal structures. The basic cutting modes of an fcc crystal are presented in Fig. 2.10b. As can be seen, both the interatomic distances of the surface atomic layers and the in-plane coordination number of the surface atoms vary as a function of the cutting mode. The 111 plane of the fcc crystal shows the same hexagonal symmetry as the 001 plane of the hcp crystal. Such crystal phases are energetically favourable and often form if crystals are formed by solidification from melt or by evaporation since they exhibit a high packing density and the highest possible in-plane symmetry.

It is mentioned for the sake of completeness that the energy of the topmost atomic layer at the crystal surface is energetically non-identical to an atomic layer of same direction within the crystal simply due to the absence of atoms in one direction. This may (although not necessarily does) lead to the spontaneous modification of the surface atomic layer of a metal. The two possible modification modes are the *surface relaxation* (where the distance between the topmost and next atomic layers changes only) and the *surface reconstruction* (where the packing density and the symmetry of the surface atomic layer also differ from that of the atomic layer of similar direction within the crystals). During surface reconstruction, the atomic layer at the surface typically shows a larger packing density and/or a higher symmetry than the atomic layer underneath. Surface reconstruction is a common phenomenon during electrochemical potential cycling of otherwise inactive noble metals.

2.9.2 Defects on Crystal Surfaces and Within Crystals

Perfect single-crystal surfaces are very difficult to prepare. A miscut of 0.5° means that a *step edge* is formed at about every 115th atom; or, in other word, the width of a *terrace* is 115 atomic layers along the surface. The atoms at the edge of the terrace are identical in this case. Although the orientation of a slightly miscut layer can also be expressed with the usual *hkl* Miller indices (like 0-1-115 in the above example), this is never used since the overwhelming majority of the atoms at the surface are in the same position and show the features of one particular low-index plane. Such crystals can be used to study the behavior of a crystal plane with nearly no compromise.

If the slight miscut of a crystal is bidirectional, the overall picture is similar to the case of unidirectional miscut, but the atoms at the edge of the terrace are no longer identical but *kink positions* are also produced (see Fig. 2.11). The kink position is often called the *half-crystal position* since the coordination number of the atom in the kink position is the half of that of an atom within the crystal. The kink position is also special in the sense that the incorporation of a newly arriving atom along the terrace edge into the kink position reproduces the same surface feature. If we count the atoms in the crystal, addition of an atom to or removal of an atom from the kink position leaves the number of the surface atoms unchanged but modifies the atoms in the bulk by 1. The kink position is the typical site where near-equilibrium growth can takes place, should this happen by vapour condensation or by electrodeposition. Similarly, the near-equilibrium evaporation and the electrochemical dissolution both proceed by removal of an atom from the kink position.

Crystal surfaces may also have various defects. If the crystal itself is perfect but the surface exhibits irregularities, one can find terrace edges and kink atoms along these edges. Beside these unequal surface sites, adatoms and atomic vacancies may also form that can accumulate as clusters and vacancy clusters, respectively. The combination of such surface imperfections can lead to faceting during the growth and the deviation of the real surface planes from that defined by the crystal orientation.

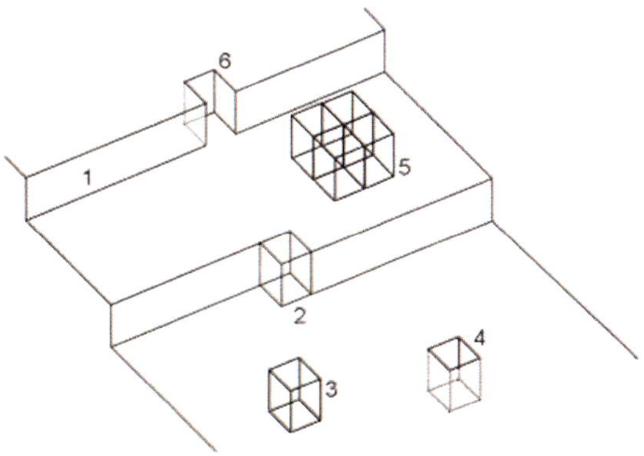

Fig. 2.11 Schematic view of a crystal face with terraces and various defects. 1: terrace edge, 2: atom in the kink position, 3: adatom, 4: surface vacancy, 5: adatom cluster, 6: terrace edge vacancy

The surface concentration of the various defect sites on a single-crystal surface in equilibrium is simply determined by their energy of formation. It is to be emphasized that the imperfections of the surface of a perfect crystal may move dynamically and may not be imaged if the time scale of the imaging method used is higher than the transformation time of the imperfection.

If the crystal itself is also imperfect, it can contain *dislocations*. The concentration of these imperfections strongly depends on the prehistory of the metal (like mechanical deformation). *Edge dislocations* (shown in Fig. 2.12) are characterized by the inclusion/loss of a complete atomic plane from a particular line of atoms along a low Miller index direction. When an edge dislocation reaches the surface, the symmetry of the atomic arrangement is locally violated.

Another typical defect is the *screw dislocation* (shown in Fig. 2.13) where the dislocation center cannot be circumferred by making steps along the same (or equivalent) direction. Screw dislocations are also common growth centres that do not annihilate as the growth proceeds but may also give rise to the occurrence of facets aligning to the original surface by essentially retaining the texture (shown in Fig. 2.14). The

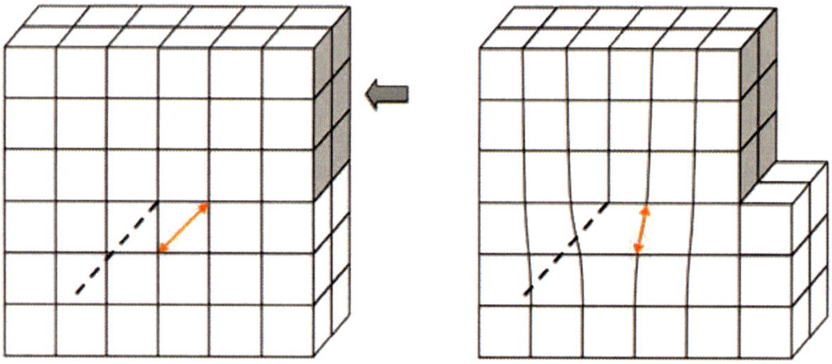

Fig. 2.12 Formation of an edge dislocation by the distortion of a perfect crystal. **Left**: the perfect crystal; **right**: a crystal with edge dislocation. The dashed line shows the so-called Burgers vector of the dislocation, and the coloured double arrow shows the original (left) and final (right) position of the new nearest neighbours as a result of the dislocation formation

Fig. 2.13 Crystallographic representation of a screw dislocation

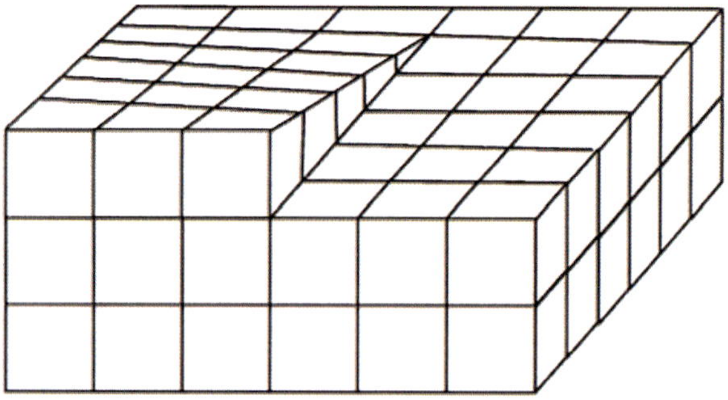

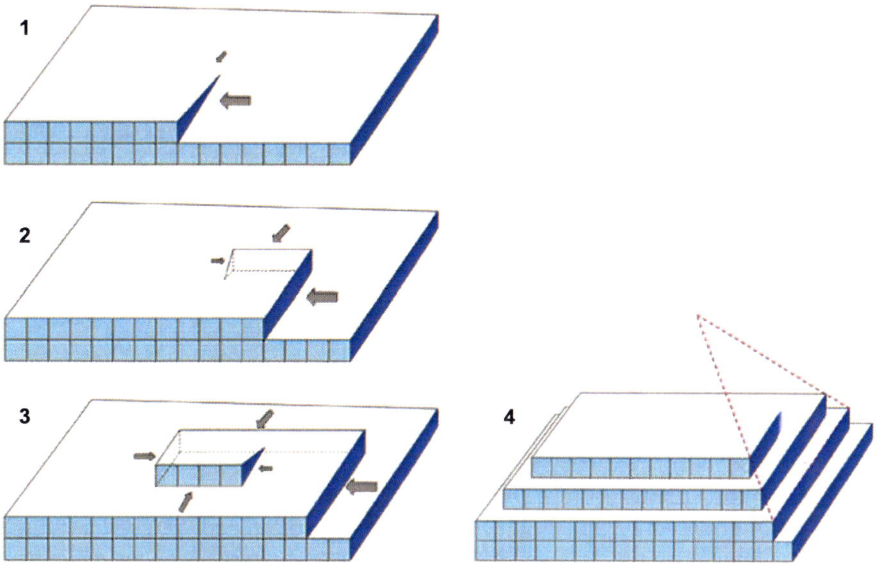

Fig. 2.14 Crystal growth along a screw dislocation. Arrows indicate the relative rate of the growth from various directions. Dashed line in phase 4 of the growth process shows the formation of a facet as the crystal growth proceeds

crystal growth along a screw dislocation is particularly interesting since the growth direction (i.e., the normal of the mean surface) and the crystal planes being formed are different.

Finally, it has to be mentioned that alloys, compound semiconductors and electrically conducting other metal compounds (like oxides of chalcogenides) can form more complicated phase structures, but their surface structure and growth follow the same principles as those that are used to describe metallic elements.

2.10 Nucleation During Electrochemical Phase Formation

For the comparison of the electrodeposition with another electrochemical reaction taking place on a noble metal, a very important contemplation issue has to be clarified. When the electrochemical reactions of soluble compounds are studied without the formation of a new phase, the electron conductor (used as the working electrode), which is an inert material like noble metal or carbon, can be characterized prior to the experiment. In such cases, electrode properties such as the texture, surface roughness and, surface composition, seldom vary. However, when a metal is either deposited or dissolved, one encounters an ever-varying surface. Although the *in situ* characterization of such a dynamically changing surface is seldom possible, it has to be kept in mind that many features of the experimental data can be explained by such variations. In the forthcoming chapters, an overview will be given concerning structural and composition effects that accompany metal deposition and dissolution.

The nucleation of a phase always precedes its growth. In this chapter, two issues will be dealt with: first, the general features of the nucleation of a deposit on a foreign substrate at the atomic level; secondly, the evolution of the grain structure from the substrate-dominated grain structure to the steady-state growth. The major trends in steady-state electrocrystallization, especially in relation to their solution chemistry and the current density, will be discussed in Sect. 2.11.

2.10.1 Nucleation and Growth Modes

When a foreign material is deposited onto a crystalline surface, two parameters together play a crucial role in the determination of the nucleation mode. These parameters are: (i) the interaction strength of the deposit atoms to the substrate atoms, as referred to the interaction strength between the atoms of the deposit in the bulk; and (ii) the relationship of the nearest neighbour distances of the atoms of the contacting phases. The latter is often summarized by a ratio called the relative misfit: $\Delta d/d_S$ where d_S is the nearest neighbour atomic distance in the bulk of the substrate and Δd is the difference of the bulk nearest neighbour distances of the two contacting phases. The temporal evolution of the deposit in the three major growth modes is summarized in Fig. 2.15. The trends highlighted in Fig. 2.15 have a general validity, whether or not the nucleation is assisted by electrochemical discharge or not (e.g., also for evaporation and sputtering).

During the Volmer–Weber type growth, E(Su-Me) < E(Me-Me), Su and Me denote the substrate and deposit atoms, respectively, and E refers here to the adhesion energy (strength of interaction). In such a case, regardless of the relative misfit of the contacting phases, the growth starts with three-dimensional nucleation, and the formation of a continuous deposit, if any, takes place via the coalescence of the

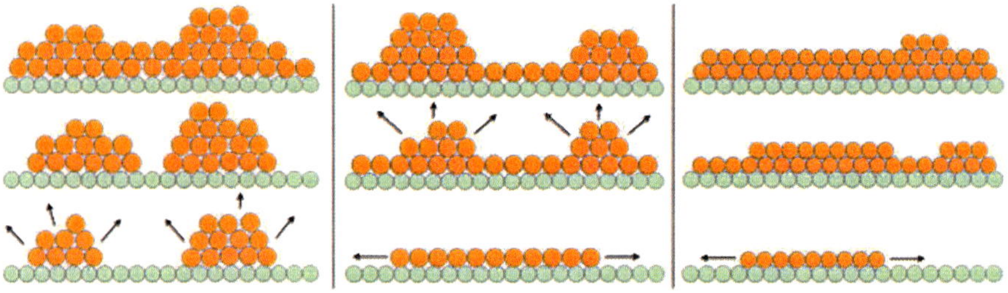

Fig. 2.15 Schematic representation of the tree major nucleation and growth modes. From left to right: Volmer–Weber, Stransky–Krastanov and Frank–van der Merwe type layer formation; from the bottom to the top: temporal evolution of the deposit. Light and dark circles indicate the substrate and the deposit atoms, respectively. Redrawn after Fig. 2.3 of Ref. [9]. Copyright (2004), with permission from Elsevier

islands produced in the nucleation period. It is the Volmer–Weber type growth that should be kept in mind when the nucleation takes place on a defect of the surface instead of a perfect crystal plane. In such cases, the large difference in the activation energy of the nucleation favours the defect sites whose surface density can also be taken as the upper limit of the number of grains.

For the Stransky–Krastanov type nucleation, the role of the interaction energy is decisive: $E(\text{Su-Me}) \gg E(\text{Me-Me})$, and hence, a monoatomic coverage of the deposit can form at the substrate, despite the relative misfit is definitely not zero. In other words, the strong interaction between the substrate and the deposit stabilizes the first (potentially incomplete) atomic layer of the deposit. Later, this tends to grow further as if islands were formed on the first atomic layer, and the stress due to the misfit is relaxed in the upcoming atomic layers, finally leading to the natural relaxed atomic distances of the deposit crystals. The Stransky–Krastanov-type growth is an intermediate mode between the island-like Volmer–Weber type growth and the heteroepitaxial growth. The latter is called the Frank–van der Merwe type nucleation, where not only the $E(\text{Su-Me}) > E(\text{Me-Me})$ relationship holds but the relative misfit is very close to zero.

The analysis of the aforementioned nucleation and growth modes is essential for the understanding of electrochemical methods that are used for the study of the phase formation. When a Volmer–Weber type nucleation takes place, the temporal evolution of the surface density of the newly formed crystallites is a key parameter. The density of the grains plays a crucial role in the variation in the deposition current density until the surface of the substrate is completely covered with the deposit. When the substrate coverage becomes full at all points, the deposit has a significant mean thickness, and the surface roughness may also differ considerably from that of the bare substrate due to the uneven growth.

The number (or surface density) of the newly formed grains always has to be studied on the time scale of the experiment. If all grains form in a very early phase of the experiment and the number of nuclei is practically constant for the rest of the observation period, one can speak about an *instantaneous nucleation*. In contrast, it is also possible that the age of the newly formed grains has a wide distribution in the full time scale of the experiment, which is equivalent to that the grain nucleation has an even probability throughout the observation. This is the case of the *progressive nucleation*. The two extreme cases of the nucleation can be regarded as a result of the same nuclei formation rate law with different rate constants:

$$\frac{\mathrm{d}N}{\mathrm{d}t} = -A(N - N_0) \tag{2.22}$$

where N_0 is the number (or density) of the surface sites that are suitable for the nucleation of a new grain, N is the actual number (density) of the grains at the surface. The parameter A is the nucleation rate constant, which can be taken as invariant in the first approximation only since it may vary also as a function of the near-substrate concentration of the reactants that is consumed by the growth of the already existing grains. The solution of the 2.21 differential equation is:

$$N = N_0\left[1 - \exp(-At)\right] \quad (2.23)$$

If we accept Eq. 2.21 as the nucleation rate law, then two extreme cases can be distinguished. If $A \gg (\Delta t)^{-1}$, we get that $N = N_0$. Here, Δt is the period between two consecutive observation events during the experiment (or, in other words, $1/\Delta t$ is the sampling frequency). The result means that the sites that are appropriate for nucleation are saturated with nuclei from the very beginning of the process, which is the definition of the *instantaneous nucleation*. However, if $A \ll (\Delta t)^{-1}$, $N = N_0At$ is obtained, indicating an even nucleation probability throughout the entire observation period, as defined for the *progressive nucleation*.

The study of the Volmer–Weber type nucleation in electrochemical systems usually requires electrolyte solutions that exhibit much lower concentrations than baths used in the galvanic industry (at most a few tens of mmol dm^{-3} versus about 1 mmol dm^{-3}, respectively). This is because the instantaneous or progressive nature of the nucleation is already hidden once the newly-formed grains coalesce. Also, the substrate used in the nucleation study has to be structurally rather incompatible with the deposit in order to fulfil the conditions at which this type of nucleation can take place at all.

The deposit formation usually does not start at a small overvoltage (as referred to the equilibrium potential of the Me^{z+}/Me electrode) because of the energy barrier of the nucleation. The typical experimental conditions include the potentiostatic polarization with a potential step to a value so that the deposition becomes diffusion-limited. Under such conditions, a spherical diffusion field develops around all grains, as shown in various representations in Fig. 2.16.

Although Fig. 2.16 indicates well the overlap of the diffusion field of the grains as time passes, the limits of the schematic image should also be kept in mind (e.g., perfect hemispherical growth in spite of the laterally uneven supply of the precursor, disregarding the crystalline nature of the deposit and the possible directional preference of the crystal growth etc.). As the grains grow and the solution becomes depleted with respect of the precursor ions, the diffusion field is planarized, and the same transport conditions are obtained as during the Cottrell experiment.

The chronoamperometric transient in the early phase of the process bears information on the nature of the nucleation. Here a few works of Sharifker are referred to [14, 15] without the deep mathematical details. The treatment of the chronoamperometric data includes a normalization of the current density with respect to the peak current density ($j^* = \{j/j_{MAX}\}^2$). The time also has to be dealt with in a dimensionless form, using the time belonging to the current density maximum as a normalization parameter ($t^* = t/t_{MAX}$). Typical results for the major nucleation modes are presented in Fig. 2.17. The concave or the convex nature of the $j^*(t^*)$ function near $t^* = 0$ is usually regarded as a diagnostic criterion for the instantaneous and progressive nucleation, respectively. Chronoamperometric transients for more complex nucleation cases are also available in the literature (see Chap. 5 or Ref. [3]).

In contrast to the Volmer–Weber type growth, the initial stage of the phase formation can be completely different if the substrate–deposit interaction is strong enough to stabilize an at most monoatomic layer of the deposit atoms on the substrate. In

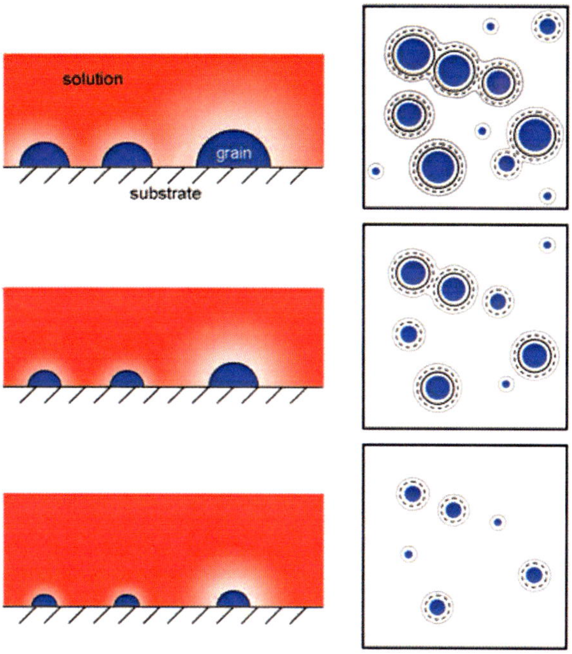

Fig. 2.16 Grain growth on a surface as presented by using the example of the progressive nucleation. Figures from the bottom to the top show the growth of the nuclei with the temporal evolution of their diffusion field. **Left:** cross-sectional view; the colour of the solution indicates the level of depletion (light colour around the grain refers to the large degree of depletion with respect to the reactant). **Right:** top view of the surface with the growing nuclei and the depleted solution around them. The thickness of the line (from thin dotted through dashed to thick solid line) indicates the degree of the solution depletion

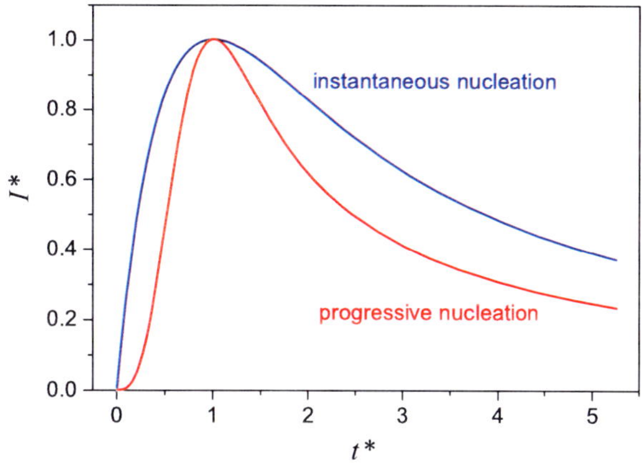

Fig. 2.17 Chronoamperometric transients for the instantaneous and progressive nucleation if the deposition process is diffusion limited. For the explanation of the quantities at the axes, see the text. Redrawn after Fig. 5 of Ref. [14]

this case, the formation of the initial layer of the deposit atoms takes place at more positive potential than the equilibrium potential of the Me^{z+}/Me system. This is called *underpotential deposition* (UPD). A few important key features of the UPD phenomenon are given below:

- UPD is not simply a feature of a substrate–deposit pair but it is also highly specific for a particular crystal plane of the substrate since the interaction of the atoms depends, beside the quality of the atoms themselves, on the surface atomic configuration. It is possible that one crystal face of a substrate can accommodate a monoatomic layer of a deposit, but another crystal plane of the same substrate cannot.

- UPD can be studied with substrates that are stable in contact with the solution and themselves do not dissolve (typically noble metals). The deposit which makes the UPD layer is less noble than the substrate. The UPD layer should form in the stability range of the solvent, otherwise voltammetric observations are not possible because of the background current of the solvent decomposition. Although it is theoretically possible that the substrate–deposit interaction itself provides favourable condition of the UPD with non-noble substrates, such cases have not been identified because of the technical difficulties.
- A UPD layer is always ordered. The special periodicity of the UPD layer along the substrate surface is at most a few times of the nearest neighbour distance of the substrate surface atoms.
- The maximum theoretical coverage of a UPD layer is 1 (as referred to the number of the surface atoms of the substrate), but the theoretical maximum cannot be achieved in each single system. However, partial UPD layers may also form. In some systems, UPD layers with more and more densely-packed adatom configurations can form at more and more negative electrode potentials prior to the formation of a bulk deposit.
- The modification of the nature of the atoms at the solid surface by the UPD process is often accompanied with a change of the coverage with adsorbed anions. The change in the surface coverage with adsorbed anions leads to a change of the surface charge, too (see Sect. 2.5 on the capacitive effects). Hence, the charge passing through an electrode during the formation of an UPD layer is often not indicative of the total surface coverage. Instead, the charge corresponding to the reduction of the ions of the resulting UPD layer must be corrected with the charge corresponding to the change in surface coverage with co-adsorbed (or desorbed) anions, too. The electrode potential where a UPD layer can form may also depend on the nature of the anions present in the solution since the co-adsorbed anion layer has an energy contribution to the stabilization of the UPD layer.
- As the coulometric data referring to the atomic density of a UPD layer may misestimate the real coverage due to the charge corresponding to anion co-adsorption, local imaging methods (like scanning probe techniques) also show the topmost species only, should they be the UPD atoms of the co-adsorbed anions. Therefore, the identification of the coverage and the complete structure of a UPD layer is often a very involved task.

A detailed description of the theoretical background of the UPD phenomenon and a collection of the substrate plane–deposit pairs exhibiting UPD can be found in Ref. [6]. A former issue of this book series was fully devoted to the discussion of UPD [16]. The interested readers are advised to study these works. Below, a few important practical aspects of the study of UPD are summarized only that can facilitate the understanding of the UPD-based nanostructure formation.

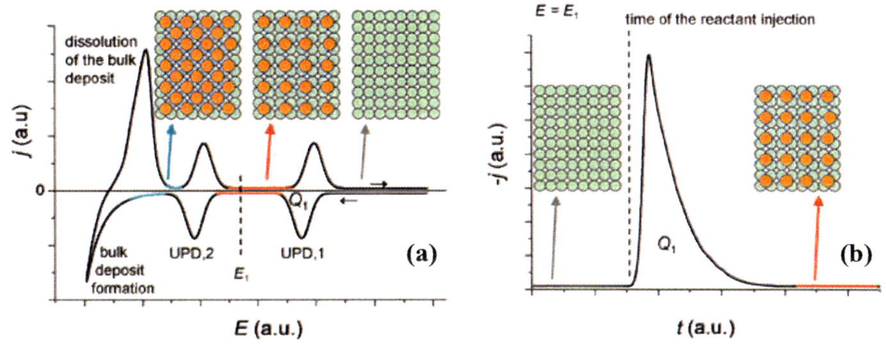

Fig. 2.18 **a** Schematic cyclic voltammogram for a UPD process taking place by the formation of two different packing densities. Arrows and coloured intervals in the voltammograms indicate the intervals of the relevant surface coverage as shown in the insets. **b** Potentiostatic current transient during the formation of a UPD layer. Quantities in the two graphs indicate the appropriate potentials and charges that belong to the current peak (CV) and transient (chronoamperogram)

The investigation of the UPD requires a very clean system and well-cut single crystals as substrate. The concentration of the reactant is mostly within the 0.1–5.0 mM range. When UPD is studied with a potential scan (like cyclic voltammetry), the formation of the stages of the UPD layer manifests itself as a relatively sharp current peak in the voltammogram. The charge corresponding to each peak refers to the assembly of all processes taking place (cation discharge and anion co-adsorption). A schematic voltammogram for a two-stage UPD layer formation and its relation to the surface coverage and the formation potential of a bulky deposit are presented in Fig. 2.18a.

In the practical application of a UPD process, there are cases when the potential is fixed and the reactant to be reduced as an UPD material is added to the system. In this case, a current transient can be measured. Figure 2.18b shows the result of this method, and the relation to the working potential and the charge transient is also compared to those measured in the cyclic voltammogram.

2.11 Major Factors of the Grain Structure of Electrodeposited Metals

The steady-state electrochemical growth of a metal can be elucidated from the atomistic picture described in Sect. 2.9. When the deposition takes place at a low rate, the formation rate of new crystals is negligible, and typical growth centres as terrace edges and screw dislocations serve as incorporation points of the discharged ions. The term “low rate” has no absolute scale but can be compared to the exchange current density of the metal in contact with the solution containing its ions. If the deposition rate as compared to the exchange current density is small, the dominant process is the growth of existing crystals, and the surface density of the adatoms at the crystal planes differs from its equilibrium value to a small extent only. In parallel, it is also true that the surface concentration of the precursor ions of the growing metal is just slightly smaller that its bulk concentration, even though the deposition leads to a little depletion near the cathode surface (see the bottom graph on the right-hand side of Fig. 2.6).

A "thumb rule" for the interrelation of the crystal size and the deposition conditions was first outlined by Dini [2] and later completed by Winand [17]. According to Dini's concept, essentially all factors lead to a reduction in the crystal size of a metal being deposited which decrease the surface concentration of the precursor metal ions. In contrast, if a modification of the deposition conditions leads to a better replenishment of the precursor ions, an enhancement of the crystal size is expected. In accord with these considerations, the increase in either the stirring rate or the temperature as well as the application of a more concentrated solution of the salt of the precursor metal results in a grain coarsening. The increase in the deposition rate, should it be achieved by the application of either a more negative deposition potential or a larger cathodic current density, leads to a grain refinement (i.e., smaller crystallites).

A factor that requires the consideration of additional effect is the application of chemicals in the plating bath that modify the deposition. The compounds applied for such purposes do not make a distinct compound family since inorganic salts and organic materials with a large variety of functional groups belong to these bath components. The common name of these compounds, *additives*, refers to that they are often a minor component in the plating bath. Additives themselves may also undergo electrochemical reactions and their decomposition products may incorporate into the deposit, too; nevertheless, their decomposition is a less important process. Additives are somewhat similar to catalysts and inhibitors in the sense that they modify a process without being transformed. Additives are usually classified on the basis of the primary impact of their application. Brighteners, levelling agents and stress relievers are the most common types of additives. In spite of the different target parameters, levelling agents and brighteners exhibit similar impact mechanisms.

The common mechanism by which additives modify metal deposition is that they adsorb at the typical growth centres. Since the atoms at the near-equilibrium growth centres of the metal crystal surface are the most exposed ones to the solution and least surrounded by atoms of the crystal, the adsorption of a molecule with one or several free non-bonding electron pair(s) is more likely on these crystal surface features. The adsorption of the additives reduces the probability that the crystal growth proceeds but increases the nucleation of new crystals from the adatoms that form during the discharge of the metal ions.

It is important at this point to highlight the impact mechanism of the additives for both potentiostatic and galvanostatic deposition. If potentiostatic conditions are applied and the concentration of additives is increased, the current decreases since the same driving force can lead to the discharge of less ions (the deposition is hindered). The discharge of the metal ions will result in an increase in the surface concentration of the adatoms whose dissolution rate also increases. As a result, the application of additives leads to a smaller current density and a more frequent nucleation of the new crystals. Since the rate of the process decreases, the additives have an inhibitory effect on the deposition. In galvanostatic mode, the additives act similarly, and the driving force (overpotential) must increase to maintain the same process rate, and the concomitant consequences on the grain refinement are the same as for potentiostatic mode.

Winand's treatment of the parameters impacting metal deposition can unify the two major trends, the one related to the transport and the other in connection with the crystal growth and nucleation. This approach leads to a two-dimensional diagram as presented in Fig. 2.19.

The horizontal axis in the diagram is related to the current density. In order to make it closer to a dimensionless representation (and hence, independent of the actual bath and metal deposited), the current density on this axis can be replaced with a j/c ratio. An alternative substitution method is to use the j/j_{LIM} parameter, j_{LIM} being the limiting current density. Nevertheless, j_{LIM} is not surely known for a process

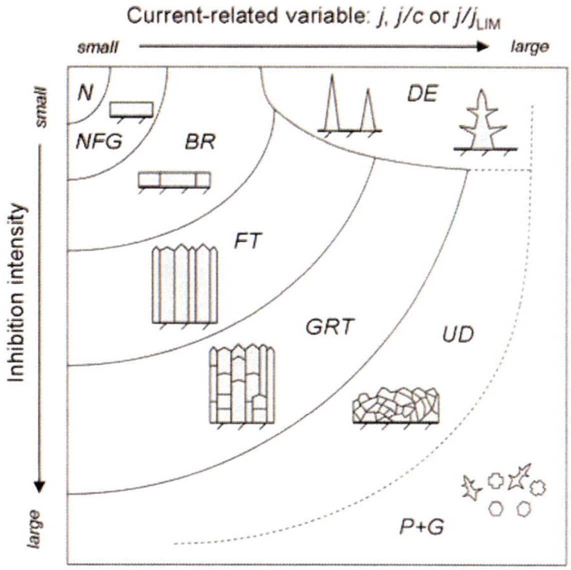

Fig. 2.19 Relationship of the applied current density, the inhibition efficiency and the typical crystallite size, shape and orientation in the deposit. *N*: no deposit formation; *NFG*: nucleation-free growth along the growth centres that serve as deposition sites near the equilibrium (screw dislocations and terrace edges); *BR*: base reproduction type growth where the deposit conserves the crystallographic orientation of the substrate; *FT*: field-induced texture (where the texture tends to differ from that of the substrate); *GRT*: grain-refined textured growth (where a dominant texture can be found, but the growth takes place via frequent re-nucleation along the growth direction); *UD*: unoriented (non-textured) dispersion type deposit; *DE*: dendritic growth; $P + G$: powder formation accompanied by gas evolution. The relative importance of the domains in the diagram may vary depending on the deposit material. Concept adapted from [17]. The figure was redrawn from Fig. 7.15 of Ref. [8]. Copyright (2008), with permission from Elsevier

in which the metal deposition is accompanied with a side process like hydrogen evolution. Therefore, no wonder that the diagram parameters are always given rather qualitatively than in a strictly quantitative manner.

The ordinate in the Winand diagram indicates the inhibition efficiency, which is even more ill-defined than the parameter related to the current density. Although the impact of various additives is difficult to compare, it is plausible that the increase in the adsorption bond strength to the surface is related to a larger inhibition effect. Also, the increase in the additive concentration (at least, below the saturation concentration) leads to a larger coverage of the surface, and hence, it increases the deposition efficiency. All these trends can be phrased also with alternative terms: a larger surface coverage with additives and a larger residence time of an additive molecule at the surface is related to an enhanced inhibition effect.

In spite of the lack of the fully quantitative treatment, the Winand diagram plays a very important role in the formation of our contemplation on the simultaneous treatment of the bath operation parameters in relation to the deposit properties. One of the major messages of the diagram is that the bath operation at high current density may drive the planar metal coating growth to a regime where dendrites are developed, especially when the process is accompanied with gas evolution, too. Another important piece of information to remember is that by using an additive-rich bath, there is more chance to obtain a fine-grained deposit than with a non-inhibited bath. It is to be noted that complexing agents play a very similar role to other additive types. The excess free molecules with complexing capability can also attach to the surface and block the growth. Therefore, it is very typical that baths with complexed metal salts yield a finer grain structure under otherwise similar conditions than non-complexed baths of the same metal.

It is indispensable to mention that the modulation of the deposition current may have an immense impact on the grain structure. *Pulse plating* usually leads to a very fine grain structure. The principle of pulse plating is that a current pulse is applied for a short time, often with much larger current density than the mass transport-limited current density corresponding to the steady-state deposition. Due to the high current density, the surface concentration of the adatoms will be high that favours the nucleation as opposed to the growth of the already existing crystals. The on-time is fairly short (typically falls in the 1–100 ms range) and the pulsating depletion layer is much thinner than the depleted solution layer during d.c plating. After the on-time, an off-time follows the deposition period when the concentration of the precursor ions near the substrate relaxes and approaches the original (bulk) concentration. The repetition of this period, the on-time and the upcoming off-time, leads to a deposit in which the nucleation zone is not restricted to the close vicinity of the substrate but is extended to the entire deposition time, which results in a fine-grained structure. Another advantage of the pulse plating is that the parts of the surface with different accessibility can be plated in a much more even manner than with d.c. plating, simply because the thickness of the depleted solution layer follows the shape of the work piece. The theory and practice of pulse plating are available in various excellent reviews [18, 19].

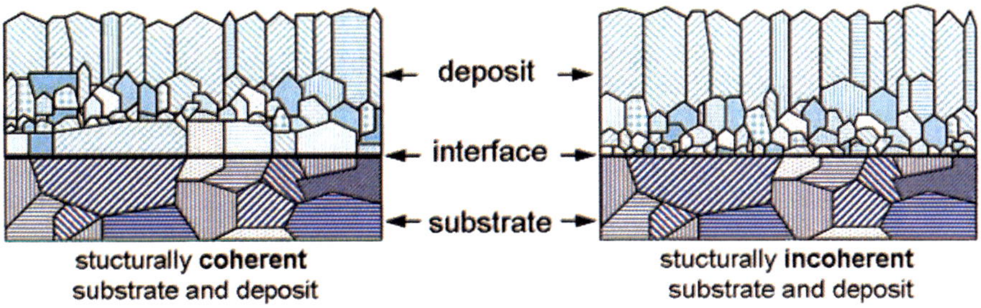

Fig. 2.20 Cross-sectional view of the deposit growth on a structurally coherent (**left**) and incoherent substrate (**right**) if the deposit has a dominantly columnar structure with a pronounced texture. Redrawn after Fig. 1.54 of Ref. [9]. Further cases are also discussed in Sect. 1.4 of Ref. [9]

Figure 2.18 offers a schematic picture on the grain structure during a steady-state electrocrystallization process. However, the complete grain structure of a deposit is the result of the substrate–deposit interaction (in the nucleation period and shortly thereafter) and the deposit–solution interaction (in the steady-state). Since the stable grain structure (size and orientation) may vary during the growth, a transient may take place as the deposit thickness increases. The most important basic cases of the grain structure evolution are presented in Fig. 2.20 as the cross-sectional view of the deposit.

If the deposit is structurally coherent with the substrate, its crystals can grow, and the original substrate–deposit boundary is difficult to distinguish in the cross-sectional cut of the work-piece. However, as the preferred growth direction under the specific plating conditions does not necessarily coincide with the texture (crystal orientation) of the substrate, new grains can nucleate and the growth mode changes. In such cases, a nearly epitaxial mode is followed by a nucleation zone and a forthcoming deposit with a dissimilar grain structure.

When the deposit is structurally incoherent with the substrate, new crystals must nucleate already at the very beginning of the deposition process. The initial nucleation zone is composed of fine-grained crystals. It is a common trend in the layer formation, should the growth process be of electrochemical nature or not, that the grain size in the initial nucleation zone scales with the deposit thickness. As the layer growth proceeds, the steady-state grain structure sets in and crystals characteristic of the bulk (concerning the orientation, the shape and the size) appear.

It is seldom mentioned in electrochemical textbooks that the layer growth may, and in almost all cases does, lead to a change in the surface roughness. The definition of the root-mean-square surface roughness is:

$$w(l) = \sqrt{\langle (h_i - \langle h \rangle)^2 \rangle} \tag{2.24}$$

where $w(l)$ is the surface roughness observed over the length scale l, h_i is the height of the surface during the ith observation, and brackets < > denote the mean value of the argument. The surface roughness is not a universal property of the surface since it depends on which length scale it is measured. The surface roughness characterization of a surface is based on a series of roughness measurements on a wide range of length scales, and the growth process is described with a series of length scale-dependent measurement for subsequent phases of the layer growth. A review of this field is recommended for further reading on the topic [20].

2.12 Composition Aspects of Alloy Electrodeposition

2.12.1 Selection of the Variables

When polarography is discussed in an electroanalytical handbook, the reduction of metal ions of various kind on mercury surface is often taken as a superposition of parallel processes whose current densities are additive and which do not influence each other. During the formation of an alloy, such thumb rules cannot be created. In most cases, codepositing metals often impacts both the rate and the deposition mechanism of each other. Due to the interaction in the solid phase, the Nernst equation is not suitable to assess the potential range in which a particular metal can be deposited as an alloy component. The thermodynamic relationship of the alloying free energy on the deposition potentials can be found elsewhere [21]. Below, a phenomenological picture of the codeposition process is offered that is hoped to give an efficient help for the beginner electroplater.

First of all, it is essential to realize that the number of possible variables in metal codeposition is quite large, and these variables are not independent. One can choose either the concentration of the salt of a particular metal, the ratio of the concentration of metal salts in the solution or the total metal ion concentration as a technical variable while keeping any other concentration-related variable constant. Concerning the electrical parameter of the deposition, either the electrode potential or the deposition current density can be taken as the independent variable. Beside these variables, other parameters such as the temperature and the hydrodynamic conditions have to be constant. For the general discussion of the codeposition modes, the natural choice is as follows. The total metal ion concentration in the solution is fixed and the metal ion concentration ratio is treated as $c_1/(c_1 + c_2)$; i.e., with a similar expression as the mole fraction in a binary solid. As an electrical parameter of the deposition, the current density is chosen as a technical variable of the experiment. It is important that the current density cannot be larger than the mass transport-limited deposition rate of either of the metal ions if it is present in the solution alone. With this parameter selection, both axes of the composition diagram ranges from 0 to 1, the abscissa indicating the relative solution concentration of the ions of the less noble metal and the ordinate referring to the mole fraction (y) of the same component in the deposit. The line in a composition diagram where $c_1/c_2 = y_1/y_2$ is called the *reference line*.

2.12.2 Basic Codeposition Modes

The most commonly used classification of the codeposition modes of the metals with respect to the comparison of the composition of the bath used and the resulting alloy stems from Brenner who first summarized this topic [1]. Brenner's classification is still the basic approach nowadays. The codeposition modes have been established for metal pairs. For alloys with more than two components, the retention of the mutual codeposition preferences is a principle that often helps to design alloy composition. The basic codeposition modes in accord with Brenner's definition are as follows:

- Equilibrium codeposition
- Regular codeposition
- Irregular codeposition
- Anomalous codeposition
- Induced codeposition.

The first four out of the above mentioned five codeposition modes can be conveniently displayed in a composition diagram as shown in Fig. 2.21a. Equilibrium, regular and irregular codeposition together are also termed as normal codeposition, indicating that the metal whose ion has a more positive deposition potential on its parent metal dominates the deposit as compared to its ion concentration ratio in the bath. Occasionally, transition from one deposition mode to another may also

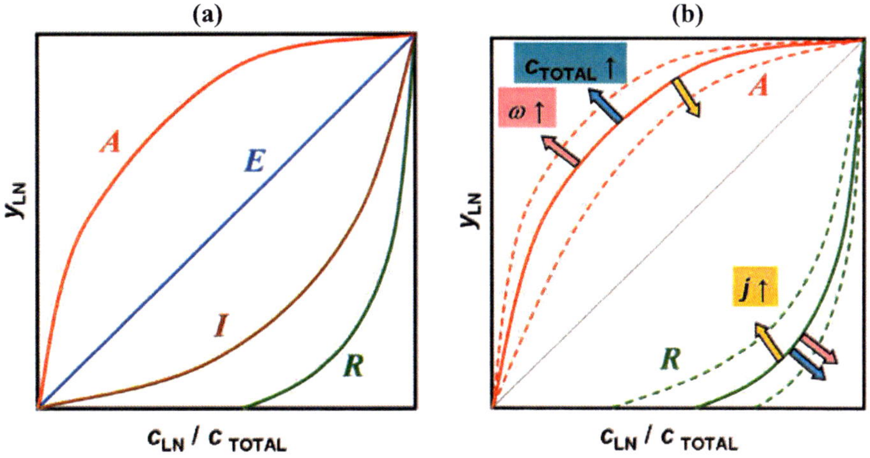

Fig. 2.21 **a** Codeposition diagram with composition lines of the most common codeposition modes: *E*: equilibrium, *R*: regular, *I*: irregular, *A*: Anomalous. The diagram refers to fixed current density, agitation conditions, total metal ion concentration and temperature. **b** Variation of the composition functions as a result of the change in the experimental conditions: *j*: total current density, c_{TOTAL}: total metal ion concentration, *ω*: rotation rate of the electrode (or solution agitation intensity). The grey diagonal line is the so-called reference line. *y* on the ordinate of both diagrams indicates the mole fraction, and the LN index refers to the less noble component

occur with the change in the concentration. For instance, the codeposition mode can be regular as long as the more noble component is deposited alone, but once the ion concentration ratio makes the codeposition possible, the mole fraction of the less noble component in the deposit may increase as if the codeposition mode was anomalous. Below, only the main codeposition mode will be dealt with shortly.

Equilibrium codeposition is characterized with that the incorporation ratio into the solid is equal to the component ratio in the solution, and hence, the composition line coincides with the reference line. The term also includes that the deposit is in equilibrium with the solution, so there is no cementation process if the deposit is left in contact with the solution that it was deposited from. Obviously, the standard potentials of the constituent metal ion/metal systems must be nearly identical, otherwise the independent equilibria could not be set at comparable metal ion concentrations. This codeposition mode is very scarce. Brenner [1] brings only a few examples for this deposition mode (Cu–Bi and Pb–Sn by using baths with appropriate anions), and it practically never appears in the modern literature.

Regular codeposition takes place when the components practically do not influence the incorporation of each other into the resulting deposit. This deposition mode looks like as if only the ion transport in the solution were responsible for the partial current densities of the components. This may happen in two distinct cases: (i) The deposit components are neither miscible nor they can form a metastable alloy but a granular mixture is obtained with (nearly) clean phases of the codeposited metals. Typical examples are, by listing always the more noble element first: Ag–(Fe,Co,Ni), Pb–(Co,Ni), Au–(Fe,Co,Ni). (ii) The components are miscible and indeed form an either thermodynamically truly stable or metastable alloy also during the codeposition process but the A–A, A–B and B–B atomic interactions are energetically so close that it cannot overwrite the deposition preference originating from the list of the standard potentials of the metal ion/metal redox systems. The most important examples include the Cu–Ni and Ag–Cu systems.

For regular codeposition, if the solution is rich enough for the more noble component, it can be deposited as a pure metal. If the metal ion ratio in the solution changes and the concentration of the more noble metal is too small to account for the current density used for the deposition, the less noble metal starts codepositing. This is the apparent starting point when the composition function leaves the abscissa of the composition diagram. It is important to note that the regular one is the only means of codeposition when a single component can be plated as a pure element, and this is always the more noble one.

For both *irregular codeposition* and *anomalous codeposition,* the interaction between the components is strong enough to achieve the deposition of the less noble component also in the case when the transport of the more noble component would be sufficient to provide a large enough flux so that it could be deposited alone in the absence of the less noble component. The difference between the two modes, concerning the composition, is that for anomalous codeposition the deposit is always more rich with respect to the less noble component than the solution. Metal pairs characterized with irregular codeposition mostly include complex ion systems such as the Cu–Zn from a cyanide bath, while the anomalous codeposition is valid for any pair among the Fe–Co–Ni–Zn group. The common feature for these categories is that the metals indeed form an alloy.

For a better understanding of the impact of the deposition conditions on the deposit composition, it is worthwhile of studying Fig. 2.21b. It shows that the increase in the supply of the *preferentially deposited metal*, should it be either the more noble one (for regular and irregular codeposition) or the less noble one (anomalous codeposition), results in the increase of the mole fraction of this metal in the deposit. In contrast, when the abundance of the ions of the preferentially deposited metal decreases in the vicinity of the cathode because of, e.g., the increase in current density, an adverse effect can be seen. Figure 2.21b yields a guideline for composition modulation during the electrodeposition from a single solution by means of agitation and current density modulation.

Composition diagrams usually refer to a steady-state deposition process which is not influenced by the transition that takes place at the beginning of the deposition. In the initial transition period of the alloy deposition, the concentrations (and hence, also the concentration gradients) of the precursor ions change until the deposition process is stabilized. In the near-substrate zone of the deposit, the mole fraction of the preferentially deposited metals is always larger than in a steady-state deposit. At room temperature and in a stagnant solution, the thickness of the transition zone ranges up to 200 nm. Figure 2.21b can also be used for assessing the impact of the initial transient on the composition: the composition line for a very thin deposit moves in the same direction as if the mass transport war stimulated. The explanation is that at the beginning of the deposition process, the near-substrate solution composition is closer to the bulk one than in the steady-state.

The impact of pulse plating on the composition diagram can be elucidated on the basis of the same train of thought that was applied for the initial transient of a d.c. deposit. As a result of the frequently intermitted current, the solution depletion near the cathode is counteracted by the mass transport during the off-time. Hence, it is straightforward that the application of current pulses modifies the composition the same way as discussed for the initial zone or for an enhanced bath agitation (provided that the d.c. current density is the same as the on-time current density during the pulse plating). Pulse plating can also be regarded from the view-point of the deposit composition as if only the beginning of a d.c. deposit was plated a large number of times.

For induced codeposition, no curves can be displayed in the conventional composition diagram like Fig. 2.21 in the entire solution composition range. The reason is that the metal whose deposition is induced by the other one cannot be deposited alone (at least, not with an appreciable current efficiency). This takes place when iron-group metals (that play the role of the *inducing component*) are codeposited with oxoanion-forming elements (metals like Mo and W and also non-metallic elements like P) or germanium. As the ratio of the ions of the induced metal increases, an alloy is formed, and the initial part of the composition diagram looks like for irregular codeposition. However, as the concentration of the induced metal increases in the deposit, the current efficiency decreases, and a limiting deposit composition can be seen. Finally, the current efficiency drops so much that no deposit is produced, and the composition function line ends in the middle of the composition diagram. Interestingly, these systems are the most prone to develop amorphous deposits as the concentration of the oxoanion-forming element in the deposit increases.

2.12.3 Structural Consequences of Alloy Formation

Even if a metal pair exhibits an equilibrium alloy, there is no guarantee that a particular alloy can be synthesized by electrochemical codeposition. The general trend is that the closer the deposition potentials of the constituents are, the easier to deposit a real alloy with no segregation but with miscibility at the atomic scale. The miscibility at the atomic scale is often promoted by the application of a complexing agent that makes a more stable complex with the ions of the more noble element to be deposited. Hence, the difference in the deposition potentials is diminished, which is a favourable factor in the tendency of codeposition.

Beside the deposition potential, the ordering and the unit cell size are also important factors in the electrolytic alloy formation. The larger the unit cell size and the more ordered an alloy must be, the less likely that it can be synthesized by direct electroplating without an annealing step afterwards. Obviously, alloys with a random atomic arrangement tend to form the most easily, in accord with the non-equilibrium nature of the electrodeposition process. Another general trend is that the increase in the number of components results in a decrease in grain size of the deposit.

Finally, it must be mentioned that phase mixtures are often deposited when the composition of the deposit is close to an equilibrium phase transition composition of the system even if the equilibrium phase transition takes place at a definite composition without a miscibility gap. In such cases, electrodeposition leads to a phase mixture in a 10–15 atomic percent wide composition range. This is related to the non-equilibrium nature of the electroplating method. The general experience is that the formation range of a phase is extended with respect to its equilibrium stability range. It is a direct consequence of the deposition of a phase mixture that the grain size exhibits a minimum in the mixed-phase deposition regime, simply because the competitive growth of the various crystals hinders the growth of each other and minor changes in the local deposition conditions often lead to the nucleation of new crystals. The nanocrystalline nature of the deposit in the phase mixture composition range also contributes to the widening of the formation zone of a crystalline form as compared to the equilibrium phase diagram.

2.13 Behaviour of Metals During Anodic Polarization

When a non-noble metallic element is positively polarized, it can actively dissolve if the surface is not covered by a protecting passive layer. This can be observed for a number of metals if the solution does not contain anions with which the metal of interest forms a weakly soluble compound (Ag, Cu, Pb, Bi, Cd, Co, Zn etc.). Depending on the properties of the metal, a passive layer may also form during the anodic polarization (like Cr). The passivation is observed as the fall of the current as the electrode potential increases. The potential regime of the passivity may be followed by transpassive dissolution if the ions formed are of higher valency than during the active dissolution at less positive potentials. If the metal is covered by a passive layer already in the native form, the active dissolution may be missing if the media does not contain appropriate anions that facilitate the break-up of the passive layer (like chloride ions for Ni or fluoride ions for Ti and Zr). If the metal remains passive also at fairly positive potentials, the increase in the current can be attributed to the decomposition of the solvent. Typical steady-state anodic polarization curves of various metal types can be seen in Fig. 2.22.

It has to be considered that, similarly to metal deposition, the dissolution of the metals can also lead to morphological changes. It is common that during the active dissolution when fully soluble products are formed the surface of the metal changes drastically. The most active dissolution centres are typically the grain boundaries where the position of the atoms differs from the equilibrium one, and hence, their energy is larger than in the bulk. The grain boundary can also be the location of the impurity accumulation which promotes the anodic dissolution. Hence, a disintegration process can take place, similar to the intergranular corrosion.

In the growth of the oxide layer on a metal, the driving force differs from the active metal dissolution or the initial formation of the passive layer. For the same reason, the exponential law corresponding to electrochemical activation does not hold. For compact passive layers, the most common case is that the layer grows up

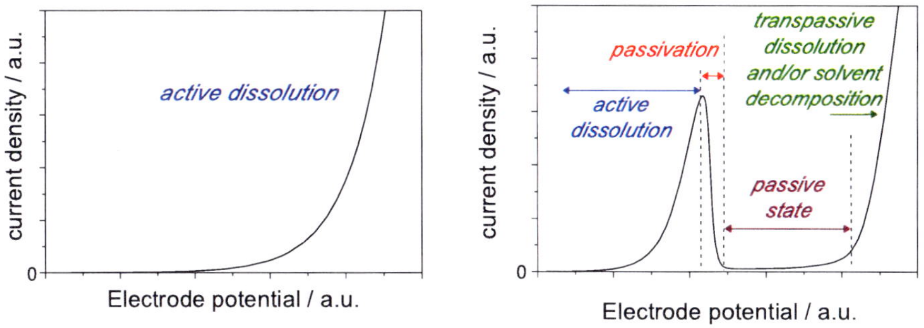

Fig. 2.22 Typical anodic polarization of metals. **Left**: active dissolution only, **right**: active dissolution followed by passivation and transpassive dissolution

to a threshold value as long as the electrical field within the passive layer is capable of leading to ionic motion within the passive layer. As the layer grows to such a thickness at which the electrical field falls below this limit, the growth stops. The critical electrical field of provoking ionic mobility within the passive layer is in the range of 10^7 V/cm. In experiments of electrochemical growth of metal oxides, the total cell voltage can be in the 15–200 V range, where the potential drop on the counter electrode is negligible as compared to the total cell voltage. For this reason, in such experiments it is customary to report the total cell voltage and no reference electrode is used.

For oxide formation, various potential regimes can be identified for a number of metals (such as Al). For relatively small anodization voltage, a compact oxide layer can be produced. At higher voltages, an oxide with large porosity and regular pore size can take shape. At extremely large voltages, the stress in the oxide becomes so large that the oxide layer is delaminated from the surface (burn-down of the oxide), and hence, no compact layer can form. The potential intervals where these phenomena take place usually depend on the solutes of the anodization media.

The anodic behaviour of metal alloys is usually much more complicated that that of metallic elements. There are several alloys that dissolve actively altogether as if they were of a metal of single component. This is common, for instance, for the Fe–Co–Zn and Cu–Zn alloy families. Even for these alloys, the composition, and consequently, the specific crystallographic form of the actual alloy strongly determines the electrode potential at which the dissolution starts. This makes the possibility of an indirect phase analysis of thin metallic layers (typically deposits), being often termed as *anodic linear sweep voltammetry* (ALSV). In ALSV experiments, a positive-going sweep is run with the alloy as working electrode, and the dissolution of the grains of various alloy phases takes place in different potential regimes and manifests itself as peaks of waves in the voltammograms. The charge ratio corresponding to the dissolution of various alloy phases corresponds to the phase composition of the metal layer. Since ALSV is a transient method, it can be used for thin layers only but not for bulk metals since it requires the complete dissolution of the specimen.

Another type of anodic behaviour can be obtained if the dissolution of a real alloy cannot take place as if it were a single component. If the potential regimes corresponding to the anodic dissolution of the element of an alloy differ very much and the mobility of the atoms of the more noble element is high enough, *anodic dealloying* may happen. During dealloying, the atoms of the less noble alloy component are oxidized and hence leave the metal, while the atoms of the more noble element remain in elemental form. As a result of dealloying, the remaining metallic structure does not correspond to the atomic arrangement of the more noble element in the parent alloy. Rather, the atoms of the remaining metal are rearranged within a distance determined by the mobility of these atoms. This rearrangement also diminishes the excess energy of the system due to the newly produced large metal surface. The resulting structure is highly porous. A typical alloy family leading to dealloying upon positive polarization is Au–Cu alloys.

Finaly, it must also be mentioned that passivity of alloys also differs from the passivity of metallic elements in the sense that the ratio of the constituents of the passive layer does certainly not correspond to the ratio of these elements in the bulk alloy. The components forming an oxide with larger stability are more abundant in the passive layers that those having oxides with lower stability. The passivation of the alloy can be regarded as the partial dealloying of the surface layer and the subsequent formation of a compact oxide from the remaining component. A typical example of passive alloy with dissimilar surface oxide composition as compared to the bulk alloy is stainless steel where chromium is essential in thc formation of a stable protecting oxide layer.

References

1. Brenner A (1963) Electrodeposition of alloys. Principles and practice. Academic Press, New York and London
2. Dini JW (1993) Electrodeposition. The materials science of coatings and substrates. Noyes publications, Park Ridge, New Jersey, USA
3. Gamburg YD, Zangari G (2011) Theory and practice of metal electrodeposition. Springer, New York, Dordrecht, Heidelberg, London
4. Schlesinger M, Paunovich M (2010) Modern electroplating, 5th edn. John Wiley and Sons Inc., Hoboken, New Jersey
5. Kanani N (2004) Electrodeposition—basic principles, processes and practice. Elsevier, Oxford-Amsterdam
6. Budevski E, Staikov G, Lorentz WJ (1996) Electrochemical phase formation and growth. An introduction to the initial stages of metal deposition. VCH, Weinheim–New York–Basel–Cambridge–Tokyo
7. Kiss L (1988) Kinetics of the electrochemical metal dissolution. Elsevier, Amsterdam–Oxford–New York–Tokyo
8. Plieth W (2008) Electrochemistry for materials science. Elsevier, Amsterdam-Oxford
9. Watanabe T (2004) Nano-plating. Elsevier, Oxford
10. Bard AJ, Inzelt G, Scholz F (eds) (2012) Electrochemical dictionary, 2nd edn. Springer, Heidelberg
11. Láng GG, Barbero CA (2012) Laser techniques for the study of electrode processes. In: Scholz F (ed) Monographs in electrochemistry. Springer, Heidelberg, Chapter 1
12. Hamann CH, Hamnett A, Vielstich W (2007) Electrochemistry, 2nd edn. Wiley-VCH, Weinheim, Chapter 4.3, pp 182–202
13. Damaskin BB, Petrii OA (2011) J Solid State Electrochem 15:1317–1334
14. Scharifker B, Hills G (1983) Electrochim Acta 28:879–889
15. Scharifker BR, Mostany J, Palomar-Pardavé M, González I (1999) J Electrochem Soc 146:1005–1012
16. Oviedo OA, Reinaudi L, García SG, Leiva EPM (2016) Underpotential deposition. From fundamentals and theory to applications at the nanoscale. In: Scholz F (ed) Monographs in electrochemistry. Springer, Berlin
17. Winand R (1994) Electrochim Acta 39:1091–1105
18. Puippe JC, Leaman F (1986) Theory and practise of pulse plating. American Electroplaters and Surface Finishers Society, Orlando
19. Hansal WEG, Roy S (2012) Pulse plating. Eugen G Leuze Verlag KG, Bad Saulgau
20. Schwarzacher W (2004) J Phys Condens Matter 16:R859–R880
21. Plieth W, Lorenz WJ, Staikov G (2004) J Solid State Electrochem 8:941–946

Chapter 3
Experimental Methods in Characterization of Nanosystems

3.1 The Nature of Such Methodological Overviews

The present book is dedicated to the electrochemical methods that can yield nanostructures. One may ask the question why a methodological chapter should precede the main part of the entire book discussing the core phenomena. The answer is simple: a serious knowledge of the fundamentals of characterisation techniques is necessary for their correct application and also for the ability to critically read and understand works in which these techniques have been used.

Therefore, the methodological introduction is by far not a formal part of a book. If a researcher is not aware of the basis of a method, she/he will not strive for its application, and the information yield of this method cannot be exploited.

Nowadays, even average-level publications use a handful of auxiliary methods for characterizing the samples synthesized in an electrochemical laboratory. The knowledge of the possible characterization methods is crucial for both the in-depth understanding of the publications of other researchers and to become an author who can perform cutting-the-edge works.

While the awareness of the characterization means cannot be emphasized enough, it must be also admitted that no electrochemical laboratory can afford the maintenance of the entire range of instruments that is necessary for the characterization of the electrochemically produced materials. What general-purpose electrochemistry laboratories can sometimes operate as non-electrochemical characterization facility of their own includes at most a scanning probe facility, a simple diffractometer or a photoelectrochemical workstation. Everything else can be reached through some means of cooperation, and the special knowledge concerning the operation conditions and information yield of the auxiliary instruments comes from the cooperation partners. This also means that the maintenance and operation of each instrument suitable for the characterization of electroplated nanostructures make the know-how of a different profession. For being able to ask the right question related to our reaction product, electrochemists also must be familiar with at least the key features of the possible characterization techniques, otherwise they were not be acknowledged

L. Péter, *Electrochemical Methods of Nanostructure Preparation*,
Monographs in Electrochemistry, https://doi.org/10.1007/978-3-030-69117-2_3

as competent cooperation partners of the operators of a specialty characterization instruments.

Since dozens of books are available of each single characterization technique to be mentioned later, this overview cannot offer more than a basic introduction by clarifying a few crucial features of some techniques that give the reader a sufficient basis for further autodidactic self-improvement. The focus will be on the essential qualitative understanding, the elemental interaction mechanism of the instruments with the object to be studied and the nature of information to be gained. For detailed information on the study of sample properties, two general [1, 2] and one specific electrodeposition-related [3] handbooks are recommended.

3.2 Non-destructive Analysis Methods with Irradiation

3.2.1 Classification of the Irradiation-Based Methods

Figure 3.1 presents a scheme that helps categorize a large number of non-destructive analysis methods that are seldom mentioned in line with each other due to the immense difference in the nature of the incoming beam, the elementary interaction process of the radiation and the target, and very importantly, the peculiarities of the necessary instrumentation. Tables 3.1 and 3.2 list the overviews of the available methods applying electromagnetic and particle beams, respectively. Those which are of highest relevance in the study of solid structures are detailed in the forthcoming chapters.

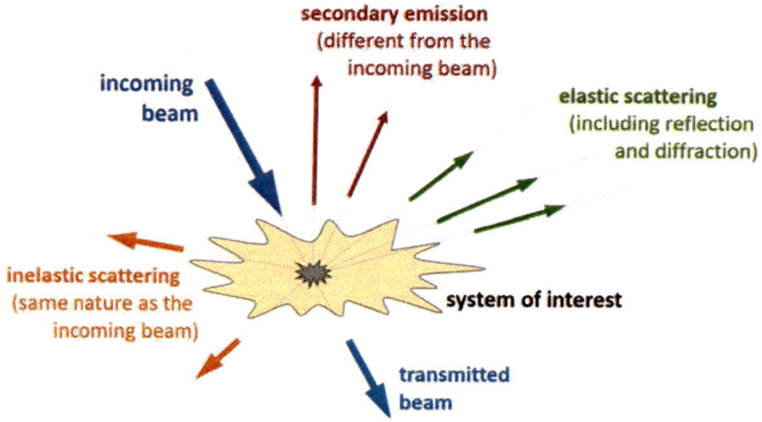

Fig. 3.1 Scheme of the irradiation-based non-destructive analysis methods. Specific techniques with the nature of the incoming beam, the elementary interaction process of the beam with the target and the information yield are listed in Tables 3.1 and 3.2

Table 3.1 Overview of irradiation-based non-destructive analysis methods applying electromagnetic irradiation. Methods are listed roughly in the order of increasing beam energy

Name of the method (abbreviation, if any)	Range (wavelength or photon energy)	Interaction of the beam with the specimen	Signal detected (with some details of the detection)	Information yield of the method
Spectrophotometry	IR/visible/UV	Absorption	Transmitted beam (transmission or absorption, detected in the beam direction)	Light absorbance, absorption intensity; if absorbance is known: concentration
Photoluminescence spectroscopy	IR/visible/UV	Absorption and forthcoming emission at lower photon energy	Spectrum of the emitted light (not in the direction of the incoming beam)	Decay of excitation, band structure, electron energy levels
Raman spectroscopy	IR/visible/UV	Excitation of various vibration/rotation modes of molecules	Spectrum of the inelastically scattered light (not in the direction of the incoming beam)	Vibration/rotation energy related to covalent bonds
Dynamic light scattering method (DLS)	Visible	Light scattering	Scattered beam intensity and its fluctuation (out of the original beam direction)	Hydrodynamic size of the scattering entities (like colloidal particles dispersed in a fluid)
Ellipsometry	IR/visible	Reflection at layer boundaries, absorption in the layers	Intensity and polarization of the reflected light as a function of the incidence angle and wavelength; $\Theta-2\Theta$ geometry	Layer thicknesses, complex refractive index of the material of the layers
Ultraviolet photoelectron spectroscopy (UPS)	UV (20–30 eV)	Ionization of gas-phase molecules	Flux of the electrons emitted with various energies (out of the incoming beam direction)	Ionization energy, band structure (of molecules primarily in gaseous phase)

(continued)

Table 3.1 (continued)

Name of the method (abbreviation, if any)	Range (wavelength or photon energy)	Interaction of the beam with the specimen	Signal detected (with some details of the detection)	Information yield of the method
X-ray photoelectron spectroscopy (XPS)	X-ray (1–2 keV)	Ionization of the atoms of a solid material (in the near surface zone) followed by electron/X-ray emission	Flux of the electrons emitted with various energies (out of the incoming beam direction)	Surface composition, oxidation state of the atoms detected
X-ray fluorescence (XRF) spectroscopy	X-ray (λ: 1.2–0.09 nm; up to 60 keV)	Core-level ionization followed by electron hopping and emission of characteristic X-ray	X-ray intensity as a function of either energy or wavelength (out of the incoming beam direction)	Elemental composition (insensitive to oxidation state), for sodium and elements of higher atomic number
X-ray reflectometry (small-angle X-ray scattering)	X-ray	Reflection at layer boundaries	Radiation scattered elastically in a small angle	Layer structure, interface roughness
X-ray diffractometry (XRD)	X-ray (λ: 0.01–10 nm in general; for metals: 0.07–0.18 nm)	Diffraction (interference of beams elastically scattered by atoms of identical crystal planes)	X-ray (preferably of the same wavelength, excluding fluorescence)	Interplanar spacing of the atoms in the direction of the scattering vector, texture, grain size (from peak broadening)
X-ray absorption near-edge structure (XANES)	X-ray (about 0.2–35 keV)	Absorption followed by electron emission from the core shells; interference of the ejected and scattered electron wave with the incoming X-ray beam	Transmitted beam intensity as a function of X-ray photon energy near the absorption edge (about 5 to 200 eV)	Oxidation state and coordination environment of the absorbing atom

(continued)

Table 3.1 (continued)

Name of the method (abbreviation, if any)	Range (wavelength or photon energy)	Interaction of the beam with the specimen	Signal detected (with some details of the detection)	Information yield of the method
Extended X-ray Absorption Fine Structure (EXAFS)	X-ray (about 0.2–35 keV) (often measured together with XANES)	Same as XANES	Transmitted beam intensity as a function of X-ray photon energy higher than about 200 eV from the absorption edge	Element-specific local structure: interatomic distances and coordination numbers (up to the 2nd coordination sphere)
X-ray tomography	X-ray (λ: 0.01–10 nm)	Absorption of the radiation	Transmitted intensity (measured in transmission mode) as a function of the radiation direction	Computer-reconstructed internal structure with a resolution down to about 1 μm
Mössbauer spectroscopy	γ (nucleus-specific; for ^{57}Fe: 14.4 keV)	Recoilless resonance absorption of the γ-photon by the nucleus	Absorption (transmitted beam) as a function of the target velocity	Isomer shift, quadrupole splitting, hyperfine interaction (oxidation state, chemical and magnetic environment, resp.)

Table 3.2 Overview of irradiation-based non-destructive analysis methods applying irradiation with some sort of particles

Particles in the beam	Name of the method (abbreviation)	Particle properties and interaction of the beam with the specimen	Signal detected (with some details of the detection)	Information yield of the method
Electron	Auger electron spectroscopy (AES)	Ionization of the atoms of a solid material (in the near surface zone) followed by electron/X-ray emission	Flux of the electrons emitted with various energies (out of the incoming beam direction)	Surface composition, oxidation state of the atoms detected
	Electron energy loss spectroscopy (EELS)	0–1 kV electrons: plasmon excitation (low energy, <~50 eV), inner-shell ionization (higher energy)	Transmission mode, electron yield as a function of energy	Band structure of a solid (low energy loss), chemical composition (high energy loss) up to the 3d metals
	Energy dispersive X-ray spectroscopy (EDS)	≤30 keV (in SEM), 60-300 keV (in TEM); core-level ionization followed by electron hopping and emission of characteristic X-ray	X-ray intensity as a function of the photon energy (out of the incoming beam direction)	Elemental composition (insensitive to oxidation state), from elements of higher atomic number than boron; sensitivity: ~0.2 at.%
	Wavelength-dispersive X-ray spectroscopy (WDS)	≤30 keV (in SEM)	X-ray intensity as a function of the photon energy (out of the incoming beam direction)	Same as EDS but with about two orders of magnitude larger sensitivity
	Scanning electron microscopy (SEM)	≤ 30 keV, emission of both secondary and backscattered electrons (with lower and higher energies, resp.)	Electron yield	Surface morphology (primarily from secondary electrons, low escape depth), atomic number contrast (electrons of higher energy)

(continued)

Table 3.2 (continued)

Particles in the beam	Name of the method (abbreviation)	Particle properties and interaction of the beam with the specimen	Signal detected (with some details of the detection)	Information yield of the method
	Electron backscattered diffraction (EBSD)	$\leq$ 30 keV (in SEM), diffraction-type interference of the backscattered electrons	Angular distribution of the backscattered electron intensity	Orientation of the crystals (near-surface zone), crystallite size, texture, orientation distribution of grains
	Transmission electron microscopy (TEM)	60–300 keV, scattering on samples with a thickness less than about 100 nm	Electron beam intensity	Grain size, grain orientation, structure at the atomic scale
	Electron diffraction (mostly in TEM)	60–300 keV, samples with a thickness less than about 100 nm	Electrons (preferably those scattered elastically) as a function of diffraction angle	Interplanar spacing of the atoms in the direction of the scattering vector, texture, grain size
Positron	Positron annihilation spectroscopy (PAS)	Various (thermalized positrons take part in the forthcoming annihilation process)	γ emission of the $e^- + p^+$ annihilation process (511 keV)	Annihilation of positrons entrapped in atomic-size cavities such as vacancies etc. yield information on the defect structure at the atomic scale
Neutron	Neutron radiography	Thermal neutron beam ('white beam' with a wide energy distribution); neutron absorption	Transmission intensity	Inner structure of macroscopic objects containing neutron-absorbing elements (B, Cd, H)
	Small-angle neutron scattering	Neutron beam (wavelength between about 0.3 and 2 nm); elastic scattering on objects larger than about 1 nm	Scattered neutron intensity at low angles (typically up to a few degrees)	Shape and size of objects on the nanometer to micrometer range; interparticle structure

(continued)

Table 3.2 (continued)

Particles in the beam	Name of the method (abbreviation)	Particle properties and interaction of the beam with the specimen	Signal detected (with some details of the detection)	Information yield of the method
	Neutron diffractometry	Neutron beam (wavelength: 0.01–0.2 nm); diffraction (interference of beams scattered elastically by atoms of identical crystal planes)	Intensity of scattered neutron beam as a function of the scattering angle (and/or energy of the scattered neutrons)	Atomic structure; texture, grain size and residual stress
	Neutron activation analysis	Neutron absorption in the nuclei to form radioactive isotopes	Gamma emission with element-specific energy upon decomposition of unstable nuclei	Elemental composition
α particles	Rutherford backscattering spectroscopy (RBS)	Elastic scattering (on atomic nuclei) and inelastic scattering (when passing through a large thickness of material)	Particle yield, typically at a fixed angle as a function of energy or vice versa	Composition depth profile

3.2.2 *Diffraction Methods*

The theory of diffraction methods is based on the quantitative description of the scattering of incoming beams on some scattering centres. The nature of scattering is fully elastic in this case. The complete theory of scattering events leading to diffraction patterns is available in various works. Here, we can offer a simplified summary only that gives the relationship of the diffraction angle and the lattice plane distances but does not treat the intensity of the diffracted beam. This is usually enough if one would like to correlate the structure of a sample with that of a crystalline material described in various databases but the identification of the structure of a yet unknown crystalline material is not required. The latter task usually needs an expert specialized for structural studies.

The scheme in Fig. 3.2 shows the conditions of the positive interference of the beam scattered on the atoms of neighbouring crystallographic planes for two orientations of the same two-dimensional crystal. It is to be emphasized that the method is sensitive to lattice plane distances in the direction of the change of the wave number vector. The suitable wavelength for the study of a structure is in the range of the

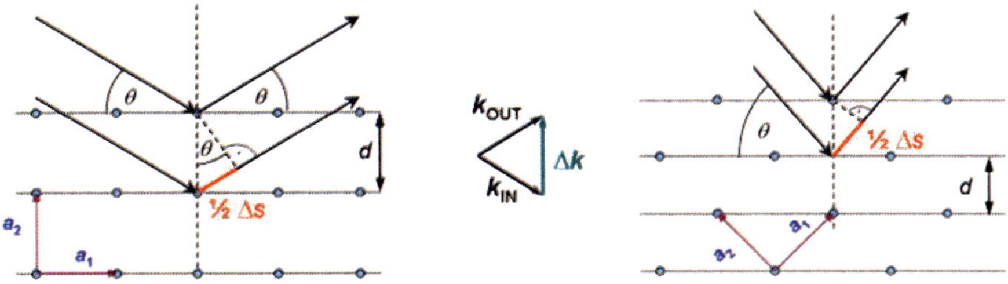

Fig. 3.2 Scheme of the diffraction conditions for the 01 and 11 planes of a two-dimensional lattice (left and right, respectively). The red lines (identical in length) indicate the half of the path difference. The comparison of the two positions of the lattice shows that as the lattice plane distance decreases, the positive interference conditions are fulfilled at higher diffraction angles. In the middle, the wave number vectors of the incoming and outgoing beams are shown with the wave number vector difference

distances of the planes that diffract the beam. The relationship between the incidence angle of the beam (θ), the lattice plane distance (d) and the wavelength of the diffracted beam (λ) is described by Bragg's equation:

$$2d \sin \theta = n\lambda, \tag{3.1}$$

where n is a positive integer.

The meaning of the wavelength is obvious for X-rays. For an electron beam, the wavelength can be calculated from the de Broglie equation:

$$\lambda = \frac{h}{m_e v} = \frac{h}{\sqrt{2m_e eV}}, \tag{3.2}$$

where h is the Planck constant (6.6×10^{-34} J s), m_e is the weight of the electron (9.1×10^{-31} kg), e is the elementary charge (1.6×10^{-19} C), v is the actual electron velocity and V is the acceleration voltage that leads to the velocity v (non-relativistic approximation).

Diffraction studies are performed with two typical configurations called single crystal diffraction and powder diffraction (see Fig. 3.3). For single crystal diffraction, the position of both the source and the object studied is fixed, while the direction of the observation of the diffracted beam varies. In this case, the diffracted beams of the various planes of the *same* crystal are observed as a function of the diffraction angle. In the powder diffraction geometry, some of the high Miller index planes of *different* crystals of non-identical orientations give the subsequent reflections corresponding to the varying diffraction angle. Here, the source and the detector are aligned in the arms of a goniometer and two objects are rotated in a synchronized manner (either the source and the detector or the object studied and the detector; Bragg–Brentano or θ–2θ geometry). For a truly random orientation of a large assembly of crystals (i.e., the powder), a large majority of the crystals never fall in the right diffraction

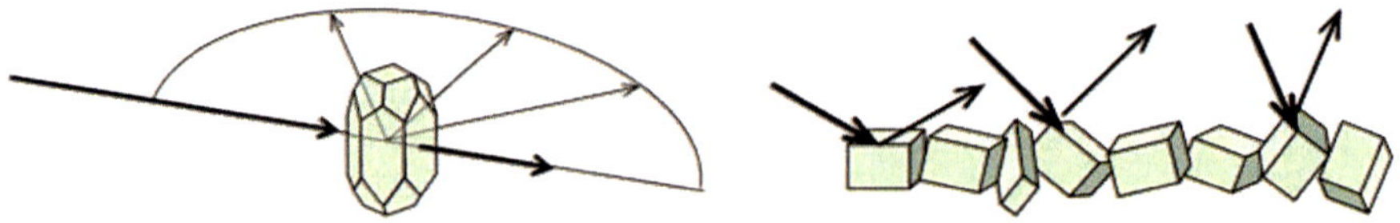

Fig. 3.3 **Left**: Diffracted beams coming from the same crystal in different directions in the single crystal diffraction geometry. **Right**: Beam diffracted by the crystals of various orientations in the powder diffraction geometry (showing also crystals for which the diffraction conditions are never fulfilled in the Bragg–Brentano geometry)

angle. However, powder diffraction can serve as a good standard for the study of a coating or other solid material as well. The reason for this is that in a solid, the high Miller index planes are typical in the preferred directions (parallel or perpendicular to the surface) because these planes have the higher surface density of atoms.

Laboratory X-ray diffractometers are equipped with sources of fixed wavelength. Copper cathodes have a reasonably long Kα radiation ($\lambda = 0.154056$ nm), which gives good resolution in registration of most of the metallic materials. Another well-established standard source for powder diffraction is cobalt ($\lambda = 0.1789$ nm) that has high importance in the study of iron-based materials in which the Cu Kα radiation causes a high unwanted fluorescence. Where both powder and single crystal diffraction studies are of importance, an optimal single X-ray source is Mo Kα ($\lambda = 0.071073$ nm) that can also be used for substances of smaller interatomic distances (organic materials). If the dimension of the scattering object is extremely small and a high luminosity source is required, synchrotron-based sources are preferred that are usually equipped with accurate wavelength selectors.

Apart from the establishment of the interplanar distances, diffractograms bear information on the size of the coherently scattering objects. The smaller the size of the coherently scattering zones, the wider diffraction lines are obtained. By using a single high-intensity line, the grain size (l) can be estimated with the Scherrer equation:

$$l = \frac{K\lambda}{\beta \cos\theta} \tag{3.3}$$

where K is a numerical constant called shape factor (whose value is usually taken as 0.9) and β is full width at half maximum corrected with the instrumental line broadening. Since the small grain size is not the sole origin of line broadening but stress in the sample also has a contribution. The parallel treatment of both above mentioned effects are taken into account in the Williamson–Hall method. This is based on the fact that strain-induced and size-related line broadenings differ in distinct ways on the diffraction angle. The usual form of the final equation is ($C\varepsilon$ being the strain-related parameters):

$$\beta \cos\theta = \frac{K\lambda}{l} + C\varepsilon \sin\theta \tag{3.4}$$

By displaying $\beta\cos\theta$ as a function of $\sin\theta$, the true grain size can be obtained from the intercept. Since the graphs thus obtained are often very scattered, it is practical to fit the entire diffractogram as a whole by including all effects related to the line broadening simultaneously to all peaks. This method called convolutional multiple whole profile (CMWP) fitting requires high-resolution devices for recording the diffractograms that are not commonly available for routine tests.

A parameter often used for the characterization of materials from its X-ray diffraction pattern is the calculation of its relative texture index (*RTI*). The calculation of the relative texture index of a particular line (referred to with the *hkl* Miller indices) involves a double normalization of the intensity (I) of each specific peak with respect to the total line intensity of the measurement it is derived from (see M in the lower index) and then with respect to the same parameter given for the powder diffractogram of the same material (see REF in the lower index), N being the number of diffraction peaks considered:

$$RTI(hkl) = \frac{1}{N}\frac{I(hkl)_{\mathrm{M}}}{I(hkl)_{\mathrm{REF}}} \Bigg/ \frac{\sum_{hkl} I(hkl)_{\mathrm{M}}}{\sum_{hkl} I(hkl)_{\mathrm{REF}}} \tag{3.5}$$

If the orientation of many grains can be measured at the same time (i.e., with electron backscattered diffraction in a scanning electron microscope), so-called pole figures can be constructed that show a quasi-continuous orientation distribution function of the grains in coloured maps.

3.2.3 Methods Involving Ionization and Used for Chemical Analysis

There are a number of different methods that all are based on the ionization of atoms followed by an element-specific emission. Although these processes are seldom discussed as a single family and the corresponding methods use very different instrumentation, their scheme is presented with a unified approach in Fig. 3.4.

The energy balance of the deep-level ionization shown in the first step of the scheme in Fig. 3.4 is easy to express if the ionized state is not excited. Namely, we can write that

$$E_{\mathrm{IN}} = E_{\mathrm{ION}} + E_{\mathrm{KIN}} \tag{3.6}$$

by calling the energy of the incoming beam E_{IN}, the specific ionization energy of the atom E_{ION} and the kinetic energy of the electron leaving E_{KIN}. The ionization energy depends on the element and the electron shell where the emission takes place from, but the exact energy of the electron leaving is affected by the chemical environment of the atom ionized. The larger the oxidation number of the atom ionized, the higher

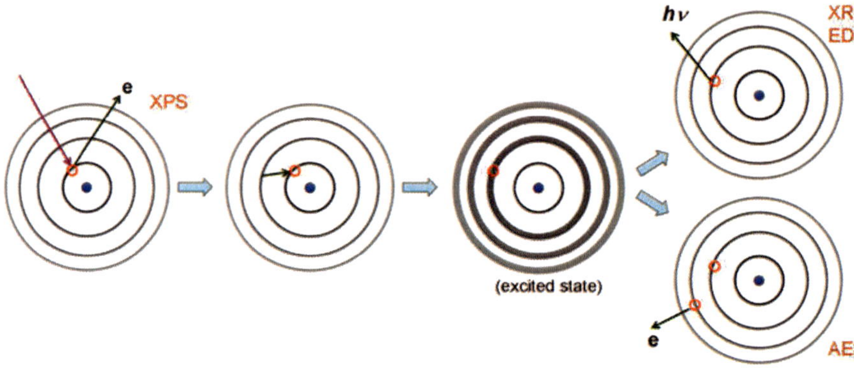

Fig. 3.4 Scheme of the core-level ionization processes leading to element-specific emission that can be used for chemical analysis. The irradiation in the first stage can be performed by photons or electrons. The electron shells depicted by the rounds of grey colours are shown schematically and the size of the circles does not indicate the actual radius ratios. Little red circles at the electron shells indicate the electron vacancy. The energy releasing process in the third step is shown separately for the sake of clarity, although it cannot be separated from electron hopping to fill the deep-level vacancy

the ionization energy (from the same shell). This can be used for chemical analysis as well as the estimation of the oxidation state in XPS which is a surface-sensitive method due to the small escape depth of the electrons (usually less than 10 nm). The resolution of the electron energy spectrometers used in XPS is a fragment of an electronvolt, whereas the ionization energy difference due to the change of the oxidation state may exceed 2 eV. Therefore, the analysis of the chemical state is well possible, including the splitting of the peaks obtained due to the spin state of the final electron configuration.

The energy of the electron leaving in the first step of the scheme in Fig. 3.4 depends on the energy of the incoming beam. However, the ionized state left behind after the first ionization step is characteristic of the element only, irrespectively of how the ionization took place in the previous step. This is why the forthcoming emission during the filling of the deep-level electron vacancy can be used for chemical analysis like XRF (where the first ionization is due to X-ray irradiation) and EDS/WDS (which is usually used in SEM instruments and involves an electron beam-induced ionization in the first step). Due to the features of the instruments, the detection methods and the accompanying side processes the sensitivity of these methods are very different, although the elemental processes are the same. The resolution of the methods based on the detection of X-ray emission is lower than that needed to distinguish the oxidation state of the atom emitting the radiation. If the chemical state of the specimen is well defined (like for metals used as cathode in an X-ray tube), the emission energy is also fixed and can be used as standard. Since the penetration depth of the electrons in SEM is restricted to at most 1 μm while that of the X-ray radiation is much larger, EDS/WDS can yield an analysis restricted to about the same layer, while XRF yields results much closer to bulk values, having much less restrictions on both the incidence and escape depths of the relevant radiations.

The release of energy after the deep-level ionization is also possible by the emission of another electron when the deep-level vacancy is filled. The electrons emitted in this process are called Auger electrons and the name of the associated measurement is the Auger electron spectroscopy (AES). Unlike that of photoelectrons, the energy of the Auger electrons does not depend on the energy of the original ionization source, which is a distinctive feature of these electrons in the spectrum. For other features like resolution and energy dependence on the oxidation state, AES electron spectrometry exhibits a large analogy to XPS.

The nomenclature of the emitted radiation differs a bit, depending on the particle emitted. The emitted primary electrons are named after the sign of the shell they come from (K, L, ...). If another electron fills up this vacancy and the leftover energy is emitted by electromagnetic radiation, the first letter indicates the level that was filled up, and the forthcoming Greek letter refers to the relative position of the shell from which the electron jumps. For instance, K_α (or K_β) means that the vacancy on the K shell is filled up by another electron coming from the neighbouring (or the second following) electron shell. The denotation of the Auger electrons emitted is composed of three Latin letters: the shell at which the primary ionization took place, the shell from which the vacancy is filled up and that from the Auger electron is emitted (e.g., KLL).

3.3 Thermal Analysis

In thermal analysis, a predefined temperature program is applied and the behaviour of the sample is detected by using various methods. The properties detected as a result of the annealing can be the structure of the sample, its weight (thermogravimetry, TG), the heat flow during the annealing process (differential thermal analysis, DTA) or the gaseous products emitted during the annealing. Each method can be performed either in an inert atmosphere or in oxidizing media (air or pure oxygen). Thermal measurements are typically performed in a controlled gas stream.

In the simplest case, the sample having a metastable structure undergoes some structural relaxation and phase transformation upon annealing. The structure of the sample can be detected with conventional diffractometric analysis at room temperature after particular periods (maximum temperatures and/or annealing times) after which the samples are cooled again. In highly specific devices designed for *in situ* structural observation, the diffraction pattern can be recorded at high temperature, too.

While the diffractometric analysis of a sample series can give information on the processes taking place, the threshold temperature of the structural transformation can seldom be established due to the limited number of samples treated. Since the structural transformations are accompanied by some heat effect, its detection during the annealing process offers a convenient tool to detect a collateral effect of the structural transformation as a quazi-continuous function (i.e., measured with frequently collected data point during the annealing process, not only by detecting an integral effect of the entire annealing process). The reason why it is called a *differential* analysis method is that the devices that can measure the heat effect of the process are equipped with two identical sample holders. One of the sample holders (or chambers) contains the sample to be tested, and the other is loaded with an equal portion of standard material for comparison that does not undergo any transformation upon annealing. The temperature of both sample holders is recorded as the device is annealed. The temperature measurement is usually made with thermocouples. Since the heat effect of the transformation taking place in the sample causes a deviation in temperature as compared to the standard, it can be compensated by modulating the heating programs in order to equilibrate the temperature of the sample holder pair (differential scanning calorimetry, DSC). What is measured is the compensation heating power that directly yields the excess energy absorbed/released during the annealing process. Since practically no transformation takes place instantaneously, the key parameters established are the onset temperature of a process and the integral heat effect of the transformation that belongs to a peak (wave). The faster the heating rate, the wider the temperature range over which the thermal peak of the same process is smeared out.

Annealing processes are often accompanied with a weight change, too, when the sample has a gas emission during a decomposition process. The weigh change can be measured by observing the weight of the solid residue of the sample, often in parallel to the recording of the heat effect. The correlation of the weight loss (gain) and the heat effect is a tool of high performance to identify the processes taking place.

There are various ways to analyse the emission product accompanying the annealing processes. Devices that can be attached to the outflow of a thermal analysis instrument include chromatographs and mass spectrometers. Neither of the above-mentioned devices can perform an absolute quantitative analysis, although their relative signal intensities can be calibrated. Nevertheless, the information yield of such an analysis refers primarily to the quality of the gas released.

A fully quantitative analysis during a heat treatment is much easier if the product to be released is a one-component material and its quality is well known. This is the case when the absorbed hydrogen is emitted upon annealing. Degassing devices designed to rate hydrogen emission are calibrated with heat conductivity detectors since they are very sensitive to light molecules appearing beside a heavier carrier gas (mostly argon). With this method, hydrogen solubility can be measured.

3.4 Mechanical Tests

Four major forms of mechanical tests are noteworthy of mentioning here: tensile, hardness, wear and adhesion tests. Tensile tests are exact in the sense that coefficients defined in physical formulae can be derived from them, but the other three test methods are based on conventions and are more comparative.

Tensile tests are primarily used for bulk or self-supporting materials. Therefore, coating can be studied either after the removal of the substrate or on a substrate that is much thinner (and weaker) than the coating itself. The appearance of the stress vs. strain curves is presented in Fig. 3.5. For essentially all materials, the initial section of the stress–strain curve is linear (see the curve in Fig. 3.5 till the point P). Here, Hook's law holds for the deformation process. The elastic deformation zone includes a section where the proportionality between the stress and the strain is no longer valid but the deformation is still elastic, allowing a fully recovery of the deformation if the stress is eliminated (see the curve in Fig. 3.5 till the point P). Further deformation is accompanied with the occurrence of plastic deformation. Brittle materials tend to break nearly at the end of the elastic deformation regime, while materials that tend to flow may have a long plastic deformation regime where strain hardening takes place. The maximum on the stress–strain curve is called the ultimate strength (U in Fig. 3.5) after which further deformation leads to necking (where the character of the curve is determined by the uneven narrowing of the cross-section of the test specimen). It is very common, especially for soft materials and viscous fluids, that the stress–strain curve exhibits a strong dependence on the deformation rate. For metals, the stress–strain curves are usually measured at a relatively small strain rate where the deformation rate does not affect the stress–strain curve.

Hardness is the quantity defined for rating the resistance of a material to plastic deformation. Hardness tests are based on procedures that involve indentation. Various modes of hardness tests differ mostly in the shape of the tool that causes the indentation imprint and in the load applied (marco, micro and nanoindentation, in the order of decreasing load, leading to decreasing indented area). For the most commonly used hardness test method in metallurgy (e.g., the Vicker's hardness, resulting in the empirical quantity HV), the definition is related to the load applied (L) and the mean diameter (d) of the indented area hence produced:

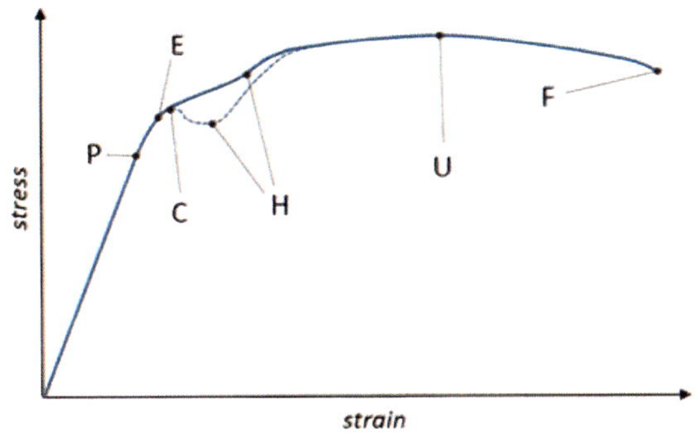

Fig. 3.5 Stress–strain curve. P: proportionality limit of the deformation until the curve is linear. E: elastic deformation limit. C: cross-section of the sample change evenly (plastic deformation starts). H: hardening due to the strain starts. U: ultimate strength of the specimen. F: fracture. The slope between C and H depends on the material. Fracture of brittle materials takes place at E

$$HV = KL/d^2 \tag{3.7}$$

where K is a predefined constant and the units are chosen so that the quantity HV is rendered to be dimensionless. The schematic result of a hardness test is shown in Fig. 3.6.

The problems related to hardness tests are numerous. For any given load, the hardness increases rapidly by diminishing the diameter of the indented area, and this deviation becomes more pronounced at small loads. Also, hardness values are not an absolute measure of a material property due to the so-called work hardening taking place during the deformation process. Therefore, hardness test is to be carried out under strictly defined circumstances and is to be used as comparative indicators. For a hardness test of a thin layer (coating), the key rule is that the indentation depth should not exceed the third of the total layer thickness.

Wear tests show a great variety. In one family of the wear tests, a fluid media is used as a carrier of abrasive particles that hit the surface of the test specimen. The contact pressure between the particles and the surface results from the kinetic energy of the fine particles transported. Depending on the carrier, such test can be divided to either gas blast erosion or fluid jet erosion tests. The major parameters to be controlled are the type of both the fluid and the particles, the size and impact velocity of the particles, and the impact angle of the particle stream. The quantitative result of these tests is the erosion loss of the test specimen, while the examination of the worn surface may give a hint on the wear mechanism.

Another family of wear test is based on the relative motion of surfaces that contact each other. The field that deals with the behaviour of contacting and interacting surfaces displaced relative to each other is called tribology. Three-body abrasion test

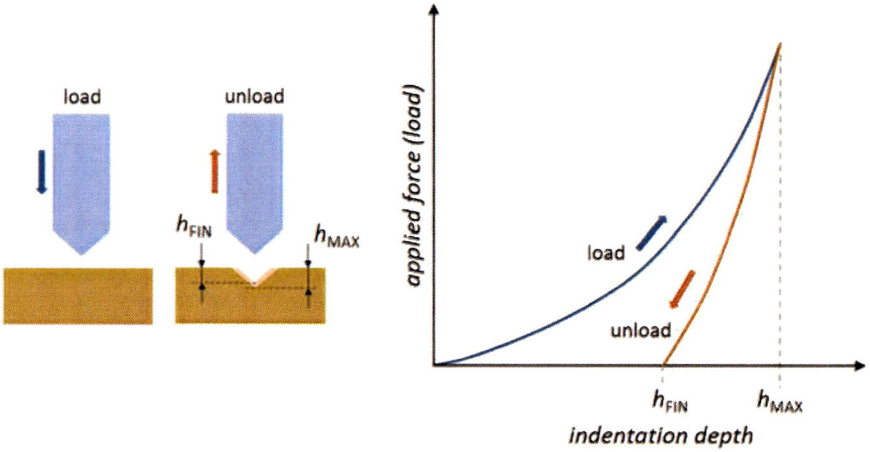

Fig. 3.6 Scheme of an indentation-based hardness measurement and the corresponding load versus indentation depth curve. The final indentation depth, h_{FIN}, is related to the plastic deformation, while $h_{MAX}-h_{FIN}$ is the elastic part that is relaxed during the unloading of the indentation tool

uses abrasive particles between the surface studied and the counterbody used in the test, under either dry or wet conditions. Two-body sliding wear tests are of the most common type for laboratory-scale wear test or relatively small objects. The relative displacement of the test specimen and its counterbody is realized by a rotating motion of either of the contacting bodies, and the test is standardized by fixing both the sliding distance and the pressure (force) acting between the two contacting surfaces. These methods can be further classified on the basis of the shape of the counterbody, such as ball cratering, pin on wheel and thrust washer test. If the driver of the instrument that provides the power for the relative motion of the two facing surfaces is equipped with a suitable feedback system, the friction coefficient can be measured simultaneously from the force required to maintain the desired relative motion. Several tribological methods are defined by various national standards.

Wear resistance is strongly related to other properties of the specimen. As a thumb rule, the wear resistance increases with hardness. The reduction of surface roughness leads to improved wear resistance through a lower friction coefficient.

Adhesion tests include a number of methods of various complexity and are often combined with other sorts of impact. In all cases, a load applied to the substrate–coating system to provoke the detachment of a surface layer from the substrate. The simplest qualitative adhesion test is the well-known scotch tape method. A number of qualitative test methods include mechanical impact such as bending, twisting, deep drawing, grinding, filing or local cupping. Other adhesion tests are based on heat treatment exposing either the entire specimen or applying a local heat shock. In these tests, the deformation difference of the coating and its substrate is of importance, either due to the hardness/tensile strength difference of the mismatch of the heat expansion coefficients. Hence, these tests are not of general validity but refer to specific conditions that correspond to the load that the coating is expected to withstand to. If the coating layer is flexible enough for rolling, a quantitative peel-off test can be carried out by delaminating the entire coating layer by simultaneously measuring the force required or the work performed during a complete delamination. In both cases, the result is quantitatively elucidated and can be normalized (force and work with respect to the width of the layer strip and to the entire surface area, respectively).

3.5 Scanning Probe Methods

In scanning probe techniques, the surface of a solid specimen is scanned with a sharp tip along a raster. The recorded parameter is a physical quantity that is converted to a colour-scaled image which shows the local variation of the physical property measured. The lateral dimension of the scanned area is at most a few micrometres. The lateral resolution depends on the specific technique and is mainly determined by the tip shape and size. The vertical resolution of all methods can be at the sub-nanometre scale, which is due to the precise motion control technique based on piezoelectric drives.

Scanning tunnelling microscopy (STM) applies atomically sharp tips. The measured parameter is the tunnelling current between the conductive surface of the test specimen and the tip. The tunnelling current depends exponentially on the distance between the tip and the surface and is linearly proportional to the energy-dependent electronic density of states of the sample surface. Since the density of state strongly varies at much smaller scale than the size of a single atom, the atomic structure of a conducting surface can be mapped with subatomic resolution. Specific features on single-crystal surfaces detectable with STM include surface defects, steps and terrace edges, and binding mode and configuration of adsorbed molecules including their conformation. For surfaces that are not atomically smooth, the estimation of surface roughness is a prominent application mode of STM. The constant-current operation is typical for STM. In electrochemical environment (i.e., under solution and with electrode potential control), combined apparatus with potentiostatic control of the specimen surface is required in combination of the potential difference regulation between the tip and the surface scanned by an integrated high-sensitivity electronic device. In vacuum or in air, the side of the tip is of no importance, while in electrochemical STM mode, it has to be covered nearly to the very end of the tip so that to reduce parasite current as much as possible. A variation of STM is scanning tunnelling spectroscopy where the potential difference between the tip and the specimen is also modulated, and the detection of local tunnelling conductance yields a local excitation spectrum.

Atomic force microscopy (AFM) applies tips of 5–25 nm end radius. Since the measured parameter is the force acting between the tip and the specimen surface, the tip does not have to be electrically conductive. The most common material for commercially available tips is silicon nitride for which the half-angle of the tips is around 20°. Sharper ceramic tips with a half-angle of a few degree and curvatures down to the 1-nm-level are available with special manufacturing techniques, and custom-grown carbon nanotubes can also be applied as tip material. The basic configuration of an AFM instrument is shown in the left scheme of Fig. 3.7.

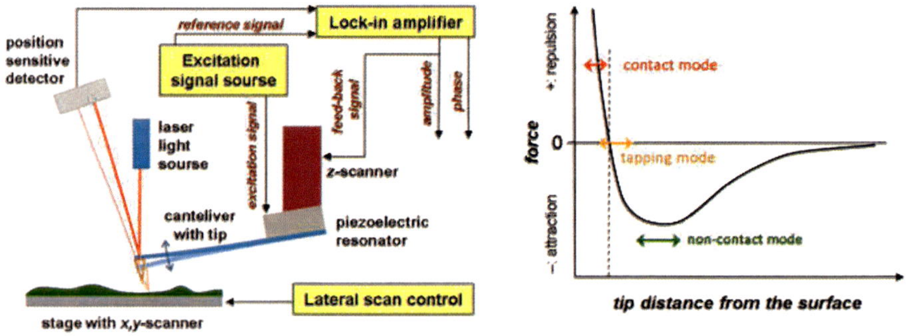

Fig. 3.7 **Left**: Scheme of the apparatus used for atomic force microscopy. **Right**: Force–distance function for the tip–specimen pair (approach curve)

The operation mode of an AFM apparatus depends on in which interaction regime of the tip and the test specimen is utilized (see Fig. 3.7, right). In non-contact mode, the interaction is attracting, and the tip oscillation is around the distance of maximum attracting force. In the tapping mode, the tip touches the surface at the end of its motion towards the surface, while in the contact mode, the tip contacts the specimen surface throughout the entire test, and hence, interaction turns to be repulsive. The danger of damaging the surface increases in the order of the list of operation modes described above.

For the observation of surface topography, the measurement of force in the direction of the surface normal is of prime importance. Nevertheless, the measurement of lateral forces is also possible, which allows one to test the adhesion strength of namometric objects or even molecules to the surface. By using magnetic tips, we arrive at magnetic force microscopy (MFM) that measures the magnetic interaction between the tip and the surface and accounts for local field strength.

3.6 Corrosion Studies

The reason why corrosion test must be dealt with is that the original version of several sorts of these tests used to be developed for materials much simpler than the nanostructures synthesized nowadays. Therefore, the blind application of some corrosion tests may produce misleading results.

Non-electrochemical corrosion tests involve the exposure of the specimens studied to a certain media (salt solution, steam etc.) without either controlling or measuring the electrode potential of the samples. The absolute rate of general corrosion can be determined from the weight loss of the specimen. The corrosion performance, however, is defined somewhat differently from the rate of general corrosion. For instance, the time elapsed until the appearance of red rust is an important parameter for rating of relative corrosion performance of various coatings on iron-containing specimens. Other corrosion test methods are designed to assess the inclination of a material to some specific mode of corrosion. One of these methods involves the reaction with $FeCl_3$ solution and the areal density of cavities formed bear information on the inclination to pitting corrosion. Many non-electrochemical test methods are standardized, and various fields of industry develop their on corrosion test protocols. These tests are applicable for nanostructured materials in the case only when the samples comply with the requirements of the standards (for instance, a nanocrystalline coatings of large surface area are suitable for such tests, while individual nanoparticles are not).

Concerning electrochemical corrosion studies, some methods apply a large potential perturbation. The most common method is to record Tafel polarization data for both the cathodic and anodic direction as referred to the corrosion potential. Although the manner as corrosion test is displayed is quite analogous to that presented in Fig. 2.5b for an equilibrium reaction, there are fundamental differences between the two measurements. While in an equilibrium reaction, the cathodic and anodic currents correspond to the same reaction proceeding in opposite directions, the anodic and cathodic arms belong to *different* reactions in the corrosion studies: oxidation of the metal (anodic dissolution) and reduction of the oxidizer (mostly O_2 or H^+). Obviously, the Tafel slopes of the two partial reactions are independent of each other. The corrosion potential is set to the value where the anodic and cathodic partial reactions take place at an equal rate. In an ideal case, the two Tafel lines have the intercept at the corrosion potential, and the current density corresponding to the intercept yields the corrosion rate.

The reason why one has to be particularly cautious with the large-amplitude perturbation method is that it may induce irreversible changes in a material of complex structure, which makes correct corrosion rate estimation impossible. A few possible pitfalls of the large-amplitude polarization studies are as follows. (i) When the polarization curve is recorded with a potential sweep, the capacitive current contribution causes a current misestimating, which has the highest relative error around the corrosion potential where the current is presumably zero. Therefore, the corrosion potential as assessed from a sweep is often misestimated. (ii) If the polarization curve is recorded with a single sweep from the cathodic end potential to the anodic one, the accumulation of the cathodic reaction product may be oxidized in the anodic part, hence falsifying the current measured. (iii) If the cathodic and the anodic parts of the polarization measurement are performed on an identical specimen, the cathode reaction may irreversibly modify the surface, and the anodic part is recorded for a surface much different from the original surface to be tested. A great care must be taken so that the measurement conditions indeed refer to the material of interest in the state to be studied.

While the small-amplitude perturbation methods eliminate a few of the above mentioned problems, some others may arise. A common method is to record a slow-rate sweep in the narrow (say, ± 10 mV) neighbourhood of the corrosion potential and to approximate the polarization curve with a line. The inverse slope of the line, $(\mathrm{d}j/\mathrm{d}E)^{-1}$, is called the polarization resistance. If the Tafel slopes of the cathodic and anodic partial reactions are known, the corrosion rate can be calculated (even though the polarization resistance of similar materials is also suitable to estimate a relative order of the corrosion rate of various specimens).

Another small-amplitude perturbation method is the measurement of electrochemical impedance that involves sinusoidal potential perturbation of various frequencies. For electrochemical impedance spectroscopy (EIS), the high-to-medium frequency part of the impedance spectrum yields a semicircle on the Nyquist plot whose diameter can be associated with the charge transfer resistance. Even though the charge transfer resistance, $\left(\partial j/\partial E\big|_{c_i,\theta,..}\right)^{-1}$, has a different meaning than polarization resistance, it is often used for quantification of the corrosion resistance. While the application of the charge transfer resistance if more or less correct for rating the relative corrosion rate of samples of the same family, the entire impedance spectrum is often fitted with the parameters of some equivalent circuits that cannot be validated. Hence, the border between the correct application of the impedance method and the arbitrary and heuristic "equivalent circuit engineering" is very narrow. It is to be remembered that a kinetic scheme unambiguously defines the impedance response of a system that can be often characterized with an equivalent circuit, while the good fit to an equivalent circuit is not at all a diagnostic criterion for a reaction mechanism.

References

1. Ida N, Meyendorf N (eds) (2019) Handbook of advanced nondestructive evaluation. Springer Nature Switzerland AG, Cham
2. Morrison K (2019) Characterisation methods in solid state and materials science. IOP Publishing, Bristol
3. Kanani N (2004) Electroplating—basic principles. Processes and practice, Elsevier, Oxford-Amsterdam

Part II
Nanostructured Materials Obtained by Using Non-structured Substrates of Large Surface Area

Chapter 4
Ultrathin Layers

4.1 Layer Preparation Methods Based Solely on UPD Processes

4.1.1 Atomic Layer Deposition Processes

Atomic layer deposition (ALD) is a method originally developed for coating surfaces with layers that are produced in a reaction of surface-bound species with some gaseous molecules. The principle of the method is that functional groups that can react with each other belong to different phases; i.e., to either the surface layer or to the molecules in the gas phase. Therefore, the reaction between them cannot proceed until the full consumption of the gaseous reactant but is limited to the number of active surface sites.

An example for atomic layer deposition processes is the formation of SiO_2 layers on an activated surface. Assume that the surface of a solid sample is already terminated with silicon-containing molecular fragments that are hydrated; therefore, the surface is dominantly covered with ≡Si–OH groups (the ≡ sign indicates that the other three covalent bonds of the Si atom belong to atoms below the topmost atomic layer). If $SiCl_4$ molecules arrive at the surface, they react with the surface-bound entities. The reaction leads to the elimination of HCl, and another Si-containing atomic layer is attached to the surface. Depending on the new bonds formed, the freshly chemisorbed species will be the terminal part of a $(\equiv Si–O–)_x SiCl_{4-x}$ surface site, where $1 \leq x \leq 3$. If then the surface site can react with water molecules, the chlorine atoms are exchanged to –OH groups again. The latter process is accompanied by HCl elimination, similarly to the previous reaction. With the formation of the –OH groups at the surface, the starting state is reproduced. With the repetition of the above described cycle, the number of atomic layers can be controlled by the number of the cycle repetition. Similarly to the formation of the ALD SiO_2 layers, various covalent and ionic metal oxide layers can be produced with the reaction of metal halogenides and water vapour, among which Al_2O_3 and TiO_2 have the

L. Péter, *Electrochemical Methods of Nanostructure Preparation*,
Monographs in Electrochemistry, https://doi.org/10.1007/978-3-030-69117-2_4

largest technological relevance. Nitride layers can be produced by the reaction of metal chlorides and ammonia, while elemental metallic layers can also be produced at higher temperatures in various catalytic processes. Mixed oxides and multilayered oxide structures are also available with ALD with a thoughtful combination of the precursor compounds. Various reviews of this topic are available for interested readers [1–3].

An important feature of the ALD process is that the reactions leading to the formation of the surface layer are *surface-are-limited*.[1] This means that once the surface is covered with the newly introduced reactant, the process stops and the thickness increment in each cycle is constant, regardless of the excess of the available reactant in the gas phase. A direct consequence of the surface-area-limited nature of the ALD process is that it leads to a highly conformal deposit on the surface, practically independently of the shape of the workpiece. Beside the variability of the ALD process concerning the surface layer composition, the conformal nature makes it ideal to cover objects with a complex shape, especially as compared to physical methods where the deposit is grown by a process having a preferential direction (like evaporation and sputtering).

The experimental setup of ALD devices includes reservoirs of the reactants, a tank for an inert gas, a pump system and valves. The entire system can be automated. The reaction of the surface with one of the reactants is followed by the removal of the reactant gas, flushing of the system with an inert gas and introduction of the new reactant. Since the full coverage of the surface requires a minor amount of the reactant, the pressure is usually much below the atmospheric conditions. The exposure time must be high enough so that the reactant gas can reach all surface sites of the workpiece even at the bottom of the cavities and pores (it is which usually happens within a few minutes). Hence, the ALD process is quite time-consuming. Another drawback of the ALD process is that the leftover precursor materials introduced to the system in a particular cycle cannot be used in the next cycle, which makes the process inefficient with respect to the usage of the precursor material.

An ALD-based process using liquid phases is the so-called successive ionic layer adsorption and reaction (SILAR) method. This is used mostly for the deposition of oxide and chalcogenide materials by immersing the substrates alternately in reactant-containing and rinsing solutions. The sequence of the steps in the SILAR processes is essentially the same as for the ALD method with the difference that ALD always takes the advantage of the reaction of molecules while SILAR uses dissolved ions as reactants in a liquid phase. A deeper insight into the current state of SILAR-based research is available in various reviews [4, 5].

---

[1] Throughout this book, the term "surface-area-limited" will be used in order to indicate that we mean the limitation of the number of active sites, not only the zone of the reaction. When the term "surface-limited" is retained for sake of the compliance with the literature (i.e., in the name of some methods that are commonly used elsewhere), the maximum extent of the reaction is also determined by the surface area, but it will not be mentioned again.

4.1.2 *Layer-by-Layer Electrodeposition Based on ALD*

The electrochemical analogue of the ALD and SILAR processes is referred to with various names in the literature. Hereinafter, it will be named as electrochemical atomic layer deposition (EC-ALD). The name electrochemical atomic layer epitaxy (EC-ALE) is also common, especially if the deposit grows in a manner to build up a coating structurally coherent with the substrate. In the latter case, the substrate is typically a well-defined single-crystal face. The EC-ALD method was pioneered by John Stickney from the early 1990s on [6, 7].

The operation of an EC-ALD workstation is based on principles similar to those of ALD and SILAR devices: all have reservoirs for the reactants and for the rinsing fluid, hence ensuring that only the desired reactant can reach the sample surface at a given time. The EC-ALD workstation operation is also sequential. In contrast to some physical and chemical layer deposition methods, the big majority of the EC-ALD processes work at room temperature, which makes the operation of the device rather simple.

An important difference as compared to either ALD or SILAR processes is that the reactions taking place are not spontaneous in the sense that the substrate (i.e., the working electrode) has to be polarized to the desired electrode potential. This makes it indispensable that the workstation is equipped with the usual electrodes of an electrochemical cell (working, auxiliary, and reference electrodes), and the setting of the electrode potential must be precisely synchronized with the pumping system. The scheme of a possible EC-ALD workstation configuration is shown in Fig. 4.1.

Some other variations for the setup of the EC-ALD systems are also available from the literature. It is possible to use one single pump connected to the outflow of the electrochemical cell, but the regulation of the valve system remains the same also for such setups [15–17]. Regardless of the details of the feed regulation, the EC-ALD setups use thin-layer cells. Even if the volume of the solution used is as small as possible, still a large excess of the component to be deposited is used. The surface density of atoms is around 2×10^{-9} mol cm^{-2}. If the cell volume is only 0.1 ml and the solution concentration is 1 mmol dm^{-3}, the cell contains 10^{-7} mol reactant, which provides approximately 50-fold excess of the reactant for a working electrode of 1 cm^2 surface area.

While the working and the counter electrodes are situated face-to-face as parallel plates in the thin layer cell, the reference electrode is usually placed in the outflow of the cell. Although this is not ideal from the viewpoint of the electrical regulation of an electrochemical cell, it does not impact negatively the desired process. This is because the length of a deposition section of a component is long enough so that the current decays by the end of the pulse, which means that the completion of the build-up of a new layer means equilibrium conditions with zero current corresponding to the reaction of interest in an ideal case. Hence, the potential measurement in equilibrium is free of any ohmic drop. For the good control of the electrode potential also during the rinsing periods, it is important that the rinsing solution(s) contain(s) an appropriate supporting electrolyte that may also vary depending on the composition

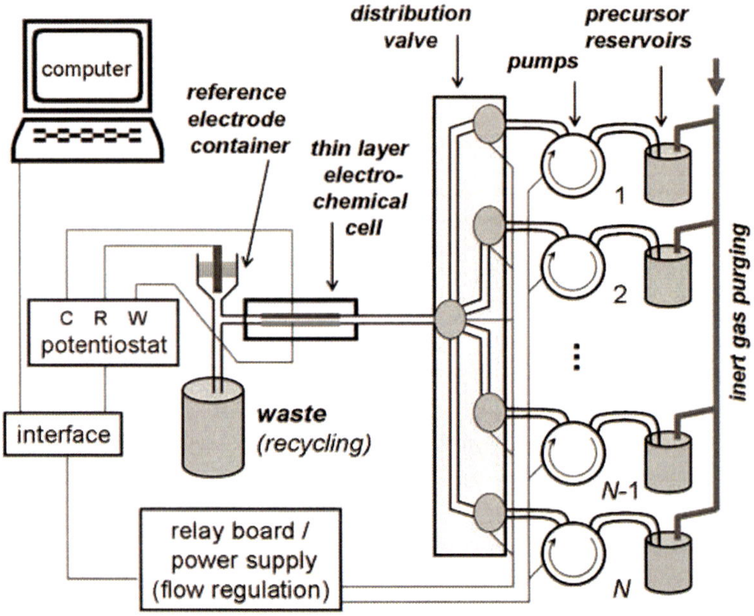

Fig. 4.1 Configuration suitable for EC-ALD processes: multiple-pump system allocating a pump for each inlet port of the distribution valve. The figure was drawn on the basis of the configurations published in various works [8–14]

of the layer previously deposited. The separate deposition of each element and the achievement of the equilibrium eliminate the simultaneous mass transport control that always occurs when bulk deposits form in a single step from mixed solution.

The surface-area-limited nature of the EC-ALD process is provided by the fact that in each step, one single UPD layer is deposited only and no bulk deposition is allowed. This imposes a very strong prerequisite towards both the reactant and the surface: each new component to be deposited as a layer of at most monoatomic coverage must form this layer under underpotential deposition conditions [18]. For the sake of simplicity, the expression "monoatomic" will be used in this chapter for indicating that the *thickness* of the newly forming layer does not exceed *one atomic layer*, even though the *coverage* with the deposit atoms as referred to the number of the atoms on the substrate surface can be less than 1, as it is usual for UPD layers.

If the formation of a monoatomic layer is not possible, a step for the post-treatment of the deposit layer may be necessary to strip off the excess material and to leave a monoatomic layer behind. The EC-ALD operation with aqueous solution imposes a strong limitation on the available deposit composition. As opposed to ALD where the synthesis of the oxide of a number of metals is feasible and the resulting layer is mostly amorphous, EC-ALD is suitable for depositing *layered semiconductors* only with typically a high degree of crystallinity.

4.1.3 Optimization of the EC-ALD Processes

It is essential that the first atomic layer exhibits a UPD process on the substrate. It may occur that the components forming the alternating atomic layers undergo UPD process on each other, but only one of them can form a surface-area-limited monoatomic layer on the substrate. This was demonstrated for indium telluride deposition onto GaAs [19]. In such cases, it is essential that the first component to be deposited can cover the substrate in a surface-area-limited manner, which has a consequence on the right order of the layers to be deposited. The same situation occurs when NiS is deposited onto silver substrate [20].

The primary optimization of the EC-ALD process involves the establishment of the potential regions of the UPD process. When a single-crystalline substrate is used as working electrode, a monoatomic layer of one of the components (A) is deposited onto the working electrode. Then, the solution is exchanged under controlled working electrode potential for introducing the precursor compound of the other component (B), and cyclic voltammetry is used to establish at which potential this component forms a UPD layer with the desired coverage. This preliminary screening of the suitable deposition potential of both components on the other component as substrate has to be completed ("A" on "Su/B" and "B" on "Su/A", Su being the substrate) [7]. Figure 4.2 presents an example for the preliminary experiments and the establishment of the suitable deposition potentials. In the example given in Fig. 4.2, both components are deposited with a reductive UPD process from an oxidized precursor material.

The graphs in Figs. 4.2a, b, compare the reduction of Te(IV) on bare Pt and Sb-covered Pt surface, respectively. It can be seen that the peak intensities change, but the shift in the peak positions is fairly small. Since the UPD coverage of Pt with Sb is less than 1, UPD of Te on the Sb-modified Pt surface is at least partly due to the Te–Pt interaction. The situation is very similar to the other pair of curves; i.e., Sb on either Pt or Te-modified Pt. The diagnostic criterion for the appropriate deposition potential in the initial EC-ALD cycles is as follows. If the cathodic-going sweep is reversed at the right potential, the formation of the UPD layer is already complete, but the forthcoming anodic-going sweep does not exhibit the stripping peak of the *bulk* deposit (which is indicated as A3 in all graphs in Fig. 4.2). Since the deposition of semiconductor components is seldom electrochemically reversible (i.e., the onset of the dissolution potential of the bulk phase is more positive than the onset of the formation of the same phase), it is advised that a number of cyclic voltammograms are run with varying cathodic limit to establish the onset of the formation of the bulk phase. This approach is particularly important when there is also a current contribution due to hydrogen evolution.

The reactants for the preliminary study of an EC-ALD process shown in Fig. 4.2 were in the oxidized state, Te(IV) and Sb(III). Hence, the study of the reactant on Pt could be performed by running cyclic voltammograms starting at the positive potential limit. However, when the substrate is pre-covered with the UPD layer of one of these components, it is important to keep an accurate potential control when the solution is changed, so that the surface layer is well conserved for the study of the other component. This is why the starting potentials are indicated in the graphs b, and d, of Fig. 4.2. It is also to be observed that the anodic limit of these measurements is always more positive than the stripping peak of the UPD layer of either of the system components (i.e., that of the topmost atomic layer or that of the other one that was used for the surface modification). For this reason, such curves cannot be recorded repeatedly, but a new electrode preparation procedure has to be carried out by depositing the surface-modifying layer. In rare cases, the stripping of the two components of a bilayer can be distinguished in anodic linear voltammograms (e.g., for InAs as seen in Fig. 10 of [21]).

When comparing the UPD peak potential for the bare and pre-coated substrate, the shift can be either positive or negative. As pointed out by Xiao et al. [23], the shift of the peak potential is a result of various competing effects. Since the interaction between the components of semiconductor compounds is essentially always stronger that the strength of their interaction with the substrate, one can expect a positive

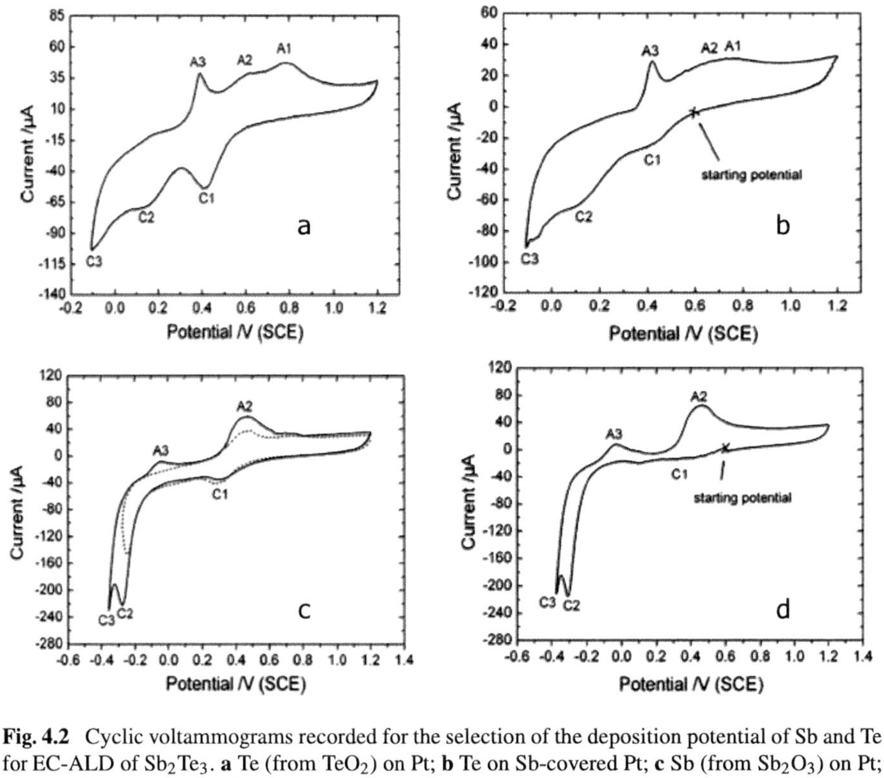

Fig. 4.2 Cyclic voltammograms recorded for the selection of the deposition potential of Sb and Te for EC-ALD of Sb_2Te_3. **a** Te (from TeO_2) on Pt; **b** Te on Sb-covered Pt; **c** Sb (from Sb_2O_3) on Pt; **d** Sb on Te-covered Pt. All data were recorded in 0.1 $MNaClO_4$ solution of pH = 1.5 at a sweep rate of 10 mV s^{-1}. C1 and C2 are surface-area-limited (UPD) peaks, while C3 indicates the onset of the bulk deposition. Figures reprinted with permission from [22]. Copyright (2006) American Chemical Society

potential shift if the peak position on the pre-coated substrate is compared to that of the bare substrate. However, the UPD layer of one of the components also modifies the work functions, hence giving rise to a contact resistance. If the latter effect is dominating, the peak shift is negative. This was found for Bi deposition on Se-covered Pt [23] and also for Bi deposition on Se-covered Au [22].

The deposition of the metallic component of an EC-ALD deposit (mostly Cu, Zn, Ga, Cd, In, Sn, Hg, Pb, Bi) is a reduction process with no exception since the precursor material of these components contain the cation (i.e., an oxidized form) of these elements. By running a potential sweep from positive to negative potentials, working regimes of 'no deposit' → 'UPD layer(s)' → 'bulk deposit' follow each other, which dissolve electrochemically in the reverse order during an anodic sweep. For some chalcogenide elements, the deposition in a reductive pulse is also possible when the precursor compound contains an oxoanion of these elements in the solution ($H_xSeO_3^{(2-x)-}$, $H_xTeO_3^{(2-x)-}$ and $H_xAsO_3^{(3-x)-}$ from SeO_2, TeO_2 and As_2O_3 as solute, respectively). However, the atomic layer of a chalcogenide element can also be deposited with an oxidative pulse if the precursor material is the hydride of these elements such as H_2S (or Na_2S) as demonstrated first for CdS by Colletti et al. [24], and the oxidative S deposition became predominant later. In the case of oxidative deposition, the order of the potential regimes of the surface-area-limited and bulk deposition region is the opposite than for cations.

The selection of the UPD peaks is often carried out by varying the concentration of the precursor material when cyclic voltammograms are recorded [11, 25]. However, as the concentration of the precursor material is in the sub-millimolar range, the position of the UPD peaks as observed in the cyclic voltammograms may vary, especially at a high sweep rate [26]. Since the position of the stripping peaks does not change with the solute concentration, this shift of the UPD peak can be attributed to a sluggish kinetics of the UPD layer formation, to mass transport effects or both. Apart from the concentration change, a shift in the cathodic end potential of the cyclic voltammogram is often employed, which leads to a constant stripping charge for UPD peaks and a growing bulk deposition peak as the negative potential limit increases.

The diagnostic method for selecting the suitable deposition potential of component "A" on an electrode pre-covered with a monolayer of the component "B" is the stripping of component "A" under circumstances that leave the layer of "B" intact (or after the stripping of component "B"). If we take into account one bilayer only, the total charge corresponding to the stripping peak of the topmost component usually shows a plateau region as a function of the deposition potential of this component. The independence of the stripping charge from the deposition potential of the component is the indication of the surface-area-limited nature of the process. A deviation from the constant charge indicates potential-dependent coverage because of either the formation of a sub-monolayer coating or that of a bulk deposit. This optimization method of the deposition potential of both components was well exemplified by Loglio et al. [27] whose work will be used below for reference. In this work, CdSe was deposited by applying reduction to produce the monolayer of Cd and Se from solutions containing Cd^{2+} and SeO_3^{2-} ions, respectively. The Cd layer formation on Se-covered electrode was studied by oxidative stripping of Cd (reaction: $Cd \rightleftarrows Cd^{2+} + 2e$). The quantitative analysis of the Se UPD on Cd, however, required the reductive stripping of Se (reaction: $Se + 2e \rightleftarrows Se^{2-}$) because the oxidative stripping would have led to the oxidation of the Cd layer, too. The width of the potential interval where a monolayer could be produced varied from one component to the other, yielding intervals of about 90 and 40 mV width for Cd and Se deposition, respectively. It seems to be a general experience that the optimal potential interval for the deposition of the metallic component is wider than that of the chalcogenide component. For EC-ALD of bismuth telluride, Zhu et al. found [28] a nearly 300 mV electrode potential range for the monolayer formation of Bi, while Te deposition with an appropriate coverage required a more accurate choice of potential within a ~70 mV potential interval. When CuS is deposited by EC-ALD, the Cu deposition potential can be chosen in a 180 mV potential interval [29].

The choice of the deposition potentials can be also verified by observing the stoichiometry of the deposit. If the desired ratio of the element is found, the deposition potentials can be accepted. As shown for CdTe deposition [30], the stoichiometry of the deposit, observed as a function of the Cd deposition potential, showed a plateau region in the same Cd deposition potential range where the coverage also had a plateau as a function of the same independent variable, although the coverage per cycle for Cd in the optimal potential interval was less than one monolayer.

It is to be stressed that the optimization of the UPD deposition potentials of atomic monolayers or bilayers on a noble metal substrate (i.e. normally used in preliminary experiments) is not a guarantee that the deposition of successive layers (bilayers) can take place in a similar way at large cycle numbers. The explanation is that the properties of the first bilayer on a foreign substrate can be substantially different from the surface of a bulk deposit. As described by Wade et al. [31] for InAs and InSb, the application of the same deposition potentials that result in a well-defined monolayer on a noble metal electrode may lead to the stop of the entire deposition process later. In contrast, if the deposition potentials are chosen by 200–300 mV more negative than the potentials of the UPD range on the starting substrate, the deposit keeps growing in the subsequent cycles at a reasonable rate. In such a case, however, excessive deposition currents may occur in the first few tens of cycles due to the initial roughening that is related to the three-dimensional nucleation instead of a mere UPD at the beginning. The deposition potentials chosen from the cyclic voltammograms run on the bare substrate are often not sufficiently negative to achieve the growth of the deposit with large cycle numbers. This points out that "the potentials derived from voltammetry on the substrate are simply a good first approximation for the potentials needed for at most one cycle, but do not represent the potentials needed to deposit the elements onto each other" [32]. In many cases, especially for multicomponent structures, it appears that there is no potential set that would work both in the initial phase and in the steady-state [33].

It is not only the chemical nature of the surface that can lead to a change in the deposition potential suitable to produce full monolayer coverage. Since most of the deposits prepared by EC-ALD are semiconductors, the potential drop across the semiconductor layer of increasing thickness also varies because of the occurrence of a space charge layer. Based on this concept, Vaidyanathan et al. [34] suggested a cycle-number-dependent potential programme with successively decreasing deposition potentials. The decrease rate of the deposition potential that was found to lead to complete monolayers varied from one material to another. For the deposition of In_2Se_3, the deposition potentials were about 400 mV more negative after some 30 cycles than that established for the bare substrate [34] and had to be decreased further as the deposit grew. The principle of the continuous deposition potential decrease with cycle number was applied for a variety of compounds. For PbSe, a 100 mV negative shift in the potential proved to be sufficient after 20 cycles without any further decrease [35], and a steady-state deposition sequence could also be achieved for Bi_2Se_3 [23]. In the case of Sb_2Te_3, both negative [22, 36, 37] and positive shifts [38] in the deposition potential were applied in various works of the same group. For PbTe, a positive shift of the Te deposition potential was necessary to maintain a single monolayer per cycle ratio [39]. Although this finding was empirically well confirmed, no explanation was given why a deposition potential correction of the opposite direction had to be employed.

If the deposition of one component on the topmost atomic layer of the other component cannot reliably lead to monolayer coverage but the nucleation of a bulk material is likely at all possible potentials, then another stripping pulse can be applied. This case is common when Te is a component of the system to be deposited. In such a case, one can take advantage of the cathodic stripping of bulk Te. Anodic stripping of bulk Te may not be possible since this treatment is likely to damage the metal atom layer underneath. However, the cathodic stripping of Te leading to Te^{2-} will surely leave both alternating atom layers underneath *and* one Te surface monolayer intact, while the excess of Te is removed. Suitable potential and solution exchange programmes either without or with a cathodic stripping pulse are shown for the EC-ALD of Bi_2Te_3 in Fig. 4.3. The cathodic stripping pulse may also have another role when the reduction of the metal takes place also from an oxoanion and the "atomic"

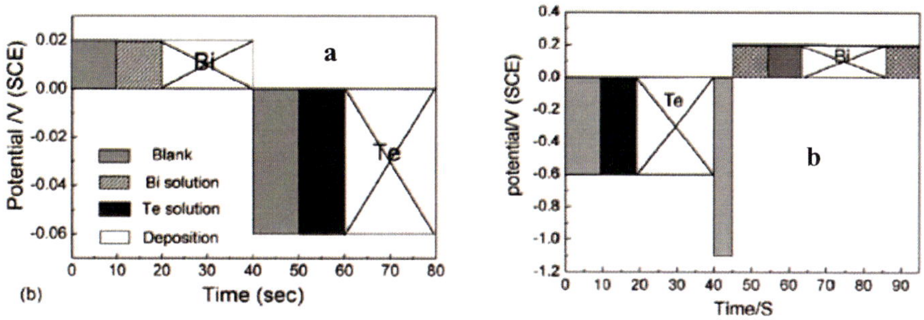

Fig. 4.3 Various potential programs and solution exchange sequences for an EC-ALE deposition of Bi_2Te_3. **a** two-step program without a stripping pulse [41], **b** three-step program including a cathodic stripping pulse for the removal of excess Te [11]. Reproduced from the references indicated. Copyright (2005)/both works/, with permission from Elsevier [11] and from the American Chemical Society [41]

layer produced in the relevant EC-ALD step cannot lead to a pure metal layer. This is the case for $MoSe_2$ when the reduction of the molybdate anion leads to a molybdenum oxide with lower oxidation state whose reduction to molybdenum and the release of the residual oxygen takes place during the reductive stripping of the leftover Se [40].

Should the sequential deposition of the atomic layers onto each other be a surface-area-limited process, the current after the deposition of a complete atomic layer usually does not return to zero but a residual current remains. A few typical chronoamperometric records are presented in Fig. 4.4. A current decay is usually found when the achievement of a constant background current indicated the completion of a monolayer [11, 25, 30, 42–44]. Since the current corresponding to the formation of a UPD layer is small, the residual current often amounts to more than 20% of the peak current at the beginning of a UPD step [28, 36, 37, 40, 41, 45–47], which makes a current monitoring-based pulse length optimization rather difficult. It occurs only in ideal cases that the current drops to a near-zero level during the rinse with blank solution [23, 39]. The combination of the chronoamperometric measurement and EQCM frequency change detection is a useful approach to obtain more information on an EC-ALD system [48].

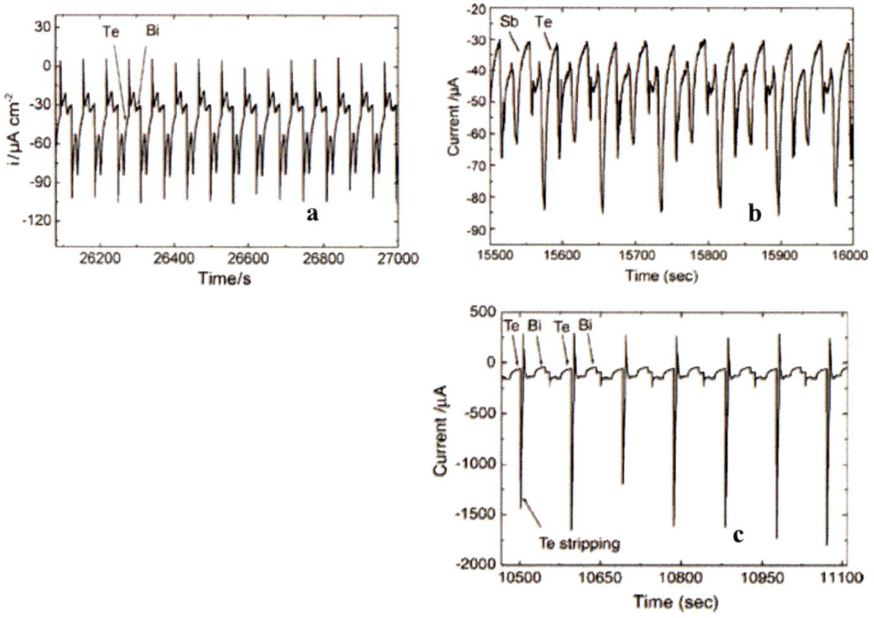

Fig. 4.4 Chronoamperometric records for EC-ALD processes of various materials and pulse sequences. **a** Bi_2Te_3 [38] and **b** Sb_2Te_3 [37] deposited with two deposition pulses with no stripping; **c** Bi_2Te_3 deposited with a stripping pulse after the Te deposition [11]. Figures were reproduced from the references indicated above. Copyright (2008) **a** and **b** and (2005) **c** with permission from Elsevier

Table 4.1 Summary of the works dealing with EC-ALD of binary compound semiconductors. For the more electronegative element, the oxidation state of the atoms in the precursor compound is indicated in parenthesis

	S (−2)	As (+3)	Se (+4)	Sb (+3)	Te (+4)
Ni	[20]				
Cu	[29]		[49, 50]		
Zn	[51]		[52–54]		[55]
Ga		[56]			
Mo			[40]		
Cd	[9, 24, 51, 57–61]		[9, 27, 42, 62]		[6–10, 15, 30, 63–69]
In		[21, 31]	[34]	[31]	[19]
Sn			[25]		
Sb					[22, 36, 37, 70]
Hg			[32, 71]		[72]
Pb	[73]		[35]		[39]
Bi			[23]		[11, 28, 38, 43, 74]

The composition of EC-ALD deposits reported so far is summarized in Table 4.1, where the oxidation state of the source compound of the chalcogenide element is also indicated.

4.1.4 *Characterization of the Composition, Structure and Semiconductor Properties of EC-ALD Layers*

Regardless of either the number of steps during a cycle or the choice of the potential programme, a common diagnostic criterion of the surface-area-limited nature of the deposition process is that the sample grows proportionally with the cycle number [10, 19, 21, 24, 51–53, 60, 67, 69, 71]. The proportional growth can be measured either by the thickness or by the charge consumed during the stripping-off treatment of the deposit. It is important to note that the growth rate as measured with the thickness increment with cycle number is a function of the crystallographic orientation of the deposit. In a stripping experiment, the charge for the cathodic and anodic stripping can be measured independently, hence also checking the stoichiometry of the composition.

For characterizing the composition of the EC-ALD layers, two major *ex situ* methods are applied. The composition of the layer can conveniently be studied by EDX in an electron microscope. The proper application of EDX can yield a quantitative result on the composition [8]; however, it does not account for the chemical state of the surface. Auger electron spectroscopy (AES) and X-ray photoelectron spectroscopy (XPS) are both suitable to check the composition as long as the thickness is within the depth sensitivity range of the electron spectroscopic analysis. In addition, electron spectroscopy is also sensitive to the chemical state of the surface, and the desired layer chemistry can also be checked. Both AES and XPS have been applied successfully to characterize EC-ALD layers since the introduction of the EC-ALD concept [7, 20, 63], and even electrochemical cells directly connected to the antchamber of the high-vacuum equipment have been elaborated [56].

The electrochemical stripping analysis method can also be made suitable for the analysis of the deposit after a few EC-ALD cycles. The discussion of these procedures below closely follows the description of Innocenti el al. [21, 29, 51, 73, 75]. The forthcoming discussion refers to binary deposits for sake of simplicity but the procedure was shown to work also for ternary deposits with two metallic components [76]. For analysing the charge corresponding to the stripping of either of the components in a binary deposit, one needs two series of samples, and each of them will be used for the quantitative assessment of one component only. As a first step of the experiment, one of the components has to be stripped off from the EC-ALD deposit. The metallic component can be stripped off with an anodic treatment and vice versa, the chalcogenide component is stripped off as negative ions produced with a cathodic treatment. It is obvious that the atoms remaining on the surface have to be rearranged and will behave thereafter as a deposit formed from a single element. Then, the solution containing the ions of the already dissolved component is removed, and it may also be necessary that a solution with dissimilar composition should be used for dissolving the component still on the surface of the working electrode. During the rinsing/solution exchange process, the electrode potential has to be well controlled so that the coating component still present remains intact. Then, a linear sweep is run in the appropriate direction in which the charge corresponding to the stripping peak is proportional to the amount of the deposit. This study has to be carried out with two series of samples, each series containing samples with different EC-ALD cycle numbers. A representative set of curves for the stripping analysis of CuS layers is shown in Fig. 4.5.

Attention has to be drawn to a few features of the stripping curves of EC-ALD deposits. When the chalcogenide component (mostly sulphur) is stripped off with a linear potential scan, two peaks are observed. If the curves are regarded in the sweep direction, the surface area of the first peak changes with EC-ALD cycle number but the second does not, should the second be more negative [29, 51, 73, 76] or more positive [21]. This is because the stripping treatment first removes the bulk of the deposit, and the UPD layer of the deposit can be dissolved only later. This is in good agreement with the fact that sulphur can form a UPD layer on the substrate used in the above-mentioned studies. The curves shown in Fig. 4.5 have such a character. However, stripping of the metallic component exhibits one stripping peak in cases when this metal does not tend to form a UPD layer on the substrate used (In [21], Pb [73] and Cu [29] on Ag(111), also the curves in Fig. 4.5). However, if the metallic component of the EC-ALD deposit also exhibits UPD on the substrate, a voltammogram with two peaks can be obtained for its oxidative stripping with the

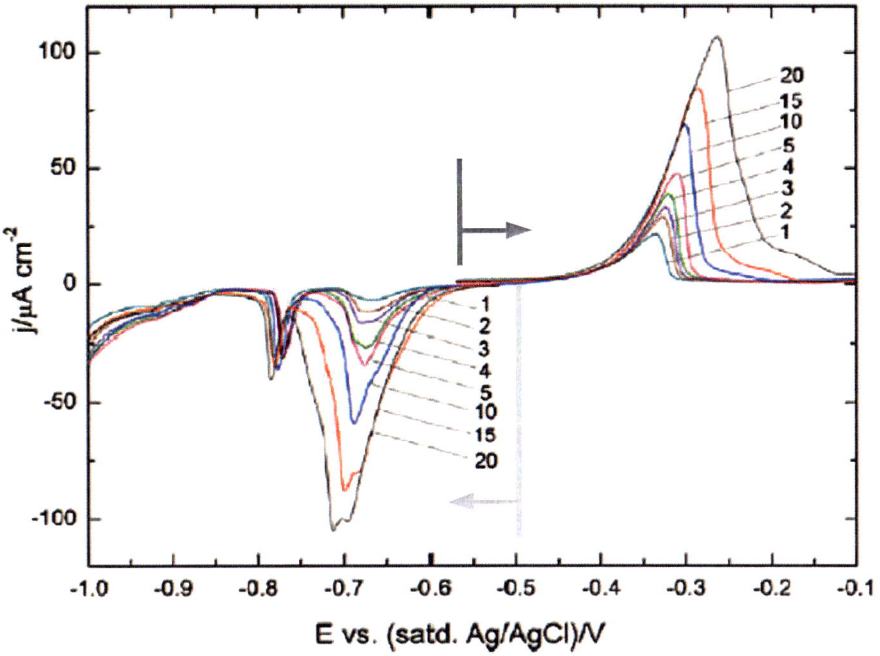

Fig. 4.5 Stripping analysis of two series of CuS EC-ALD layers deposited onto Ag(111) surface. Numbers next to the curves indicate the cycle number. Arrows indicate the sweep direction for the groups of curves. Anodic stripping: oxidation of Cu; cathodic stripping: reduction of S. Reproduced from [29]. Copyright (2011), with permission from Elsevier

same peak area vs. cycle number characteristics as for the chalcogenide component (e.g., Cd on Ag(111) [75]).

Beside the composition analysis of the deposits, structural studies play an important role to verify the atomic arrangement produced with EC-ALD. Low-energy electron diffraction (LEED) is appropriate to detect the arrangement of the atoms of the first ordered layer on a single-crystal surface [7, 56, 63]. Although LEED is sensitive to the topmost atomic layer only, it proved to be particularly useful to clarify the atomic arrangement of the deposit on various crystal surfaces [64]. Similar results were also achieved with STM studies [62, 65, 68]. *In situ* STM with atomic-scale resolution is also applicable for EC-ALD [66].

Concerning the structure of the samples, X-ray diffraction can be applied in the glancing angle mode because of the small scattered intensity if the deposit thickness is small [9, 13, 15, 23, 31, 33, 34, 45, 53, 69]. The conventional Bragg–Brentano geometry is feasible for thicker deposits obtained with a fairly large cycle number [10, 30, 36, 37, 39, 42, 74]. Deposit structures identical to the known crystal structures of the same material were found for all such studies. X-ray photoelectron diffraction [58] and *in situ* X-ray analysis [77, 78] are quite unique but already adapted tools for the structural study of EC-ALD layers.

SEM images recorded for EC-ALD layers indicate that although the deposition method itself was designed as a layer-by-layer process, the nature of the deposit turns into a granular one as the cycle number increases to a few dozens or more. This trend was observed for a variety of layer composition, even if the substrate is a well-defined single crystal face [8, 9, 51]. The deviation from the planar layer-by-layer nature of the process can be understood by considering the stress originating from the misfit between the atomic structure of the substrate surface and that of the relaxed structure of the deposit. The granular growth sounds even more natural if the substrate is polycrystalline [11, 15, 22, 28, 36, 37]. The morphology of the deposit is also sensitive to the composition of the solutions applied for the deposition processes [79]. When the substrate does not allow a uniform nucleation of the EC-ALD coating, the deposit may be intentionally granular [80] or dendritic [55]. In both latter cases, the process seems to begin in a similar manner to an instantaneous nucleation since the size distribution of the resulting features is narrow. This results in a situation that even if the growth starts on a porous material like a nanowire array, the size of the deposit will be similar, regardless of the longitudinal position of the EC-ALD-produced dot on the nanowire [81]. This is well in accord with the nature of the atom-by-atom deposition process.

The granular nature of the deposit can obtain a quantitative assessment with a height distribution of the deposit if AFM is used for surface imaging [21, 29, 44]. AFM is also suitable to characterize the evolution of the surface roughness with cycle number. It proved to be common that the surface roughness starts increasing from the early stage of the growth process. Depending on the deposit, the surface roughness may either tend to saturate after a few hundred cycles [59] or keep increasing without saturation [82]. The way as the surface roughness increases upon growth also depends on the deposition potential of each atomic layer [42]. AFM was also used to keep track of the change in the surface morphology of EC-ALD deposits after exposure to air due to oxidation [73].

The major goal of depositing EC-ALD layers is to exploit their semiconductor properties. As a primary parameter, the band gap of the semiconductor layer deposited has been determined for a number of samples [23, 30, 32, 34, 37–39, 41, 48, 71, 72, 80]. The determination of the band gap is based on the following equation:

$$\alpha(h\nu)h\nu = A(h\nu - E_{\mathrm{G}})^n \tag{4.1}$$

where h is the Planck constant, ν is the photon frequency, α is the absorption coefficient at the photon energy specified in the argument, E_{G} is the energy gap and A is a proportionality constant. The exponent n is equal to ½ for a direct band gap and 2 for an indirect band gap. The graphical method of the determination of the direct band gap is shown in Fig. 4.6.

The experience related to the band gap found for thick EC-ALD semiconductor layers is that it approaches the bulk value (or that measured for single-crystalline reference samples). However, at small layer thicknesses, a blue shift is often found [39], which is due to the quantum confinement.

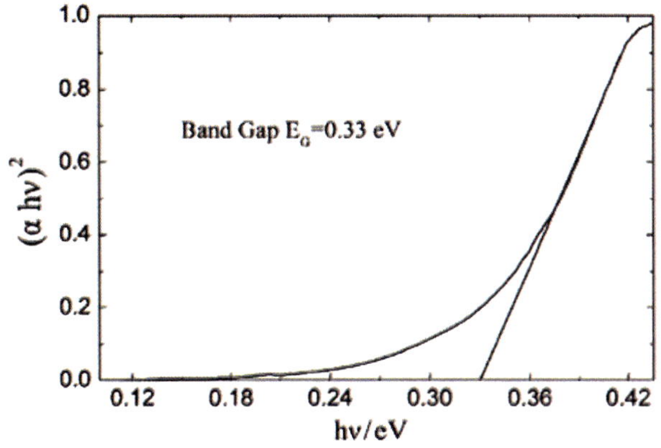

Fig. 4.6 Determination of the band gap from the plot depicting the square of the product of the absorption coefficient multiplied with the photon energy as a function of the photon energy. Data refer to a Bi_2Te_3 EC-ALD layer and the straight line is a tool for extrapolation. Reproduced from [38]. Copyright (2008), with permission from Elsevier

4.1.5 Multicomponent and Superlattice Structures Obtained with EC-ALD

The major driving force of combining more than two elements in EC-ALD layers comes from the need that this way the semiconductor properties of the deposits can be tuned. This goal is often termed as *band gap engineering*. The application of several chemical elements in the EC-ALD layers can be carried out by layering more than two types of atomic layers onto each other in a specific order or even by packing various EC-ALD deposits onto each other where each segment in the layered structure contains several atomic layers. A visualization of the above mentioned material families can be seen in Fig. 4.7. The main difference between the images shown in Figs. 4.7b, c, is that although the former is a material with dissimilar atomic layers, the layer alternation is so frequent that individual phase properties of the materials with composition AC and BC cannot manifest themselves but the properties obtained will be a weighted average of them. However, for the structure shown in Fig. 4.7c, the goal is to produce another interface, hence modifying the electron transfer between the topmost semiconducting layer and the substrate.

The deposition of multicomponent EC-ALD materials requires a very similar setup as the one shown in Fig. 4.1, keeping in mind that the number of the storage tanks for the solutions of the precursor materials has to be varied in accord with the number of the types of the atomic layers, and this may necessitate the multiplication of the blank rinsing solution tanks, too. The basic optimization viewpoints for multicomponent EC-ALD deposits are exactly the same as for binary ones. An important difference is that the atoms of the apparently buried layers are so close to the deposit/solution interface that they can be stripped off in the presence of the component of the second atomic layer if it is different. For instance, it was found

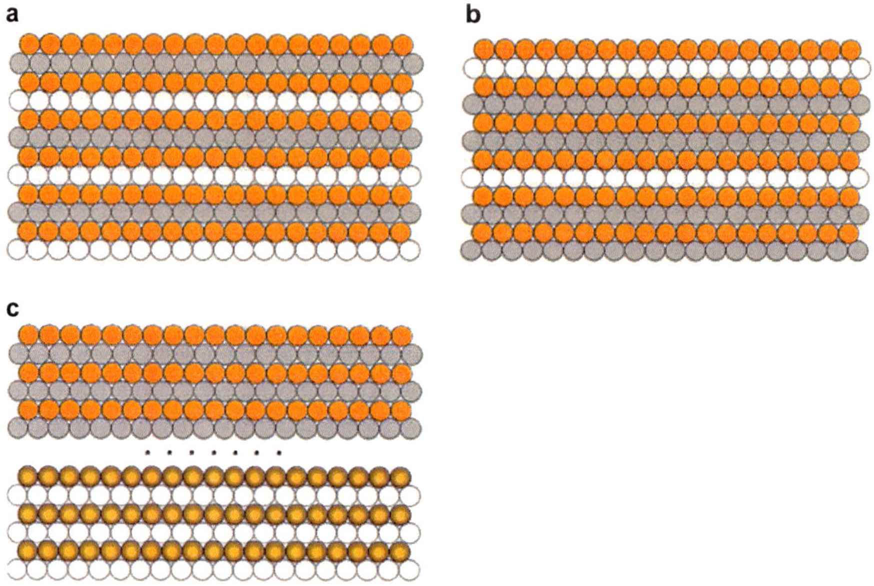

Fig. 4.7 Various structures obtained with EC-ALD with different atomic arrangements. **a** (ACBC) $\times N$ structure of an ABC_2 type compound, **b** $[(AC)_k(BC)] \times N$ structure as the approximation of an $A_xB_{1-x}C$ material with $x = k/(k+1)$; **c** $(AB) \times N/(CD) \times M$ structure with separate phases. In the notation, A through D mean the elemental components, x is the atomic fraction of an element within a mixed layer, while k, N and M refer to the number of repetitions

for sequential deposition of Hg and Cd on Te intermediate atomic layers [48] that at the potential needed for Hg UPD onto the topmost Te layer, the Cd atoms under the covering Te atomic layer are yet somewhat reactive. This fact has various consequences. If the solution containing Hg^{2+} ions is left to react with the Te-terminated Cd/Te layer structure, a part of the Cd layer is replaced with Hg atoms, as evidenced with quartz crystal microbalance experiments. The replacement of the underneath Cd layer can also take place if the potential is controlled during the introduction of the Hg^{2+}-containing solution. Such reactions may substantially complicate the optimization of multicomponent EC-ALD systems. Another ternary system where the atomic ratio in the deposit was significantly different from the designed value was $Cu_xSn_yS_z$ [83]. Here, the deposition mode was in accord with the scheme shown in Fig. 4.7b. As much as an order of magnitude difference was found between the experimental Cu/Sn atomic ratio and that calculated from the layer numbers applied in the EC-ALD process. This discrepancy was explained with structural rearrangement and porosity formation, but the spontaneous stripping and replacement process cannot be excluded either.

It is also a consequence of the difference in UPD and stripping potentials of the otherwise similar components that the deposition of one single component can be optimized in a particular step; therefore, mixed atomic layers cannot be produced with

EC-ALD. In turn, the layer-by-layer deposition mode can result in nanostructured materials whose bulk counterparts with *both* the same mean composition and atomic layer structure cannot be synthesized by any other means. This opportunity offers a great perspective for the EC-ALD method.

In EC-ALD materials corresponding to the layer structure shown in Fig. 4.7a, one can find materials that mimic the structure of bulk semiconductors in which the atomic layers are spontaneously ordered. Prominent examples for such materials are $CuInS_2$ [84] and $CuInSe_2$ [45, 85]. While $CuInS_2$ proved to be nearly stoichiometric, $CuInSe_2$ was found to be relatively poor for In. An even more difficult pursuit was to obtain $CuIn_xGa_{1-x}Se_2$ [47]. For this material, the Cu/(In + Ga) = 1 ratio could be achieved with a 1:5 ratio of the Cu vs. In + Ga deposition periods, and the Ga/(In + Ga) ratio was only 0.25 for a 1:1 ratio of In and Ga deposition pulses. This composition discrepancy clearly indicates how the optimization becomes more and more involved with the increase of the number of elemental components. Cu_2ZnSnS_4 films were also reported as quaternary EC-ALD layers [13], although the composition of the as-deposited material was quite far from the stoichiometric ratio.

Figure 4.7b shows the case when every other atomic layer in the sequence is the same (atomic layers of even number in Fig. 4.7b), and the ratio of the numbers of the other two layer types is practically arbitrary (but is often equal to the ratio of small integers). In such structures, it is possible to deposit two different metallic layers with one chalcogenide element. Examples for such materials are $Cu_xZn_{1-x}S$ [86], $Cd_xZn_{1-x}S$ [76, 82], $Cd_xZn_{1-x}Se$ [44, 76], $Pb_xSn_{1-x}Se$ [87] and $Hg_xCd_{1-x}Te$ [48]. $Ge_xSb_yTe_z$ [88] is special in the sense that the Ge and Sb are of different valencies, and hence, the composition cannot be described simply with a component replacement, unlike in the previous examples. Also, it is possible to match one metal with two chalcogenide element like CdS_xSe_{1-x} [89].

Finally, it has to be mentioned that non-identical semiconductor materials can also be deposited onto each other by the EC-ALD method. Here, it must be stressed that every component of the system packed onto each other is composed of a large enough number of atomic layers so that the individual phase properties, especially the band gap of the particular component, can manifest themselves and contribute to the overall functionality of the system. The principle of the deposition order here is to create a structure with gradually varying band gap from the substrate to the topmost phase where the light absorption is the most intense. By creating such structures, the light absorption can be widened to nearly the entire wavelength range of the visible light and also to enter the UV region. The gradual change of the band gap results in a diminished recombination of the photogenerated electron–hole pair, which leads to enhanced photocurrents. The desired cascade-type multiple heterogeneous junctions are illustrated in Fig. 4.8. Superlattice-like architectures with composition of PbSe/PbTe [33] and PbTe/CdTe [90] up to 15 atomic bilayers of each component have been reported, and fairly thick sandwich structures with a $Ag/(S/Cu/S/Zn)_{20}/(S/Cd)_{40}/S$ layer sequence [91] have also been described. However, the favourable influence of the sandwich structure for the absorption properties has been demonstrated only for $TiO_2/CdSe/Cu_2Se$ [92] and $TiO_2/CdSe/CdTe$ [81] deposits.

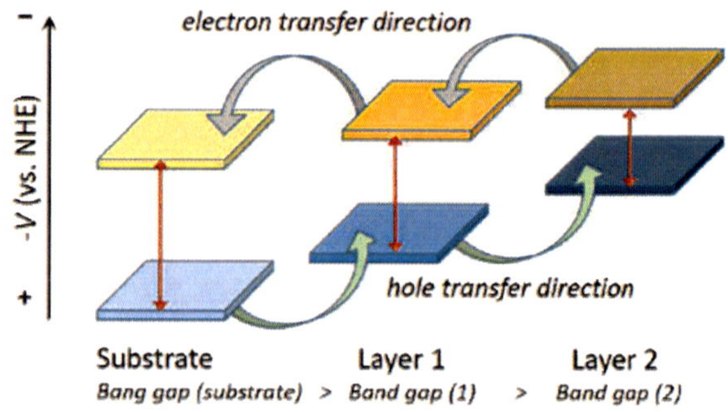

Fig. 4.8 Principle of the adjustment of the band gap of the neighboring semiconductor trilayer based on their band gap properties for obtaining nanoscale architectures with a wide light absorption wavelength range and a large photocurrent. Redrawn after Scheme 2 of [92]

As it was shown above, the EC-ALD process on semiconductor surfaces has a great significance because of the improved functionality. Indeed, the layer-by-layer electrochemical deposition that was originally demonstrated as a heteroepitaxial process with small atomic bilayer numbers has entered into the field of engineering by applying more advanced substrates. Titanium dioxide as substrate was already mentioned above, which is often used as a nanorod array for increasing the effective surface area of the specimen towards light absorption [80, 81, 93]. More recent pursuits for a better substrate choice in order to enhance the photoelectrochemical performance include various composite materials, mostly composed of polymers with either carbon nanotubes or graphene [49, 54, 55, 94].

4.1.6 Various Nanostructure Deposition Processes Based on UPD Principles

It is possible to produce a pair of atomic layers composed of different elements from a single solution on a suitable substrate if the underpotential deposition of one element on another and the corresponding UPD potentials make it feasible. The prominent example of such a system is Au(111)/Ag/Cd [95]. The solution contains both Ag^+ and Cd^{2+}, the former being present in a very small concentration. UPD of Ag on Au(111) takes place at a fairly positive potential, and the Cd UPD on Ag(111) that can grow epitaxially on Au(111) goes on at a much more negative potential. The growth of the Ag/Cd bilayer can be produced by polarizing the Au(111) electrode to the UPD regime of Ag, and then the potential is swept fast to the UPD regime of Cd on Ag(111). Since the Ag^+ concentration is only a few tens of μM, as opposed to $c(Cd^{2+}) = 1$ mM, the diffusion-limited Ag deposition rate is so small as compared to the Cd adlayer formation rate that the growth of a bulk Ag layer cannot take place. Beside the atomic bilayer formation, a significant interdiffusion was also detected in the same system [96].

The method named as selective electrodesorption-based atomic layer deposition (SEBALD) is a modification of the conventional EC-ALD process for a controlled deposition of nanosized metallic clusters [75, 97]. This method was elaborated for depositing metal clusters on a surface where UPD of these metals does not take place and even at potentials slightly negative of their reversible potential value, no effective control of the layer thickness and/or morphology is possible due to the occurrence of the bulk deposition. This is the case, for instance, for the iron-group metals on silver surface.

In the first step of the SEBALD process, a non-metallic monolayer is produced at the surface, just like in any other EC-ALD procedure. In the studies establishing this method, it was a sulphur monolayer on Ag(111), that can later be modified with metals that are otherwise not possible to deposit onto silver surface with an atomic-level control. The EC-ALD cycles then lead to the growth of a MeS deposit with controlled thickness (Me: Cd, Fe, Co, Ni or an alloy of the iron-group metals). As a final step, sulphur is stripped off with a fairly negative cathodic pulse, when the remaining metal atoms are rearranged at the surface and form clusters in a size range that is not accessible with a direct deposition. The size and the areal density of the metal clusters left behind by the electrodesorption of sulphur are tunable by the thickness of the EC-ALD layer of the MeS compound. This surface modification method may gain important role in the elaboration of mixed catalysts [97] where the synergetic effect of nanoscale metal clusters has an importance for various elementary steps of reactions with a complicated mechanism.

As opposed to the SEBALD-type process where the deposition cycles were followed by a single dealloying step at the end, the EC-ALD process can also be completed with a cathodic stripping peak in each cycle. Here, the EC-ALD deposit has the metal/chalcogenide/metal atomic layer sequence near the surface when a cathodic stripping pulse is applied to remove the chalcogenide atomic layer partly buried down beneath the topmost metal atom layer. Thereafter, the *chalcogenide deposition—metal deposition—chalcogenide removal* steps can be repeated in the further cycles. The method named by the authors as a "bait and switch" surface-area-limited reaction [98] was suitable to produce a continuous film from germanium, a metal whose deposition is usually self-limited (see Sect. 4.3.1) by using the Te content of EC-ALD GeTe deposits as a sacrificial component.

A great advantage of the EC-ALD process is that the deposition conditions of the constituents can be tuned absolutely independently of each other, hence gaining a large degree of freedom in the pursuit of the successive formation of complete monolayers of alternating components on each other. However, the separation of the deposition steps in time is not necessary if the following conditions are fulfilled: (i) The goal is not to form a layer of as even thickness as possible, but some thickness fluctuation is allowed (which, in turn, occurs also for EC-ALD layers, as it was shown above, and does not necessarily impact the functionality negatively). (ii) The UPD of either of the components onto the monolayer of the other components is possible from a properly chosen solution, hence being able to apply one single bath. Such processes have to be operated under potentiostatic control in order to provide the right stoichiometry of the deposit. Although the process rate is rather limited due to the low concentration of the solution, it may be acceptable for the deposition of a nanoscale layer.

The above deposition mode can be called accumulative underpotential deposition. It was shown to be feasible for the deposition of, e.g., stoichiometric PbS from one single solution, leading to crystals of micrometre-range lateral dimensions [99]. Another attempt to deposit PbS in a single step was based on a hot saturated solution of *in situ* formed PbS particles [100]. In the latter study, the deposition took place in a UPD manner from the small-concentration solution from which the whole particles themselves did not incorporate into the deposit but served as the source of the precursors by saturating the solution via their dissolution. The accumulative underpotential deposition from a single solution is also feasible if the solution concentrations, like in conventional baths for alloy deposition, are not limited by the low solubility of the precursor(s). This method was employed for a variety of binary (CdTe [101], CuTe [102] and PbTe [103]) and ternary deposits ($Bi_2Te_{3-y}Se_y$ [26], $Bi_{1-x}Sb_x$ and $Bi_{2-x}Sb_xTe_3$ [79]).

4.2 Combination of UPD with Other Surface-Area-Limited Processes

4.2.1 Surface-Limited Redox Replacement Processes

It is commonly known that once a relatively non-noble metal (like iron) is immersed into the solution of the ions of a more noble metal, an exchange (or displacement) reaction takes place between these metals, such as $Fe + Cu^{2+} \rightleftarrows Fe^{2+} + Cu$ in the above example. Such processes are also called cementation in the technical literature. The cementation process takes place spontaneously in the presence of the suitable reactants and it does not require an external current source. If the metal deposited cannot cover conformally the one that is displaced, the process does not stop until one of the reactants is fully consumed and is accompanied with the occurrence of porosity in both the displaced and deposited metals.

Surface-limited redox replacement (SLRR) is a similar process to cementation concerning the driving force of the process. The limitation comes from the fact that the component to be displaced (i.e., the sacrificial component) covers the substrate surface in a thin layer which is often restricted to a UPD layer. Hence, the displacement reaction has to consume an at most monoatomic layer. Therefore, the accomplishment of the displacement process results in a similarly thin layer of the displacing metal on *surface average*; i.e., the atomic distribution of the displacing metal is not certainly as surface conformal as that of the displaced layer. The diversified aspects of SLRR processes and the deposits thus produced were recently reviewed by Dimitrov [104].

The deposition of a single atomic layer by SLRR can be carried out with subsequent immersion into different solutions coupled with a suitable electrode polarization, as it was described in the papers on the elaboration of the SLRR technique [105, 106]. This approach requires a thorough control of the experiments by maintaining an inert atmosphere for avoiding the unwanted oxidation of the sacrificial metal. Although the absence of oxygen is a prerequisite for all possible configurations, the requirements for providing a non-reactive environment is the most difficult if sequential immersion is applied.

The build-up of an SLRR layer can also be organized into cycles provided that a suitable UPD layer can form again onto the metal deposited in the previous SLRR process. This newly-formed UPD layer can be stripped off again by a more noble metal. Due to the high demand for the exclusion of the mutual solution contamination, the method with subsequent immersions is impractical for several cycles. This results in the situation that in the technical sense, the apparatus to deposit several sequential SLRR monolayers is analogous to that used in EC-ALD. The first step of the cycle is the deposition of a UPD layer, but the next one is a replacement process where the thickness does not increase. Instead, the composition of the surface changes, and no electrochemical control is applied in this step (the process goes on under open circuit conditions) except for monitoring the electrode potential. The big majority of the SLRR studies apply this type of instrumentation. The apparatus involving a solution replacement is easy to use also for the deposition of an alloyed layer if the solutions used in the subsequent displacement half-cycles contain the ions of various growing metals [107]. In such a case, the UPD of the sacrificial metal has to undergo on the surface of both depositing metals.

Another approach for SLRR is based on a single solution in which the ions of the sacrificial and the depositing metals are present in different concentrations [108–114]. Although the control of the deposit composition is less accurate than for separated solutions, should this mean an alternating immersion or a flow cell approach, the simplification offers the opportunity of the repeated application of the same bath without a high consumption of rinsing solution. The deposition of the UPD layer of the sacrificial metal takes place during a relatively short potential pulse with kinetic control, and then the replacement process is left to take place in a much longer time with the diffusion control of the reaction. The concentration ratio of the sacrificial and the deposited metal is typically 2–100 for the one-pot method. The larger the ratio of the apparent deposition rate constants of the sacrificial metal to the deposited one, the lower concentration ratio of their ions can be allowed. The incorporation of the sacrificial metal into the desired layer was occasionally found [109], which may necessitate an anodic dealloying step after the completion of the deposition cycles. By omitting the dealloying step, the incorporation of the sacrificial metal into the deposited one can also be used intentionally for producing alloyed layers.

The production of the sacrificial UPD layer to be displaced is also possible in an electroless manner [115, 116], although the mechanism of the deposition process is purely electrochemical. In this case, a simple galvanic couple is created by connecting the substrate to be covered with a so-called "executive" metal on which a dissolution process takes place to the same extent as the UPD layer formation on the substrate. The driving force of the deposition of the sacrificial metal is hence provided by creating an electrochemical cell instead of applying an electronic power source. Electroless SLRR is also possible if the substrate can absorb hydrogen which is applied as a sacrificial material in the deposition half-cycle [117]. Of course, the electrical connection has to be interrupted when the solution carrying ions of the deposited metal is introduced into the cell.

Concerning the composition of the deposits obtained with an SLRR process, these are mostly pure metals, unlike for EC-ALD. Hence, the significance of SLRR does not come from the fact that a material with special chemical composition is deposited, but from the very small mean thickness of the layer that is at the same time surface-conformal. Unlike the conventional electrodeposition of metals that seldom leads to monoatomic layers (see Chap. 2.10), the deposition by SLRR is much more similar to an ALD process in its surface-conformal nature. At the same time, the production of special non-equilibrium atomic structures of multilayers is also possible (see later).

The control of SLRR processes, driving the displacement half-cycle reaction strictly under open circuit condition, can be based on the observation of the electrode potential. Kinetic models include the substrate–deposit and deposit–deposit atomic interactions, adsorption isoterms and Gibbs free energy changes of the reactions [118]. Kinetic studies showed [119] that the rate of the replacement reaction can be elucidated in terms of surface limitation at a concentration of the ions of the growing metal in the millimolar range, while the transport limitation is in effect when this concentration is lowered to the level of 0.01 mM. The typical replacement time is a few tens of seconds for the mM concentration range, which increases inversely proportional to the concentration in stagnant solutions. The $E(t)$ functions obtained could be fitted with the appropriate isoterm, taking also into account the order of the replacement reaction. Figure 4.9 shows typical surface coverage vs. electrode potential and electrode potential vs. displacement time functions.

In a later work of the Brankovic group [120], *in situ* surface reflectivity measurements were also applied for following the kinetics of the displacement process, and an excellent agreement was found between the theoretical and experimental values. In the latter work, surface reflectivity data were taken advantage in the calculation of kinetic coefficients under various circumstances.

Although the general goal is to achieve a complete replacement of the sacrificial metal, the partial replacement may also be a goal [114]. To this end, the cut-off potential should be set so that the potential decay is yet incomplete, and a significant part of the sacrificial metal remains on the surface.

4.2.2 Substrates and Displacement Pairs in SLRR Processes

The majority of the studies refer to noble metals as substrates (Au, Pd, Ru), many of them having a single-crystalline character or at least a preferred orientation like Au(111) for sputtered gold. Beside noble metal substrates, it is worthwhile of mentioning that the SLRR process has also been tested for materials where the "classical" UPD process could not be observed. In the study of Fang et al. [121], the substrate was a sputtered Co film, and the sacrificial metal was Pb that mediated the deposition of Cu. Although the Pb UPD peak was not very clear, the Pb-to-Cu replacement cycles were successful, and the technique suggested was suitable to

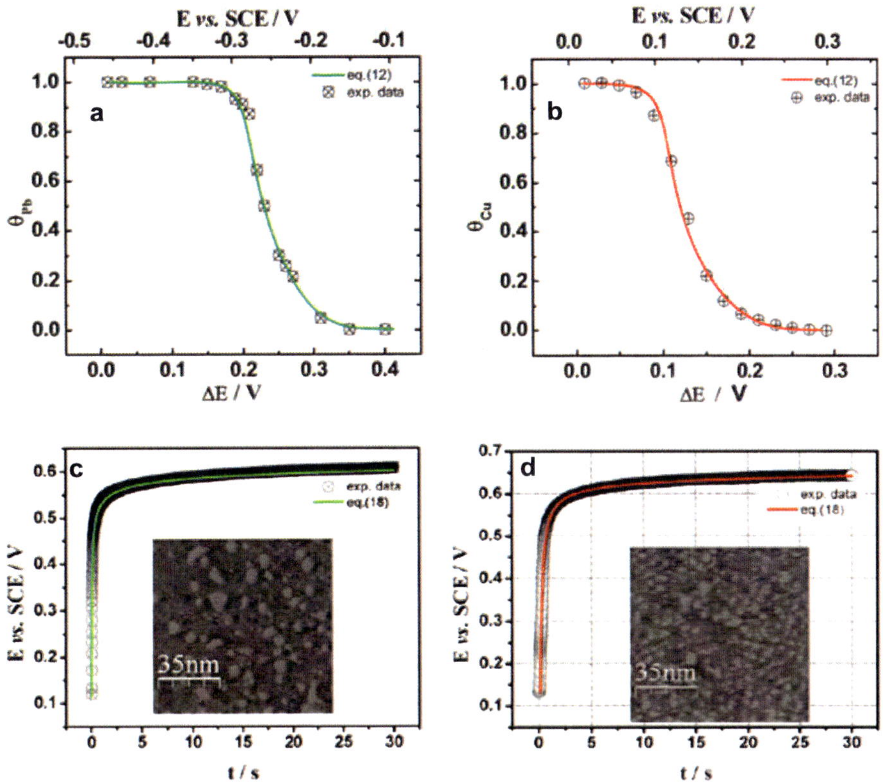

Fig. 4.9 Typical working curves for UPD and SLRR of Pb and Cu on Au(111) substrate. **a** and **b** UPD coverage versus electrode potential (the potential relative to the onset of the bulk phase formation in the bottom axis); **c** and **d** Replacement of the monolayer of Pb and Cu, respectively, with 10^{-3} M $[PtCl_6]^{2-}$ + 0.1 M H_2SO_4 solution with 1000 rpm electrode rotation rate. For equations, see the reference work. Reproduced from [119]. Copyright (2011), with permission from Elsevier

avoid corrosive conditions that prevail in direct Cu deposition. In a study applying porous substrates such as carbon paper or nickel foam [122], the optimization details of the SLRR process by showing voltammetric curves were not specified at all; nevertheless, the replacement cycles resulted in a fairly even deposit.

SLRR-type growth of metallic layers on semiconductors is also an important goal, also when Si is covered with some type of barrier layer used in microelectronics. An example is when Cu is deposited by Pb-sacrificed SLRR on Ta/TaN-covered Si [123]. On the TaN surface, the direct deposition of both Cu and Pb was observed at more negative potentials than on their parent metal due to the nucleation barrier. However, in the anodic-going part of the cyclic voltammograms, the dissolution of neither of the metals could be observed because the TaN layer is oxidized easier than the deposit layers, and hence, the metal becomes electronically separated from the substrate. Hence, the occurrence of the UPD process is observed only in a peak of

Table 4.2 Sacrificial/growing metal pairs used in SLRR studies

Sacrificial metal	Growing metal					
	Pt	Au	Pd	Ru	Ag	Cu
Cu	[109, 124–137]	[129, 138]	[16, 122, 139–143]	[133]	–	–
Pb	[109, 112, 114, 124, 128, 144]	[110, 120]	[16]	–	[145]	[121, 123, 146, 147]

the cathodic-going scan. Nevertheless, this was enough for the optimization of the Pd-sacrificed SLRR of Cu layers.

The most common sacrificial and growing metal pairs used in SLRR studies are summarized in Table 4.2. Although some processes are formally equivalent concerning the stoichiometry, they do not necessarily lead to the same layer morphology. The comparison of the impact of the sacrificial metal in the $2Me + Pt^{4+} \rightleftarrows Pt + 2Me^{2+}$ reaction showed [124] that the surface roughness changes slowly with cycle number if Me = Pb, while a progressive surface roughness development can be observed for Me = Cu. For the latter, the uncertainty of the valency of dissolving Cu may be a reason for the roughening as discussed later. Figure 4.10 presents representative results for the comparison of the Cu-to-Pt and Pb-to-Pt replacement processes, where the change of the current in the resulting Pt surface volatmmograms is indicative of the surface area development of the deposits.

It has long been the pursuit to eliminate heavy metals from SLRR processes. A nice example is to use Zn as sacrificial metal to deposit Cu [148]. In order to observe a relatively clear UPD peak of a non-noble metal on another, the solution has to be alkaline in order to suppress the hydrogen evolution; nevertheless, the parasitic current for hydrogen evolution was still a problem in 100 mM NH_3 solution. The major goal here is to keep the surface roughness of Cu deposit much lower than what is accessible with direct Cu electrodeposition. Recently, evidence for the applicability of the SLRR process has been demonstrated for the deposition of metals with quite negative standard potentials. A prominent example is the Zn-mediated deposition of Co [149], based essentially on the same solution chemistry as in [148]. The full coverage of a Ru substrate used was claimed to be achieved only after five deposition cycles, which is consistent with the potential decay observed during the stripping period. Mn was also tested as a sacrificial metal [147]; however, the voltammetric results for the Mn UPD are quite uncertain and the details of the process are yet to be elaborated.

An attempt was made also for using a Ni layer of controlled thickness as sacrificial metal [150]. Ni UPD on the Au substrate was not evidenced; however, a protocol was developed for depositing a thin enough (approximately two monolayer thick) Ni layer that makes it possible to deposit nearly one single atomic layer of Pt if Pt(IV) is used as precursor material. We have to keep in mind that such calibration-based processes are also possible, although they did not gain much attention so far.

Although the majority of the sacrificial components in SLRR processes are metals, it is also possible to use adsorbed hydrogen atoms for the sacrificial UPD layer

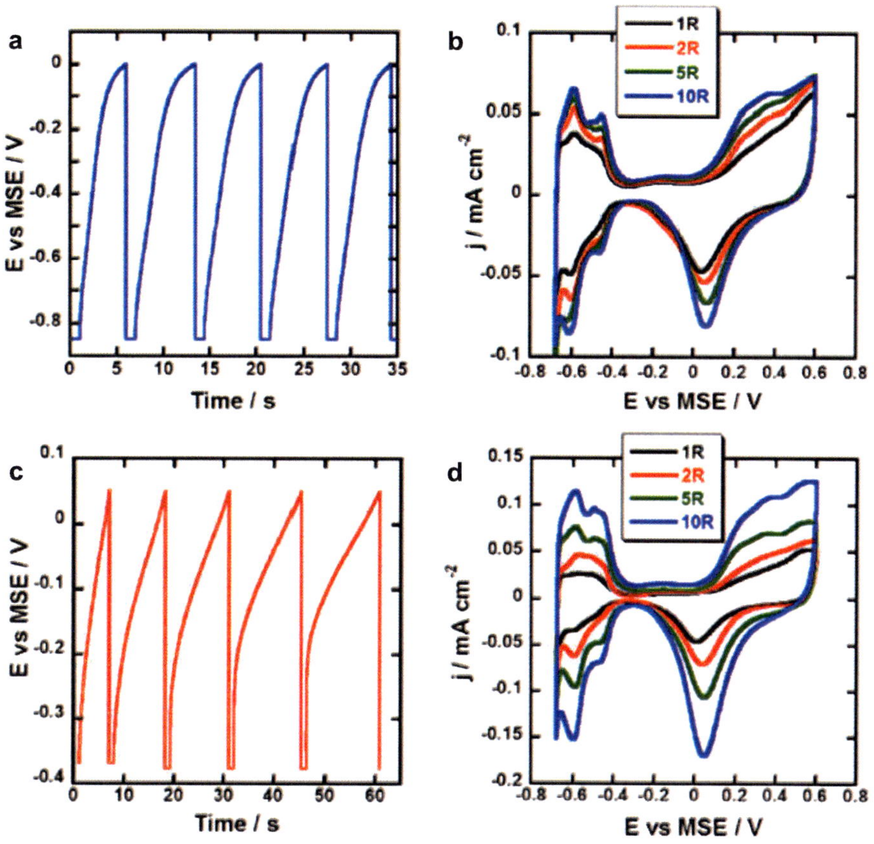

Fig. 4.10 Pt film growth on Au by successive SLRR cycles ('R') with the single solution method. **a** and **c** Potential transients during Pt deposition via SLRR of Pb UPD (at $E = 0.85$ V) and Cu UPD (at $E = 0.38$ V) vs. MSE, respectively. **b** and **d** Cyclic voltammograms of the resulting Pt films in 0.1 M H_2SO_4 solution obtained with 50 mV s^{-1} sweep rate as a function of SLRR cycle number. Reused from [124] with Creative Common License (https://creativecommons.org/licenses/by/4.0/)

component if the deposited metal is either Pd or Pt, due to 1:1 the adsorption capability of the hydrogen atoms on these metals. The solution for the production of the UPD hydrogen layer is usually acidic. While Pt only adsorbs but does not absorb hydrogen [111], Pd dissolves hydrogen fairly easily, and the absorbed hydrogen can also take part in the replacement process as a sacrificial material when the Pd deposit is thick enough for exhibiting hydrogen absorption capability similar to bulk Pd [113].

4.2.3 Structural Aspects of the SLRR Processes

The advantage of the SLRR-based process is obvious for cases when the atomic layer of the sacrificial metal can be replaced with a similarly even distribution of the growing metal. If the process cannot be driven along an atom-by-atom replacement manner, the growing metal will form clusters instead of layers, similarly to the nucleation-initiated growth during direct deposition (see Fig. 2.15 and the related discussion), and the advantage of the subtle growth process is hence annulled. It has to be clearly seen that the displacement of the sacrificial layer can indeed take place via a so-called local cell mechanism, which means that the location of the oxidation and reduction process can be well separated due to the conducting nature of the substrate (as it is nearly exclusively seen for corrosion processes on inhomogeneous metals or in environments with uneven oxygen supply). An atomic-scale scheme of the two types of replacement mechanism is given in Fig. 4.11.

The stoichiometry of the displacement system has a major impact on the growth mode, the key parameter being the ratio of the number of the sacrificial to growing

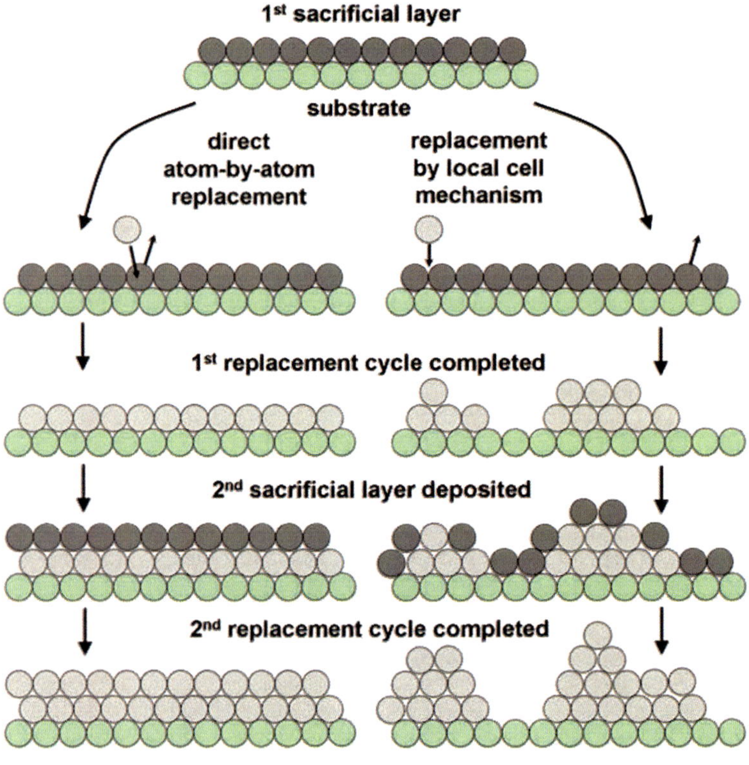

Fig. 4.11 Result of the monolayer displacement process depending on the replacement mechanism. For more details, see also [104] from which Fig. 21 was used for constructing the scheme

metal atoms, n_S/n_G. If this ratio is less than one (for instance, Pb replacement by a Pt(IV) compound as solute), the direct exchange mechanism is more likely, whereas for $n_S/n_G < 1$ (e.g., for $Pb + 2Ag^+ \leftrightarrows 2\ Ag + Pb^{2+}$), the growing layer is rendered to be uneven. The clarification of the stoichiometry of the system sometimes requires a great care. For instance, Cu displacement by Pt(IV) that is present as a $[PtCl_6]^{2-}$ complex was shown to lead to the formation of Cu(I) as a chloride complex [137]. Nevertheless, with a different anion that does not stabilize Cu(I) but provides that Cu(II) is the sole soluble species, copper behaves as a divalent sacrificial metal.

Not only does the morphology of the layer deposited in the SLRR process depends on the nature of the sacrificial/depositing metal pair but the technical implementation and the kinetic parameters of a displacement system are also important. These factors determine together which mechanism depicted in Fig. 4.11 takes place in a particular reaction. If the displacement reaction is very fast and the fill-up of the flow cell with the reactant is relatively slow, an overplating takes place at the cell ingress [16, 146]. By tuning the composition of the solution of the growing metal, especially by applying a suitable complexing agent for slowing down the displacement process, this discrepancy could be eliminated [142, 146]. This draws the attention to a complex view of the SLRR process and to the importance of the optimization of the chemistry of the reactions. It was also shown that the nucleation density and the cluster surface area are a complicated function of the technical parameters of the SLRR process [132], in which the initial UPD coverage of the sacrificial metal and the concentration of the ions of the growing metal play a crucial role.

For monolayer-wise growth, a general question is the commensurability of the atomic structures of the substrate and the deposit. If they have the same structure and the lattice misfit is small, heteroepitaxial growth can take place, at least for the first few monolayers like for Pd on Pt [140].

Although the general view on the SLRR process is that the first monolayer is already a fully reduced metal layer, this point is questioned by some studies. The doubt concerning the reduction is typical for Pt whose ions appear to react in a rather sluggish way during the replacement half-cycle of the SLRR process. The partial reduction of Pt(IV) in the Cu replacement reaction [127] was explained with a partial reduction of the Pt(IV) species. This leads to the presence of both O and Cl in the surface layer, which was evidenced by EXAFS data on the basis of bond lengths. The results of another polarization-dependent total-reflection fluorescence X-ray absorption fine structure study of a Pt monolayer on Au(111) prepared with Cu monolayer-mediated SLRR showed that the structural data are consistent with the presence of $[PtCl_4]^{2-}$ species at the surface [136]. The planar tetrachloro complexes appeared to lay on the gold surface, each atoms of the complex being in touch with the surface gold atoms at a Pt coverage equivalent to 0.45 as referred to surface gold atoms. Since the electrolyte solutions used in this study did not contain excess chloride ions but only as a component of the H_2PtCl_6 precursor material of the Pt-layer, the impact of the insufficient rinsing of the electrode prior to the transfer to the surface investigation system could be ruled out. The results indicate that the simple view on the replacement process is to be scrutinized. The finding shown above is in agreement with another study on Pt SLRR in the sense that the adsorption of Pt chlorocomplexes was found to adsorb on the electrode that was reduced only later in the period meant for the UPD layer formation, hence contributing to an unexpectedly fast growth of the deposit [128].

The variation in surface roughness of SLRR layers with deposition cycle number is somewhat similar to EC-ALD layers. When direct exchange takes place, the surface roughness remains close to that of the substrate for a couple of SLRR layers. In this case, the surface roughness increment is very low, and a nearly planar deposit is obtained even if the same metal is deposited as three-dimensional nuclei during d.c. deposition (like Pd with Cu sacrificial layer [143]). However, after a number of cycles, roughening can take place, and the metal growth becomes faster than what can be anticipated in the monolayer-wise growth process. The deposit is more prone to develop a significant roughness as compared to the substrate if there is a lattice mismatch between the deposit and the substrate. For instance, Au SLRR on Au substrate causes much smaller surface roughening than the same process on a Pt substrate [110]. The threshold thickness where roughening takes place varies from one material to another. For H-adsorption-mediated Pd growth, the turnover was found to be around 20 deposition cycles [113]. Concerning the SLRR growth of Pt, it is the common experience for both classical [109, 112, 124, 128] and electroless [116] processes that Pb as a sacrificial metal promotes a nearly even Pt layer growth, while a Cu replacement leads to a progressive growth of the surface roughness practically from the first cycle (see also Fig. 4.10).

Although it is not clearly evidenced, it is logical to assume that alloy layers obtained with SLRR by using different deposited metals in the subsequent cycles exhibit neither fully random nor completely ordered structure. In accord with the original concept [12], such deposits may work as a bifunctional catalyst due to the atomic arrangement at the surface that does not exist in conventional alloys. Mixed Pt/Ru atomically layered catalysts proved to have beneficial properties for methanol oxidation [12]. It is likely that for alloys produced with alternating replacement with various growing metals at uneven cycle ratio, the minority component forms two-dimensional island-like features in the matrix composed of the majority component [107]. For a regular alteration of the components deposited from one growth layer to another, an atomically layered structure was presented [129]. This picture shows the most probable (dominant) type of atomic arrangement; nevertheless, the roughening process is likely to induce irregularities in the deposit.

The inclusion of an intermediate layer between the substrate and the deposited layer strongly impacts the deposit structure. Graphene-covered Au leads to a different structure of Pt than the pure gold substrate [135], mostly due to the compressive strain of the compact atomic spacing of the graphene sheet. Interestingly, the inclusion of an intermediate graphene layer proved to be transparent in the sense that both UPD on gold was possible to produce the intermediate layer and the Pt replacement could be carried out in the usual way.

4.2.4 Application of SLRR Deposits

The major motivation to study SLRR processes has been the efficient utilization of noble metals as catalyst since the invention of the technique. On the one hand, heterogeneous catalysts must always exhibit a surface layer of favourable catalytic properties, irrespectively of the bulk composition. Hence, the most efficient application of a relatively expensive catalyst is its deposition as a monoatomic surface layer. This is exactly what SLRR does. On the other hand, however, a monoatomic coverage of a substrate material with foreign guest atoms creates a different environment at the atomic scale from the bulk catalyst. Therefore, catalytic properties can also be tuned with a proper *substrate–guest atom* match, creating catalysts that cannot exist in the bulk form.

It is an important feature of the SLRR process that even sub-monolayer coverage can be achieved that is very hard to create by direct d.c. deposition methods. The control of the coverage of the sacrificial UPD layer (through its deposition potential) is a convenient manner to carry out such a regulation [130, 132, 141]. It was found that catalysts with such sub-monolayer thickness have different catalytic effects than the bulk counterparts of the same materials; moreover, the catalytic activity in the ultralow coverage state is outstanding as compared to the bulk form. Such effects were found for the oxidation of methanol [130, 141], formic acid [112, 124] and ethanol [131] and for the reduction of oxygen [134] on sub-monolayer Pt or Pd SLRR films. In a similar manner, a sub-monolayer Pt SLRR layer enhanced the hydrogen adsorption–desorption rate of Pd films (that were also obtained with the SLRR route) [17]. It is interesting to note that not only a sub-monolayer Pt coverage on Au can yield an excellent catalytic activity but vice versa; i.e., a sub-monolayer-thick Au deposit can also enhance the catalytic activity of Pt [138]. This again points out the complex nature of the atomic-scale interaction in catalysis.

The SLRR process is applicable not only for ultrathin layer formation from one element but also for the preparation of a mixed layer if the process has the suitable stoichiometry. This was exploited for the preparation of mixed Pd–Pt layers [151]. In the first step, Au was covered with a Pt layer of sub-monolayer coverage by taking advantage of the $2Cu + Pt^{4+} \rightleftarrows 2Cu^{2+} + Pt$ process that leaves behind half as many atoms on the Au surface than that produced in the preceding Cu UPD process. Then, the replacement cycle was repeated with Pd as the growing metal. The even distribution of Pd and Pt was evidenced with scanning TEM–EDX at the nanometre scale, and the mixed catalyst proved to be advantageous as compared to their pure Pt- and Pd-based counterparts in formic acid oxidation [151].

SLRR is a very appropriate technique to study the transient from a surface layer to a bulk-like material. Unlike for direct deposition with d.c. current when the nucleation phenomenon may lead to bulk-like grains also at very small nominal coverages, SLRR leads to layers of extremely well-defined even thickness. This feature of the method was exploited for the study of hydrogen absorption in Pd layers [140]. While no hydrogen absorption was found for a single monolayer, the absorption contribution increased gradually as the nominal coverage increased to several monolayers, in parallel with the change of hydrogen adsorption characteristics due to Pd atomic terraces and their edges.

For tuning the catalytic properties of Pd and Pt, layered systems were synthesized by using 8–24 subsequent cycles first for one of the components and then for the other [152]. The deposits were tested for oxygen reduction reaction. Another multilayer approach to synthesize nanostructured catalysts was to deposit a Pd/Ru sandwich structure with coating *ex situ* synthesized Pd particles with Ru [153] or by performing 15 SLRR cycles for each metal on a foreign substrate [133]. In these studies, the surprising finding was that the hydrogen absorption/desorption into/from the Pd layers/particles was accelerated, although they were nearly fully buried under another metal that does not adsorb/absorb hydrogen. Although no firm conclusion could be drawn whether the elimination of the Pd/H adsorption state or some competing absorption mechanism was responsible for the unexpected acceleration, these examples show the nature of nanostructures that is difficult to elucidate on the basis of bulk properties.

The study of Pt–Pb SLRR layers is particularly interesting since these metals form alloys with various compositions (Pt_3Pb, PtPb, $PtPb_4$), but the atomic arrangement in the SLRR layers is probably completely different from that of the ordered crystalline phases, and, in addition, they can also by synthesized as layers having a non-equilibrium composition. The study of Pt–Pb SLRR layers with Pb mole fraction less than 0.1 [114] indicated an immense decrease of the hydrogen adsorption, much higher than expected from the density of Pb atoms at the surface, which was explained by the electronic interaction of the alloying element with Pt. For CO oxidation, alloyed Pt–Ru SLRR layers proved to exhibit an outstanding activity [107].

It is to be taken into account that the atomic spacing of the thin layer deposited on a foreign substrate may substantially differ from that in the bulk form, which is due to the stress and surface relaxation. This is also true for layer produced with SLRR. Intentional modification of this spacing with intermediate layers like graphene has a great perspective in the purposeful development of catalysts [135].

It is also a common feature of catalysts that they are applied in powder form or as nanoparticles. Therefore, various techniques have been proposed for creating deposit layers with even thickness by SLRR also on particles. The obvious case is when a particle is formed *in situ* on a surface, possibly also by electrodeposition [138, 145] or electroless deposition [154], which makes it possible to perform the SLRR process without another step of particle immobilization. If the preparation process starts with *ex situ* synthesized nanoparticles, the immobilization can be based on a standard process in which the substrate surface is first modified with an aminoalkyl trimethoxysilane compound. The substrate surface is hence terminated with an assembly of $\equiv$Si–R–NH_2 chains, which immobilize gold nanoparticles efficiently. The potential of the immobilized nanoparticles can be controlled, making it possible to subject them SLRR cycles without a particle loss [125].

Various procedures have been developed without a chemical immobilization of particles for SLRR. The preparation of catalyst-coated nanoparticles is also possible with a sedimentation process in which the nanoparticles are collected on a horizontal graphite cathode. This configuration provides a sufficient electrical contact between the particles in a stagnant solution where SLRR was shown to be effective at least at the monolayer-level coverage [131]. Due to the accessibility of the deep regions of the sediment, the SLRR reaction time for nanoparticle sediments was about two orders of magnitude larger than for planar electrodes. A better mechanical control on the assembly of nanoparticles is provided if they are pressed against each other with small-porosity filters, introducing a current lead to the same chamber. This compacted flow cell makes it possible to exchange the working solution around the nanoparticles without loosing the potential control [153].

Another way for increasing the surface area of a catalyst is to apply a porous material as substrate [126, 139]. Dealloyed Ag–Au alloys proved to be an ideal catalyst support when the internal surface was modified with SLRR-produced Pt [144], taking advantage of the conformal deposition characteristics of the SLRR process.

It is not only the efficiency but also the degradation of a catalyst can be studied by using SLRR layers. When a catalyst degrades, the corrosion rate can be estimated from the outflow of a reactor, but this method requires the measurement of very small concentrations, which is often a problem. When ultrathin layers are studied, the loss of the catalyst can be followed with the surface analysis of the catalyst itself since the substrate underneath is a different material than the ultrathin catalyst layer. This approach has been taken advantage of for Au-supported Pt catalysts formed with SLRR [112].

Finally, it has to be mentioned that catalysis is by far not the only field of the potential application of SLRR-wise metal deposition. In microelectronics, the ever-decreasing size of circuit elements due to the ultra-large-scale integration (ULSI) may require the deposition of even metallic layers, which can be carried out in a feasible manner by SLRR [155].

4.3 Non-UPD Deposition of Ultrathin Metallic Layers

4.3.1 Electrodeposition of Ultrathin Layers with Self-limiting Processes

It is needless to say that any deposition process can lead to the formation of some nanoscale objects if the deposition is stopped early enough after the start of the process. However, an interruption of the deposition process shortly after the onset of the deposition does not necessarily result in a structure that is different from the bulk and what deserves the name "nanostructure". Therefore, the common feature of the cases discussed in Sect. 4.3 is that some particular observation which is normally not observed in the bulk form of the same material can be made that is closely related to the nanoscale nature of the deposits.

A large family of nanocoatings where the small layer thickness occurs naturally is the group of so-called self-limited electrodeposit. The self-limiting growth is common for electropolymerization if the resulting coating is insulating. A well-known example is the oxidative electropolymerization of phenol in aqueous solution that leads to a coating of about 20 nm limiting thickness. This reaction is often used for sealing the side surface of STM tips to be used under an electrolyte solution in order to avoid parasitic reactions of the solution component at the relatively large surface of the side wall of the STM tip, leaving the end of the tip uncoated. Like any other surface-area-limited process, self-limiting electropolymerization of phenol is suitable to achieve conformal coatings on porous materials [156]. Self-limiting electropolymerization of other organic monomers can yield a similar limiting thickness to that of poly(phenylene oxide) [157]. However, if the monomer to be polymerized is a metal complex, the limiting thickness achievable by electropolymerization can be tuned in a wider interval. This was exemplified in a study dealing with electropolymerization of tris(5-amino-1,10-phenanthroline) metal complexes with various central metal atoms, Fe and Ru [158], leading to limiting thicknesses of cca. 50 and 2 nm, respectively. A self-limiting growth process takes place during the electrochemical oxidation of thiourea in a two-step process [159], where the electrically insulating nature of the sulphur layer is responsible for the stop of the layer growth, similarly to the insulating polymers.

It has a great technological importance that platinum exhibits a self-limited deposition process under suitable conditions. The major goal here, similarly to SLRR processes, is the efficient use of the expensive catalyst material, partly in combination with the enhancement of the catalytic effect due to the deposit–substrate interaction at the atomic level. The principle of the platinum self-limited electrodeposition is as follows [160–162]: Platinum is deposited in a dilute solution of a platinum salt. With an appropriate choice of the electrode potential, the discharge of the Pt(II) or Pt(IV) species at the cathode surface is followed by the adsorption of hydrogen on the Pt atoms. Hence, although the deposition of platinum itself is not a UPD process, the follow-up UPD of the hydrogen atoms on the newly-formed Pt surface prevents the further reduction of the platinum salts. If the potential is then switched to a sufficiently positive value where the surface of the working electrode is cleaned and the adsorbed hydrogen atoms are oxidized, the surface becomes receptive again to the deposition of a new Pt atomic layer. The voltammetric behaviour of the system and the current response to a suitable potential pulse programme is presented in Fig. 4.12.

The analysis offered by the authors of the original study on the self-limitation of the Pt deposition stressed that the origin of the self-limiting nature of the process is due to the change in the double layer structure caused by the H adsorption on Pt atoms. The H-terminated Pt surface at potentials more negative of the potential of zero charge makes the proton reduction possible (protons access the negatively charged surface easily), while anionic Pt complexes such as $[PtCl_4]^{2-}$ are strongly repelled by the Coulombic interaction, hence completely quenching the metal deposition process.

Although the self-limiting Pt deposition proved to be a linear function of the cycle number in the two-pulse plating process, it turned out that the surface coverage is also a function of the substrate. Au(111) surface allows one monolayer (ML)

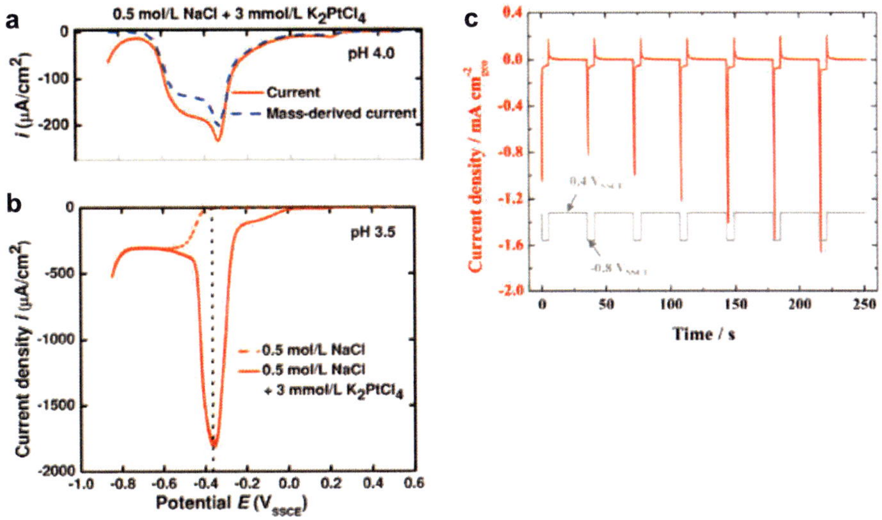

Fig. 4.12 Voltammetric and chronoamperometric results related to self-limiting Pt growth. **a** Measured current density during a 2 mV/s cathodic-going scan in a dilute Pt(IV) solution of the indicated composition (solid line) and the current density equivalent to the frequency change of the EQCM used in the same experiment (dashed line). **b** Comparison of the behaviour of the system with (solid line) and without (dashed line) the Pt(IV) compound as measured in cathodic-going cyclic voltammograms, indicating the identical behaviour in the hydrogen adsorption and hydrogen evolution potential interval. **c** Current density response (upper curve) as a result of the potential pulse applied with the wave form shown in the lower curve. The current density decay during the short pulses at $E = -0.8$ V indicates the self-limited growth of Pt. Adapted with permission from Refs. [160, 161]. Copyright (2012) and (2015), AAAS and American Chemical Society, respectively

platinum atom per cycle, but two atomic layers per cycle can be obtained on Ni [162]. Concerning the features of the low-coverage platinum layers, the bimetallic and ensemble effects associated with sub-monolayer coverage of Pt and resulting in enhanced catalytic efficiency were in excellent agreement with those obtained with SLRR concerning the oxidation of organic compounds [161].

Similarly to the deposition of platinum, other processes leading to a self-limited electrodeposition are also based on the product produced from the solvent. As it was shown for the reduction of germanate ions [163], the first process involves three electrons per Ge(IV) species, leading to a Ge(OH) adsorbate, which is reduced to an H-terminated Ge atom (GeH_{ads}). This surface species can block the further reduction of the germinate ions (GeO_3^{2-}) probably through a mechanism discussed above for the self-limiting reaction of Pt (negatively charged surface and repulsion of the precursor anion).

For the self-limiting growth of transition metals (Ni, Co and Fe), the hydroxide ions play a key role in the inhibition of the further growth of a layer with a limited thickness. When the reduction process of the precursor Me^{2+} ions is carried out in their dilute aqueous solution [164], a sufficiently negative potential has to be achieved to produce an effective protective hydroxide layer, where the alkalization of the vicinity of the cathode is due to the hydrogen evolution from the unbuffered solution of pH = 3. When the experiment involves a potential scan, a regular cyclic voltammetric peak of the metal ion reduction can be seen, then the decay of the negative current is reversed as the onset potential of the water decomposition is reached. This is the potential where the growth of the metal layer stops, as shown by the EQCM signal observed in parallel to the voltammetric scan. Therefore, we encounter the paradox situation that at relatively small cathodic overvoltage a continuous growth of a metal layer can be seen, while at more negative potential, where the driving force of the reduction of the metal ion is higher, the process stops. The resolution of the controversy is possible if the blocking effect of the hydroxide adlayer on the electrode is taken into account. The self-limitation is clearly a consequence of a finely-tuned system where the transport of the protons/hydroxide ions towards/from the cathode surface, respectively, is too slow to eliminate the blocking layer of hydroxide ions from the freshly-deposited metal surface. The thickness of the thus produced nickel layers proved to be a function of the Ni^{2+} concentration in the solution.

Water can be a key component of the self-limiting metal electrodeposition process also in the case when it is not a major component of the electrolyte solution. In a common deep eutectic solvent, the 2:1 ratio mixture of choline chloride and urea, the passivation of the nickel deposition was found [165]. Beside the observation of the cathodic passivation itself, the common origin of the self-limitation with the previously discussed process was evidenced by the necessity of water for the blocking of the metal deposition, the occurrence of the self-limitation at a sufficiently negative potential only and the transformation of the deposit morphology from granular to surface-conformal. The regulation of the solution composition and especially the buffer capacity of the solvent was unimportant here since the dissolution of the nickel(II) hydroxide in the deep eutectic solvent does not take place. An indirect evidence for the passivation process is that upon the switch of the electrode potential to the Ni dissolution regime, the anodic process is retarded due to the presence of the passivating layer that was previously formed in the cathode process.

4.3.2 *Atomic-Scale Observation of the Initial Phase of the Layer Growth*

As it was discussed in Chap. 2.10.1, nucleation of a new phase on a foreign substrate can take place through various deposition modes. Although the Frank–van der Merwe type nucleation and growth mode would theoretically be the most suitable layer formation pathway to deposit ultrathin layers, this seldom takes place in practical systems. The Volmer–Weber type nucleation can also lead to layer-like deposits, though with less uniform thickness distribution, if the growth of the newly-formed nuclei is not isotropic but the lateral growth is highly preferred. Detailed analysis of the role of the anisotropy on the coalescence of the grains and the formation of continuous layers can be found in the works of Guo and Searson [166–168]. Here, we will deal with the systems only whose behaviour was described the best at the atomic scale. Detailed overviews of the topic were offered by Allongue and Maroun [169, 170].

In many aspects, the appropriate measurement conditions for the *in situ* observation of the early stage of electrodeposition are similar to UPD-based processes (e.g., small cell volume, concentration of the electroactive species in the mM concentration range, ultrapure conditions with the removal of oxygen with an inert gas purging, application of a precisely-cut single crystal with well-defined surface orientation as substrate). The small concentration is necessary so that a sufficiently small nucleation density and a slow enough deposition process are provided, hence the changes caused by the deposition process can be easily followed with the available time resolution of the measurement. The current density is often of the $\mu A\ cm^{-2}$ level and the corresponding growth rate is in the monolayer/minute range or even less. The amount of deposit depends on the time of potentiostatic polarization in a nonlinear manner, typically showing a decrease with deposition time. Although the data obtained for dilute solutions are not transferable for fairly concentrated solutions used in conventional plating systems, the observations yield a valid guidance for the phenomena to be considered under different circumstances. The need for the potentiostatic control of the experiment is evident. The above mentioned conditions are valid for the *in situ* magnetization studies and will not be mentioned again. Many studies apply *in situ* observations of the atomic structure of the deposits and magnetic properties in parallel. The general aspects of the atomic structure will be dealt with in this chapter; however, where the observation of the structure at the atomic scale is indispensable for the elucidation of the magnetization behaviour, the structural data will be mentioned therein.

Concerning the observation method of the atomic arrangement, one possibility is the application of the scanning tunnelling microscopy (STM). In this case, the electrode potential of the tip and of the substrate must be controlled independently, hence the bipotentiostatic control of the experimental setup is needed. The side of the STM tip is insulated for reducing the parasitic current as much as possible.

Most of the observations of the atomic structure of the non-UPD type deposits refer to Au(111) or Cu(111) as substrate and Ni, Co and Fe as host metal. The general experience is that the growth of the host metal starts at the kink positions or at the step edges of the substrate. This nucleation is followed by the growth of an epitaxial atomic layer. The thickness where a full coverage of the substrate can be achieved is the higher, the larger the overpotential of the deposition. Representative examples for the growth of subsequent atomic layers as observed by STM are shown in Fig. 4.13.

For the most commonly used Au(111) substrate and the growth of nickel, the small overpotential of the deposition leads to a place exchange of the host and substrate metal atoms, while an exchange-free nucleation at the step edges takes place at higher overpotentials [171]. This behaviour was not observed for other metals deposited; however, Ni deposited on Ag(111) showed a very similar character [173]. It was shown also for the Au(111)/Ni system that a monoatomic coverage can be achieved without the nucleation of the second atomic layer. In the latter case, if the growth rate

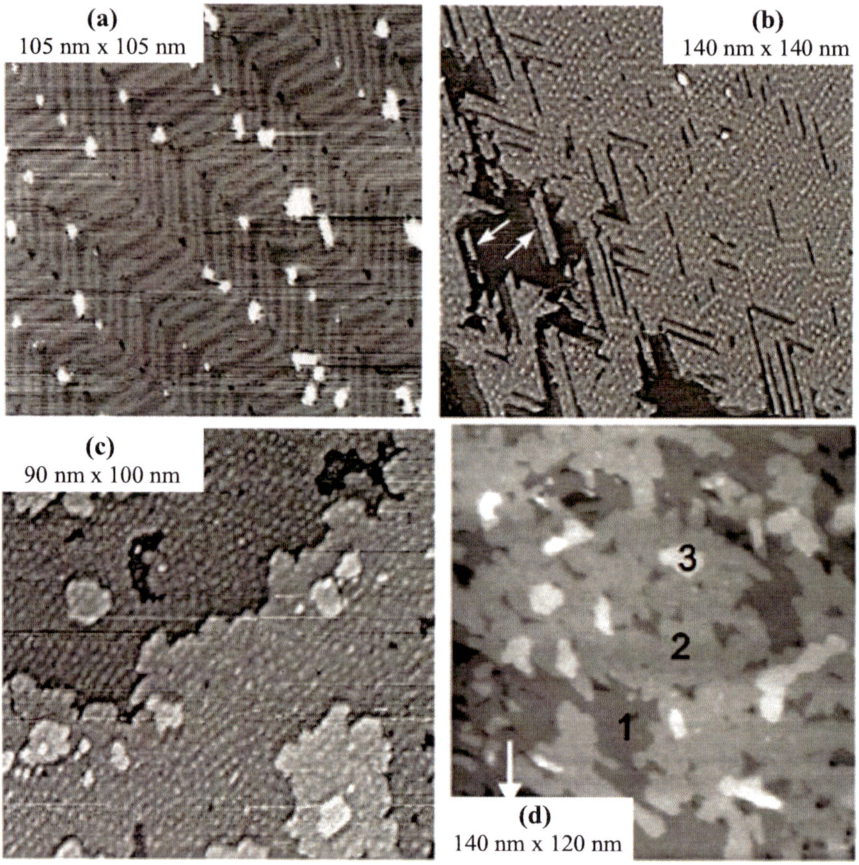

Fig. 4.13 Representative STM images of the layer growth on Au(111) substrate. **a** Ni deposited from dilute sulfate–chloride solution, nucleation of the islands of monoatomic layers at the elbows of the reconstructed surface; **b** Ni deposited from nitrate solution, with nearly full coverage of the first atomic layer; **c** Ni deposition at higher overpotential; **d** Fe deposition, with numbers indicating the deposit thickness in monolayers. For **c** and **d** lighter colours belong to areas where larger number of atomic layers can be found. Reprinted from [171] (**a** through **c**) and [172] (**d**). Copyright (1997) and (2004), with permission from The American Physical Society and Elsevier, respectively

of the layer was small enough, both the growth direction of the need-like features of the deposit and the remaining pinholes in the first layer were aligned along the major crystallographic direction of the Au(111) substrate [171]. The deposition of Co is only somewhat similar to nickel. The nucleation starts at the step edges and in the elbow of the reconstructed Au(111) surface, but later the deposit atomic structure strongly depends on the anions [174]. The nucleation of the deposit grown at small overpotential on reconstructed Au(111) can be taken as typical since it was observed also for Ru [170].

An important finding of the Ni deposition studies was that the place exchange of the substrate atoms at the step edges with the deposit atoms plays a significant role in the nucleation of the deposit, in particular at low overpotentials. This is somewhat unexpected in the sense that Ni and Ag are immiscible as shown by the equilibrium phase diagrams. The occurrence of the place exchange between the atoms of metals that show zero miscibility in the bulk indicates that bulk properties may be misleading when they are used for interpreting nanoscale phenomena.

It has to be noted that the deposition of even similar metals cannot be treated uniformly. For instance, Co growth on Au was found to start with a biatomic layer instead of monoatomic ones [174, 175], unlike the deposition of nickel.

The coverage level where the next atomic layer is nucleated at the surface of an atomic terrace is a function of essentially all deposition parameters (substrate/deposit pair, overpotential, concentration of the components etc.). The nucleation sites of the second and further atomic layers are no longer related to the terraces of the original substrate but rather to some stacking fault anomalies of the previous deposit layer. The quantitative treatment of the growth can be carried out if the coverage of each layer plane is given as a function of the mean deposit thickness, as shown in Fig. 4.14, which also shows a significant difference of the Ni coverage obtained for two systems, showing a typical two-dimensional and three-dimensional nucleation and growth pattern (Fig. 4.14a, b, respectively).

The fill-up of the last pinholes of the first atomic layer is preceded by the nucleation of the second atomic layer on the top of the first one. In general, the larger the overpotential of the deposition, the less complete is the first atomic layer when the nucleation of the second atomic layer starts. The second atomic layer often contains fewer faults than the first one. This is due to the fact that the first atomic layer is

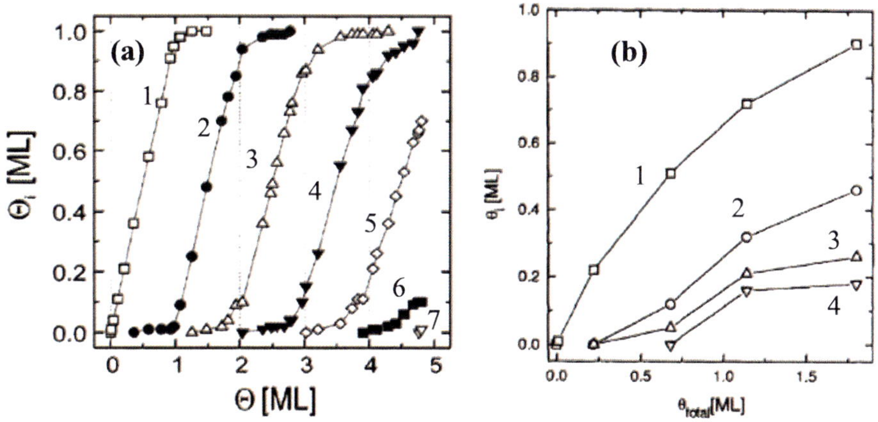

Fig. 4.14 Occupation of ith atomic layer of Ni from the substrate surface(Θ_i) as a function of the total coverage for **a** Au(111) and **b** Cu(111) substrates. Solution in both cases: 10^{-2} M H_3BO_3 + 10^{-3} M HCl + 10^{-3} M $NiSO_4$, overvoltage: $\eta = 150$ mV; numbers next to the curves refer to the layer number i. Reprinted from [171] **a** and [176] **b**. Copyright (1997) /both works/, with permission from The American Physical Society and Elsevier, respectively

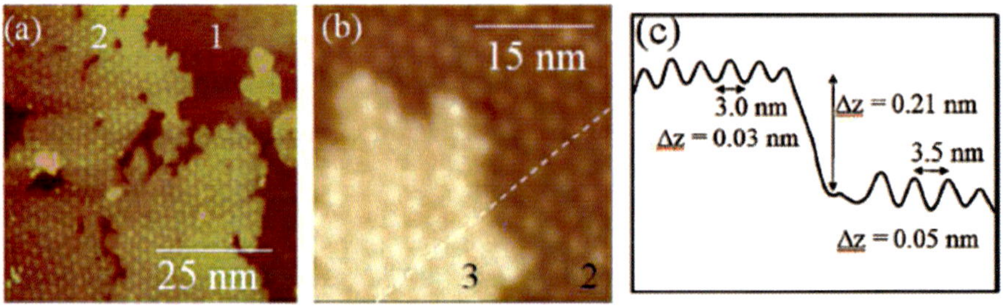

Fig. 4.15 *In situ* STM images (**a, b**) collected for Pt(111) coated with 1.5 and 2.3 ML of Co at −0.73 V (vs. Ag/AgCl) in 0.1 M $KClO_4$ + 1 mM HCl + 0.04 M $CoCl_2$ solution. Panel (**c**) shows the height profile along the dotted line marked in panel **b**. The Moiré pattern observed is due to the uneven periodicities and corrugation heights of the upcoming layers, as revealed by the profile shown in panel **c**. Reproduced from [177]. Copyright (2013), with permission from Elsevier

much stressed as opposed to the upcoming ones because of the lattice mismatch of the substrate and the deposit. The strain of the deposit relative to its equilibrium structure in the bulk form is gradually relaxed upon the growth.

As a consequence of the small difference between nearest neighbour atomic distances of the neighboring layers and the identical symmetry of the atomic arrangements in these layers, a so-called Moiré pattern is observed when the atoms of the adjacent atomic layers are sketched from the top view. This can be seen for layer pairs at both the substrate/deposit boundary and also within the deposit if unrelaxed/relaxed layer pairs are compared. Figure 4.15 shows an example for the STM observation of the modified interatomic distances in the first deposit layers as well as the Moiré pattern due to the difference in the lateral atomic distances.

Deposit features such as nucleation characteristics, thickness range for the nearly layer-by-layer growth and the turnover thickness into a three-dimensional growth mode varies from one system to another. All these parameters are a function of both the chemical environment (e.g., anions present) and the electrode potential. A compact summary of the results obtained for various substrate/metal pairs is shown in Table 4.3 (for elemental substrates only).

While the concentration of the ion of the deposited metal is in the millimolar concentration range in most of the cases, the application of the solutions with concentrations in the micromolar range can open a way for the study of the electrode potential dependence of the structure without the change in mass transport impact. This is simply due to the fact that with such concentrations, the deposit always forms in a diffusion-controlled manner, and the deposit structure is merely a function of the electrode potential. The size of the islands formed under such circumstances depends on the mobility of the adatoms, which is tuned with the electrode potential and partly with the co-adsorbing anions. This method was applied for the study of Au deposition onto Pt(111) surface [184]. The trend found was that for chloride-free solution, the island density decreased by applying more negative deposition potentials. However, solutions containing also chloride ions resulted in a growth mode when the already existing Pt terraces grew by the newly deposited Au atoms. The

Table 4.3 Atomic-scale observations of ultrathin electrodeposited layers unrelated to UPD processes

Substrate (method)	Deposit	Solution	Overvoltage	Growth characteristics	References
Au(111) reconstructed (STM)	Ni	$NiSO_4 + H_3BO_3 + HCl$ $NiSO_4$ $Ni(NO_3)_2$	$\eta > 80$ mV	Layer-by-layer growth up to 5 ML (all solutions)	[171]
	Fe	$FeSO_4 + K_2SO_4 + H_2SO_4 + KCl$	$\eta > 200$ mV	Fe forms bcc structure at $d > 3$ ML; uniaxially strained deposit	[178]
		$FeSO_4 + K_2SO_4 + H_2SO_4 + KCl$	$\eta > 200$ mV	Simultaneous growth of a few atomic layers; 2 ML fcc then bcc crystal structure	[172]
	Co	$CoSO_4 + KCl$ $CoSO_4 + KSCN$	$\eta < 300$ mV	Biatomic Co coverage at first, then layer-by-layer growth Co UPD, then nanometer-sized Co islands	[174]
		$CoSO_4 + K_2SO_4 + H_2SO_4 + KCl$	$\eta = 270$ mV $\eta = 570$ mV	Biatomic Co coverage at first, then layer-by-layer growth up to 5ML Biatomic Co coverage at first, then three-dimensional growth	[179]
Au(111) reconstructed (AFM)	Co	$CoSO_4 + CoCl_2 + H_3BO_3$	$\eta = 180$ mV $\eta = 680$ mV	Smooth Co film Granular Co islands	[180]
Au(111) with sub-ML Pd, unreconstructed (STM)	Ni	$NiSO_4 + K_2SO_4 + H_2SO_4 + KCl$		Pd pre-coating prevented the surface reconstruction, well-ordered Ni monolayer obtained	[181]
Au(100) (LEED)	Pd	$H_2SO_4 + HCl + PdCl_2$	$\eta > 100$ mV	Layer-by-layer growth up to 30 ML	[182]

(continued)

Table 4.3 (continued)

Substrate (method)	Deposit	Solution	Overvoltage	Growth characteristics	References
Ag(111)	Ni	$NiSO_4 + H_3BO_3 + HCl$	$\eta > 160$ mV	Layer-by-layer growth with a wider height distribution then for Au(111)	[173]
Cu(111)	Ni	$H_3BO_3 + HCl +$ $NiSO_4$ or $Ni(H_2NSO_3)_2$	145 mV $\leq \eta \leq$ 180 mV	Three-dimensional islands (all growth potentials, any solution)	[176]
Cu(100) Pb ML (STM)	Ni	$NiCl_2 + HCl$		Pb monolayers shifts the growth mode from the Stransky–Krastanov to the Frank–van der Merwe mode	[183]
Pt(111) (STM)	Au	$H_2SO_4 + KAuCl_4$ $H_2SO_4 + KAuCl_4 + HCl$	200 mV $\leq \eta \leq$ 900 mV	Monolayer islands with island surface density decreasing with potential; 3D island growth from $d > 2$ ML	[184]
	Co	$KClO_4 + HCl + CoCl_2$	$\eta \approx 100$ mV	Disordered first layer in the UPD potential range, then layer growth up to 5 ML	[177]
Pt(111) covered with Cu ML (STM)	Co	$KClO_4 + HCl + CoCl_2$	$\eta > 100$ mV	Nearly layer-by-layer growth of 2–3 atomic layers simultaneously up to 12 ML	[185]

growth with larger thickness shows an example for the Stransky–Krastanov type growth above the thickness of two monolayers.

The sequential build-up of structures comprising various atomic layers yield an opportunity to tune the deposit structure at the atomic level. For instance, a submonolayer of Pd on Au(111) was found to impede the surface reconstruction, hence providing an immense change in the deposition conditions [181]. The difference in Ni deposition on reconstructed and unreconstructed Au(111) surface was attributed to a larger extent to the difference in the atomic structure of the surface than to the modified chemical environment caused by the Pd adatoms.

Although STM observations yield very spectacular images on both atomic arrangement and distances of the deposit surface, STM is by far not the sole method that can reveal the deposit structure at very small thickness. X-ray diffraction as applied in the grazing incidence mode (GIXRD) can also reveal the in-plane atomic distances, although such investigations require a high-intensity source (like synchrotron beamline) due to the small number of atoms detected. This method has been applied from the early phase on the investigation of ultrathin electrodeposited layers. GIXRD was successfully used to indicate the structural transformation of Co deposited onto Cu as a function of the layer thickness [186]. As often seen for substrate/deposit systems with a misfit, the pseudomorphic (epitaxial) near-substrate layers are not reproduced after several layers, and a relaxed structure is obtained as the deposit grows.

Ultrahigh vacuum methods like low-energy electron diffraction (LEED) can be used *ex situ*. For example, an LEED study of electrodeposited Pd on Au(100) revealed an expanded Pd deposit (as compared to the relaxed bulk Pd atomic distances) up to at least 6 ML coverage, and the bulk lattice distances were achieved near the coverage of 20 ML. Interestingly, the three-dimensional island growth started only at 30 ML. The relatively large threshold thickness for the three-dimensional island growth was attributed to the small lattice misfit (4.5%). The difference between the atomic distances between the bulk Pd and thin Pd layers is to be taken into account for the explanation of the modified catalytic activity (see examples of Sect. 4.2, too).

It is worthwhile of mentioning that the study of the dissolution of ultrathin layers clearly revealed that the dissolution is by far not the reversal of the deposition in the sense of the order of immobilization and removal of the atoms. Even in the case when a layer-by-layer growth could be observed for several monolayers (like Ni on Au), the dissolution induced etched pits that also served as dissolution sites, similar to the already existing step edges [171]. The formation of pits was also found for the dissolution process of Co [179] where the fragmentation of the initial biatomic layer was even more pronounced.

4.3.3 Magnetization of Ultrathin Electrodeposited Layers

Bulk magnetism originates from the parallel spin orientation in an assembly of species that are paramagnetic as stand-alone entities. Ferromagnetism in metallic elements can be observed when the exchange coupling interaction is capable of ordering the spins of the electrons in the metal in a parallel manner. It is obvious that there must be a size limit where the exchange interaction in the growing phase can give rise to a long-range magnetic order. This is particularly true for ultrathin magnetic layers [187] where differences in thickness at the sub-monolayer scale can fundamentally influence the magnetic behaviour. The magnetization occurring during the deposition of the ultrathin layers is by far too small for a vibrational magnetometer. Therefore, early *in situ* studies are based on the observation of the magneto-optical Kerr effect or alternating field gradient magnetometry. Experimental setups are available in a number of works [186, 188–191]. Results of the early studies were summarized by Allongue and co-workers in various reviews [192, 193]. Recently, *in situ* magnetic observation of ultrathin electrodeposited films was made possible in a SQUID magnetometer device [194].

If an *in situ* magnetic measurement is carried out, the electrochemical cell has to be placed into the gap of an electromagnet. This means that the gap of the electromagnet has to be larger than for a usual measurement of a tiny sample, which limits the maximum field available. In any case, the maximum available field has to be larger than the saturation field of the samples. A field of a few hundred oersteds proved to be sufficient for the observation of square-shaped magnetization curves [186, 191, 195], but a few kOe maximum field is also available with a suitable configuration [174, 179, 196, 197]. The external magnetic field has to be adjusted as in-plane (longitudinal arrangement) and perpendicular-to-plane (polar arrangement) with respect to the substrate, which usually means a horizontal and a vertical direction of the magnetic field, respectively. As a special example for *in situ* observation of magnetism, a Mössbauer spectroscopy setup has also been implemented [198].

Magnetic properties of the ultrathin layers are studied with solutions in which the concentration of the metal ions is typically a few mM (but at most 40 mM). The supporting electrolytes used were typically sulphate salts or boric acid, the pH being slightly acidic only (pH = 3…5). The comparison of the deposits obtained from acidic and alkaline solutions was demonstrated for Co deposition [199], and the role of the deposition conditions in the magnetic behaviour was found to be crucial (see later). The impact of additives has not been tested in magnetization-oriented experiments. Since an interference of the hydrogen evolution can always occur during the deposition of the magnetic metals due to their negative deposition potential, the equivalent thickness of the deposit cannot be determined from the cathodic charge. Instead, the equivalent layer thickness is determined after the magnetic measurement from the anodic charge consumed for the layer dissolution [195, 196], occasionally corrected for the background current obtained with blank solution [188]. The accuracy of the determination of the equivalent layer thickness was estimated to be 0.02–0.1 monolayers (see [188] and [196], respectively). For eliminating the negative impact of the hydrogen evolution, some attempt was made to apply non-aqueous solvent already in the pioneering work of the field [195], but it did not gain a further interest and later works used aqueous systems only. Due to the small charge required for the deposition of the thin layers, the geometry of the electrochemical cell is not of importance, and hence, the counter and the reference electrodes were usually placed in side compartments of the cell, hence providing access to the working electrode for the magnetic observation.

Substrates in these experiments are either single crystals (Cu or Au) or highly textured noble metal surfaces (like sputtered gold with a dominant (111) orientation). When gold was used as substrate, it was found that the appropriate surface reconstruction is indispensable for the bulk deposition of the metals of interest [174, 180, 196, 197]. When a capping layer was applied for protecting the magnetic layer (i.e., for an *ex situ* study after completing the electrodeposition), a flow cell arrangement is necessary [200]. The construction principles of such cells are quite identical to those listed in Sect. 4.1.

The magnetic behaviour of ultrathin magnetic layers is generally very similar to their counterparts produced with high-vacuum processes, while the experimental setup costs much less. An advantage of the electrochemical experiment as opposed to high-vacuum ones is that the deposition and the dissolution both can be carried out *in situ*, and hence, the growth and the dissolution can be compared [188]. Electrodeposition exhibits a variable that is mostly very restricted in high-vacuum devices; namely, the deposition rate which is controlled through the overpotential of the deposition. As it will be discussed later, it has a strong impact on the magnetization behaviour of ultrathin layers.

An important factor of interest in the ultrathin magnetic layers is the out-of-plane magnetization. For a planar magnetic object, the minimization of the magnetostatic energy favours the in-plane magnetization. However, this is not true for an ultrathin layer that mostly exhibits a threshold thickess of a few atomic layer below which the magnetization is perpendicular to the layer plane. This effect is highly sought in magnetic data storage where the miniaturization of the bit areas requires the occurrence of perpendicular magnetization.

Since magnetization stems from the exchange coupling of the electron spins in a particular layer, the surface state of this layer is of high importance. The situation is similar to the ratio of volume and surface energy in any other nanosystem. As the surface interaction can influence the electron density at the boundary layer, the occurrence of either the magnetization itself or its direction depends on the atomic environment of both the bottom (substrate side) and the top (solution or covering layer side) of the deposit. A valid comparison of the results obtained from magnetization measurement is possible only if these factors are taken into account, as shown in Table 4.4 that summarizes the features of ultrathin electrodeposited layers by focusing on their magnetization. It cannot be stressed enough that once a magnetic layer is in contact with an electrolyte solution, the adsorbed anions and/or the oxidation state of the surface is also an important determining factor of the overall magnetization process. An example for the turnover of the spontaneous magnetization direction as a function of the chemical environment of ultrathin magnetic layers is shown in Fig. 4.16.

Among electrodeposited magnetic metals, cobalt gained an outstanding interest. This is because Co has two crystalline forms, hcp (thermodynamically stable bulk form at room temperature) and fcc (stable as a bulk form above 422 °C). Electrodeposition can be driven so that either of the Co crystalline form is obtained, and

Table 4.4 Thickness corresponding the onset of magnetization (d_{ON}) and that corresponding to the onset of spontaneous in-plane magnetization direction (d^*) in electrodeposited metals as obtained with *in situ* observations. Data completed after Table 1 of Ref. [201]

Substrate	Deposit	d_{ON}/ML	d^*/ML	Conditions	References
Au(111)	Ni	6	N/A		[192]
	Co	1.6	6.2–7.2	Cu-capped	[179, 192]
			1.6 20	SO_4^{2-}–solution solution, pH = 8.5	[199]
		N/A N/A	1.5–2 4–5	Cl^- –solution SCN^- –solution	[174]
		2 N/A	8 2	$\eta = -0.68$ V, Cu-capped $\eta = -0.18$ V, Cu-capped	[180, 196]
		N/A	>8	Au-capped	[200]
	Fe	1	2 N/A	SO_4^{2-}–solution Cu-capped	[172, 192, 197]
Ag(poly)	Co	<5.5	<15		[198]
Cu(001)	Co	N/A	1.5	SO_4^{2-}–solution	[186, 188]
		1.5	2	SO_4^{2-}–solution	[195]
	Fe	2	8	SO_4^{2-}–solution	[195]

this opens a new opportunity for tuning the magnetic properties of the ultrathin Co deposits. It was observed already in a very early study that the spontaneous magnetization direction depends on the deposition potential [196]. This was attributed to the formation of hcp Co islands at high overpotential, although no direct evidence was available for this assumption. The occurrence of the in-plane magnetization at low deposition overpotential is in agreement with other observations when the deposition potential was not thought to be of high importance [188, 189, 195]. The hcp versus fcc ratio proved to be a decisive factor for the thickness range below about 100 nm, above which the deposition rate and the solution used was much less crucial [202]. The analysis of the contribution of various anisotropy factors to the spontaneous magnetization direction can be found in the relevant literature and is much beyond the scope of this work.

Data in Table 4.4 indicate magnetization directions of the electrodeposited layers under stagnant conditions. However, the electrochemical setup makes it possible to tune the magnetic properties of the deposit *in situ* in the electrochemical cell after achieving the desired thickness, also in a reversible manner while changing the electrode potential forth and back. This manipulation with an electrodeposited layer opens the way of electrochemically tunable magnetism. The magnetism of ultrathin Co layers can vary as a result of either reductive [203] or oxidative [192] electrochemical treatment as well as due to adsorption of neutral molecules like CO [203]. Figure 4.17 shows the variation of the magnetization curve during a potential pulse treatment between potential values corresponding to virtually zero current density and intense hydrogen evolution for CO-covered ultrathin Co surface. It can

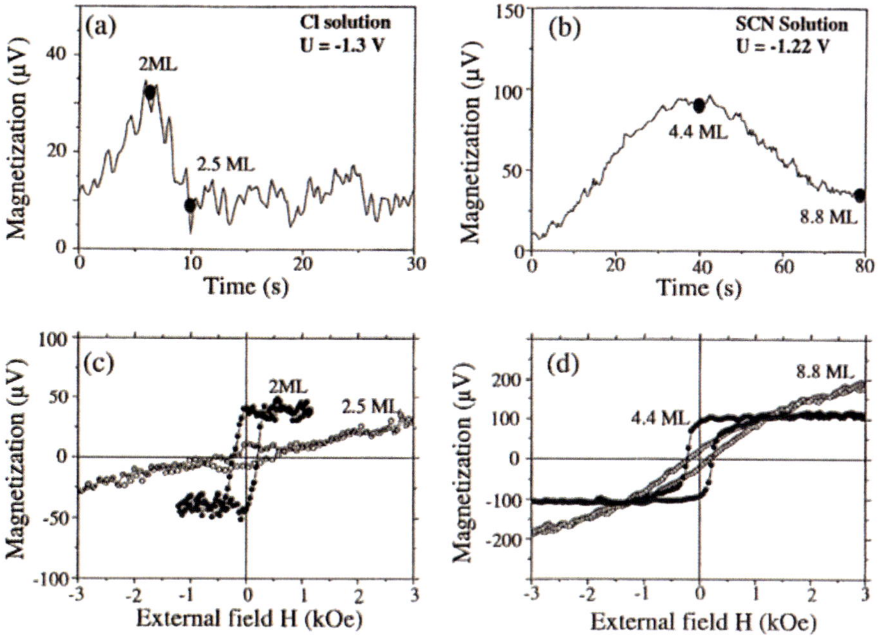

Fig. 4.16 *In situ* recorded magnetization curves obtained in the presence of Cl^- ions (left column) and SCN^- ions (right column) in the solutions. Top row: transient *M(t)* curves, bottom row: *M(H)* hysteresis loops for different Co thickness (ML: monolayer-equivalent thickness). The corresponding times are labelled in *M(t)* curves using the same figures. Deposition potentials are indicated in the top figures; magnetization is indicated in the unit of the measurement gauge. Reprinted from [174]. Copyright (2000), with permission from Elsevier

be well seen that, practically regardless of the potential applied for the provocation of the hydrogen evolution at a sufficient rate, the hydrogen-covered surface of the ultrathin CO-covered Co layer exhibits in-plane magnetization, as opposed to the perpendicular magnetization when no hydrogen is evolved. The reversibility of the hydrogen-induced magnetization reorientation is remarkable.

Another study applying the stress induced by the hydrogen absorption for the modification of the properties of thin electrodeposited Co films was based on a fully electroplated Au/Co/Pd structure [175]. In this case, the hydrogen absorption capability of the topmost layer provides a platform for reversible hydrogenation. Due to the limited lateral expansion possibility of the top Pd layer due to its mechanical couple with the Co layer underneath, hydrogen absorption leads only to relaxation perpendicular to the surface. However, this effect is enough for a reversible magnetization change fully parallel to the change in the hydrogen content of the Pd layer that is regulated by the electrode potential.

Surface-area-limited oxidation of an ultrathin Co film is possible in alkaline solutions. The alternating surface coverage with adsorbed hydrogen atoms and hydroxide ions is accompanied with a reversible change of magnetization direction.

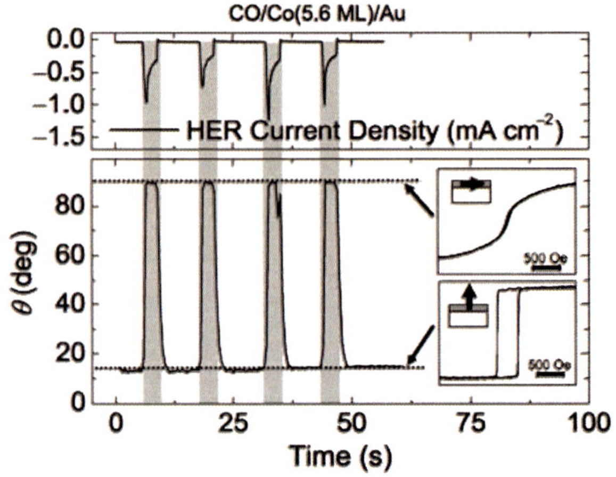

Fig. 4.17 Influence of the hydrogen evolution current (top graph) on the magnetization easy axis angle θ (with respect to the film normal) for a 5.6 monolayer thick CO-covered Co film (bottom graph) as a function of time during four consecutive potential steps between −1.15 V (no hydrogen evolution) and −1.3 or −1.4 V (first and second pair of potential steps, respectively). The two *M-H* curves in the inset of the lower panel correspond to near-zero and high hydrogen evolution rates, respectively. Reproduced with permission from Ref. [203]. Copyright (2012) American Physical Society

The summary of the magnetization change as a result of the surface redox treatment of an ultrathin electrodeposited Co layer is presented in Fig. 4.18.

It was also demonstrated with *in situ* magnetic measurement of ultrathin electrodeposited cobalt layers that the magnetic moment per cobalt atoms differs from the bulk value [205]. The latter study could particularly exploit the opportunity of electrodeposition coupled with magnetization measurement; namely, the continuous, repetitive, and background-free monitoring of the magnetic moment of ultrathin Co films during the film growth and subsequent dissolution with an accurate adjustment of the average film thickness. The magnetic moment increment for ultrathin Co layers in electrodeposited films was in accord with that found for Co films produced with physical deposition methods [205].

It is to be emphasized that the coverage in monolayer-equivalent units is by far not enough to characterize a layer, as it was also shown in Sect. 4.3.2. The morphology of the deposit has an immense impact on the magnetization process. A combined electrodeposition and theoretical calculation study showed [206] that the edge atoms have a crucial role in the overall magnetization process, in particular in the thickness dependence of the perpendicular magnetization component. Since electrodeposition has a number of variables for adjusting the ratio of bulk, surface and perimeter atoms in the growing ultrathin islands, it can be applied in a flexible way to achieve the desired magnetic properties. A similar problem is the even vs. island-like growth. In the presence of surfactant, vacuum deposition methods (MBE, sputtering or evaporation) can be used to deposit even layers without island formation even at the

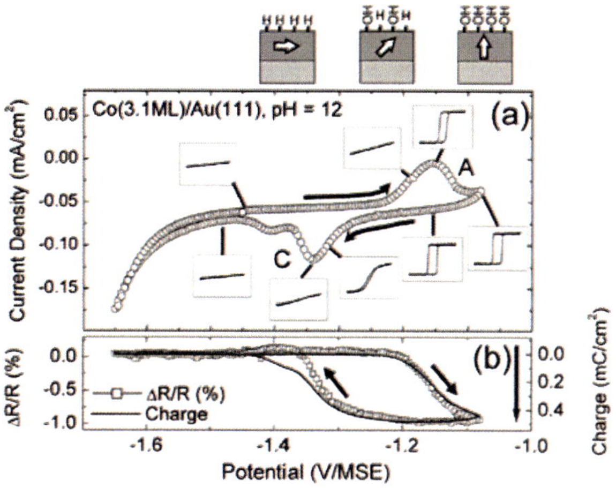

Fig. 4.18 **a** Variations of the current during a 10 mV/s potential sweep of a 3.1 ML Co film on Au(111) in 0.1 M K_2SO_4 + 0.01 M KOH (pH = 12). The insets are *M–H* magnetization curves at selected electrode potentials. Drawings on top sketch the correspondence between surface chemistry and the magnetization easy axis of the Co film during the positive potential scan. **b** Variations of relative optical reflectivity $\Delta R/R$ as a function of the electrode potential (open symbols, ordinate). The integral of the electrochemical current over time (solid line, right ordinate) is calculated after baseline correction for the water decomposition current. The charge sign was reversed for sake of easy comparison with the reflectivity. Reprinted from Ref. [204], with the permission of AIP Publishing

monolayer level. In such cases, a Pb layer is deposited first onto the substrate that behaves as a "floating" layer and the deposition of the further components takes place with the diffusion of newly arriving atoms through these layers [183]. In this study mimicking the vacuum deposition methods applying surfactants, $PbCl_2$ was applied beside $NiCl_2$ for regulating the surface morphology of the Ni deposit. In order to avoid the deposition of bulk Pb (which takes place at less negative potential than that of Ni), the concentration of $PbCl_2$ had to be as low as 0.06 mM, a level where the total Pb content of the electrochemical cell was barely enough for the formation of a monolayer, which can behave as surfactant without a further Pb growth. Beside the effect that the near-monolayer coverage of Pb reduced the hydrogen evolution rate, hence eliminating its interference with Ni deposition, it decelerated Ni deposition, too. While Pb had practically no influence at the Ni layer thickness below 10 atomic layer, further Ni growth occurred in a layer-by-layer manner, unlike in the absence of Pb. For thicker films, quasi-two-dimensional islands with a flat top layer were detected, which resulted in an increased squareness of the magnetization loop.

References

1. Leskelä M, Ritala M (2002) Thin Solid Films 409:138–146
2. George SM (2010) Chem Rev 110:111–131
3. Johnson RW, Hultqvist A, Bent SF (2014) Mater Today 17:236–246
4. Pathan HM, Lockhande CD (2004) Bull Mater Sci 27:85–111
5. Pawar SM, Pawar BS, Kim JH, Joo OS, Lokhande CD (2011) Curr Appl Phys 11:117–161
6. Gregory BW, Stickney JL (1991) J Electroanal Chem 300:543–561
7. Gregory BW, Suggs DW, Stickney JL (1991) J Electrochem Soc 138:1279–1984
8. Huang BM, Colletti LP, Gregory BW, Anderson JL, Stickney JL (1995) J Electrochem Soc 142:3007–3016
9. Colletti LP, Flowers BH, Stickney JL (1998) J Electrochem Soc 145:1442–1449
10. Flowers BH, Wade TL, Garvey JW, Lay M, Happek U, Stickney JL (2002) J Electroanal Chem 524–525:273–285
11. Zhu W, Yang JY, Gao XH, Bao SQ, Fan XA, Zhang TJ, Cui K (2005) Electrochim Acta 50:4041–4047
12. Mkwizu TS, Mathe MK, Cukrowski I (2010) Langmuir 26:570–580
13. Valdes M, Modibedi M, Mathe M, Hillie T, Vazquez M (2014) Electrochim Acta 128:393–399
14. Chen Y, Wang L, Pradel A, Ribes M, Record MC (2014) J Electroanal Chem 724:55–61
15. Villegas I, Napolitano P (1999) 146:117–124
16. Sheridan LB, Czerwiniski J, Jayaraju N, Gebregziabiher DK, Stickney JL, Robinson DB, Soriaga MP (2012) Electrocatal 3:96–107
17. Jagannathan K, Benson DM, Robinson DB, Stickney JL (2016) J Electrochem Soc 163:D3047–D3052
18. Oviedo OA, Reinaudi L, García SG, Leiva EPM (2016) Underpotential Deposition. From Fundamentals and Theory to Applications at the Nanoscale. In: Scholz F (ed) Monographs in Electrochemistry. Springer, Berlin
19. Kim JY, Stickney JL (2008) J Phys Chem C 112:5966–5971
20. Loglio F, Innocenti M, Jarek A, Caporali S, Pasquini I, Foresti ML (2010) J Electroanal Chem 638:15–20
21. Innocenti M, Forni F, Pezzatini G, Raiteri R, Loglio F, Foresti ML (2001) J Electroanal Chem 514:75–82
22. Yang J, Zhu W, Gao X, Bao S, Fan X, Duan X, Hou J (2006) J Phys Chem B 110:4599–4604
23. Xiao C, Yang J, Zhu W, Peng J, Zhang J (2009) Electrochim Acta 54:6821–6826
24. Colletti LP, Teklay D, Stlckney JL (1994) J Electroanal Chem 369:145–152
25. Qiao Z, Shang W, Wang C (2005) J Electroanal Chem 576:171–175
26. Köse H, Bicer M, Tütünoğlu C, Aydın AO, Şişman I (2009) Electrochim Acta 54:1680–1686
27. Loglio F, Innocenti M, D'Acapito F, Felici R, Pezzatini G, Salvietti E, Foresti ML (2005) J Electroanal Chem 575:161–167
28. Zhu W, Yang JY, Hou J, Gao XH, Bao SQ, Fan XA (2005) J Electroanal Chem 585:83–88
29. Innocenti M, Bencistà I, Bellandi S, Bianchini C, Di Benedetto F, Lavacchi A, Vizza F, Foresti ML (2011) Electrochim Acta 58:599–605
30. Venkatasamy V, Jayaraju N, Cox SM, Thambidurai C, Happek U, Stickney JL (2006) J Appl Electrochem 36:1223–1229
31. Wade TL, Vaidyanathan R, Happek U, Stickney JL (2001) J Electroanal Chem 500:322–332
32. Venkatasamy V, Mathe MK, Cox SM, Happek U, Stickney JL (2006) Electrochim Acta 51:4347–4351
33. Banga D, Kim YG, Stickney J (2011) J Electrochem Soc 158:D99–D106
34. Vaidyanathan R, Stickney JL, Cox SM, Compton SP, Happek U (2003) J Electroanal Chem 559:55–61
35. Vaidyanathan R, Stickney JL, Happek U (2004) Electrochim Acta 49:1321–1326
36. Yang JY, Zhu W, Gao XH, Fan XA, Bao SQ, Duan XK (2007) Electrochim Acta 52:3035–3039
37. Zhu W, Yang JY, Zhou DX, Xiao CJ, Duan XK (2008) Electrochim Acta 53:3579–3586

38. Zhu W, Yang Y, Zhou D, Xiao C, Duan X (2008) J Electroanal Chem 614:41–48
39. Banga DO, Vaidyanathan R, Xuehai L, Stickney JL, Cox S, Happeck U (2008) Electrochim Acta 53:6988–6994
40. Tsang CF, Ledina MA, Stickney JL (2017) J Electroanal Chem 793:242–249
41. Zhu W, Yang JY, Zhou DX, Xiao CJ, Duan XK (2008) Langmuir 24:5919–5924
42. Mathe MK, Cox SM, Flowers BH, Vaidyanathan R, Pham L, Srisook N, Happek U, Stickney JL (2004) J Crystal Growth 271:55–64
43. Zhu W, Yang JY, Gao XH, Hou J, Bao SQ, Fan XA (2005) Electrochim Acta 50:5465–5472
44. Loglio F, Telford AM, Salvietti E, Innocenti M, Pezzatini G, Cammelli S, D'Acapito F, Felici R, Pozzi A, Foresti ML (2008) Electrochim Acta 53:6978–6987
45. Banga D, Jarayaju N, Sheridan L, Kim YG, Perdue B, Zhang X, Zhang Q, Stickney J (2012) Langmuir 28:3024–3031
46. Banga D, Perdue B, Stickney J (2014) J Electrochem Soc 161:D141–D146
47. Ramasamy M, Jung CY, Yeon YB, Lee CW (2017) J Electrochem Soc 164:D1006–D1014
48. Venkatasamy V, Jayaraju N, Cox SM, Thambidurai C, Stickney JL (2007) J Electrochem Soc 154:H720–H725
49. Li J, Kou H, Jiang Y, Lu D, Zheng Z, Wang C (2012) J Solid State Electrochem 16:3097–3103
50. Zhang L, He W, Chen XY, Du Y, Zhang X, Shen Y, Yang F (2015) Surf Sci 631:173–177
51. Innocenti M, Pezzatini G, Forni, Foresti ML (2001) J Electrochem Soc 148:C357–C362
52. Pezzatini G, Caporali S, Innocenti M, Foresti ML (1999) J Electroanal Chem 475:164–170
53. Zhang X, Shi X, Wang C (2009) J Solid State Electrochem 13:469–475
54. Jia L, Kou H, Jiang Y, Yu S, Li J, Wang C (2013) Electrochim Acta 107:71–77
55. Jiang X, Kou H, Li J, Yu S, Du Y, Ye W, Wang C (2012) J Solid State Chem 194:336–342
56. Villegas I, Stickney JL (1992) J Electrochem Soc 139:686–694
57. Boone BE, Shannon C (1996) J Phys Chem 100:9480–9484
58. Cecconi T, Atrei A, Bardi U, Forni F, Innocenti M, Loglio F, Foresti ML, Rovida G (2001) J Electron Spectr Related Phenom 114–116:563–568
59. Innocenti M, Cattarin S, Cavallini M, Loglio F, Foresti ML (2002) J Electroanal Chem 532:219–225
60. Salvietti E, Loglio F, Innocenti M, Cavallini M, Facchini M, Pezzatini G, Raiteri R, Foresti ML (2007) Electrochim Acta 52:6034–6040
61. Carlà F, Loglio F, Resta A, Felici R, Lastraioli E, Innocenti M, Foresti ML (2014) J Phys Chem C 118:6132–6139
62. Lister TE, Stickney JL (1996) Appl Surf Sci 107:153–160
63. Suggs DW, Stickney JL (1993) Surf Sci 290:362–374
64. Suggs DW, Stickney JL (1993) Surf Sci 290:375–387
65. Goetting LB, Huang BM, Lister TE, Stickney JL (1995) Electrochim Acta 40:143–158
66. Hayden BE, Nandhakumar IS (1998) J Phys Chem B 102:4897–4905
67. Forni F, Innocenti M, Pezzatini G, Foresti ML (2000) Electrochim Acta 45:3225–3231
68. Lay MD, Stickney JL (2004) J Electrochem Soc 151:C431–C435
69. Perdue B, Czerniawski J, Anthony J, Stickney J (2014) J Electrochem Soc 161:D3087–D3092
70. Venkatasamy V, Shao I, Huang Q, Stickney JL (2008) J Electrochem Soc 155:D693–D698
71. Mathe MK, Cox SM, Venkatasamy V, Happek U, Stickney JL (2005) J Electrochem Soc 152:C751–C755
72. Venkatasamy V, Jayaraju N, Cox SM, Thambidurai C, Mathe M, Stickney JL (2006) J Electroanal Chem 589:195–202
73. Fernandes VC, Salvietti E, Loglio F, Lastraioli E, Innocenti M, Mascaro LH, Foresti ML (2009) J Appl Electrochem 39:2191–2197
74. Yang J, Zhu W, Gao X, Bao S, Fan X (2005) J Electroanal Chem 577:117–123
75. Innocenti M, Bellandi S, Lastraioli E, Loglio F, Foresti ML (2011) Langmuir 27:11704–11709
76. Loglio F, Innocenti M, Pezzatini G, Foresti ML (2004) J Electroanal Chem 562:117–125
77. Foresti ML, Pozzi A, Innocenti M, Pezzatini G, Loglio F, Salvietti E, Giusti A, D'Anca F, Felici R, Borgatti F (2006) Electrochim Acta 51:5532–5539

78. Giaccherini A, Russo F, Carlà F, Guerria A, Picca RA, Cioffi N, Cinotti S, Montegrossi G, Passaponti M, Di Benedetto F, Felici R, Innocenti M (2018) Appl Surf Sci 432:53–59
79. Bicer M, Köse H, Şişman I (2010) J Phys Chem C 114:8256–8263
80. Feng S, Yang J, Liu M, Zhu H, Zhang J (2012) LiG, Peng J, Liu Q. Electrochim Acta 83:321–326
81. Li W, Yang J, Zhang J, Liu M, Luo Y, Xiao Y, Fu L (2015) J Electrochem Soc 162:D137–D141
82. Innocenti M, Cattarin S, Loglio F, Cecconi T, Seravalli G, Foresti ML (2004) Electrochim Acta 49:1327–1337
83. Caporali S, Tolstogouzov A, Teodoro OMND, Innocenti M, Di Benedetto F, Cinotti S, Picca RA, Sportelli MC, Cioffi N (2015) Sol Energy Mater Sol Cells 138:9–16
84. Lin S, Shi X, Zhang X, Kou H, Wang C (2010) Appl Surf Sci 256:4365–4369
85. Kou H, Zhang X, Jiang Y, Li J, Yu S, Zheng Z, Wang C (2011) Electrochim Acta 56:5575–5581
86. Innocenti M, Becucci L, Bencistà I, Carretti E, Cinotti S, Dei L, Di Benedetto F, Lavacchi A, Marinelli F, Salvietti E, Vizza F, Foresti ML (2013) J Electroanal Chem 710:17–21
87. Lin S, Zhang X, Shi X, Wei J, Lu D, Zhang Y, Kou H, Wang C (2011) Appl Surf Sci 257:5803–5807
88. Liang X, Jayaraju N, Thambidurai C, Zhang X, Stickney JL (2011) Chem Mater 23:1742–1752
89. Foresti ML, Milani S, Loglio F, Innocenti M, Pezzatini G, Cattarin S (2005) Langmuir 21:6900–6907
90. Banga D, Perdue B, Stickney J (2014) J Electroanal Chem 716:129–135
91. Berretti E, Cinotti S, Caporali S, Cioffi N, Giaccherini A, Di Benedetto F, Foresti ML, Montegrossi G, Lavacchi A, Vizza F, Picca RA, Innocenti M (2016) J Electrochem Soc 163:D3034–D3039
92. Chong B, Zhu W, Liu Y, Guan L, Chen GZ (2016) J Mater Chem A 4:1336–1344
93. Zhang J, Yang J, Liu M, Li G, Li W, Gao S, Luo Y (2014) J Electrochem Soc 161:D55–D58
94. Yu S, Jiang Y, Wang C (2013) Electrochim Acta 114:430–438
95. del Barrio MC, Salinas DR, García SG (2007) Electrochem Commun 9:1382–1387
96. del Barrio MC, García SG, Salinas DR (2009) Electrochim Acta 55:451–457
97. Innocenti M, Zangari G, Zafferoni C, Bencistà I, Becucci L, Lavacchi A, Di Benedetto F, Bellandi S, Vizza F, Foresti ML (2013) J Power Sources 241:80–86
98. Liang X, Zhang Q, Lay MD, Stickney JL (2011) J Am Chem Soc 133:8199–8204
99. Öznülüer T, Erdoğan I, Şişman I, Demir Ü (2005) Chem Mater 17:935–937
100. Alanyalıoğlu M, Bayrakceken F, Demir Ü (2009) Electrochim Acta 54:6554–6559
101. Şişman I, Demir Ü (2011) J Electroanal Chem 651:222–227
102. Aydın ZY, Abacı S (2017) J Solid State Electrochem 21:1417–1430
103. Erdoğan IY, Öznülüer T, Bülbül F, Demir Ü (2009) Thin Solid Films 517:5419–5424
104. Dimitrov N (2016) Electrochim Acta 209:599–622
105. Brankovic SR, Wang JX, Adzic RR (2001) Surf Sci 474:L173–L179
106. Brankovic SR, Wang JX, Adzic RR (2001) J Serbian Chem Soc 66:887–898
107. Jayaraju N, Banga D, Thambidurai C, Liang X, Kim YG, Stickney JL (2014) Langmuir 30:3254–3263
108. Viyannalage LT, Vasilic R, Dimitrov N (2007) J Phys Chem C 111:4036–4041
109. Fayette M, Liu Y, Bertrand D, Nutariya J, Vasiljevic N, Dimitrov N (2011) Langmuir 27:5650–5658
110. Mitchell C, Fayette M, Dimitrov N (2012) Electrochim Acta 85:450–458
111. Nutariya J, Fayette M, Dimitrov N, Vasiljevic N (2013) Electrochim Acta 112:813–823
112. Fayette M, Nutariya J, Vasiljevic N, Dimitrov N (2013) ACS Catal 3:1709–1718
113. Achari I, Ambrozik S, Dimitrov N (2017) J Phys Chem C 121:4404–4411
114. Mercer MP, Plana D, Fermín DJ, Morgan D, Vasiljevic N (2015) Langmuir 31:10904–10912
115. Ambrozik S, Rawlings B, Vasiljevic N, Dimitrov N (2014) Electrochem Commun 44:19–22
116. Ambrozik S, Dimitrov N (2015) Electrochim Acta 169:248–255
117. Cappillino PJ, Sugar JD, El Gabaly F, Cai TY, Liu Z, Stickney JL, Robinson DB (2014) Langmuir 30:4820–4829
118. Mkwizu TS, Cukrowski I (2014) Electrochim Acta 147:432–441

119. Gokcen D, Bae SE, Brankovic SR (2011) Electrochim Acta 56:5545–5553
120. Bulut E, Wu D, Dole N, Kilic H, Brankovic SR (2017) J Electrochem Soc 164:D159–D168
121. Fang JS, Lin LY, Wu CL, Cheng YL, Chen GS (2017) J Electronic Mater 46:6677–6684
122. Modibedi RM, Mathe MK, Motsoeneng RG, Khotseng LE, Ozoemena KI, Louw EK (2014) Electrochim Acta 128:406–411
123. Fang JS, Chen JH, Chen GS, Cheng YL, ChinTS (2016) Electrochim Acta 206:45–51
124. Al Amri Z, Mercer MP, Vasiljevic N (2016) Electrochim Acta 210:520–529
125. Jin Y, Shen Y, Dong S (2004) J Phys Chem B 108:8142–8147
126. Liu P, Ge X, Wang R, Ma H, Ding Y (2009) Langmuir 25:561–567
127. Cheng S, Rettew RE, Sauerbrey M, Alamgir FM (2011) ACS Appl Mater Interfaces 3:3948–3956
128. Jayaraju N, Vairavapandian D, Kim YG, Banga D, Stickney JL (2012) J Electrochem Soc 159:D616–D622
129. Mkwizu TS, Mathe MK, Cukrowski I (2013) J Electrochem Soc 160:H529–H546
130. Zhang X, Choi I, Qu D, Wang L, Lee CWJ (2013) Int J Hydrogen Energ 38:5665–5670
131. Loukrakpam R, Yuan Q, Petkov V, Gan L, Rudi S, Yang R, Huang Y, Brankovic SR, Strasser P (2014) Phys Chem Chem Phys 16:18866–18876
132. Gokcen D, Yuan Q, Brankovic SR (2014) J Electrochem Soc 161:D3051–D3056
133. Sheridan LB, Yates VM, Benson DM, Stickney JL, Robinson DB (2014) Electrochim Acta 128:400–405
134. Qu D, Tao Y, Guo L, Xie Z, Tu W, Tang H (2015) Int J Electrochem Sci 10:3363–3371
135. Abdelhafiz A, Vitale A, Joiner C, Vogel E, Alamgir FM (2015) ACS Appl Mater Interfaces 7:6180–6188
136. Yuan Q, Takakusagi S, Wakisaka Y, Uemura Y, Wada T, Ariga H, Asakura K (2017) Chem Lett 46:1250–1253
137. Gokcen D, Bae SE, Brankovic SR (2010) J Electrochem Soc 157:D582–D587
138. Bi X, Wang R, Ding Y (2011) Electrochim Acta 56:10039–10043
139. Kiani A, Fard EN (2009) Electrochim Acta 54:7254–7259
140. Hossain MA, Cummins KD, Park YS, Soriaga MP, Stickney JL (2012) Electrocatal 3:183–191
141. Kazemi R, Kiani A (2012) Int J Hydrogen Energ 37:4098–4106
142. Sheridan LB, Gebregziabiher DK, Stickney JL, Robinson DB (2013) Langmuir (29):1592–1600
143. Sheridan LB, Kim YG, Perdue BR, Jagannathan K, Stickney JL, Robinson DB (2013) J Phys Chem C 117:15728–15740
144. McCurry DA, Kamundi M, Fayette M, Wafula F, Dimitrov N (2011) ACS Appl Mater Interfaces 3:4459–4468
145. Zhang G, Kuang Y, Liu J, Cui Y, Chen J, Zhou H (2010) Electrochem Commun 12:1233–1236
146. Thambidurai C, Gebregziabiher GK, Liang X, Zhang Q, Ivanova V, Haumesser PH, Stickney JL (2010) J Electrochem Soc 157:D466–D471
147. Fang JS, Sun SL, Cheng YL, Chen GS, Chin TS (2016) Appl Surf Sci 364:358–364
148. Venkatraman K, Gusley R, Yu L, Dordi Y, Akolkar R (2016) J Electrochem Soc 163:D3008–D3013
149. Venkatraman K, Dordi Y, Akolkar R (2017) J Electrochem Soc 164:D104–D109
150. Rettew RE, Guthrie JW, Alamgir FM (2009) J Electrochem Soc 156:D513–D516
151. Qin SJ, Zhao Y, Peng F, Chen XQ, Pan GB (2015) RSC Adv 5:92757–92761
152. Motsoeneng RG, Modibedi RM, Mathe MK, Khotseng LE, Ozoemena KE (2015) Int J Hydrogen Energ 40:16734–16744
153. Benson DM, Tsang CF, Sugar JD, Jagannathan K, Robinson DB, El Gabaly F, Cappillino PJ, Stickney JL (2017) ACS Appl Mater Interfaces 9:18338–18345
154. Yu Y, Hu Y, Liu X, Deng W, Wang X (2009) Electrochim Acta 54:3092–3097
155. Gebregziabiher DK, Kim YG, Thambidurai C, Ivanova V, Haumesser PH, Stickney JL (2010) J Crystal Growth 312:1271–1276
156. Deheryan S, Cott DJ, Muller R, Timmermans M, Heyns M, Vereecken PM (2015) Carbon 88:42–50

157. Fischer AE, McEvoy TM, Long JW (2009) Electrochim Acta 54:2962–2970
158. Bélanger S, Stevenson KJ, Mudakha SA, Hupp JT (1999) Langmuir 15:837–843
159. Zhang L, Huang H, Xia Y, Liang C, Zhang W, Luo J, Gan Y, Zhang J, Tao X, Fan HJ (2017) J Mater Chem A 5:5905–5911
160. Liu Y, Gokcen D, Bertocci U, Moffat TP (2012) Science 338:1327–1330
161. Ahn SH, Liu Y, Moffat TP (2015) ACS Catal 5:2124–2136
162. Liu Y, Hangarter CM, Garcia D, Moffat TP (2015) Surf Sci 631:141–154
163. Zhao F, Xu Y, Mibus M, Zangari G (2017) J Electrochem Soc 164:D354–D361
164. Wang R, Bertocci U, Tan H, Bendersky LA, Moffat TP (2016) J Phys Chem C 120:16228–16237
165. Cherigui EAM, Sentosun K, Bouckenooge P, Vanrompay H, Bals S, Terryn H, Ustarroz J (2017) J Phys Chem C 121:9337–9347
166. Guo L, Searson PC (2008) Langmuir 24:10557–10559
167. Guo L, Searson PC (2010) Electrochim Acta 55:4086–4091
168. Guo L, Oskam G, Radisic A, Hoffmann PM, Searson PC (2011) J Phys D: Appl Phys 44:443001(1–12)
169. Allongue P, Maroun F (2006) Curr Opin Solid St M 10:173–181
170. Allongue P, Maroun F (2006) J Phys Condens Matter 18:S97–S114
171. Möller FA, Kintrup J, Lachenwitzer A, Magnussen OM, Behm RJ (1997) Phys Rev B 56:12506–12518
172. Gündel A, Devolder T, Chappert C, Schmidt JE, Cortes R, Allongue P (2004) Physica B 354:282–285
173. Morin S, Lachenwitzer A, Möller FA, Magnussen OM, Behm RJ (1999) J Electrochem Soc 146:1013–1018
174. Cagnon L, Gündel A, Devolder T, Morrone A, Chappert C, Schmidt JE, Allongue P (2000) Appl Surf Sci 164:22–28
175. Maroun F, Reikowski F, Di N, Wiegmann T, Stettner J, Magnussen OM, Allongue P (2018) J Electroanal Chem 819:322–330
176. Lachenwitzer A, Vogt MR, Magnussen OM, Behm RJ (1997) Surf Sci 382:107–115
177. Kuo YC, Yen PY, Chen W, Chen S, Yau SL (2013) Electrochim Acta 112:831–837
178. Jurca HF, Damian A, Gougaud C, Thiaudière D, Cortès R, Maroun F, Allongue P (2016) J Phys Chem C 120:16080–16089
179. Cagnon L, Devolder T, Cortes R, Morrone A, Schmidt JE, Chappert C, Allongue P (2001) Phys Rev B 63:104419(1–12)
180. Bubendorff JL, Beaurepaire E, Mény C, Bucher JP (1988) J Appl Phys 83:7043–7045
181. Lecadre F, Damian A, Braems I, Maroun F, Allongue P (2015) Surf Sci 631:135–140
182. Pinheiro ALN, Zei MS, Luo MF, Ertl G (2006) Surf Sci 600:641–650
183. Lin CL, Wu AW, Wang YC, Tseng YC, Tsay JS (2013) Phys Chem Chem Phys 15:2360–2367
184. Sibert E, Ozanam F, Maroun F, Behm RJ, Magnussen OM (2004) Surf Sci 572:115–125
185. Yen PY, Chen S, Tu HL, Wu H, Yau SL (2011) J Phys Chem C 115:23802–23808
186. Schindler W, Koop T, Kirschner J (1999) Z Phys Chem 208:93–106
187. Vaz CAF, Bland JAC, Lauhoff G (2008) Rep Prog Phys 71:056501(1–78)
188. Schindler W, Kirschner J (1997) Phys Rev B 55:1989–1992
189. Schindler W, Koop T, Hofmann D, Kirschner J (1998) IEEE Trans Magn 34:963–967
190. Poduska KM, Morin S (2003) Rev Sci Instr 74:4723–4727
191. Mangen T, Bai HS, Tsay JS (2010) J Magn Magn Mater 322:1863–1867
192. Allongue P, Maroun F (2007) Electrodeposition of two-dimensional magnetic nanostructures on single crystal electrode surfaces. In: Staikov G (ed) Electrocrystallization in nanotechnology. Wiley-VCH, Weinheim, Germany, pp 217–241
193. Allongue P, Maroun F, Jurca HF, Tournerie N, Savidand G, Cortès R (2009) Surf Sci 603:1831–1840
194. Topolovec S, Krenn H, Würschum R (2015) Rev Sci Instr 86:063903(1–6)
195. Schindler W, Schneider O, Kirschner J (1997) J Appl Phys 81:3915–3917

196. Bubendorff JL, Beaurepaire E, Mény C, Panissod P, Bucher JP (1997) Phys Rev B 56:7120–7123
197. Gündel A, Morrone A, Schmidt JE, Cagnon L, Allongue P (2001) J Magn Magn Mater 226–230:1616–1617
198. Dézsi I, Fetzer Cs (2007) Electrochem Commun 9:1846–1849
199. Borges JG, Prod'homme P, Maroun F, Cortès R, Geshev J, Schmidt JE, Allongue P (2006) Physica B 384:138–140
200. Prod'homme P, Maroun F, Cortès R, Allongue P, Hamrle J, Ferré J, Jamet JP, Vernier N (2007) J Magn Magn Mater 315:26–38
201. Péter L, Bakonyi I (2011) Electrodeposition as a fabrication method of magnetic nanostructures. In: Nasirpouri F, Nogaret A (eds) Nanomagnetism and spintronics. World Scientific, Singapore, pp 89–120
202. Bubendorff JL, Mény C, Beaurepaire E, Panissod P, Bucher JP (2000) Eur Phys J B 17:635–643
203. Tournerie N, Engelhardt AP, Maroun F, Allongue P (2012) Phys Rev B 86:104434(1–5)
204. Di N, Kubal J, Zeng Z, Greeley J, Maroun F, Allongue P (2015) Appl Phys Lett 106, 122405(1–4)
205. Topolovec S, Krenn H, Würschum R (2016) J Magn Magn Mater 397:96–100
206. Geshev J, Gündel A, Zaharieva I, Schmidt JE (2012) Appl Phys Lett 101:132407(1–4)

Chapter 5
Compositionally Modulated and Multilayered Deposits

5.1 On the Electrodeposition of Nanolaminated Materials

Compositionally modulated alloys (CMA) exhibit an intentional composition modulation along the surface normal (i.e., in the growth direction). In general, any coating with uneven composition can be classified as CMA. The CMA coating will be called hereinafter a *multilayer* when the interface between the layers of distinct composition is relatively sharp and no graded composition zone occurs along the growth direction with a thickness that is comparable to that of either of these layers. For such deposits, the term *compositionally modulated multilayer coating* (CMMC) is commonly found in the literature. It has to be noted that deposits with layers of alternating structural properties are also called multilayers, even if composition modulation does not necessarily occur here. In such cases, mostly the crystallite size of the deposit is the modulated parameter.

With the development of the field, various reviews were published either with a general approach [1–5] or focusing on a particular physical property [6] of the multilayered deposits. Sections 5.2 and 5.3 will present the basic deposition methods of CMA materials, while their specific properties will be detailed in Sect. 5.4.

5.2 Multiple-Bath Methods

5.2.1 Common Features of the Multiple-Bath Methods

The advantage of the application of multiple-baths for multilayer electrodeposition is that the chemical environment in each deposition step is completely unrelated to that of the other steps. The application of fully independent solutions for the deposition of each layer in a multilayer structure yields a special degree of freedom that makes it possible to optimize each single bath used separately, should it be used to produce a metallic element or an alloy. It can be particularly important to use the

L. Péter, *Electrochemical Methods of Nanostructure Preparation*,
Monographs in Electrochemistry, https://doi.org/10.1007/978-3-030-69117-2_5

appropriate additives for each type of layers, which are highly specific for the metal to be deposited.

Since the contamination of either of the baths used for depositing a layer with the components of the other bath(s) is highly undesired, the electrode has to be rinsed between the deposition steps. Rinsing is accompanied with a loss of the plating bath (called drag-out) as well as with the production of a large volume of waste solution. The large amount of waste solution produced is the common feature of all multiple-bath methods, should they be based on an alternating electrode immersion or a solution flow.

5.2.2 *The Sequential Immersion Method*

Historically, the dual bath technique was the first electrochemical method to produce metallic multilayers [7], although the thickness of the constituent layer was in the micrometre range in the first trials. The basis of this technique is that the cathode is moved between separated solutions and, hence, it is exposed to the baths containing the ions of various metals alternatingly. This chapter summarizes the methods applying stagnant solutions between which the cathode is transferred.

The mechanical transfer of the sample makes the system required quite complex. A manual sample transfer imposes a very tight limit for the number of layers to be produced, even though the composition of the layers (and also that of the bath) may vary arbitrarily, allowing to plate a large variety of desired layer permutations. The available thickness of the layers was assessed to be at least 25 nm when the dual bath technique was elaborated and automated [8], but nowadays much lower layer thickness values also appear in dual immersion-plated samples [9–11].

The electrical control mode for the sequential immersion method is exclusively galvanostatic. This can be rationalized with the simplicity of the cell construction as well as with the fact that baths are usually optimized for a specific current density rather than for a particular electrode potential.

It has long been believed to be a disadvantage of the alternating immersion methods that the sample being transferred cannot be held under potential control, which may lead to the oxidation of the sample in the ambient atmosphere. Although it is theoretically possible to provide an inert gas atmosphere for the entire sample transfer process (see the subsequent immersion method in Chap. 4.2.1), it is seldom applied if the layers are thicker than a single atomic layer. This means that layers with as small a thickness as a few nanometres can hardly be protected sufficiently upon the transfer step, and the impact of the environment can be the more significant, the thinner the layer.

Even if the oxidation in the ambient atmosphere is eliminated, another type of corrosion process may occur when the sample is transferred into the solution of the salt of the more noble metal. Taking the example of Co and Cu, when the Co-covered specimen is immersed into the Cu^{2+}-containing solution, the exchange reaction (also termed as displacement or cementation) takes place. The process can be formulated as

$Co + Cu^{2+} \rightleftarrows Cu + Co^{2+}$, which clearly leads to a deviation of the layer thicknesses as well as their ratio from their designed values. Hence, the synchronization of the current source applied for the deposition of a layer with the immersion process is recommended if the deposit can undergo a cementation reaction.

It is to be emphasized that the displacement process can take place also when the surface is not fully covered with the more noble metal and its deposition rate as measured with the current density is smaller than that allowed by the mass transport of the reagent (in the above example, Cu^{2+}). The Cu^{2+} ions not consumed by the current applied can freely react with the uncovered Co, hence superimposing a cementation reaction onto the deposition. Further examples for this phenomenon and the thoughtful application of the cementation reaction will be given in Sect. 5.3.2.

The view on the possible corrosion damage during the sample transfer was, however, strongly denied by the authors of an early work of the field dealing with Ni/Cu multilayers [12]. Both Cu and Ni proved to be relatively inert during the rinsing and immersion processes, and the possibly forming monolayer-thick oxide did not prevent an epitaxial growth of the subsequent layer (which might also be caused by the reduction of these oxides when the cathodic current pulse was applied in the next bath). As small as 5 nm thick layers were achieved by retaining a continuous layer structure as seen in the cross-sectional TEM images [12]. The displacement of Ni did not seem to be a problem for multilayers produced for mechanical studies [13]. However, it has to be noted that the resistance of both Cu and Ni to both corrosion and cementation is rather exceptional among the metal pairs considered so far.

A few examples of the cross-sectional imaging of a few dual bath-plated multilayers can be seen in Fig. 5.1. In the two cross-sectional images of the Ni/Cu multilayers, it can be seen that the relative undulation of the layers increases as the periodicity decreases. This will later be seen also for multilayers obtained with the single-bath method.

Originally, the alternating immersion method is rendered to be a tool for multilayer deposition with fairly large layer thicknesses where the influence of the layer degradation during the sample transfer does not impact significantly the topmost layer exposed to the atmosphere. For producing deposits with micrometre-scale periodicity and with a limited number of bilayers (<10 repeat periods), the alternating immersion method can be used economically and is applied still nowadays [14]. The application of the dual bath technique for corrosion research is quite widespread. The thickness of the constituent layers of a multilayered deposit does not have to be in the nanometre range but can increase up to several micrometres, although thinner layers down to about 20–40 nm proved to be optimal. This case when the less noble metal (or the one that is not preferentially deposited) is required as a pure metal without any impurity. Typical layer sequences in anticorrosion coatings are Zn/Co [15], Zn/Ni [16–18] or Zn–Fe/Ni [19].

In magnetization-related studies, one of the motivations was to study layer compositions that are difficult or impossible to obtain with the single-bath method. In one group of such materials, a Co–Zn alloy makes either the magnetic layer (in Co–Zn/Cu multilayers [9]) or the non-magnetic layer (in Co/Co–Zn multilayers [10, 11]). Another pursuit was to obtain $Ni_{80}Fe_{20}$ (Permalloy) as a constituent without

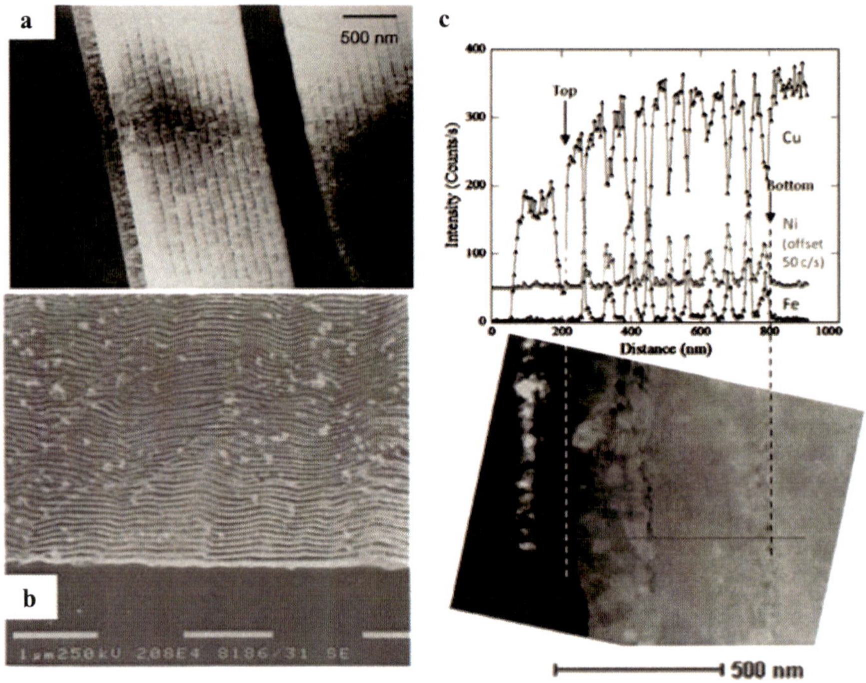

Fig. 5.1 Examples for the observed layer structure of a few samples prepared with the dual bath method. **a** TEM image of a CoPtP/Cu multilayer with 100 nm and 20 nm CoPtP and Cu layer thicknesses, respectively [20]. **b** Cross-sectional SEM image of a Ni/Cu multilayer obtained by using a sulphamate and a sulphate bath for Ni and Cu deposition, respectively; neither of the baths contained growth-modifying additives [12]. The scale bar is 1 μm. **c** Cross-sectional TEM image (bottom) and the corresponding EDS map (top) of a NiFe/Cu multilayer with approximately 100 nm periodicity. The EDS map indicates the parallel change of the Fe and Ni signals as well as their counteroscillation with the Cu signal [21]. Republished from the papers indicated in the figure references. Copyright (2004), (1994) and (2011) for **a**, **b** and **c**, respectively; with permission from Elsevier (**a** and **c**) and from The Electrochemical Society **b**. Permission conveyed through Copyright Clearance Center, Inc.

a Cu impurity in Ni–Fe/Cu multilayers [21]. In the latter case, the optimization of the small coercivity of the permalloy layer could be spoiled with the Cu impurity, which would certainly be present in the alloy of the less noble metal if the laminated structure is deposited from a single bath. In these magnetization-related studies, the layer thicknesses reported were much lower than those in the typical multilayers obtained with the dual bath method (e.g., d(Co–Zn) = 12 nm and d(Cu) = 3 nm [9]; d(Co) = 4 nm and d(Co–Zn) = 3 nm [10, 11]). This was necessary for making it possible to compare the thickness obtained from the Faraday low with that deduced from magnetization measurement. The periodicity below 10 nm was low enough so that multilayer satellite peaks could be observed for these multilayer samples in the X-ray diffractograms, although the peak intensity was very small. It was only a

Ni–Fe/Cu multilayer for which a TEM study was published, indicating a periodicity of about 50 nm (see also as Fig. 5.1c) [21]. It was found for each system containing Co–Zn layers that the intermixing at the layer interfaces has a large impact on the magnetization behaviour of the samples. Ni/Cu multilayers could also be obtained with the sequential immersion method, which was the subject of a ferromagnetic resonance study [22].

In the deposits containing both Co and Pt, the common goal is to achieve a high magnetic anisotropy and a perpendicular magnetization [23, 24]. Since pure Co cannot be deposited in a bath containing Pt, the sequential immersion approach can be verified. In a study of Co/Pt deposits, the mean Co layer thickness was varied in nearly monoatomic steps [23], which is a true challenge in the sequential immersion method. High-quality cross-sectional TEM images showing the layer structure could be obtained for repeat periods of at least 30 nm [24]. The magnetic properties were measured as a function of both the mean Co layer thickness and the Co deposition potential. The anisotropy energy of the deposit varied in the same manner as MBE-grown counterparts (i.e., linearly increasing anisotropy energy with decreasing Co layer thickness), although the anisotropy energy values were different [23].

For the deposition of Co–Pt–P/Cu multilayers, the aim was to produce a thick deposit containing the hard-magnetic Co–Pt–P layer by retaining its perpendicular magnetic anisotropy [20, 25]. Without an "interrupting" layer, the growth of the hard-magnetic layer with a large thickness is not possible because of the grain coarsening and the development of a texture with in-plane easy magnetization direction (hence loosing the perpendicular anisotropy). Here, the materials properties of the Cu spacer layer were not of importance but the Cu layer merely served as a fresh substrate for the fine-grained Co–Pt–P deposition.

There are various other fields where the multiple-bath method gained some role. In the case of Bi–Te/Bi–Se multilayer thin films with about 100 nm layer thicknesses, the individual Bi_2Te_3 and Bi_2Se_3 crystal structures were both observed [26]. Although the Seebeck coefficient of the laminated structure was not sensitive to the periodicity, the electrical conductivity could be tuned, which impacts the figure of merit of the deposit as a thermoelectric material. As the Bi–Te/Bi–Se multilayer structure cannot be obtained from a single solution, this is also the case for metals that are typically deposited from ionic liquids. Since codeposition from ionic liquids is yet fully unexplored, metals plated from such media can be used for multilayer build-up with the sequential immersion method only. The trial of obtaining Cu/Ta multilayer showed an incomplete layer structure but was certainly an innovative step towards obtaining non-conventional multilayered structures [27].

The possibility of plating more than two layer types in the multiple immersion method was seldom exploited. An example is shown in the attempt of replacing electroplated $[Cu/Sn]_N$ interconnects with a $[Cu/Ni/Sn/Ni]_N$ quadruple repeating structure for modifying the intermixing properties of the easily miscible Cu/Sn interface [28]. This study showed that the presence of a Ni layer of about 70 nm between the Cu and Sn layers gives rise to the formation of a ternary nanolayer due to the room-temperature ageing of the layered system.

An interesting combination of the double-bath and single-bath methods was applied for plating [Cu/Co/Cu/Ni–Fe]$_N$ multilayers [29]. The two baths used were as follows. One bath was optimized for the deposition of the Ni–Fe layer to obtain the low-coercivity material $Ni_{80}Fe_{20}$ (Permalloy) without either Co or Cu impurity. The typical thickness of this layer was only 2.8 nm. The second bath contained the Co and Cu salts from which pure Cu and a Co-rich Co–Cu alloy could be deposited. The Cu layer thickness was 3.6 nm, and it could protect the Co-rich layer during the immersion process. The major goal was the deposition of well-separated magnetic layers with strongly different coercivity.

5.2.3 Cell Configurations with a Rotating Cathode

The alternating immersion variant of the multiple-bath method can be improved by avoiding the removal of the cathode from the liquid and retaining the alternating contact with the varying plating baths at the same time. Such a method was implemented by the application of a rotating cathode disc in a cell with two plating chambers. The substrate was turned into the way of electrolyte jets in the order *Bath 1—Rinsing solution and wiper blade 1—Bath 2—Rinsing solution 2 and wiper blade 2* [30, 31]. This rotating cathode configuration is shown in Fig. 5.2. The residence time of any part of the cathode in a particular solution is fixed, and the current passed through the cathode in each phase of the process can be controlled with two independent current sources. The cross-talk between the two current sources can be diminished if the wiper blades efficiently provide a physical separation of the two solutions (which prevents their intermixing, too). If both current sources are operated continuously, the currents applied in each phase of the process also determine the

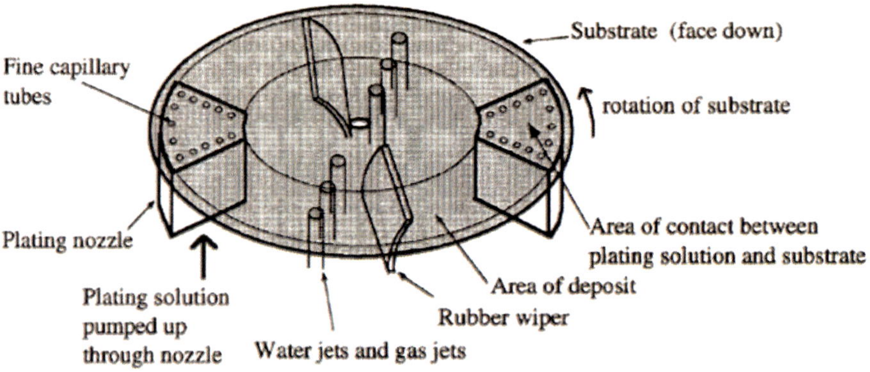

Fig. 5.2 Working electrode, plating nozzle, substrate cleaning jet and wiper arrangement of a rotating ring electrode used for multilayer deposition. Republished with permission of The Electrochemical Society, Inc. from Ref. [31]. Permission conveyed through the Copyright Clearance Center, Inc.

thickness ratio of the subsequent layers, while the thickness of each layer is determined by the residence time above the particular solution and the current applied at that position. Another means of the thickness regulation is the application of an off-time period at either of the plating positions, and the alternating operation of the current sources is also possible.

Since the electrode is in contact with some of the solution during the entire process (apart from the short periods when it passes above the separator of the cell chambers), the inertness of the atmosphere can be ensured more easily than with the subsequent immersion method. Nevertheless, the wiping blades between the cell chambers exert a mechanical force onto the deposit that can lead to an unwanted wear load.

The typical rotation rate of the rotating electrode system presented in Fig. 5.2 was 10–30 rpmin. It could be used successfully to plate Ni/Cu and Ni/NiP$_x$ multilayers, but its application for Co/Cu multilayers failed. The reason obviously lies in the high inclination of the relatively aggressive copper electrolyte to displace cobalt deposited in the previous rotation phase of the cathode.

A rotating cylinder-type multichamber deposition apparatus was also reported [32]. Here, the cathode is a cylinder with vertical axis, and the sections aligned along the surface of the cylinder in the direction of rotation include polishing chambers after both deposition steps. The operation principles reported were the same as for the rotating disc system (simultaneous or alternating on-time periods for the two deposition chambers, variation in each current density), and even the typical rotation rate was in the same range. The advantage of the inclusion of the grinding chambers was that the roughening of the deposits with the increase in the thickness could be efficiently counteracted, and shiny deposits up to a millimetre thickness could be achieved for Ni/Cu multilayers. Polishing during the preparation of the deposit improved the layer structure significantly, even though the process was optimized for about 100 nm minimum layer thicknesses.

5.2.4 *Bath Change by Using Flow Cells With Multiple Solution Inlets*

It is often desired to retain the independence of the baths used for depositing each layers and providing well-controlled electrode surface conditions at the same time. All these conditions can be fulfilled by applying a setup presented earlier in Fig. 4.1 (including the permanent potential control with appropriate potential steps when solutions are exchanged, exclusion of the mixing of the baths, absence of oxygen etc.). Although the electrolyte pumping apparatus shows a large commonality, the systems used to plate layers thicker than a monolayer also have different features. When the layer to be produced is plated in a process which is not surface limited, the concentration of the electrolyte solution is normally much larger than in the case when monolayers are produced. This also means that the solution exchange in the flow cell leads to a significant loss of electrolyte solution, even if the pumping is stopped for the deposition periods (which were reported to make the optimization of the deposition of the layers easier [33]). Therefore, the electrochemical cells used are thin-layer cells, similarly to the EC-ALD studies.

Secondly, the choice of the electrode potential during the deposition of the subsequent layers is different from that applied in the EC-ALD system. When a UPD layer is formed, the strong interaction between the alternating atomic layers ensures that the stripping of the layers takes place at much more positive potentials than the deposition potential of either of these layers. However, in multilayer systems, one of the constituent layers usually has a much more negative onset potential of dissolution than the onset potential of the deposition of the other metal. This is the case, for instance, for the deposition of a Co/Cu multilayer in which the Co dissolution can take place about 500 mV more negative potential than the onset of Cu deposition (in solutions containing no complexing agents). This means that the cathode potential always has to be negative enough to avoid the dissolution of the least noble metal, whichever deposition potential would be acceptable for the other metal on an inert substrate. The synchronization of the electrode potential change with the solution exchange is a delicate task for variable-bath deposition systems. Nevertheless, such optimization for nanometre-scale layer thicknesses has not been reported so far in detail in any available publication.

It is clear from the nature of the multiple-bath methods, should they be implemented with either an alternating cathode immersion or a solution exchange, that one does not have to take into account any interference of the reagents that are not present near the electrode at the same time. Therefore, if metallic elements form the subsequent layers, the codeposition mode of these elements is of no importance, and either of the layers can be produced as pure metal.

Examples for the flow cell-based dual bath electrodeposition include the plating of Ni/Cu and Ni/Fe [33] and Co/Cu [34] multilayers. The lowest layer thickness tested for the Ni/Cu and Ni/Fe systems was 50 nm, and the only observation made was the occurrence of the natural crystal structure of each layer type in the X-ray diffractograms [33]. During the preparation of the Co/Cu multilayers, the capability of the alternating solution method was fully exploited and the lowest nominal layer thickness was 1 nm [34]. Although the layer structure was assessed to be imperfect by the authors in the latter case, the composition modulation was clearly shown by the magnetoresistance effect measured for the layered samples.

5.3 The Single-Bath Method

5.3.1 Impact of the Codeposition Mode of the Components on the Composition Modulation During Two-Pulse Plating

The single-bath method is a very widespread tool for obtaining compositionally modulated coatings. The large number of reports on this method does not allow a complete overview of each single-layer combination and bath classification. Instead, the principles of the method will be discussed in depth, and representative examples will be presented.

The basic tool of the multilayer deposition is the so-called two-pulse plating, in which either the potential or the current is varied between two fixed values. The current is cathodic during both pulses. The features of the codeposition modes discussed here are elucidated with most attention on two-pulse plating, but their consequence on many other deposition modes is straightforward.

When the goal is the deposition of an alloy, it is generally desired that the deposition potential of the alloy constituents is brought close enough, which is mostly done by the application of a selective complexing agent that is preferably attached to the ions of the more noble (MN) metal, hence decreasing its deposition potential. For the single-bath deposition of a layered system where a pure MN layer is required, however, it is necessary that the components can be deposited in distinct potential intervals. This means that from the viewpoint of the difference between the onset potentials of the deposition of the components, the chemistry of the solutions used for plating alloyed and layered systems must fulfil different criteria.

If all components of a CMA coating are present in a single solution, the prerequisite of obtaining a composition modulation is that the deposit composition must vary with one of the plating parameters. In principle, the dependence of the composition on the temperature or on the solution agitation intensity could also be used for composition modulation. While the change of temperature cannot be instantaneous, the modulation of either the flow rate or the electrode rotation was seldom applied for obtaining CMA deposits. However, the transient width between the subsequent layers in flow-modulated deposits is weakly predictable. Therefore, the nearly exclusive manner of the electrochemical composition modulation is based on the variation of the *electrical parameter* of the deposition, which can be either the current density or the electrode potential. It has to be emphasized that the regulation of these two parameters is not always interchangeable even if their interrelation is mutually unambiguous (which means that the polarization curve of the system is monotonous). Hence, the choice of the appropriate regulation conditions requires a deep understanding of the electrochemical background.

The composition of an alloy usually varies with the current density, regardless of the codeposition mode of the components. The mole fraction of the preferentially deposited component(s) increases when the current density decreases. This is true for all codeposition modes included in the composition diagram in Fig. 2.21, but the rule is also fulfilled for induced codeposition. For induced codeposition, the induced metal can be classified as the preferentially deposited one in accord with the above definition, at least in the current density range where continuous deposit layers can be obtained.

Figure 5.3 shows schematically the polarization behaviour of a system with two metals exhibiting the regular codeposition mode. Since the nobility scale is a thermodynamic order of the standard electrode potential of the metal ion/metal systems, the order of the partial polarization curves may somewhat vary because they are also influenced by kinetic factors. It is fulfilled in all cases that the deposition of the less noble (LN) metal takes place at a more negative potential than that of the MN metal. If the layers to be deposited have to differ significantly in their composition, the concentration of the ions of the MN metal has to be small enough so that its deposition becomes mass transport limited already at the onset potential of the deposition of the LN metal. In this case, the composition can be calculated as indicated in the figure and in the figure caption. If the deposition efficiency is smaller than 1, j_{dep} should be the effective current density.

Line 3 in Fig. 5.3 indicates a $j(E)$ curve that can be commonly recorded for systems of regular codeposition by using a cathodic-going linear sweep on an inert electrode. If the deposition of the LN metal on the MN one takes place with a nucleation barrier, the codeposition starts at more negative potential than the reversible potential of the

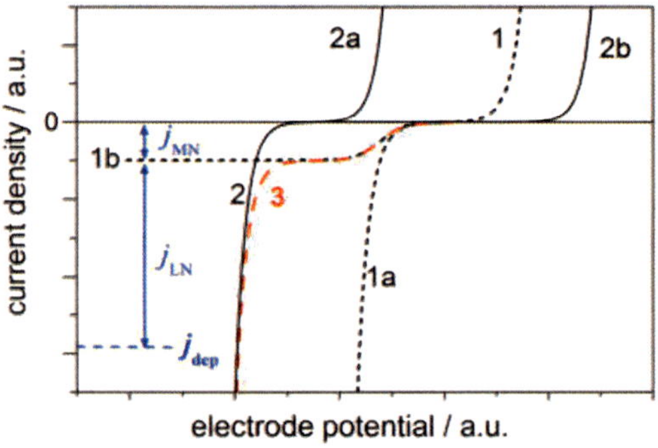

Fig. 5.3 Relative position of the parts of the polarization curves of a binary metal system exhibiting regular codeposition. **1**: Anodic partial polarization curve of the MN metal; **1**a: Cathodic partial polarization curve of the MN metal at high concentration of its ion (no mass transport control); **1**b: Cathodic partial polarization curve of the MN metal at small ion concentration when a diffusion-limited deposition can take place (plateau region). **2**: Cathodic partial polarization curve of the LN metal; **2**a: anodic partial polarization curve of the LN metal if the deposition is nearly reversible (the exchange current density is high); **2**b: anodic partial polarization curve of the LN metal if the dissolution is hindered (the deposition is not reversible). **3** (red line): Cathodic polarization curve of the mixed system when the concentration of the ions of the MN metal is small (diffusion-limited deposition) and the nucleation of the LN metal on the MN one is not hindered. Arrows and the corresponding current density notations indicate the relationship of the current densities and the deposit composition: is $y(\mathrm{MN}) = j(\mathrm{MN})/\{j(\mathrm{MN}) + j(\mathrm{LN})\}$ if the current efficiency is 100%

LN metal. In this case, the increase in the cathodic current density is faster than that expected from the exponential rate law until the previously deposited MN metal is fully covered with the LN one.

The anodic behaviour of the system after reversing the sweep direction is not displayed because it substantially depends on a number of factors such as (i) the relative position of the anodic partial polarization curves of the two metals, (ii) the thickness of the MN metal on the deposit which may form when the deposition of the LN metal stops in the anodic-going sweep. Below the trends concerning the above-mentioned factors will be given.

In many cases, the onset of the dissolution potential of the LN metal is more negative than that of the MN metal. This is the case for systems such as Co/Cu and Co/Pd (in sulphate baths), Fe/Ag and Co/Ag (in perchlorate baths) and Ni/Cu (in chloride baths). However, the onset of the dissolution of the LN metal can be more positive than that of the MN metal if the LN metal is relatively inert; this is very typical for Ni/Cu and Ni/Ag systems when halogenide ions are not present (i.e., sulphate and perchlorate baths, respectively) and also occur for the Co/Pb system (in an acetate bath).

It cannot be stressed enough that when cyclic voltammetry is used for the characterization of the deposition and dissolution process, the elucidation of the anodic-going sweeps is much more difficult than that of the cathodic-going ones. The reasons for the difficulties are diverse and range much beyond the relative positions of the partial polarization curves of the constituents [35]. When the anodic-going sweep (usually the 2nd segment of the CV curve) is run, the potential reaches again a value where the MN metal is deposited alone. Depending on how easily the MN metal can cover the previously deposited layer of graded composition (where composition grading comes from the variety of electrode potentials scanned before), the dissolution of the LN metal may or may not remain hidden. The "covering ability" of the MN metal is generally the larger, the slower is the sweep, the less negative is the cathodic limit of the polarization, the larger is the concentration of the MN metal and the stronger is the structural coherence of the deposit. As it was shown for the deposition of Co/Cu multilayers, the cyclic voltammetry is often insufficient to find the right deposition potential of the MN metal of multilayered samples [35]. Similar results were achieved when the quartz crystal microbalance was used to minimize the unwanted dissolution of the LN metal [36].

It is very important to clarify whether the current or the potential control (i.e., galvanostatic /G/ or potentiostatic /P/ mode, respectively) is expected to yield a better result during multilayer deposition. Let us start the analysis with the deposition of the LN metal layer. Here, the deposition can be optimized for either of the electrical control modes. In the G mode, the layer composition can be easily calculated from the current density, as it was described for Fig. 5.3. When the LN layer is deposited in the P mode, at least two additional factors are to be taken into account. First, the current may vary in time as the concentration of the LN metal ions in diffusion layer near the cathode decreases. This results in a graded composition, which may be undesired. Secondly, the potential control requires a well-defined cell geometry. The position of the reference electrode is particularly important due to the Ohmic drop (see also Chap. 2.7). Since the current density for the deposition of an LN metal-rich layer is typically high, the resulting current is sensitive to the reference electrode position that determines the ratio of the interfacial potential drop at the cathode and the Ohmic drop occurring in the solution. The current density is determined merely by the interfacial potential drop. However, the accurate correction of the Ohmic drop is seldom carried out in works related to preparative electrochemistry, which may lead to a weak reproducibility if the conditions are not properly reported.

Concerning the deposition of the MN metal layer, the viewpoints for the choice of the regulated parameter and its setup value are different. When the anodic partial polarization curve of the LN metal is like the line 2b in Fig. 5.3 (i.e., the LN metal is relatively inert), the deposition conditions for the MN metal can be selected in a wide range of parameters, either in the G or the P mode. In this case, the dissolution of the LN metal during the deposition of the MN one will not occur. The current density can be set to any value lower than the diffusion-limited one; or, alternatively, the electrode potential can chosen in the interval of the onset of the deposition potentials of the two metals.

However, when the LN metal is not inert in the potential interval of the deposition of the MN metal, the deposition conditions of the MN metal have to be chosen very accurately so that the thickness of the LN metal can be designed with the help of the charge passed in the corresponding deposition pulse. The current control in this case is not an option, especially when the MN metal layer is to be obtained in a pure form. If the current density is too small, the LN metal dissolves; when it is too large, the LN metal is codeposited with the MN one. If potential control is applied, the "neutral point" of the LN metal can be found where the partial polarization curve of the LN metal crosses the abscissa. Here, the MN metal deposition is the only reaction, regardless of the relative position of the polarization curves.

The first method to find the potential of no reaction of the LN metal was the measurement of the equilibrium potential of the LN metal in the solution used for multilayer deposition but without the ions of the MN metal [37]. While this method proved to be satisfactory for a single-component LN metal, it cannot guarantee that the equilibrium potential of the LN metal is the same as the rest potential of the same metal in a partially depleted solution. The depletion always occur when the deposition of the LN metal is just finished, and the rest potential may vary as the mass transport leads to the alleviation of the metal ion concentrations. Therefore, a more complex optimization method was suggested on the basis of the observation of the potentiostatic current transient during the pulse for the MN metal deposition [35]. The chronoamperometric records are compared for various potentials applied in the pulse for the deposition of the MN metal (and meanwhile, the pulse parameters for the deposition of the LN metal were constant). A demonstrative collection of the chronoamperometric records is shown in Fig. 5.4. The potential optimum for the deposition of the MN metal from the viewpoints of both the conservation of the previously deposited LN metal and sharpness of the resulting interface is when the anodic transient is as short as possible, and the diffusion-limited current density sets in without observing higher cathodic current density values. The method of the analysis of the chronoamperometric transients was originally applied for the Co/Cu [35] and

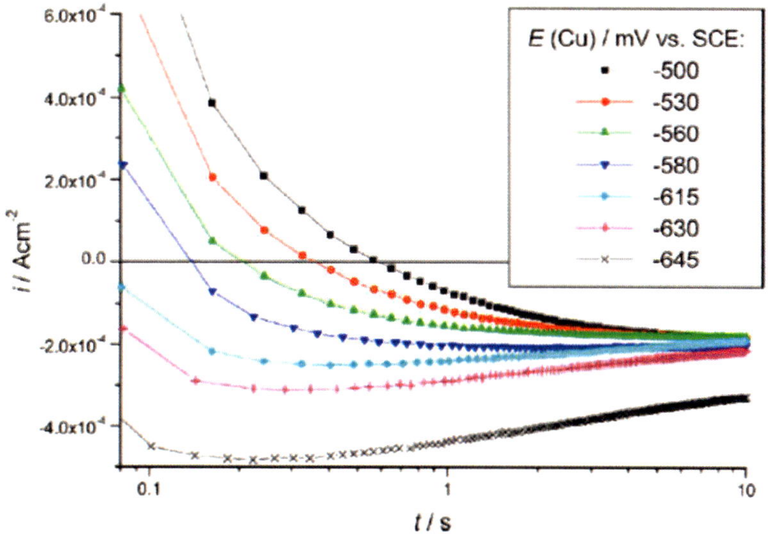

Fig. 5.4 Chronoamperometric transients recorded during the potentiostatic Cu deposition pulse after the deposition of a Co-rich Co–Cu layer. At $E = -645$ mV, the current density transient does not return to the diffusion-limited current density, which means that the Co codeposition continues throughout the potentiostatic Cu pulse. At $E = -630\ldots -615$ mV, the Co deposition does not stop at the beginning of the Cu pulse (the current density is higher than the diffusion-limited value), but later the Co deposition stops as the Cu layer covers the Co grains and the nucleation barrier does not allow any further Co codeposition. The indication of the disappearance of the Co from the layer being deposited is the return of the current density to the diffusion-limited value of the Cu deposition. At $E = -580$ mV, there is no cathodic transient and the anodic one is the shortest possible; this value can be accepted as the ideal Cu deposition potential. At $E \geq -560$ mV, the charge corresponding to the anodic transient keeps increasing, indicating a more and more significant Co loss due to its anodic dissolution. Reprinted from [35]. Copyright (2004), with permission from Elsevier

later adopted for several systems such as Co/Ag [38], Ni–Co/Cu [39], Ni–Fe/Cu [40] and the Ni–Co–Fe/Cu [41–43].

It can be seen from the above discussion that the deposition of the MN and LN metals can be optimized with different electrical control modes, i.e., with P and G modes, respectively. Such letter combinations will be used in the following chapters for identifying the deposition modes. The G/G mode offers the simplest setup (no reference electrode is necessary), while the unwanted reactions of the LN metal cannot be excluded. The P/P mode enables one to optimize the deposition conditions of the MN metal, but it also requires a good care for ensuring the reproducibility of the deposition of the LN metal. The G/P mode (indicating first the deposition mode of the high-current pulse) ensures the best conditions, but many electrochemical instruments cannot provide a seamless change between the G and P modes. The application of the G/P mode was first demonstrated by using a current limiter connected to the electrochemical cell and applying formally a P/P mode for the regulation of the potentiostat [44]. This attempt was followed shortly by the application of a potentiostat/galvanostat device that was suitable for switching the operation mode within a usual instrumental transient time (at most a few tens of microseconds) [45].

Concerning the other deposition modes (i.e., anomalous or induced), the polarization curves usually show no characteristic features (i.e., plateau) which could be indicative of the deposition rate of either of the components. Since the alloying elements in these composition modes are codeposited in the entire range of current density in which a coating can be produced at all, the measurement of the *composition–current density* relationship has to be determined experimentally. It has to be emphasized that, as it will be described in Sect. 5.4.1, a transient may also take place when the current density (or electrode potential) is modified, which is expected give rise to a difference between the composition of a nanolayer and that measured at otherwise identical deposition condition for a bulk deposit. No detailed methodical papers are available on whether the change in the deposition conditions from one layer type to the other leads to any unwanted composition change (i.e., a partial dealloying/dissolution when the high-current pulse ends and a low-current pulse begins). Hence, the choice of the electrical parameter used for the deposition seems to be less crucial for the anomalous and the induced deposition mode than for the regular one.

5.3.2 *Multilayer Formation with Single-Pulse Plating and Displacement*

The principle of the displacement-combined single-pulse method is essentially the same as for the SLRR process. The driving force of both the SLRR and the displacement-combined single-pulse deposition method is the difference in the standard potential of two metal ion/metal systems, which results in that the oxidized form of the more positive system can displace the reduced form of the more negative redox pair. The important differences are that during the multilayer formation, the sacrificial layer is much thicker than a single atomic layer, its deposition during the current pulse is not surface area-limited, the sacrificial layer is displaced by the other metal only partly so that this layer conserves its integrity when thinned by the displacing metal, and the thickness of the layer formed from the displacing metal is not limited to a monolayer either.

The concentration ratio of the ions of the LN and MN metals is similar to that applied in the SLRR processes. However, the concentration values themselves are larger. This is because a continuous layer from the LN metal can form in a short time only in the case when the concentration of its ions is high enough. The high concentration of the salt of the LN metal is necessary also because during its deposition, the mass transport control and the concomitant inclination to the formation of dendrite are undesired. For this reason, the concentration of the salt of the LN metal ranges from a few tenths of mol/litre to a nearly-saturated solution of its salt. In contrast, for achieving both a moderate rate of displacement and a small proportion of the MN metal codeposited together with the LN one, the solution is rather dilute with respect to the salt of the MN metal. In practice, this means that the concentration of the ions of the MN metal is between a few and a few tens of millimole/litre.

The structure, the mean composition and the occurrence of the composition modulation in pulse-plated materials depends on the following parameters: (i) The amount of the material deposited during a single pulse; (ii) the kinetics of the displacement process, including the mass transport of the ions of the displacing metal in the solution; (iii) the off-time between the pulses. For achieving a composition modulation along the growth direction, the following criteria are to be met: (i) The deposit thickness achieved in a single pulse should be at least several atomic layers, and the deposit must indeed form a layer. If the deposit forms segregated grains, the displacement may result in a composition modulation, but the composition gradient does not coincide with the growth direction; instead, it will result in a granular material. (ii) The displacement reaction must be kinetically possible, which can be ensured with an appropriate choice of the anions present. Additives often impact the kinetics of the displacement, while they may also incorporate into the deposit [46]. (iii) The off-time should be long enough so that a layer accumulation can take place, even in the case when the concentration of the displacing material is small in the electrolyte solution (which is the typical case). This means that, unlike in the case of the pulse plating processes of alloys when the off-time is usually at most a few tens of milliseconds, the off-time for displacement can range up to several tens of second. (iv) The displacement process has to lead to a nearly even consumption of the displaced metal and an even accumulation of the displacing metal. This condition is the harder to fulfil because both the local cell effects and the nucleation barrier may favour the separation of the dissolution and the deposition sites. It is always a concern whether the LN metal slowly dissolves at the pinholes of the MN metal cover layer [47]. If the coverage is even and no pinholes occur, the displacement process may become self-terminating, and the electrode potential returns during the off-time to the rest potential of the MN metal in the plating solution [48]. The morphological features of the displacement process may depend on the bath pH, too [49]. Cases when the pulse plating process combined with displacement lead to granular materials will be discussed in Chap. 7.

The displacement-based nanostructure synthesis processes always apply metals among which at least the most noble one (which will then displace the other(s)) is regularly codeposited with the rest. This makes it possible to measure the diffusion-limited deposition rate of this metal and compare it with the rate of displacement.

When the displacement process is kinetically not hindered, the displacement process can be taken as a superposition of the diffusion-limited deposition of the displacing (MN) metal and the dissolution of the displaced (LN) metal at the same rate [50]. This mechanism can be verified with the independent measurement of the diffusion-limited current density of the LN metal. A diffusion-limited mechanism was found for Cu displacement by Ag [47, 51] and Ni displaced by Cu [50]. When the LN metal is an alloy, the displacement can take place in a diffusion-limited manner concerning the MN metal deposition, but the displacement modifies the composition of the LN layer due to the selective displacement of the least noble component. This was found for the Ni–Co–Fe–Cu quaternary system where Cu^{2+} displaced Fe and Co and left Ni nearly unaffected [52]. The key point of the mechanism test is usually the comparison of the deposit composition with calculations either allowing or excluding the exchange process. The diffusion-limited displacement mechanism was found to be valid also for reverse pulse plating of Ni–Cu CMA coating [50]. In such a case, the rate of the dissolution of the LN metal is the sum of the reverse current density applied and the diffusion-limited current density of the MN metal.

The displacement-based multilayer deposition method was applied for the following systems (indicating the metals in the order of the displaced/displacing ones): Cu/Ag [47, 51], Ni/Cu [44, 48, 50, 53–56], Co/Cu [49, 54, 57, 58] and Ni–Co–Fe/Cu [52]. The Ni/Cu and the Co/Cu systems are worthwhile of comparison because they show distinct differences. The maximum amount of Ni displaced by Cu was found to be 4–5 atomic layers [44, 53] or even below the detection limit of the method applied [54], which is attributed to the relatively passive nature of Ni and its good alloying ability with Cu, hence leading to a conformal surface Cu layer. However, Co displacement does not lead to a self-limited Cu thickness, which may cause a significant porosity and the formation of a three-dimensional structure instead of well-ordered layer system [49, 54, 57, 58].

In spite of the fact that some physical properties of the deposits (like magnetoresistance of the Co/Cu and Ni/Cu samples) show the sign of the composition modulation at the nanometre scale, the occurrence of the true layer structure was evidenced with some direct imaging method only for layer thicknesses larger than 50 nm [57, 58]. Although the composition change of the deposit is a clear evidence of the displacement and the reaction can be followed with a quartz crystal microbalance experiment [55], we have no clear picture on the morphology of many displacement-processed deposits.

5.3.3 *Multilayer Formation with Various Electrical Waveforms*

If the single-bath method is used with a metal pair with regular codeposition, the concentration ratio of the LN and MN metals in usually high varying between 7 and 300. After finishing the high-current pulse, the deposition of the MN metal is certainly mass transport limited. If the mass transport-limited deposition is maintained during the low-current pulse, it may unfavourably impact the deposit morphology. When the relative position of the polarization curves is like 1a and 2a in Fig. 5.3, there is obviously no opportunity to deviate from the mass transport-limited deposition in the pulse leading to the deposition of the MN metal since the LN layer will be at least partly etched away. However, if the relative position of the partial polarization curves is like 1a and 2b, there is more freedom to choose the deposition conditions of the MN metal.

This deposition condition adjustment opportunity was exploited for the deposition of Ni in Ni/Cu multilayers produced for various purposes [44, 59, 60]. The common approach is to include a pause after the deposition of the Ni layer so that the concentration of the Cu^{2+} ions can increase (while possibly some degree of cementation may also take place), then the Cu deposition pulse can be adjusted in accord with the achievement of the desired sample property [44, 59]. Instead of a pause with no regulation of the electrode potential, a potentiostatic pulse was also applied with more positive potential than the onset of the Cu deposition [60]. While cementation in this case is not possible, it was speculated that this pulse may lead to the formation of a monoatomic oxide layer on Ni that facilitates the Cu deposition later, but this explanation has not obtained clear evidence.

If the dissolution of the LN metal takes place at more negative potentials than the onset of the deposition of the MN metal (curves 1a and 2a in Fig. 5.3), a pause after the deposition of the MN metals leads to displacement. However, the application of a third pulse after the deposition of the MN metal layer was assumed to be suitable for the fine-tuning of the thickness of the MN layer. This idea was tested for Co/Cu multilayers where the accomplishment of the Cu coverage at small nominal Cu thicknesses cannot be achieved with the conventional two-pulse plating. Therefore, after the deposition of a sufficiently thick Cu layer, a dissolution pulse was applied with a smaller current density than that of the diffusion-limited one of the Cu deposition, leading to a G/P/G pulse sequence [61]. The result was that the samples were relatively rich in Cu as compared to their nominal composition, which may stem from two effects: on the one hand, the dissolution of the Co layer during the anodic pulse; and, on the other hand, the excess Cu deposition with Co at the high-current pulse, which was made possible by the accumulation of the Cu^{2+} ions near the cathode during the dissolution pulse. As a whole, the G/P/G pulse sequence with anodic dissolution after the Cu deposition did not lead to a breakthrough in either the continuity of the layers or in their magnetic properties. It is to be noted that similar reverse pulse techniques are occasionally applied in the preparation of other nanolaminated coatings (like Ni_xW_{1-x}/Ni_yW_{1-y} [62]), although the reasoning was rather empirical and the reverse pulse was rationalized by the improved surface morphology without offering a detailed mechanistic picture.

Concerning the four-pulse method, it gained importance for magnetic/non-magnetic multilayers only where either the thickness or the composition of the magnetic layers varied, leading to two types of magnetic layers for odd and even number magnetic layers, respectively. This multilayer type is called the pseudo-spin-valve structure. In the four-pulse method, a cathodic current is applied, regardless of the polarization mode (G or P). The majority of the works relied on the Co/Cu system [63–65]. Here, the thickness of the magnetic layers was modulated while the current density during the deposition of the Co-rich layer was high and its Cu content was kept below a few atomic per cent. However, when the magnetic components of the LN layer were Ni and Fe, the modulation of the current density was of high importance since it regulated the ratio of the magnetic elements, hence leading to different coercivity [66]. More complicated waveforms were also tested for multiplying one of the types of the magnetic layer, while they were always separated with Cu layers. However, the desired sample properties were mostly obtained for a small number of repeat periods where the roughness increase with thickness did not counteract the desired magnetic effect.

5.3.4 Multilayer Formation with Transport Rate Modulation

Electrodeposition of a multilayer is possible with d.c. current if the transport rate in the electrolyte solution is modulated. Although the lateral homogeneity can hardly be ensured, the transport rate modulation can be carried out with the periodic operation of a simple stirrer. A demonstration of this method showed [67] that Co–P alloys with varying composition can be deposited by this manner. Since the flow rate around the cathode cannot be changed instantaneously due to the finite viscosity of the solution, sharp interface cannot be expected. The individual layer thickness achieved with this method was a few tens of nanometres.

A cell type of even cathode accessibility used for electrodepositon with transport rate modulation is based on a disc-shaped working electrode facing with a porous filter of same surface area [68]. The distance between the working electrode disc and the porous filter has to be much smaller than the diameter of the working electrode itself. The flow direction is from the pre-chamber of the cell towards the working electrode. The counter electrode can be found in the pre-chamber. If the product of the counter electrode reaction is harmful for the proper operation of the working electrode, an additional separated off-stream chamber can be provided. The schematic image of the important parts of the cell is shown in Fig. 5.5. If the aspect ratio of the cell is small, the primary current distribution is even, and so is the mass transport, therefore, the electrode configuration was named as the *uniform injection cell* (UIC). There is some analogy with the rotating disc electrode (RDE), and roughly speaking, the role of the rotation angular velocity for the rotating disc electrode is taken over by the flow rate in the thin-layer cell. The analytical expression derived for the mass-transfer-limited current density in the flow cell is analogous to the Levich equation valid for RDE, and the physico-chemical properties of the solution play

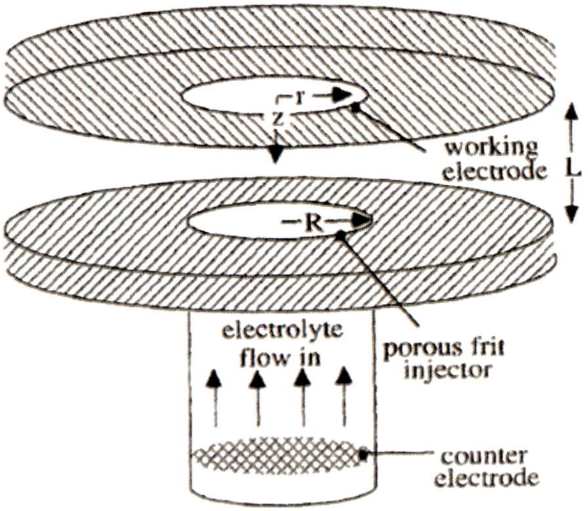

Fig. 5.5 Electrode and solution inlet configuration of the uniform injection cell. The aspect ratio of the cell has to be small for achieving a uniform primary current distribution ($L/2R \ll 1$) Republished with permission of The Electrochemical Society, Inc. from Ref. [68]; permission conveyed through Copyright Clearance Center, Inc.

the same role in the two cases [68]. The verification of the theoretical description of the cell was experimentally confirmed for steady-state cell operation [69]. The mathematical description was given for high-amplitude modulation of the flow rate including non-linear terms [70], which was found in agreement with the experiments concerning the limiting current of a single electrode reaction [71].

The UIC was used for depositing Ni–Fe CMA coatings by modulating the flow rate [72]. Both galvanostatic and potentiostatic conditions were tested, leading to the obvious result that the free (i.e., non-regulated) electrical parameter varied as the flow rate is varied between two discrete values. The decay of the non-regulated electrical parameter was comparable to the flow pulse length applied; hence, it can be assumed that the interface between the layers deposited at particular solution flow rates is not sharp. The composition modulation wavelength (Λ) was inversely proportional to the flow rate modulation frequency (ν) at a fixed current density, in accord with the formula below:

$$\Lambda = \frac{V_{\mathrm{M,av.}}\ \eta\,|j|}{2F\,\nu} \tag{5.1}$$

where the $z = 2$ substitution is due to the Me^{2+} form of the ion of both constituent metals, η is the current efficiency and $V_{\mathrm{M,av.}}$ is the average molar volume of the deposit. The smallest modulation wavelength achieved was 55 nm. The periodicity of the flow-modulated deposits was checked with a stripping analysis, but neither in-depth composition analysis nor TEM observation of the layered structure was given.

The rotating ring-disc electrode (RRDE) configuration has also been tested for the deposition of CMA structures [73, 74]. In both studies, a sinusoidal rotation rate modulation was applied, for which the complete theoretical background was published later [75]. In the composition modulation, the frequency of the rotation rate modulation is dominant, while the higher-order terms have relatively little weight. In the experimental studies [73, 74], Ni–Fe deposition was tested with rotation rate oscillation frequency of a few hundred millihertz. The smallest composition oscillation wavelength observed with potentiostatic stripping analysis was around 7 nm. Electron microscopic images and space-resolved EDX analysis was given for much larger wavelengths only. The 7-nm minimum wavelength observation was only partly due to the limited resolution capability of the stripping analysis method. The codeposition characteristics of Ni and Fe also led to decreasing mole fraction oscillation amplitudes as the modulation periodicity decreased, which made it nearly impossible to resolve the composition oscillation for rotation rate modulation frequency above 0.2 Hz. CMAs other than Ni–Fe were not tested with any flow modulation technique.

Instead of modulating the solution flow (where the electrode configuration is crucial for achieving a laterally even deposit), the transport rate can be modulated with the periodic operation of an ultrasonicator. If the sonication head of the ultrasonic device provides uniform energy dissipation in the solution, the transport rate modulation can be even along the electrode surface. This method has been demonstrated for protective coating composed of Ni–Co [76] and Ni–Sn [77] alloys. Due to the relative simplicity of the ultrasonic agitation, this method is foreseen to have bright prospects, especially if the layer thicknesses do not have to be maintained highly uniform.

5.4 Properties of Electrodeposited CMAs

5.4.1 Composition Variation in Ultrathin Electrodeposited Alloy Layers

Metal deposition as ultrathin layers discussed in Chap. 4.3 referred mostly to pure metals. In that case, the metal ion concentration was higher than a few millimole per litre, and the deposition of the ultrathin layers was not mass transport controlled. The choice of single-component deposits in the implementation of the observation technique (such as *in situ* STM or magnetometry for metal deposition) is a natural choice in general. Ultrathin alloy layers discussed in Chap. 4.3 were produced with extremely dilute solutions, and the deposition of both components was diffusion-controlled from practically the formation of the first atomic layer. In this chapter, a feature of the alloy deposition process will be presented that would make it very difficult to characterize nanoscale alloy layers produced with conventional plating solutions due to the uncertainty in the composition and in the component distribution of the resulting deposits. First, regular codeposition will be characterized, and then the composition depth profile generally occurring for any alloy plating system will be given.

The peculiar feature of the regular codeposition is that the MN metal can be deposited without the contamination of the LN metal. Hence, if the deposition of an LN/MN metal pair is started with a fixed current, the MN metal will be deposited first as long as its transport rate is high enough for accounting the current applied (see also Fig. 2.21). For estimating at what time the deposition of the LN metal has to start due to the depletion of the MN metal, the Sand equation (valid for a stagnant solution) can be applied:

$$t = \frac{\pi D z^2 F^2 c^2}{4 j^2} \tag{5.2}$$

where j is the total current density, and D (diffusion coefficient) and c (concentration) refer to the properties of the ions of the MN metal.

Sand equation indicates that for a regular codeposition system, there must be an initial layer during the galvanostatic deposition where the coating is not an alloy but is composed of purely the MN metal. Since the concentration of the MN metal ions in an alloy plating solution exhibiting regular codeposition is small and the current density is quite high, the thickness of such an initial layer is little and was not the subject of a direct observation. However, if pulse plating is applied, such a MN metal layer occurs for the initial period of each single pulse, which has an impact on the mean composition of the deposit [48]. The partial current density of the MN metal at the beginning of the pulse can be even 100%, which is followed by a decay whose rate is a function of many parameters (including the ratio of the concentration of the metal ions, that of their transport rate and the current density).

Concerning the general aspects of the unintentional composition variation during alloy plating, it was shown already in the 1960s that the mean composition of electrodeposited Fe–Ni alloys depends on the deposition time [78–80]. The reason why the composition change was discovered on Fe–Ni alloys was that the deposition of zero-magnetostriction Fe–Ni films, often called Permalloy, has been an important goal of electroplating for decades. The thickness of the near-substrate zone with a significant composition variation was assessed to be around 150–200 nm [79]. Although the observation of the mean deposit composition as a function of the deposition time is an integral method, it remained an important approach until nowadays [81, 82]. However, it is straightforward that the change of the overall composition with thickness must stem from the gradual change of the local composition. Therefore, a nanoscale composition-modulated alloy is formed unintentionally even during d.c. deposition.

Sputtering-based in-depth composition analysis methods (that are also referred to as depth profiling) are seldom suitable to detect subtle near-substrate composition changes. Sputtering-based methods always induce some roughening, and the roughness of the actual surface during the sputtering process is not predictable since it depends on the initial surface roughness and the sputtering-induced effects. Therefore, the surface during the sputtering-based analysis process practically always differs from that occurring during the deposition, which makes the composition changes far from the substrate much smeared out. Although scanning EDS for thinned samples during a TEM study can yield a composition profile with quite high resolution, this technique is seldom used, at least by far not routinely.

Near-substrate composition profile of electrodeposited samples can be obtained with a high resolution if two conditions are fulfilled. First, the substrate has to be as smooth as possible; second, the deposit has to be removed from the substrate with no damage. Once the deposit is separated from the smooth substrate, the sputtering-based depth profile analysis can be started from the "back side" of the sample, hence studying the near-substrate zone without any roughening artefact.

The technique of the sample preparation for sputtering from the reverse direction was developed by the author of this work. In a nutshell, the process involves the following steps. The substrate is a polished silicon wafer with sputtered adhesion and conducting layers (which can be Cr and Cu with as small thicknesses as 5 and 20 nm, respectively). Then, the deposit of interest is plated onto the substrate. If needed, an additional nickel supporting layer can be plated onto the deposit of interest. Then, the silicon wafer is scratched at the back side and gently broken, which makes it possible to peel off the deposit. Further details of the procedure are available in the original works [83–87]. The films thus removed from the substrate with a thickness of a few nanometre can be studied with a sputtering-based analysis method, preferably with secondary neutral mass spectrometry (SNMS) in which the sample is entirely in the vacuum chamber of the device (and hence, it is not required to withstand against a pressure difference, unlike in glow discharge devices).

The result of the studies performed for alloy deposits obtained from stagnant solutions was that the preferentially deposited component(s) of the alloy accumulated in the near-substrate zone of the deposit. This can happen because at the moment when the deposition starts, the solution composition near the cathode surface is identical to the bulk solution composition. However, as the deposition proceeds, the depletion influences the near-substrate concentration of the ions of the preferentially deposited metal to the largest extent. As the steady-state composition profile is achieved for the precursor ions, the deposition rate of the preferentially deposited metal drops. This trend proved to be valid for both regular and anomalous deposition modes of binary and ternary alloys, and the near-substrate zone in which the composition stabilization of the deposit takes place was at most 200 nm. A few typical composition depth profile curves of various alloys is the near-substrate zone is shown in Fig. 5.6. For binary alloys, there is a maximum in the mole fraction of the preferentially deposited component near the substrate. In contrast, for ternary alloys, there are subsequent maxima in the mole fractions in the order of the deposition preference of the components, and the decay of the spontaneous composition change often has a complicated character with various oscillation patterns.

The appropriate countermeasures for avoiding the spontaneous composition modulation near the substrate include either the thorough control of the hydrodynamics during the deposition [87] or the application of pulse plating instead of constant current [86]. In both cases, the initial composition change in the deposit is minimized, although the reasons for the result are different. If the cathode is rotated, the depletion of the solution with respect for the precursor ions is much depressed,

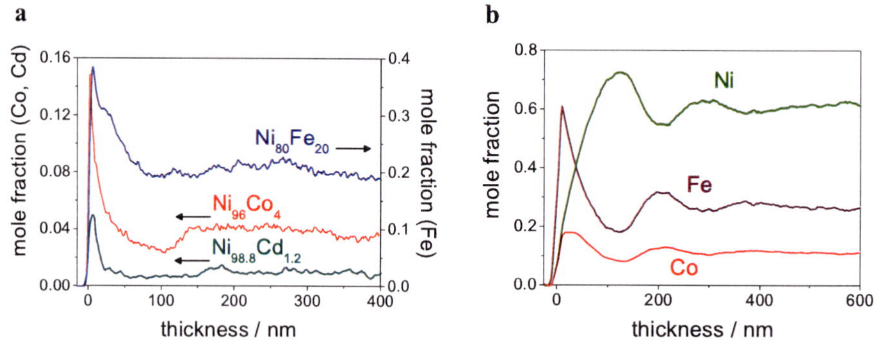

Fig. 5.6 Reverse composition depth profile of various d.c.-plated alloys near the substrate as measured with SNMS. **a** Ni–Cd, Ni–Co and Ni–Fe binary alloys (mole fraction of Ni is not shown); **b** Fe–Co–Ni ternary alloy. For all curves, zero thickness indicates the substrate/deposit interface [88]

hence allowing a much larger transport rate than for a stagnant solution. If pulse plating is applied, the transport rate of the metal ion is not enhanced; however, a long enough time period is provided so that the same metal ion concentrations are achieved by spontaneous diffusion as for the previous pulse, and this will lead to identical conditions for each pulse.

The revelation of the occurrence of the spontaneous variation of the deposit composition of d.c.-plated samples near the substrate has serious consequences on the study of nanoscale layers. First, it is obvious that graded composition deposits are obtained in most cases, which does not make it possible to relate any sample property to the composition as a single variable. Secondly, the spontaneous variation in the deposit composition bans the extrapolation of the composition of bulk deposits to small layer thicknesses. For instance, it is very convenient to analyse the composition of plated coatings of several micrometre thickness with EDS. However, one has to be aware that the composition of a nanoscale alloy layer produced technically under the same circumstances will not show the same composition as the thick deposit, and, furthermore, the composition will not be even constant at the nanoscale if d.c. plating is applied with uncontrolled hydrodynamics.

Although the zone with varying thickness shown in the above graphs ranges to a few hundred nanometres, the fast depletion of the cathode diffusion zone with respect to the preferentially deposited metal has a direct impact on nanoscale multilayers, too. As an example, Fig. 5.7 can be brought up. Here, Co–Ni–Cu/Cu multilayers

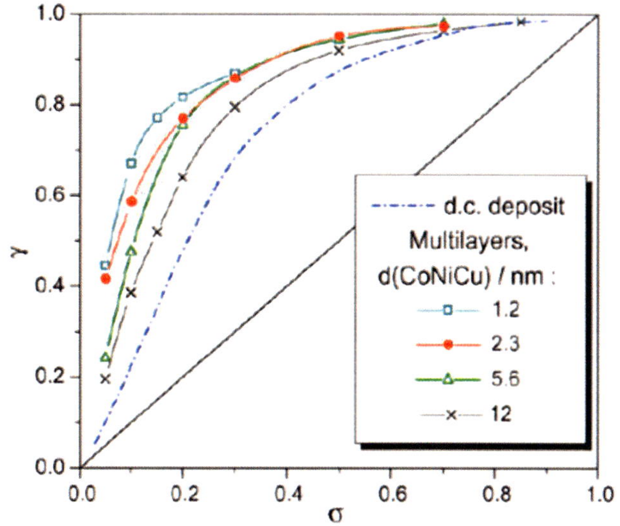

Fig. 5.7 Composition of Co–Ni–Cu/Cu multilayers prepared from various solutions with varying Co–Ni–Cu layer thicknesses. Notations: $\sigma = c(\text{Co}^{2+})/\{c(\text{Co}^{2+}) + c(\text{Ni}^{2+})\}$, $\gamma = y(\text{Co})/\{y(\text{Co}) + y(\text{Ni})\}$. Compare this graph with Fig. 2.21 and observe that only the Cu concentration/mole fraction were omitted for obtaining a quasi-binary composition diagram for a ternary system. Reprinted from Ref. [89]. Copyright (2007), with permission from Elsevier

were produced with constant and varying Cu and Co–Ni–Cu layer thicknesses, respectively. When the Co/(Co+Ni) atomic ratio is displayed as a function of the $Co^{2+}/(Co^{2+}+Ni^{2+})$ ion ratio in the solution, the composition lines deviate the more from the d.c. composition line, the thinner the Co–Ni–Cu layer.

The explanation is very straightforward in the light of the above discussion. Namely, the Cu pulse can be taken as a "break" in the Co+Ni deposition, during which the ion concentration of the less noble metal ions can relax. The shorter the deposition pulse for the layer containing Co and Ni, the least the depletion of the preferentially deposited Co can impact the layer composition. Therefore, the analysis of the mean composition of a multilayer series with varying layer thickness can reveal much useful information on the deposition preference, and, indirectly, on the composition gradient within an alloy nanolayer.

The spontaneous variation in composition along the growth direction at sub-micrometre scale can cause long-range composition fluctuations, too, but the origin of the composition change in such a case is not related to an initial transient. It was demonstrated for electrodeposited Ni–P alloys [90] that nearly random composition fluctuations occur over several tens of micrometre. The explanation was the variation in the hydrogen evolution rate, its complex interrelation with the depletion of the cathode diffusion layer and the dependence of the hydrogen evolution rate on the deposit composition.

5.4.2 Structural Features of Deposits With Modulated Composition

It is very scarce that a conformal full monolayer is produced when the deposition of the subsequent layer starts. This means that the growth of each new layer starts with a nucleation process, also in the case when the growth is heteroepitaxial; i.e., when the crystalline structure does not necessarily change with the composition modulation. In order to provide a full coverage of the previous layer, the nuclei of the newly formed layer (that are often called islands when their thickness is much smaller than their lateral diameter) must coalesce. If the nominal layer thickness is around one monolayer of even less and the components do not segregate, the resulting deposit shows a multilayer-to-alloy transition [91].

The coalescence threshold limit is around 1 nm even for metal pairs that exhibit a lattice misfit of less than 5% but it can be much larger if the conditions of the heteroepitaxial growth are not fulfilled and there is a significant lattice misfit between the neighbouring deposit layers. If the nominal layer thickness is smaller than the threshold thickness of the formation of a full layer, the result of the deposition can be a granular material. Such a structural transition can be observed in the row of Co/Cu [92, 93], Co/Ag [38] and Co/Pb [94] deposits where the lattice misfit is about 2, 14 and 39%, respectively. Co/Cu multilayers show very nice layer structures over hundreds of nanometres along the direction parallel to the substrate with layer thicknesses slightly larger than 1 nm. Co/Ag two-pulse-plated deposits show much smaller ordered zones, and Co/Pb deposits are fully granular at any nominal layer thickness.

When layers with dissimilar composition are plated onto each other with repeat periods less than the crystallite size, the composition modulation is usually independent of the crystallite size and shape of the deposit (i.e., whether the deposit is composed of large columnar crystals or equiaxed nanograins). This was well exemplified for Ni/Cu multilayers [95]. For the Ni/Cu system, the choice of the Cu deposition current density (or electrode potential) has some freedom since the partial polarization curve corresponding to the nickel dissolution has the onset at more positive potential than that of the Cu (see curve 2.b in Fig. 5.3). Therefore, the grain structure could be regulated by choosing a relatively small or large Cu deposition current density, corresponding to columnar and fine-grained structure, respectively (see the scheme in Fig. 5.8). It is interesting to note that the independence of the grain structure and the composition modulation seems to be a general trend in electroplated samples. It was also found for Ni–Co–Fe deposits showing a spontaneous composition modulation occurring due to growth instability that no correlation can be seen between the grain structure and composition modulation fronts [96].

In the range of nanoscale layer thicknesses (d < 100 nm), we can still distinguish three regimes depending on the ratio of some particular distances to the layer thickness:

1. If the layer thickness is larger than the crystallite size of the material of either of the layers, heteroepitaxy does not play a role in the formation of the total layer structure. In such cases, the nucleation probability is relatively high (which means that the deposition is often inhibited with some additives). The properties

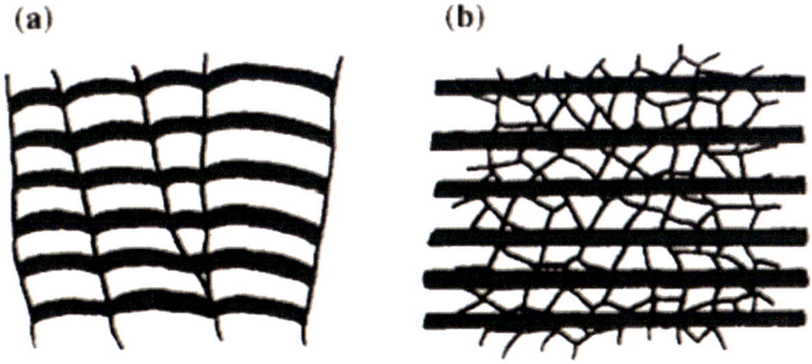

Fig. 5.8 Relative position of the composition modulation fronts and the grain boundaries in electrodeposited multilayers. Wide and nearly horizontal black and white strips refer to the subsequent layers of different composition, while thin lines correspond to the grain boundaries. The image was deducted from the structural study of Ni(20 nm)/Cu(10 nm) multilayers deposited from a citrate-sulphate bath, by varying the Cu deposition current density: **a** ~40% and **b** ~80% of the diffusion-limited current density. Reprinted from Ref. [95]. Copyright (1997), with permission from Elsevier

of the middle zone of the grains far from the layer interfaces do not differ significantly from those in a bulk deposit since the impact of the adjacent layers decays fast in the interfacial zone.

2. If the layer thicknesses are smaller than the crystallite size (at least along the growth direction) but are large enough so that the natural bulk lattice distances can manifest themselves in the middle of the layers, the heteroepitaxial growth is accompanied with an effective strain relaxation. The strain originates from the difference of the nearest neighbour distances of the bulk forms of the constituent layers. The stress-induced deformation is effective only in the close vicinity of the layer boundary.
3. When the layer thicknesses are small enough so that the natural lattice distances of the layers cannot be set in, oscillating stress and strain fields occur along the growth direction, which impact the growth of the coating.

For Case 1 of the above list, the multilayer growth is determined by the same factors as those being effective during any plating process. For Cases 1 and 2 of the above list, the X-ray diffraction pattern of the multilayer deposit shows the lines of the constituent layers independently of each other (depending on the texture, not all lines of the powder diffraction data may appear). For Case 3, however, the structural features of the individual layers do not manifest themselves, which leads to new features not seen at larger layer thicknesses. This occurs usually when the bilayer thickness is less than 20 nm [97], which will be called hereinafter as the "small layer thickness case".

If the layer thicknesses are small enough, the lattice mismatch is small and the deposition is assumed to produce truly planar layers perpendicular to the growth direction, the X-ray diffraction pattern of the multilayer changes as compared to the thick layer case. The individual diffraction lines of the layer types disappear and a single line are obtained for a particular reflection in the weighted average lattice distance of the bulk form of the constituent layers. This line merging is due to the very small thickness of the coherently scattering domains in the direction of the scattering vector. The analysis of the diffraction behaviour of a coherently grown multilayer system was accounted for in detail by Michaelsen [97], offering a thorough guideline for the diffraction study of any multilayer system.

Concerning the mean atomic distances for the "small layer thickness case", the Vegard law can be applied, which tells that the mean lattice parameter can be obtained as the weighted average of the relaxed lattice parameter of the constituent layers. However, a scrutiny of the structure tells that the in-plane and out-of-plane lattice distances are not necessarily the same. Since the atomic volume for an element can be taken constant regardless of the neighbourhood of a certain atom, the in-plane compression draws an out-of-plane dilatation and vice versa. The exact deformations can be calculated by using the relevant elastic constants. For the quantitative treatment, the reader is advised to consult some relevant literature resources [93, 98, 99]. Regardless of the stress effect, the Vegard law can be used for estimating the mean atomic distances in multilayers and the corresponding position of the lines in the XRD pattern.

Besides the merging of the diffraction lines, the secondary periodicity of the multilayer gives rise to new lines around the major diffraction peaks, which are called satellite peaks. The primary periodicity is related to the distance between the atomic planes, while the secondary periodicity is due to the composition modulation and the concomitant, though small, lattice plane distance variation from one layer to the other. The satellite peak positions are related to the modulation wavelength of the multilayer (Λ) in accord with the formula below:

$$\Lambda = \frac{\lambda}{\sin \Theta_{i,+1} - \sin \Theta_{i,-1}} \tag{5.3}$$

where λ is the wavelength of the X-ray used for the diffraction experiment, i is the order of the reflection and Θ is the angle at which the satellite peak maximum is observed. The two terms in the denominator of Eq. 5.3 refer to nearly symmetric peaks at higher and lower diffraction angles around the major diffraction line. For more details, see, e.g., Sect. 4.2 of Ref. [6]. The observation of the satellite peaks around the major diffraction line is a key tool for the assessment of the multilayer periodicity Λ with a non-destructive method for multilayers with $\Lambda < 10$ nm. A few examples for such satellite peak observations for electrodeposited multilayers are shown in Fig. 5.9. It has to be noted that satellite peaks can be observed for high-quality multilayers only where stacking and, consequently, the X-ray scattering of the layers are both coherent. Multilayers either deposited on a rough substrate or having a high undulation do not show satellite peaks.

It is commonly found that the periodicity as calculated from the satellite peak positions of the electrodeposited multilayers is either overestimated with respect to the nominal periodicity or it is significantly larger than that established with another method (e.g., EQCM weight change [100]). The explanation lies in the grain structure of the deposit. The local deposit thickness at the centre of the grains is the largest, while the growth near the grain boundary is retarded, as it is shown in Fig. 5.10. This results in a grain boundary grooving [101], which is peculiar for electrodeposited samples and does not occur for physically deposited ones (either sputtered or evaporated). Since the satellite reflection can originate only from areas where the layers are parallel to the substrate, the overestimation of the mean layer thickness calculated for the entire deposit can be easily understood.

Figure 5.10 also indicates the common observation that the roughness developing during the growth of multilayer with small layer thicknesses is *cumulative*. This means that within a grain, the deviation from the planar layer growth is the higher, the larger the distance from the substrate (faceting).

When the lateral diameter of grains is large, the multilayer growth is heteroepitaxial and the bilayer thickness is below 10 nm, the multilayer structure imposes a significant alternating stress field, which tends to relax at large sample thicknesses. One means of reducing the total mechanical energy of the system is that the composition modulation fronts will be inclined to the plane of the substrate, as it was shown in Fig. 5.10. This layer plane canting can be observed in the cross-sectional TEM

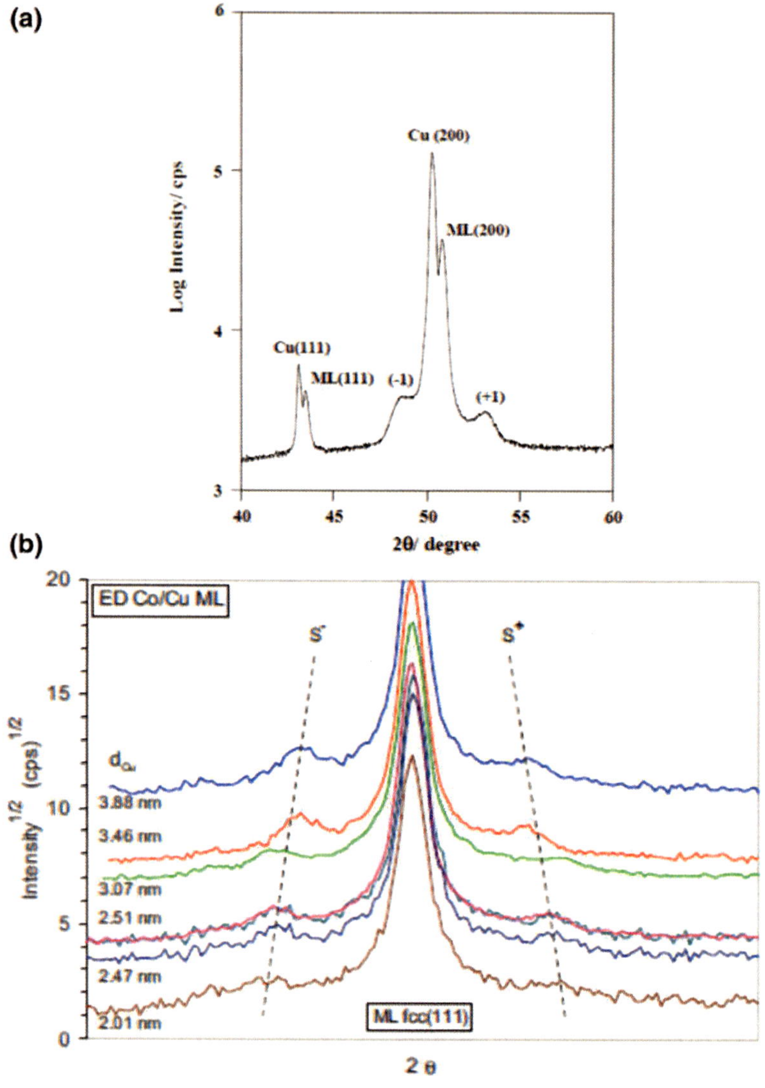

Fig. 5.9 Examples for multilayer satellite peaks on the X-ray diffractograms of electrodeposited samples. **a** Ni–Cu/Cu multilayer on Cu substrate. Peaks labelled with Cu and ML refer to the Cu substrate and the multilayer deposit, respectively. Reprinted by permission from Springer from Ref [102]. **b** Co–Cu/Cu multilayer on Si/Cr/Cu substrate where the substrate diffraction peaks do not appear in the region of the main multilayer peak. The dashed lines show the shift of the peak position with the Cu layer thickness at fixed Co layer thickness. Republished with permission of The Electrochemical Society from Ref [98] with permission conveyed through Copyright Clearance Center, Inc

Fig. 5.10 Schematic cross-sectional representation of the composition modulation in comparison with the grain boundaries in coherently grown electrodeposited multilayers. This figure is based on the observations published in Refs. [101] and [104]

images of various multilayers. Figure 5.11 shows two examples for different multilayers where the inclination of the composition modulation plane to the substrate plays a significant role in the formation of the overall sample structure.

It is to be noted that the sharpness of the composition modulation perpendicular to the layer planes is independent of the structural features of the deposit but depends rather on the deposition conditions and the difference in the composition of the neighbouring layers. The problem of the interface sharpness was studied mainly in the case of multilayers prepared with the single-bath method and small layer thicknesses (where a loss of sharpness can be detrimental for the desired physical properties). It was found in a detailed XRD study [103] that for multilayers with nearly pure layers the interfacial zone is as thin as 0.2 nm, which is in the order of the atomic distances in metals. Hence, the sharpness found was comparable to a thickness variation of one atomic plane. If the difference in the composition of the neighbouring layer is small (which can be achieved by choosing a too small current density for the deposition of the less noble layer), the structure of the multilayer may become rather ill-defined [105].

5.4.3 Mechanical Properties of Compositionally Modulated Deposits

Mechanical properties commonly tested are the internal stress, the tensile stress, the Young modulus, the hardness (mostly as defined by Vickers) and wear resistance. Although the variety of the mechanical properties is wide, essentially they are all related to the behaviour of dislocations. The major difference in the behaviour of bulk metals and multilayered ones is that in multilayers, the dislocations are pinned by the

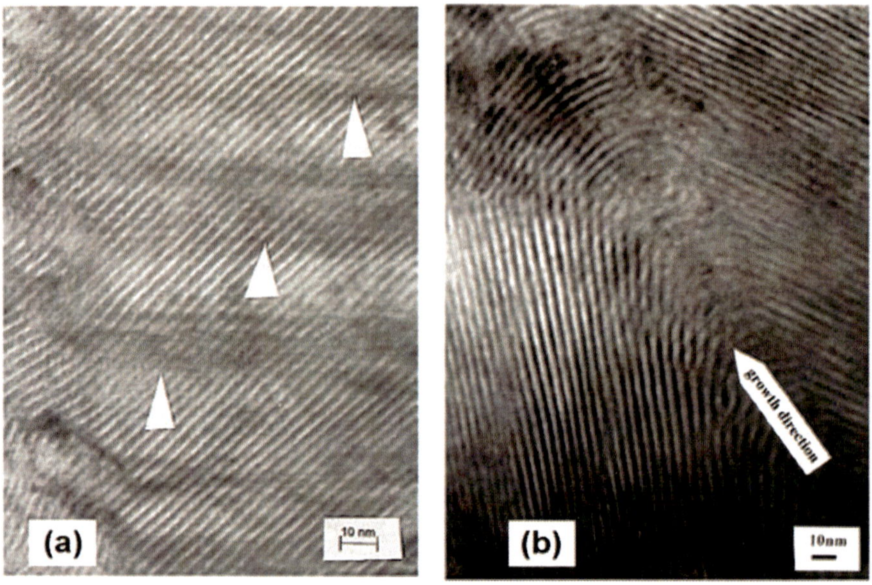

Fig. 5.11 TEM images indicating characteristic structural features in electrodeposited multilayers. **a** TEM picture of a part of a grain in a Ni–Cu(2.5 nm)/Cu(1.4 nm) sample [106]. The arrows indicate lamellar twins parallel to the substrate, while the composition modulation lines incline to both at about a 45° angle. **b** Cross-sectional TEM picture close to the top of a crystal column of a $Co_{95}Cu_5$(9.4 nm)/Cu(1.1 nm) × 1325 multilayer sample where the canting of the layer planes is the most characteristic [92]. Reprinted from Refs. [106] **a** and [92] **b**. Copyright (1998) and (2003), respectively; with permission from Elsevier

interface of the neighbouring layers of different properties. Since the propagation of the dislocation through the interface is hindered, the presence of the interface impacts how the atomic planes can slip onto each other. Starting from a bulk material and gradually thinning the constituent layers, an initial effect is that the mechanical properties are improved due to the dislocation pinning effect of the interfaces. However, at layer thicknesses smaller than about 20 nm, dislocations can no longer be embedded into one layer, which means that the advantageous factor of the CMA-type structure is lost and the mechanical properties deteriorate with a further decrease in the layer thicknesses. The physical description of the background of the deformation-related behaviour of multilayers is available in various reviews [107–111]. Although several properties of multilayered materials (wear and corrosion resistance, etc.) were found to be improved by the multilayer structure itself, some other features are highly specific for a restricted material family. For instance, an increase in deformation moduli called *supermodulus effect* can be measured for a few fcc/fcc metal pairs having (111) texture and individual layer thicknesses around 2 nm [112].

CMAs prepared from Ni and Cu take the majority of the studies dealing with mechanical properties [13, 59, 60, 100, 113–115]. Concerning the tensile stress in Ni/Cu multilayers [59, 60], the decrease in the multilayer periodicity resulted in an increase of the tensile stress. Although the actual values of the tensile stress increment as compared to the components of the multilayer are difficult to compare since the studies cited applied different thickness ratios (Ni:Cu thickness ratio was 9:1 [59] or 1:1 [60]), an approximately fourfold increment could be seen in both cases. The cross-sectional electron microscopic study of the Ni/Cu multilayer deposit as a coating on a bulk Cu tensile stress test specimen showed that the increased tensile stress was due to the barrier-type behaviour of the multilayered coating against the dislocation penetration upon the deformation [60]. The decrease in the tensile stress occurred at Ni(25 nm)/Cu(2.5 nm) and Ni(5 nm)/Cu(5 nm) for the two sample types. Vickers hardness was also found to increase with decreasing bilayer thickness, which has the same origin as for the tensile stress increment. The increment was found to be two to three times as compared to the pure Cu substrate [13, 60], which is a good agreement between the works applying different deposition methods (however, another study mentioned only 25% increment [114]). When it could be studied (i.e., when the single-bath method was used), the loss in hardness occurred at 10 nm periodicity [60]. When a gradient composition coating was compared with multilayered ones obtained under the same deposition conditions, the composition gradient was found to yield the best hardness values [115]. This was attributed to the unique misfit dislocation distribution in the gradient samples (where the dislocation distribution is nearly even, as compared to multilayered ones where they accumulate at the layer interface). The tribological behaviour of the Ni/Cu multilayers is even more difficult to compare due to both the lack of the data and the variation in the measurement protocol [100, 113, 114]. The general conclusion of the sliding wear data is that the wear resistance also increases with the decrease in periodicity down to an optimum level. The friction coefficient of the multilayer coating depends on the load, varies between the values obtained for the two pure components and approaching that of Cu with decreasing layer thicknesses.

Ni–W coatings are also in the forefront of coating research due to the large hardness of Ni–W alloys. One of the key problems related to Ni–W coatings is the development of fractures at the surface when the deposition is carried out with a constant current. This problem can be overcome by layering Ni–W alloys of various compositions onto each other, while the internal stress decreases and the hardness improves [116]. The application of a three-pulse method leads to a further improvement of the deposit properties, prominently the wear resistance, without changing the composition as compared to the two-pulse-plated deposits [62].

The multilayered Ni–W CMAs contain amorphous layers if the W content within a layer is larger than about 30 at.%; therefore, the multilayer composed of Ni–W layers of alternating composition show a crystalline/amorphous structural transition beside the composition modulation. For the Ni–P system, a similar coupled composition–structure transition can be achieved with the single-bath method [117] (for similar samples deposited with the double-bath method onto a rotating cathode, see [31]). At high-current density, the deposit is crystalline (with less than 7 at.% P in the cited work), while the lower current density was adjusted so that a Ni–P alloy with eutectic composition could be obtained (19.3 at.% P). With increasing the interface density by applying two-pulse plating with increasing frequency, the mechanical properties like hardness improved. A hardness plateau was observed when the multilayer structure could no longer be developed (at 15 nm thickness or lower). It is remarkable that the coincidence of the multilayer structure with the deviation from the trend observed at higher layer thicknesses was confirmed here with a TEM study.

5.4.4 *Corrosion Properties of Compositionally Modulated Deposits*

The composition modulation technique in coating preparation for corrosion protection was introduced in the 1990s. It was the general experience that multilayered deposits provide better corrosion protection than even layers with the same mean composition as the multilayered ones. The advantage of the multilayer structure in the corrosion resistance is evidenced for sublayers with several micrometre thickness, too; therefore, the enhanced protection performance for multilayers is unrelated to any nanoscale phenomenon that cannot be found for larger modulation length scales. The mechanism of the corrosion protection with alternating layers having different composition is presented with the help of Fig. 5.12.

In accord with the generally accepted mechanistic picture, the more noble and less noble layers mutually polarize each other; hence, the more noble layer gains a cathodic protection as long as the less noble layers already exposed to the corrosive medium completely dissolves. The growth of the corrosion cavity is decelerated by the *lateral* consumption of the less noble (sacrificial) layer, which leads to a virtually layer-by-layer dissolution of the coating. This provides longer protection for the base material than with a coating of even composition because for multilayers, the propagation of the corrosion front is periodically stopped. For a sufficient corrosion protection, the layer should be crack-free. The latter criterion usually makes it necessary to use an appropriate additive combination during the plating process.

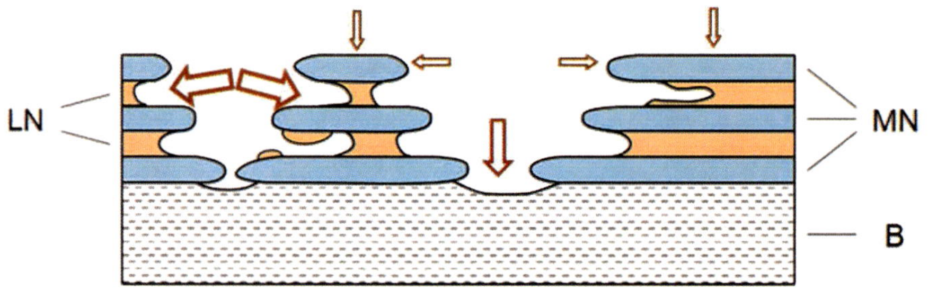

Fig. 5.12 Corrosion protection mechanism with layers of dissimilar composition. Notations: B–base metal to be protected, MN—more noble metal layers, LN—less noble (sacrificial) layers. Arrows in the cross-sectional sketch indicate the consumption rate of the various layers in proportion to their relative corrosion rate. For more details, see the text. The figure was adapted from the schemes in Refs. [17, 18, 76, 118, 119]

For the advantage of the higher corrosion resistance of the more noble layer, it has to exhibit a threshold minimum thickness even if the layer thickness undulates to a certain extent (which always happens for electrodeposited layers). Therefore, the optimum layer thickness is typically above 15 nm (but multilayer illustration images are usually given for much larger layer thicknesses only). In the literature of corrosion research, the habit is to give the layer number of the CMA multilayer within a fixed coating thickness; therefore, the layer thicknesses are often hidden among the primary experimental data. The corrosion rate of multilayer coatings can be 1–2 orders of magnitude smaller than the corresponding single-layer coating with the same mean composition.

The multilayer approach is often applied for Zn-containing coatings on mild steel by alloying Zn with an element that makes it a more efficient protecting layer at a limited cost. These alloying elements used in multilayered deposit with layer thickness below 100 nm are Co [120, 121], Fe [122] and Ni [123–125]. All these nanolayered deposits were obtained with the single-bath method. There are no sufficient structural data available to establish whether the composition modulation leads to a structural change which may have a contribution to the enhanced corrosion resistance of the coatings. For the corrosion protection efficiency, it is not necessary to produce a layered system with sharp interfaces. Therefore, beside the conventional two-pulse plating method, saw-tooth [121] and triangular [124] current–time waveforms also proved to be suitable to deposit layered coatings with high corrosion resistance. The surface roughness was found to decrease as the period of the multilayers was diminished [123, 125], which is coupled with a larger lateral homogeneity. These factors are likely to contribute to the good corrosion protection performance of the multilayered Zn-based coatings.

The deposition of multilayered Cr/Ni coating as a protecting layer was carried out from a Cr^{3+}/Ni^{2+} solution with a combination of current and solution agitation modulation [126] where the solution agitation was applied during the plating of the more noble metal (here, Ni). Even though Cr deposition is usually difficult from trivalent chromium baths, the deposits were smooth and crack-free, showing optimized corrosion protection ability at the Cr(20 nm)/Ni(50 nm) multilayer structure. The role of Ni is inverted for Ni_xW_{1-x}/Ni_yW_{1-y} multilayers where the layer with higher W content has a higher corrosion resistance. Nevertheless, the enhanced protection ability with nanoscale layer thicknesses was evidenced for this layer structure, too [119].

5.4.5 *Magnetic and Magnetoresistance Properties of Electrodeposited Multilayers*

Electrodeposited multilayers exhibiting ferromagnetism can be divided into two major groups. In the first group, both layer types in the multilayer stack are ferromagnetic. In ferromagnetic/ferromagnetic (FM/FM) multilayers, one goal can be the tuning of the coercive force of the layer stack for achieving a material with soft magnetic properties not available in the bulk form. This was demonstrated for an amorphous/crystalline soft magnetic multilayer film in which both layers are composed of Ni–Co–B alloys with different composition [127]. The layer deposited with a low-current density resulted in an amorphous deposit with about 6 wt.% boron content, while the layer deposited with high-current density was crystalline with 1% B content (in parallel, the Ni:Co ratio in the alloy also changed). The layers could be resolved with TEM down to about a 25 nm bilayer thickness, and the lowest coercivity achieved was less than 0.5 Oe with a square-shaped hysteresis loop. The study of the electrodeposited FM/FM system of Fe–Co alloys showed [128] that the deposition with two-pulse plating results in a compositionally modulated alloy even if the nominal layer thicknesses are below 1 nm. Here, both the lattice constant of the deposit and its saturation magnetization were a function of the layer thicknesses but did not depend simply on the composition. It was achieved only by the composition modulation at the nanoscale that the system approached the maximum saturation magnetization as a function of the Fe:Co ratio.

The other basic type of magnetic multilayers are composed of alternating ferromagnetic/non-magnetic layers (which will be called hereinafter as FM/NM multilayers). It was shown in the previous chapters on mechanical and corrosion properties that the thickness of the layers was orders of magnitude larger than the atomic plane distance in the crystals. For such layers, the relative fluctuation of the layer thickness (the mean deviation of the thickness divided by the mean layer thickness) was not crucial, and the desired properties could be achieved with a relative thickness fluctuation up to 10–20%. With the FM/NM multilayers, we arrive at a material family in which the layer thickness fluctuation should be suppressed as much as possible because the desired sample properties can be achieved mostly with 1–3 nm layer thicknesses. The magnetic or magnetoresistance properties are based on the so-called coupling of the neighbouring magnetic layers, which means that the relative direction of the magnetization of the neighbouring layers can be aligned in either parallel or antiparallel direction. Moreover, the parallel/antiparallel configuration can manifest itself in an oscillating manner where the coupling direction is inverted *with a ~0.5 nm change* in the NM layer thickness. Therefore, the layer thicknesses should be as even as possible with the least possible interface roughness and component intermixing, especially when magnetoresistance properties are in the focus of research.

For the ideal composition of FM/NM multilayers, the concentration of the FM ions has to be high, typically 0.5–2.3 M, while the NM metal ion concentration is at most a few tens of mM. The magnetic component is mostly Ni and/or Co, with Fe as a possible alloying element beside the previous ones. The current during the deposition of the FM metal is so high that the NM metal impurity is negligible at least concerning its impact on the magnetic properties. An NM metal mole fraction in the FM layer between 0.5 and 5 at.% is customary. However, the NM metal has to be as pure as possible in order to avoid a direct magnetic coupling between the neighbouring magnetic layers. The layer thicknesses in the relevant studies vary between 0.5 and 8 nm.

Early works related to electrodeposited FM/NM multilayers focused on the magnetic anisotropy of the layer system. The motivation was mainly the experience on thin cobalt layers; i.e., that a perpendicular magnetic anisotropy is observed at a layer thickness of up to a few atomic planes. The multilayer configuration with a suitable NM metal was hoped to give rise to a perpendicular anisotropy with high magnetization at the same time. The perpendicular anisotropy was found for both Co/Cu [129, 130] and Co/Pt [131, 132] multilayers obtained from a single bath.

Concerning the magnetoresistance in multilayered materials, the background is explained here only shortly, referring to an extensive review listing the full electrochemistry-related literature until 2010 [6]. Figure 5.13 shows the scheme of the layer structure and magnetization behaviour of a perfect multilayer system.

When two magnetic layers spaced closely to each other exhibit opposite magnetization directions, the resistance of the system is relatively large. This is because the electrons travelling from one magnetic layer to the neighbouring one retain their original spin orientation, and they undergo a spin-dependent scattering in the next layer of different magnetization direction. If an external magnetic field is applied to

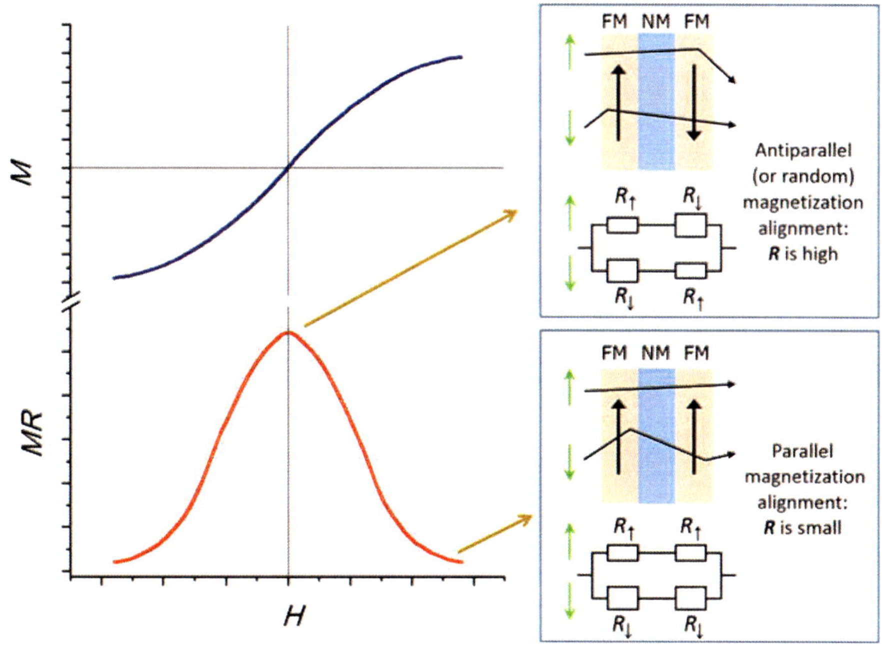

Fig. 5.13 Magnetization (*M*) and magnetoresistance (*MR*) curves of a perfect FM/NM multilayer. The right blocks show two FM layers (orange) with one NM layer in between (blue), and the arrows within the blocks indicate the direction of the magnetization of the FM layers. The equivalent circuit in the blocks represents the two-current model with parallel conduction channels for spin-up and spin-down electrons, as indicated with the coloured arrows. The resistor boxes are proportional to the resistance of the channels

align the magnetization direction of the FM layers, the spin-dependent scattering is strongly reduced, hence decreasing the resistance of the system. This phenomenon is called the giant magnetoresistance (GMR) which originates from the order of magnitude of the resistivity change but it refers to the *mechanism* of the resistivity change as well.

The magnetoresistance of a system (in general) is expressed as the resistivity change at a particular external magnetic field $R(H)$ with respect to the zero magnetic field value R_0 normalized with the latter value:

$$MR = \frac{R(H) - R_0}{R_0} \text{ and } MR\% = 100\frac{R(H) - R_0}{R_0} \tag{5.4}$$

The resistivity of a nanolayered system can be about an order of magnitude larger than that of the bulk form of the constituent metals. Therefore, it can be surprising that the magnetoresistance can as high as 50% at room temperature and 80% at liquid helium temperature [133] for Co/Cu multilayers prepared by sputtering. Such large magnetoresistance values can be measured at a Cu layer thickness corresponding to the first antiparallel coupling at ~1.0 nm. The coupling strength exponentially decreases with the increase of Cu layer thickness; therefore, magnetoresistance also decreases at the antiparallel couplings with larger NM layer thicknesses.

The first observation of the GMR in electrodeposited samples started with the Ni–Co–Cu/Cu system [134, 135], which was followed with the systematic analysis of various other element combinations. Ni–Cu/Cu [55, 56, 102, 104, 136–142] multilayers are quite simple from the preparation point of view because of the relative inertness of nickel and the insensitivity to the exact deposition condition of the Cu layer. Co–Cu/Cu multilayers obtained the greatest attention [37, 92, 93, 103, 105, 143–156], even though the dissolution of the cobalt during the Cu deposition is a prominent problem. Ni–Co–Cu/Cu multilayers remained in the focus of studies [39, 89, 157–165] because of their relatively large magnetoresistance. When iron is a component of the magnetic layer, roughening is often observed, which prevents the growth of thick multilayers. This is why the number of studies dealing with iron-containing multilayers [40, 42, 43, 166–171] is small as compared to Ni- and Co-based ones. Concerning the NM metal, Cu is by far the most common due to the small lattice misfit to either Ni or Co. FM/NM multilayers with other NM metals (Ag [38, 172, 173], Ru [174] or Au [175]) have been exemplified but were not studied often.

It has to be emphasized that many studies indicate the nominal layer thicknesses without any correction due to the displacement-like processes occurring at the pulse change (large to small cathodic current density). The optimization method described in Sect. 5.3.1 may ensure that the mean layer thickness can be calculated from the Faraday law and correct magnetoresistance vs. layer thickness data can be obtained for electrodeposited samples [156].

For the measurement of the magnetoresistance effect on electrodeposited samples, the substrate effect has to be eliminated. While physical vapour deposition methods can use non-conducting substrate, this is clearly impossible during electrodeposition.

One feasible strategy is to use a substrate that can be dissolved later. This works for Cu substrate and Ni–Co/Cu deposits since Cu can be dissolved from a Ni-rich alloy with sufficient selectivity. The substrate removal is also possible by simply peeling off the deposit if it is thick enough and self-supporting. In this case, the substrate is a flexible Ti sheet on which the adherence of a metallic deposit is week because of the native TiO_2 layer. The peeling-off method works only if a thick deposit can be grown with a small roughness, which provides that the sample is not torn apart during the removal from the substrate. Another common approach is the application of a thin evaporated metal film on a non-conducting substrate whose conductivity does not disturb significantly the magnetoresistance measurement. As the last option, semiconductor wafers are also commonly used, and their resistance is high enough so that the sample is not shunted during the resistivity test. The resistivity test on multilayer films is always carried out in a current-in-plane configuration. Depending on the direction of the external magnetic field, the longitudinal ($H \parallel I$) and the transverse ($H \perp I$) component of the magnetoresistance can be determined.

For many systems, there is a full analogy between metallurgically processed and electrodeposited samples. However, this is absolutely not true for multilayers with GMR. For evaporated, sputtered or MBE-deposited multilayers, the smooth growth of the subsequent layer could be achieved with a monoatomic surfactant metal layer (mostly Pb or Ag) that provided a smooth growth of subsequent layers that would grow otherwise as a granular film. Although there were similar attempts in multilayer electrodepositon experiments, no analogous multilayers could be produced [176, 177]. In the absence of a surfactant layer, the grain growth takes place as it is shown in Figs. 5.10 and 5.11b. This means that the layers are not parallel to the substrate, especially at large layer thicknesses, which also means that the application of an external magnetic field parallel to the substrate will not lead to a fully in-plane magnetization. This gives rise to a magnetostatic coupling between the neighbouring FM layers. This magnetostatic coupling (which is often named as Néel-type "orange peel" coupling) always leads to a parallel magnetization alignment between the FM layers. The orange peel coupling easily overwrites the coupling effect dictated by the spacing of the FM layers by the NM one, hence destroying the antiparallel alignment of the neighbouring FM layer magnetizations [101, 178]. The geometric representation of the magnetostatic coupling effect is shown in Fig. 5.14.

Indeed, the oscillatory parallel/antiparallel coupling between the neighbouring FM layers has not been evidenced for any electrodeposited FM/NM multilayer sample [156]. Instead of an oscillatory GMR as a function of the NM layer thickness, the following general trend could be seen for electrodeposited multilayers.

At small NM layer thickness, the NM layer is discontinuous, which allows direct contact (i.e., direct ferromagnetic coupling) between the neighbouring FM layers. In fact, we cannot speak about true "layers"; instead, the correct term would be "the material deposited during the subsequent high-current pulses", but this complicated phrasing is never applied. When the dominating effect is the coupling of the neighbouring FM layers via pinholes in the NM layer, the magnetoresistance character is the *anisotropic magnetoresistance* (AMR) that is a feature of the bulk ferromagnets. The pinholes in the FM layers are difficult to observe directly, especially with

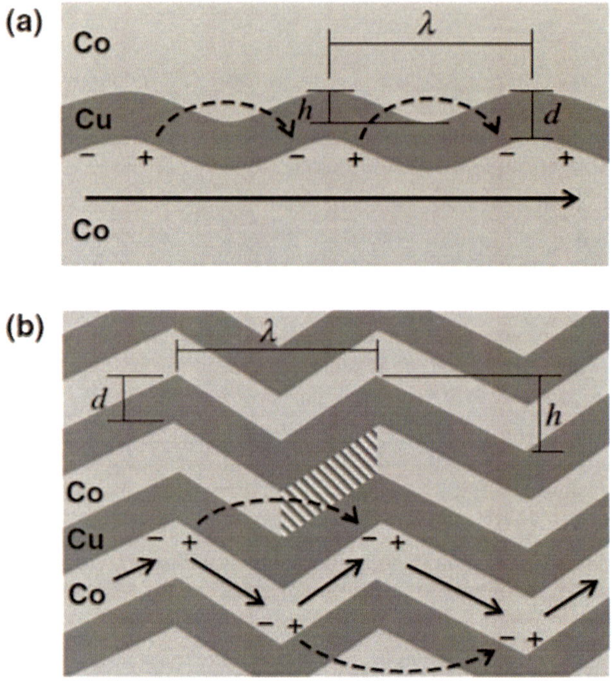

Fig. 5.14 Néel-type "orange peel" coupling in FM/NM multilayers with layer undulation. The dashed arrows show how the magnetic induction lines are diverted from the composition modulation zones in wavy (**a**) and saw-tooth-like (**b**) composition modulations. The coupling energy depends on h^2/d, which can be large enough to overcome the coupling corresponding to the distance between the FM layers (c.f. TEM images in Fig. 5.11). Reprinted from [178] with the kind permission from The Electrochemical Society of Japan

a non-destructive analysis method. Instead, the magnetoresistance measurement is used as an indirect tool to test the sufficient continuity of the NM metal layers. At the NM layer thickness where the anisotropic magnetoresistance changes to GMR and the coercive field tend to approach a maximum (plateau), we can take the pinhole density negligible [156].

From the change of AMR to GMR, the magnetoresistance first increases, then it passes through a maximum and start decreasing. The NM layer thickness at which the GMR maximum is achieved has nothing to do with the values where parallel or antiparallel couplings would be observed at perfectly smooth layers. Hence, it was rationalized that the magnetic coupling effect that is probably provided locally at small areas are modified by magnetostatic coupling effects. The GMR measured for electrodeposited multilayers is caused by the partial misalignment of the magnetization directions in the nearby FM zones, but no oscillatory parallel/antiparallel magnetic coupling occurs at the length scale of the full sample [156]. As the NM layer thickness increases, the magnetoresistance ratio starts to fall because of the decrease in the interface density.

It would sound logical to apply additives to achieve a grain refinement and hence, making the layer structure fully independent of the grain boundaries, as shown in Fig. 5.8b. However, all such attempts failed for electrodeposited multilayer films with GMR. Even if the grain refinement results in smoother layers in terms of the parallel position of the layer interfaces both with each another and with the substrate, the ultrafine-grained material has a large resistivity. This means that the zero-field resistivity of the multilayer drastically increases, which decreases the magnetoresistance ratio as defined with Eq. 5.4. This happens with the application of surfactants, stress relievers and also complexing agents. Saccharin as stress reliever, citric acid as complexing agent and tensides as wetting agents are all detrimental for the magnetoresistance properties of electrodeposited multilayers [148]. The increase in pH also impacts negatively the maximum magnetoresistance effect observed for multilayered deposits, and the pH effect is essentially independent of the multilayer composition [102, 159, 171]. This is thought to be related to the general tendency that the crystallite size of the metals grown from acidic solution is larger than from neutral and basic ones, although the logical chain between the structure and the magnetoresistance properties is much less underpinned than that of the additive effect.

As an inherent feature of the electrodeposition technique, the FM metal (which is also the LN component) always contains the NM one (which is the MN if we regard its relative nobility). Depending on the NM metal content of the FM layer, the magnetoresistance curves exhibit various shapes. An example is shown in Fig. 5.15. The NM metal content in the layer produced in the high-current pulse is determined by the $j_{\mathrm{FM}}/j_{\mathrm{Lim,NM}}$ ratio and $j_{\mathrm{Lim,NM}}$ is proportional to $c(\mathrm{NM}^{z+})$. Therefore, the increase in the NM metal content can be achieved by either increasing its concentration in the bath [154] or decreasing the current density in the high-current pulse [179]. Whichever way is chosen, the NM metal tends to segregate in the layer produced during the high-current pulse, and so does the FM metal, too. This is particularly true for Co–Cu/Cu multilayers because Co and Cu are not miscible in equilibrium. The small Co-rich fragments are no longer ferromagnetic but they exhibit *superparamagnetism.* A superparamagnetic (SPM) entity is a single-domain particle (or zone) in which the spin orientation of the magnetic atoms are aligned in a parallel manner due to the exchange coupling; therefore, the spin ordering interaction is the same as in a bulk FM metal. However, the SPM particle is so small that it cannot exhibit a fixed magnetization direction because the thermal excitation easily changes the magnetic orientation (in other words, the kT energy is much larger than the anisotropy energy of the magnetic entity). The magnetization curve related to the SPM behaviour can be described with the Langevin function:

$$L(x) = \coth(x) - x^{-1} \tag{5.5}$$

where $x = \mu H/kT$, μ being the magnetic moment of the SPM entity (the other symbols have their usual meaning). The magnetic saturation of an SPM particle requires a much larger field than that of an FM domain, which lends to a sample with

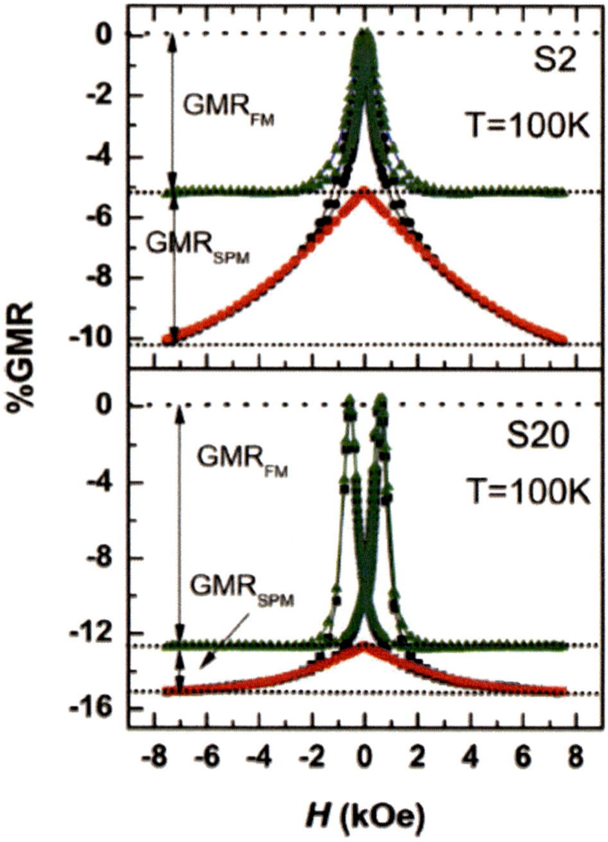

Fig. 5.15 Magnetoresistance of Co–Cu/Cu multilayers with different magnetoresistance contributions. Top: nearly equal FM and SPM contributions, bottom: large FM and small SPM contribution. Measured data are displayed with the black symbols. The red curves denote the SPM contribution in accord with the Langevin function and are shifted along the ordinate so that they follow the measured data at high magnetic fields. The green curves show the FM contribution that were obtained and a difference between the measured data and the calculated SPM contribution. Reprinted from [180]. Copyright (2015), with permission from Elsevier

SPM particles a slow saturation with the external magnetic field. This is a persistent problem for many electrodeposited multilayers.

It was first demonstrated for electrodeposited multilayers [153] that the magnetoresistance can be decomposed into a ferromagnetic and a superparamagnetic contribution:

$$MR(H) = MR_{\mathrm{FM}}(H) + GMR_{\mathrm{SPM}} L(\mu H/kT) \tag{5.6}$$

In the above equation, GMR_{SPM} is a constant and the $MR_{FM}(H)$ function saturates at magnetic field where ferromagnets are technically saturated (usually at $H < 2$ kOe).

Where magnetization was also measured in parallel to the magnetoresistance, it was shown that the SPM particle size obtained from the two measurements are the same. The significance of this decomposition procedure is that an estimate can be obtained for the size of the small segregated magnetic domains (SPM particles), and the partial FM magnetoresistance contribution of the sample can be obtained as a "residue"; i.e., the difference of the MR curve measured and that calculated as the SPM contribution. Hence, the FM part can be calculated, although there is no theoretical function shape available for the FM part of the MR curves. The FM/SPM ratio is often calculated and is taken as an indicator of the continuity and perfectness of the FM layer [38, 40, 61, 89, 154, 176, 180].

It is to be emphasized that magnetization is a volume-proportional (extensive) quantity, while magnetoresistance depends much on the relative position of all magnetic entities in the system. Therefore, large magnetoresistance values are not always obtained when the deposition parameters are tuned properly in accord with the principles described in Sect. 5.3.1. The saturation magnetization value can be large for samples in which the displacement of the magnetic layer can be identified from the deposition conditions. Also, the surface roughness of the deposit was found to play a key role in both the magnetoresistance ratio and in the superparamagnetic contribution, maximizing the GMR effect at an NM layer deposition potential where neither the composition nor the layer thickness ratio are predictable from the pulse parameters [39]. Such effects make the electrodeposited multilayers with GMR very difficult to design.

5.4.6 *Magneto-ionics*

Magneto-ionics is a new field of materials science where the ion migration is used to tune the magnetic property of a material, mostly that of a thin layer. The typical implementation of a magneto-ionic device includes a metal/metal-oxide interface (like Co/GdO_x [181]) across which oxide ions can migrate and modify both the resistance and the magnetic properties of the layered system. Since both the resistivity and the magnetization can be easily tuned with the electric charge passed, magneto-ionic devices may constitute a new class of nonvolatile memories.

The electrochemical approach to magneto-ionics [182–185] includes a thin ferromagnetic layer which can be reversibly oxidized and reduced with classical means of electrochemical polarization in a three-electrode cell. Here, the starting material is not multilayered but a bilayer structure is obtained with the device operation. The magnetic layer oxidized has to be thin (i.e., a few nanometres) so that a little charge passed the system can achieve a significant variation in the magnetic properties. The resulting oxide is not restricted so a surface atomic layer, which means that magneto-ionics differ from double layer- or adsorption-induced magnetization modulation as presented in Chap. 4.3.3 for ultrathin magnetic layers. Instead, oxidation/reduction cycles in magneto-ionic devices can be taken as a semi-bulk process.

The impact of the electrochemical treatment on the magnetization behaviour of 2.8 nm thick CoPt sample was studied in an aprotic medium ($LiClO_4$ in the equimolar mixture of ethylene carbonate dimethyl carbonate) [182]. The role of the native oxide reduction in the change of the magnetization properties was well evidenced. Although the role of the double layer charging effect in the magnetic behaviour could be excluded, no reversible magnetization change could be achieved in the non-reactive medium applied in the potential range where Co did not dissolve selectively.

A reversible change in the magnetic properties was achieved for Fe/FeO_x layer pairs [183–185]. For this system, the controlled-potential electrochemical oxidation of the Fe layer in a dilute KOH solution can lead to an oxide layer formation with a well-defined thickness of about 3–4 nm, which is equivalent to the oxidation of about 1.5 nm of metallic Fe due to the volume expansion accompanying the oxidation [183]. The relative change in magnetization is proportional to the thickness ratio of the remaining metallic Fe and the electrochemically produced oxide, and it is straightforward that the effect measured increase with a decrease of the total layer thickness. The change in magnetization measured as a function of the electrode potential and the corresponding layer structure is shown in Fig. 5.16.

The great achievement related to the electrochemical oxidation/reduction of thin Fe layers in dilute KOH solutions was that the redox transformation can take place without either a morphology change in the layer of an iron loss, at least for the limited cycle number (12) that was used for the demonstration of the reversibility effect. If the Fe layer is produced by electrodeposition, an all-electrochemical device can be constructed where the magnetization is directly governed by the sample voltage [184]. However, if the underlayer exhibits perpendicular magnetic anisotropy (like FePt as shonw in [182]), the electrochemical reduction/oxidation cycle results in a transient between rigid magnet to exchange spring magnet, respectively, hence opening the way for electrochemically driven spintronic devices.

5.4.7 Annealing Behaviour of Electrodeposited Multilayers

Electrodeposition is a non-equilibrium sample preparation method. Therefore, as-received electroplated samples often exhibit a large structural fault density which can be partly eliminated with annealing. Depending on the properties of the samples, annealing effects show a great diversity and the optimum annealing temperature may also vary from one sample to another.

Annealing at temperatures below 200 °C does not result in a significant long-range atomic motion in the solid deposit but leads only to the relaxation of dislocations and local stress. Depending on the nature of the samples, a moderate-temperature annealing may also lead to the improvement of the sample properties. This can be seen for Co/Cu multilayers with GMR properties where a gradual increase of the magnetoresistance was found for annealing temperature up to 300 °C [186]. The

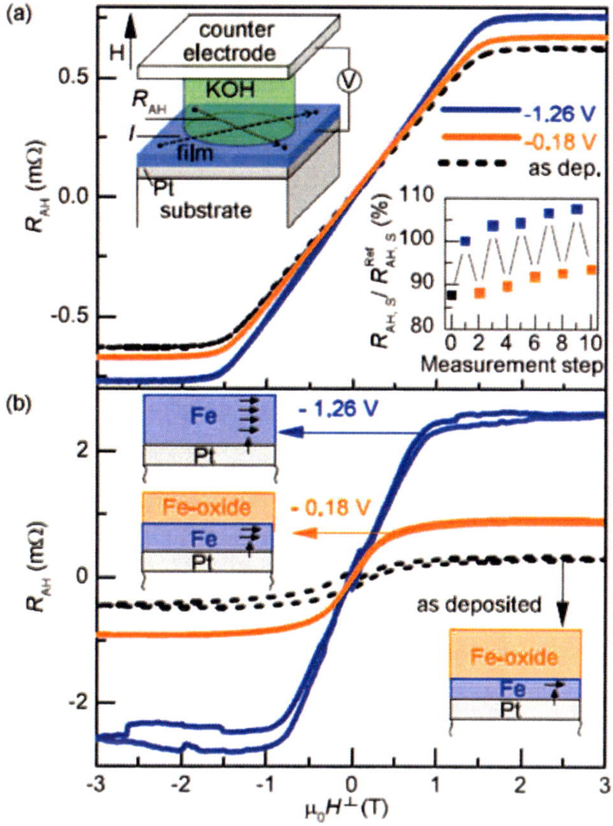

Fig. 5.16 Magnetization curves as measured by anomalous Hall effect for Fe–O/Fe films with nominal Fe thickness of **a** 10 nm and **b** 2 nm in as-deposited state and polarized in 1M KOH at −1.26 V and −0.18 V. The left inset in **a** presents the in-situ magnetization measurement geometry with both the applied current and the measured resistivity laying in the film plane. The right inset in (**a**) shows the measurement results for 10 subsequent switching experiments. Sketches in **b** visualize the film structures with arrows indicating the expected local spin orientation. Reused from [183] with Creative Common License (https://creativecommons.org/licenses/by/4.0/)

sharpening of the layer interfaces at moderate annealing conditions is followed by a collapse of the multilayer structure because of both the increase of the mutual solubility of Co and Cu at higher temperature and the recrystallization.

A partial recrystallization can have a contribution to the improvement of the mechanical behaviour of a multilayer system. Electroplated multilayer with alternating crystalline Ni-rich/amorphous Cr-rich layers showed a twofold hardness increase upon annealing at 550°C [187]. The annealing treatment was shown to lead to the formation of a nanocrystalline structure in the Cr-rich layers, while no change was detected in the Ni-rich segments. The layer structure was probably not fully metallic due to the incorporation of the solution components (mostly organic additives) into the Cr-rich layer, which is likely to have influenced the mechanical properties, too.

Finally, the Fe/Pt system must be mentioned where annealing can be indispensable for the post-treatment of electrodeposited multilayers. The importance of the annealing in the Fe–Pt system stems from that the desired $L1_0$ structure of the FePt alloys with nearly equimolar composition does not form at room-temperature deposition. Additionally, the oxygen incorporation during the electroplating process prevents the achievement of the desired magnetic properties when the equimolar alloy is deposited. For the preparation of the layer with desired composition, a very elegant solution was suggested with a multilayer approach [188]. Namely, a multilayer structure is deposited first with Fe-rich and Pt-rich layers, both of them with relatively small oxygen impurity as compared to the direct plating of FePt deposits. Then, the resulting multilayer can be annealed at a temperature where the homogenization of the multilayer takes place. Hence, the indispensable heat treatment for obtaining the desired phase structure was coupled with the sample homogenization, and the drawback of the oxygen impurity could be avoided by plating a metal-rich multilayer system as a starting material for annealing. The heat treatment accompanied with partial recrystallization was also essential for achieving a high-coercivity coating from a multilayer composed of $Fe_{75}Pt_{25}$ and $Fe_{50}Pt_{50}$ sublayers [189]. Interestingly, the samples with 15 nm layer thicknesses did not homogenize even at 650 °C annealing temperature but develop a mixed bcc/fcc crystalline system where the coercivity of the individual layers could be identified from magnetization measurement. Extremely high coercivity of 12 kOe was achieved with the combined *multilayer electroplating–annealing* approach.

References

1. Ross CA (1994) Annu Rev Mater Sci 24:159–187
2. Leisner P, Nielsen CB, Tang PT, Dörge TC, Møller P (1996) J Mater Proc Technol 58:39–44
3. Wilcox GD (2004) J Corros Sci Eng 6: Paper 52
4. Roy S (2009) Electrochemical fabrication of nanostructured, compositionally modulated metal multilayers (CMMMs). In: Schmuki P (ed) Electrochemistry at the nanoscale. Springer, New York, pp 349–376
5. Bakonyi I, Hansal WEG (2018) Trans IMF 96:237–243
6. Bakonyi I, Péter L (2010) Progr Mater Sci 55:107–245
7. Blum W (1921) Trans Am Electrochem Soc 40:307–320
8. NabiRahniy DMA, Tang PT, Leisner P (1996) Nanotechnology 7:134–143
9. El Bahraoui T, Errahmani H, Berrada A, Dinia A, Schmerber G, Cherkaoui El Mourslic F, Hajji F, Lassri H (2004) J Magn Magn Mater 272–276:e955–e957
10. El Bahraoui T, Errahmani H, Belghazi Y, Berrada A, Dinia A, Schmerber G, Lassri H, Cherkaoui El Moursli F, Hajji F (2007) J Magn Magnc Mater 316:8–12
11. Lghazi Y, Moubah R, Bimaghra I, Hlil EK, El Bahraoui T, Alleg S, Lassri H (2017) J Supercond Nov Magn 30:2641–2645
12. Haseeb ASMA, Celis JP, Roos JR (1994) J Electrochem Soc 141:230–237
13. Lamovec J, Jović V, Vorkapić M, Popović B, Radojević V, Aleksić R (2011) J Min Metall Sect B-Metall 47:53–61

14. Fu Z, Zhang Z, Meng L, Shu B, Zhu Y, Zhu X (2018) Mater Sci Eng a 726:154–159
15. Kirilova I, Ivanov I, Rashkov S (1998) J Appl Electrochem 28:1359–1363
16. Ivanov I, Valkova T, Kirilova I (2002) J Appl Electrochem 32:85–89
17. Fei JY, Liang GZ, Xin WL, Wang WK (2006) J Iron Steel Res Int 13:61–67
18. Fei JY, Wilcox GD (2006) Surf Coat Technol 200:3533–3539
19. Rahsepar M, Bahrololoom ME (2009) Surf Coat Technol 204:580–585
20. Lee KH, Kim GH, Jeung WY (2004) Electrochem Commun 6:115–119
21. Esmaili S, Bahrololoom ME, Kavanagh KL (2011) Mater Charact 62:204–210
22. Lassri H, Ouahmane H, El Fanity H, Bouanani M, Cherkaoui F, Berrada A (2001) Thin Solid Films 389:245–249
23. Hayashi Y, Kashiwabara S, Jyoko Y (1997) Physica B 239:35–40
24. Jyoko Y, Kashiwabara S, Hayashi Y, Schwarzacher W (1999) Electrochem Solid State Lett 2:67–69
25. Lee KH, Kang SW, Kim GH, Jeung WY (2004) J Magn Magn Mater 272–276:e925–e926
26. Matsuoka K, Okuhata M, Takashiri M (2015) J Alloy Compd 649:721–725
27. Mais L, Mascia M, Vacca A, Palmas S, Delogu F (2015) J Appl Electrochem 45:735–744
28. Chia PY, Haseeb ASMA (2015) J Mater Sci: Mater Electron 26:294–299
29. Jiang Y, Yao S, Zhang W (2008) Thin Solid Films 516:3210–3216
30. Goldman LM, Ross CA, Ohashi W, Wu D, Spaepen F (1989) Appl Phys Lett 55:2182–2184
31. Ross CA, Goldman LM, Spaepen F (1993) J Electrochem Soc 140:91–98
32. Wang GF, Shen LD, Dou LM, Tian ZJ, Liu ZD (2014) Int J Electrochem Sci 9:220–226
33. Briggle J, Nolidin PD, Golden TD (2010) Electroanal 22:1–5
34. Hayashi Y, Lee CG, Koo BH, Sato T, Arita M, Masuda M (2004) Phys Stat Sol (a) 201:1658–1661
35. Péter L, Liu Q, Kerner Z, Bakonyi I (2004) Electrochim Acta 49:1513–1526
36. Ghosh SK, Grover AK, Chowdhury P, Gupta SK, Ravikumar G, Aswal DK, Kumar MS, Dusane RO (2006) Appl Phys Lett 89:132507(1–3)
37. Lenczowski SKJ, Schönenberger C, Gijs MAM, de Jonge WJM (1995) J Magn Magn Mater 148:455–465
38. García-Torres J, Péter L, Révész Á, Pogány L, Kiss LF, Bakonyi I (2009) Thin Solid Films 517:6081–6090
39. Tóth BG, Péter L, Dégi J, Révész Á, Oszetzky D, Molnár G, Bakonyi I (2013) Electrochim Acta 91:122–129
40. Tóth BG, L. Péter, Pogány L, Révész Á, Bakonyi I (2014) J Electrochem Soc 161:D154–D162
41. Rajak S, Srivastava C, Bidaye AC, Ghosh SK (2014) Russ J Electrochem 50:27–31
42. Gong J, Butler WH, Zangari G (2005) Appl Phys Lett 87:012505(1–3)
43. Gong J, Butler WH, Zangari G (2005) IEEE Trans Magn 41:3634–3636
44. Meuleman WRA, Roy S, Péter L, Varga I (2002) J Electrochem Soc 149:C479–C486
45. Weihnacht V, Péter L, Tóth J, Pádár J, Kerner Zs, Schneider CM, Bakonyi I (2003) J Electrochem Soc 150:C507–C515
46. Kelly JJ, Kern P, Landolt D (2000) J Electrochem Soc 147:3725–3729
47. Tench DM, White JT (1992) J Electrochem Soc 139:443–446
48. Roy S (1998) Surf Coat Technol 105:202–205
49. Fricoteaux P, Douglade J (2004) Surf Coat Technol 184:63–68
50. Roy S, Matlosz M, Landolt D (1994) J Electrochem Soc 141:1509–1517
51. Fricoteaux P, Simon C (2005) J Phys Chem B 109:20449–20453
52. Huang Q, Podlaha EJ (2004) J Electrochem Soc 151:C119–C126
53. Meuleman WRA, Roy S (2003) Trans Inst Met Fin 81:55–58
54. Ovchinnikova SN, Maslii AI (2006) Russ J Electrochem 42:1235–1243
55. Tóth-Kádár E, Péter L, Becsei T, Tóth J, Pogány L, Tarnóczi T, Kamasa P, Bakonyi I, Láng G, Cziráki Á, Schwarzacher W (2000) J Electrochem Soc 147:3311–3318
56. Myung NV, Nobe K (2000) Plat Surf Fin 87:125–134
57. Kelly JJ, Cantoni M, Landolt D (2001) J Electrochem Soc 148:C620–C626
58. Kelly JJ, Bradley PE, Landolt D (2000) J Electrochem Soc 147:2975–2980

59. Menezes S, Anderson DP (1997) J Electrochem Soc 137:440–444
60. Kaneko Y, Mizuta Y, Nishijima Y, Hashimoto S (2005) J Mater Sci 40:3231–3236
61. Rajasekaran N, Pogány L, Révész Á, Tóth BG, Mohan S, Péter L, Bakonyi I (2014) J Electrochem Soc 161:D339–D348
62. Shreeram DD, Bedekar V, Li S, Jagtap R, Cong H, Doll GL (2017) JOM 69:2192–2198
63. Attenborough K, Boeve H, de Boeck J, Borghs G, Celis JP (1999) Appl Phys Lett 74:2206–2208
64. Attenborough K, Boeve H, de Boeck J, Borghs G, Celis JP (2000) Sensor Actuator 81:9–12
65. Seligman L, Sartorelli ML, Pasa AA, Schwarzacher W, Kasyutich OI (2001) J Magn Magn Mater 226:752–753
66. Yamada A, Shirota M, Rizal CLS, Houga T, Ueda Y (2002) J Jpn Inst Met 66:869–872
67. Fukumuro N, Nishiyama J, Shigeta K, Morimoto Y, Takagami H, Yae S, Matsuda H (2007) Electrochem Commun 9:1185–1188
68. Medina JA, Schwartz DT (1995) J Electrochem Soc 142:451–456
69. Medina JA, Sexton DL, Schwartz DT (1995) J Electrochem Soc 457–462
70. Medina JA, Schwartz DT (1996) Phys Fluids 8:2895–2905
71. Medina JA, Schwartz DT (1997) J Electrochem Soc 144:155–164
72. Medina JA, Schwartz DT (1997) Electrochim Acta 42:2679–2684
73. Leith SD, Schwartz DT (1996) J Electrochem Soc 143:873–878
74. Leith SD, Wang W, Schwartz DT (1998) J Electrochem Soc 142:2827–2833
75. Wilson JR, Adler SB, Schwartz DT (2005) 17:063601(1–11)
76. Shetty S, Mohamed Jaffer Sadiq M, Krishna Bhat D, Chitharanjan Hegde A (2016) RSC Adv 6:77465–77473
77. Shetty AR, Chitharanjan Hegde A (2017) Surf Coat Techn 322:99–107
78. Cockett GH, Spencer-Timms ES (1961) J Electrochem Soc 108:906–908
79. Doyle WD (1967) J Appl Phys 38:1441–1442
80. Lommel JM, Girard R (1967) J Appl Phys 38:1423–1430
81. Beltowska-Lehman E, Riesenkampf A (1980) Thin Solid Films 71:129–132
82. Gong J, Riemer S, Morrone A, Venkatasamy V, Kautzky M, Tabakovic I (2012) J Electrochem Soc 159:D447–D454
83. Csik A, Vad K, Tóth-Kádár E, Péter L (2009) Electrochem Commun 11:1289–1291
84. Péter L, Csik A, Vad K, Tóth-Kádár E, Pekker Á, Molnár G (2010) Electrochim Acta 55:4734–4741
85. Neuróhr K, Csik A, Vad K, Bartók A, Molnár G, Péter L (2011) J Solid State Electrochem 15:2523–2544
86. Neuróhr K, Csik A, Vad K, Molnár G, Bakonyi I, Péter L (2013) Electrochim Acta 103:179–187
87. Péter L, Vad K, Csik A, Muñíz R, Lobo L, Pereiro R, Šturm S, Žužek Rožman K, Molnár G, Németh K, Neuróhr K, Boros K, Pogány L, Bakonyi I (2018) J Electrochem Sci Eng 8:49–71
88. Csik A, Vad K, Péter L, unpublished results
89. Péter L, Pádár J, Tóth-Kádár E, Cziráki Á, Sóki P, Pogány L, Bakonyi I (2007) Electrochim Acta 52:3813–3821
90. Nee CC, Weil R (1985) Surf Technol 25:7–15
91. Kazeminezhad I, Schwarzacher W (2002) J Magn Magn Mater 240:467–468
92. Cziráki Á, Péter L, Arnold B, Thomas J, Bauer HD, Wetzig K, Bakonyi I (2003) Thin Solid Films 424:229–238
93. Cziráki Á, Péter L, Weihnacht V, Tóth J, Simon E, Pádár J, Pogány L, Schneider CM, Gemming T, Wetzig K, Tichy G, Bakonyi I (2006) J Nanosci Nanotechnol 6:2000–2012
94. Neuróhr K, Dégi J, Pogány L, Bakonyi I, Ungvári D, Vad K, Hakl J, Révész Á, Péter L (2012) J Alloy Compd 545:111–121
95. Bonhôte C, Landolt D (1997) Electrochim Acta 42:2407–2417
96. Péter L, Sturm S, Zuzek Rozman K (2018) J Electrochem Soc 165:D384–D392
97. Michaelsen C (1995) Philos Mag A 72:813–828

98. Bakonyi I, Péter L, Horváth ZE, Pádár J, Pogány L, Molnár G (2008) J Electrochem Soc 155:D688–D692
99. Bödeker P, Abromeit A, Bröhl K, Sonntag P, Metoki N, Zabel H (1993) Phys Rev B 47:2353–2361
100. Ghosh SK, Limaye PK, Bhattacharya S, Soni NL, Grover AK (2007) Surf Coat Techn 201:7441–7448
101. Shima M, Salamanca-Riba LG, McMichael RD, Moffat TP (2001) J Electrochem Soc 148:C518–C523
102. Alper M, Baykul MC, Péter L, Tóth J, Bakonyi I (2004) J Appl Electrochem 34:841–848
103. Nallet P, Chassaing E, Walls MG, Hÿtch MJ (1996) J Appl Phys 79:6884–6889
104. Cziráki Á, Gerőcs I, Fogarassy B, Arnold B, Reibold M, Wetzig K, Tóth-Kádár E, Bakonyi I (1997) Z Metallkd 88:781–789
105. Rafaja D, Schimpf C, Klemm V, Schreiber G, Bakonyi I, Péter L (2009) Acta Mater 57:3211–3222
106. Cziráki Á, Pierron-Bohnes V, Ulhaq-Bouillet C, Tóth-Kádár E, Bakonyi I (1998) Thin Solid Films 318:239–242
107. Anderson PM, Li C (1995) Nanostuct Mater 5:349–362
108. Wang J, Misra A (2011) Curr Opin Solid St M 15:20–28
109. Ren F, Zhao S, Li W, Tian B, Yin L, Volinsky AA (2011) Mater Lett 65:119–121
110. Zhou Q, Xie JY, Wang F, Huang P, Xu KW, Lu TJ (2015) Acta Mech Sin 31:319–337
111. Subedi S, Beyerlein IJ, LeSar R, Rollett AD (2018) Scripta Mater 145:132–136
112. Cammarata RC (1988) Mater Res Soc Symp Proc 103:315–325
113. Ghosh SK, Limaye PK, Swain BP, Soni NL, Agrawal RG, Dusane RO, Grover AK (2007) Surf Coat Tech 201:4609–4618
114. Qian Y, Tan J, Liu Q, Yu H, Xing R, Yang H (2011) Surf Coat Tech 205:3909–3915
115. Uemori K, Kaneko Y (2012) J Japan Inst Metals 76:309–313
116. Lee S, Choi M, Park S, Jung H, Yoo B (2015) Electrochim Acta 153:225–231
117. Müller T, Bachmaier A, Konetschnik R, Schöberl T, Pippan R (2018) Mater Sci Eng A 715:83–91
118. Rashmi S, Elias L, Chitharanjan Hegde A (2017) Eng Sci Technol Int J 20:1227–1232
119. Elias L, Chitharanjan Hegde A (2015) Surf Coat Tech 283:61–69
120. Thangaraj V, Eliaz N, Chitharanjan Hegde A (2009) J Appl Electrochem 39:339–345
121. Yogesha S, Chitharanjan Hegde A (2011) Bull Mater Sci 34:1699–1706
122. Yogesha S, Chitharanjan Hegde A (2011) J Mater Process Tech 211:1409–1415
123. Venkatakrishna K, Chitharanjan Hegde A (2010) J Appl Electrochem 40:2051–2059
124. Bhat RS, Chitharanjan Hegde A (2011) Surf Eng Appl Elect 47:112–119
125. Bahadormanesh B, Ghorbani M, Kordkolaei NL (2017) Appl Surf Sci 404:101–109
126. Etminanfar MR, Heydarzadeh Sohi M (2012) Thin Solid Films 520:5322–5327
127. Onoda M, Shimizu K, Tsuchiya T, Watanabe T (1993) J Magn Magn Mater 126:595–598
128. Rizal C, Kolthammer J, Pokharel RK, Choi BC (2013) J Appl Phys 113:113905(1–4)
129. Dariel M, Bennett LH, Lashmore DS, Lubitz P, Rubinstein M, Lechter WL, Harford MZ (1987) J Appl Phys 61:4067–4069-
130. Fedosyuk VM, Kasyutich OI (1993) J Magn Magn Mater 125:330–334
131. Georgescu V, Mazur V, Pushcashu B (2000) Mat Sci Eng B 68:131–137
132. Georgescu V, Georgescu M (2002) Surf Sci 507–510:507–511
133. Mosca DH, Petroff F, Fert A, Schroeder PA, Pratt WP Jr, Laloee R (1991) J Magn Magn Mater 94:L1–L5
134. Alper M, Attenborough K, Lane SJ, Lashmore DS, Younes CM, Schwarzacher W (1993) Appl Phys Lett 63:2144–2146
135. Alper M, Attenborough K, Baryshev V, Hart R, Lashmore DS, Schwarzacher W (1994) J Appl Phys 75:6543–6545
136. Bakonyi I, Tóth-Kádár E, Becsei T, Tóth J, Tarnóczi T, Cziráki Á, Gerőcs I, Nabiyouni G, Schwarzacher W (1996) J Magn Magn Mater 156:347–349

137. Shima M, Salamanca-Riba L, Swartzendruber LJ (1997) Mater Res Soc Symp Proc 451:419–294
138. Tóth J, Kiss LF, Tóth-Kádár E, Dinia A, Pierron-Bohnes V, Bakonyi I (1999) J Magn Magn Mater 198–199:243–245
139. Bakonyi I, Tóth J, Goualou L, Becsei T, Tóth-Kádár E, Schwarzacher W, Nabiyouni G (2002) J Electrochem Soc 49:C195–C201
140. Tokarz A, Wolkenberg A, Przeslawski T (2002) J Electrochem Soc 149:C607–C609
141. Bakonyi I, Tóth J, Kiss LF, Tóth-Kádár E, Péter L, Dinia A (2004) J Magn Magn Mater 269:156–167
142. Meuleman WRA, Roy S, Péter L, Bakonyi I (2004) J Electrochem Soc 151:C256–C261
143. Ueda Y, Hataya N, Zaman H (1996) J Magn Magn Mater 156:350–352
144. Jyoko Y, Kashiwabara S, Hayashi Y (1997) J Magn Magn Mater 165:280–283
145. Chassaing E, Morrone A, Schmidt JE (1999) J Electrochem Soc 146:1794–1797
146. Shima M, Salamanca-Riba L, Moffat TP, McMichael RD (1999) J Magn Magn Mater 198–199:52–54
147. Ueda Y, Houga T, Zaman H, Yamada A (1999) J Solid State Chem 147:274–280
148. Chassaing E (2001) J Electrochem Soc 148:C690-694
149. Péter L, Cziráki Á, Pogány L, Kupay Z, Bakonyi I, Uhlemann M, Herrich M, Arnold B, Bauer T, Wetzig K (2001) J Electrochem Soc 148:C168–C176
150. Gómez E, Labarta A, Llorente A, Vallés E (2002) Surf Coat Tech 153:261–266
151. Shima M, Salamanca-Riba LG, McMichael RD, Moffat TP (2002) J Electrochem Soc 149:C439–C444
152. Gómez E, Labarta A, Llorente A, Vallés E (2003) Electrochim Acta 48:1005–1013
153. Bakonyi I, Péter L, Rolik Z, Kiss-Szabó K, Kupay Z, Tóth J, Kiss LF, Pádár J (2004) Phys Rev B 70:054427(1–10)
154. Liu QX, Péter L, Pádár J, Bakonyi I (2005) J Electrochem Soc 152:C316–C323
155. Péter L, Rolik Z, Kiss LF, Tóth J, Weihnacht V, Schneider CM, Bakonyi I (2006) Phys Rev B 73:174410(1–10)
156. Bakonyi I, Simon E, Tóth BG, Péter L, Kiss LF (2009) Phys Rev B 79:174421(1–13)
157. Hua SZ, Lashmore DS, Salamanca-Riba L, Schwarzacher W, Swartzendruber LJ, McMichael RD, Bennett LH, Hart R (1994) J Appl Phys 76:6519–6521
158. Nabiyouni G, Schwarzacher W (1996) J Magn Magn Mater 156:355–356
159. Alper M, Schwarzacher W, Lane SJ (1997) J Electrochem Soc 144:2346–2352
160. O'Keeffe AP, Kasyutich OI, Schwarzacher W, de Oliveira LS, Pasa AA (1998) Appl Phys Lett 73:1002–1004
161. Cziráki Á, Zheng JG, Michel A, Czigány Zs, Nabiyouni G, Schwarzacher W, Tóth-Kádár E, Bakonyi I (1999) Z Metall 90:278–283
162. Kasyutich OI, Schwarzacher W, Fedosyuk VM, Laskarzhevskiy PA, Masliy AI (2000) J Electrochem Soc 147:2964–2968
163. Alper M (2002) Lect Notes Phys 593:111–128
164. Nabiyouni G, Schwarzacher W, Rolik Z, Bakonyi I (2002) J Magn Magn Mater 253:77–85
165. Zhang J, Moldovan M, Young DP, Podlaha EJ (2005) J Electrochem Soc 152:C626–C630
166. Kakuno EM, da Silva RC, Mattoso N, Schreiner WH, Mosca DH, Teixeira SR (1999) J Phys D: Appl Phys 32:1209–1213
167. Chassaing E, Nallet P, Trichet MF (1996) J Electrochem Soc 143:L98–L100
168. Huang Q, Young DP, Chan JY, Jiang J, Podlaha EJ (2002) J Electrochem Soc 149:C349–C354
169. Huang Q, Young DP, Podlaha EJ (2003) J Appl Phys 94:1864–1867
170. Kuru H, Kockar H, Alper M (2015) J Magn Magn Mater 373:135–139
171. Sahin T, Kockar H, Alper M (2013) J Supercond Nov Magn 26:825–829
172. Zaman H, Ikeda S, Ueda Y (1997) IEEE Trans Magn 33:3517–3519
173. Ueda Y, Houga T, Yamada A, Zaman H (1999) J Magn Soc Jpn 23:120–122
174. Bakonyi I, Tóth Kádár E, Cziráki Á, Tóth J, Kiss LF, Ulhaq-Bouillet C, Pierron-Bohnes V, Dinia A, Arnold B, Wetzig K, Santiago P, Yacamán MJ (2002) J Electrochem Soc 149:C469–C473

175. Rizal CLS, Yamada A, Hori Y, Ishida S, Matsuda M, Ueda Y (2004) Phys Stat Sol (C) 1:1756–1759
176. Neuróhr K, Péter L, Pogány L, Rafaja D, Csik A, Vad K, Molnár G, Bakonyi I (2015) J Electrochem Soc 162:D331–D340
177. Jafari Fesharaki M, Neuróhr K, Péter L, Révész Á, Pogány L, Molnár G, Bakonyi I (2016) J Electrochem Soc 163:D485–D492
178. Takane N, Narita H, Liu X, Arai S (2013) Electrochem 81:966–970
179. Rafaja D, Schimpf C, Schucknecht T, Klemm V, Péter L, Bakonyi I (2011) Acta Mater 59:2992–3001
180. Ghosh SK, Dogra A, Chowdhury P, Rajak S, Srivastava C (2015) J Alloy Compd 647:1098–1103
181. Bauer U, Yao L, Tan AJ, Agrawal P, Emori S, Tuller HL, van Dijken S, Beach GSD (2015) Nature Mat 14:174–181
182. Reichel L, Oswald S, Fähler S, Schultz L, Leistner K (2013) J Appl Phys 113:143904(1–7)
183. Duschek K, Pohl D, Fähler S, Nielsch K, Leistner K (2016) APL Mater 4:032301(1–10)
184. Duschek K, Uhlemann M, Schlörb H, Nielsch K, Leistner K (2016) Electrochem Commun 72:153–156
185. Duschek K, Petr A, Zehner J, Nielsch K, Leistner K (2018) J Mater Chem C 6:8411–8417
186. Uhlemann M, Gebert A, Herrich M, Krause A, Cziraki A, Schultz L (2003) Electrochim Acta 48:3005–3011
187. Huang CA, Chen CY (2009) Surf Coat Tech 203:3320–3324
188. Leistner K, Fähler S, Schlörb H, Schultz L (2006) Electrochem Commun 8:916–920
189. Liang D, Zangari G (2011) Electrochim Acta 56:10567–10574

Chapter 6
Nanocrystalline Deposits

6.1 General Considerations Concerning Nanocrystalline Deposits

6.1.1 Scope of This Chapter

In Chap. 6, the deposition characteristics and features of deposits occurring as continuous nanocrystalline (nc) coating on the surface of a substrate are summarized. The common feature of the deposition methods comprised in this chapter is that the thickness of the coating is much larger, sometimes by orders of magnitude, than the grain size of the deposit. In this sense, these deposits can be regarded as bulk nanocrystalline materials with nearly equiaxed crystals. The coatings to be discussed are mostly metallic.

There is a variety of deposits in which the crystallite size is below one micrometre at least in one direction, but the crystal structure is by far not equiaxed. When the nanocrystalline nature of a deposit is rather a consequence of either the dendritic or porous growth of the deposit, the physico-chemical parameters governing the growth mode are much different from those acting during the growth of compact layers. Therefore, these deposits are described together with other porous materials in Chap. 8.

When the thickness of the deposit is small, the crystallite size tends to scale with the thickness. In such cases, the nanocrystalline nature of the deposit is simply a consequence of the small thickness. When nanocrystals are formed in a single layer of well-segregated grains on a foreign substrate, their formation is influenced mostly by the nucleation phenomenon, and the structural incoherence with the substrate surface plays a key role in the granularity. These cases are discussed in Chap. 10 where all other sorts of the influence of the substrate inhomogeneity are also detailed.

L. Péter, *Electrochemical Methods of Nanostructure Preparation*,
Monographs in Electrochemistry, https://doi.org/10.1007/978-3-030-69117-2_6

6.1.2 About the History of the Research on Nanocrystalline Materials

It has long been known that the atomic arrangement at the grain boundary greatly differs from that in the bulk of a large crystal [1]. This difference leads to the idea to produce materials in which the majority of the atoms are at, or at least very near to, the boundary of a crystallite. This material family was called "all-grain-boundary materials" or "bulk interfacial material" by the pioneering researcher of the field, Gleiter, already before the extensive research in the field started [2, 3]. The research on nanocrystalline materials is one of the scarce cases when the desire for a new type of material was anticipated before any specific discovery, and the synthesis followed the announcement of the pursuit.

The early synthetic approaches to obtain nanocrystalline materials were based on a two-step technique involving vapour condensation followed by particle compacting. When the starting material, usually a metal, is evaporated either thermally or by an electron beam, the atoms are condensed already in the gas phase to form relatively small crystals. Therefore, the small crystals as the building blocks of a bulk nanocrystalline material are obtained with a bottom-up method, which provides that the nanocrystals themselves are relatively free from defects, unlike in the case when a bulk material is milled. By collecting these crystals and compacting them, a cluster-assembled bulk nanocrystalline material was hoped to synthesize.

However, it turned out that methods based on gas condensation coupled with powder consolidation suffer from effects that are likely to lead to false conclusions on nanocrystallinity in general. For this reason, some differences have been identified between nanomaterials processed by electrodeposition and powder consolidation [4]. The reason for the misleading results was at least twofold. First, the compactness of the thus obtained materials was not always sufficient to provide that the grains are in direct contact with each other to such extent that the role of the voids is negligible. Later, an approximately 98% compactness level could be achieved. Secondly, the inherent difficulty of the gas condensation method is that the boundary of the crystals to be compacted is always loaded with the typical impurities of the vacuum chamber. The compacting of the crystals with the adsorbates on their surface, however, leads to different materials than simply nanocrystalline ones without contaminated internal boundaries. Also, the porosity originating from the imperfect compacting methods allows the adsorption of foreign materials after the completion of the synthesis when a specimen is exposed to ambient atmosphere.

As compared to synthesis methods involving vacuum processes or mechanical treatments, electrodeposition proved to be fairly easy and technically much less demanding [5], ensuring a sufficient level of compactness of the deposits. Therefore, electrodeposited nanocrystalline materials played a decisive role in scientific debates on the properties of nanocrystalline materials by providing suitable specimens to be tested. Although the first wave of the research of nanocrystalline materials decayed until about 2005, electrodeposition of nanocrystalline materials remained a fashionable field to provide specimens for studying specific properties related to the nanocrystalline nature of the samples. The principles of the electrodeposition of nanocrystalline materials have been clarified for more than two decades. Therefore, the reviews dealing with the properties of electrodeposited nanomaterials written before 2010 can be taken as fully relevant even nowadays [6–9], also when application of nanomaterials is concerned [10], although the topic is revisited time by time [11]. Nevertheless, the number of papers in which electrodeposited nanocrystalline materials are dealt with keeps increasing since electroplating remained the simplest method to produce nanocrystalline materials; in particular, metals.

6.1.3 Electrochemical Techniques in the Deposition of Nanocrystalline Materials

Electrodeposition of bulk nanocrystalline materials can be performed for various substrate–deposit pairs regardless of the nucleation mode of the deposit (see Chap. 2.10.1). For relatively thick layers, the grain size distribution of the deposit becomes independent of the nucleation mode (as it was also shown in Fig. 2.20), and the near-substrate zone of the deposit is negligible when the properties of thick nanocrystalline deposits are concerned. When the lattice of the substrate and that of the deposit is incommensurable, the deposition starts with a layer in which the crystallite size scales with the deposit thickness until a stable growth mode is achieved. The latter example holds for microcrystalline deposits, while in this chapter, we deal with deposition processes in which the deposit is nanocrystalline throughout the entire deposit thickness.

The trends governing the deposition conditions and the grain size of the resulting deposits have been summarized in Chap. 2.11; hence, the most important factors are repeated here only. All experimental factors that contribute to the enhancement of the adatom concentration on the deposit surface lead to grain refinement. This is due to the increase of the nucleation rate.[1] All other experimental conditions that

---

[1] Although it is not in the scope of a monograph to deal with misinterpretations, it is important to mention here that the small grain size occurring during either high-current density d.c. plating or pulse plating is often explained with the electrochemical analogy of the Kelvin equation, i.e., with $\rho_{crit} = 2V_m\sigma/ze|\eta|$ (ρ_{crit} is the critical nuclei radius for a hemispherical grain, V_m is the molar volume of the deposit, and σ is the interfacial specific energy at the deposit–solution interface). The equation is the analogy of the vapour pressure change of a small-radius droplet as compared to a flat surface, and the intensive parameter analogous to the vapour pressure is the electrode potential (or the overvoltage). For more details, see Ref. [12].

It is to be clarified that the Kelvin equation refers to a situation when a deposit nucleus is at the surface of an inert foreign substrate. This state is the typical case of unstable equilibrium. Particles that are smaller than the critical nucleus size are expected to dissolve, while larger particles can grow; hence, any fluctuation drives the system away from its actual state to which the system will never return. However, plating with a high current density always represents a far-from-equilibrium experiment where the application of equations valid for equilibrium is strictly forbidden. While the Kelvin equation predicts a 5–25 mV equilibrium potential correction for grains of a few tens of atoms, plating modes applied for electrodeposition of nanocrystalline materails usually work in the drive the deposition process away from the near-equilibrium growth also lead to grain refinement. Most importantly, the application of additives has to be mentioned. Additives block the near-equilibrium growth sites (kink points, edges and screw dislocations), hence allowing adatom accumulation on the terrace plane surfaces. Since the nucleation of new crystals takes place via random processes, fine-grained deposits exhibit much weaker texture than coarse-grained ones. Another collateral advantage of the reduction of the crystallite size is that the deposits become smoother due to the lack of the preferential growth points and the suppression of the dendrite formation.

Pulse plating is often applied for obtaining nanocrystalline deposits. The advantage of pulse plating is that the current density applied during the on-time can be much larger than the mass transport-limited current density during d.c. plating. The shorter the on-time period, the larger current density can be applied; however, a sufficiently long off-time is required for the replenishment of the reactant by diffusion. The convenient calculation mode for treating the length of the on-time, t_{ON}, the duty cycle, $\theta = t_{ON}/(t_{ON} + t_{OFF})$, and the pulse-limiting current density for current-controlled deposition, j_{PG} is based on the double diffusion layer approximation [13, 14]. Here, the maximum pulse length is limited by the achievement of the zero reactant concentration at the cathode surface by the end of the pulse. The dimensionless pulse-limiting current density is obtained as

$$j_{PG}^{*} = \frac{j_{PG}}{j_{LIM}} = \left[\sqrt{\frac{4}{\pi}\frac{D t_{ON}}{\delta^2}}(1-\theta)^{3/2} + \theta\right]^{-1} \tag{6.1}$$

where j_{LIM} is the limiting current for d.c. deposition conditions and δ is the thickness of the diffusion layer.

Pulse plating is an equally suitable technique for plating both metallic elements and alloys in nanocrystalline form. The deposition preference may impact the deposition rate of the components of an alloy in a manner so that the deposition rate of the component of higher deposition preference is relatively high at the beginning of a pulse (similarly to d.c.-plated bulk alloys and CMAs, see Chap. 5.4.1). However, the pulse time during the deposition of nanocrystalline metals is usually relatively small. Hence, there is no physical evidence for any variation in the composition of the deposit produced during the time of a single pulse, although the calculation of the composition of pulse-plated deposits is based on this approach [14]. Concomitantly, the composition of pulse-plated alloys significantly differs from their d.c.-plated counterparts, should either the on-time current or the mean current be applied for d.c. plating. Also, it is to be emphasized that the grains can grow during several pulses and there is no strict correspondence between the content of the grains and the atoms deposited during a particular pulse.

---

high-overpotential regime of several hundred millivolts, and the resulting grains are composed of thousands of atoms.

Reverse pulse plating is also often applied to achieve nanocrystalline deposits. The concept of reverse pulse plating is that the dissolution pulse can further decrease the crystallite size of the deposit simply by the consumption of the already existing grains and then allowing the nucleation of new grains during the subsequent pulse. While reverse pulse plating obviously does not impact the composition of the deposits of a single metallic element, the composition of nanocrystalline alloys is sensitive to the reverse pulse parameters. Due to the selective nature of dissolution, reverse pulse plating leads to alloys enriched in the MN component since the LN component usually dissolves selectively. It is to be noted that a selective dissolution may lead to dealloying, hence producing a porous structure. Such deposits will be dealt with in Chap. 8.

For both pulse and reverse pulse plating, the pulse length is practically limited by the capacitive damping effect. This defines a practical lower limit of the pulse lengths at about 1 ms, below which the actual Faraday current gets very close to the d.c. plating (as averaged for the entire cycle). The practical upper limit of the pulse length is when the thickness of the pulsating diffusion layer becomes comparable to that prevailing during d.c. plating. The latter case is rather an intermittent d.c. plating in which the distinct features of pulse plating can no longer be exploited.

6.1.4 Properties of Nanocrystalline Materials

Volume ratio of the grain boundary region. As the crystallite size of a material is decreased, the fraction of the atoms situated in the grain boundary region increases. Hence, one encounters the enhancement of the role of the more or less disordered segments of the material where material constants valid for well-ordered crystals are no longer valid.

Figure 6.1 shows the volume fraction dependence of the regions far from the bulk zones for a boundary region thickness of 1 nm (for a different boundary region thickness, the exact volume fraction values may vary but the character of the dependencies is unchanged).

As can be seen from Fig. 6.1, the volume fraction of the intercrystalline regions drastically increases with the transition from microcrystalline to nanocrystalline materials. It is particularly interesting that the volume fraction of the triple-junction regions precedes that of the grain boundaries as the grains are refined. This is because triple junction is defined as a region where the boundary regions of various adjacent crystal pairs overlap. The saturation of the total volume fraction of the intergranular regions at small grain size indicates a virtually smooth transient from a nanocrystalline to a non-crystalline (amorphous) structure. The smooth transient clearly shows that the definition of amorphicity cannot be related to a sudden disappearance of the atomic-scale ordering; rather, it is usually related to the diffraction pattern of a material, and the term "X-ray amorphous" is used very often.

Structure of the grain boundary. Not only is the number of atoms being present in the grain boundaries is important but the structure of the grain boundary may

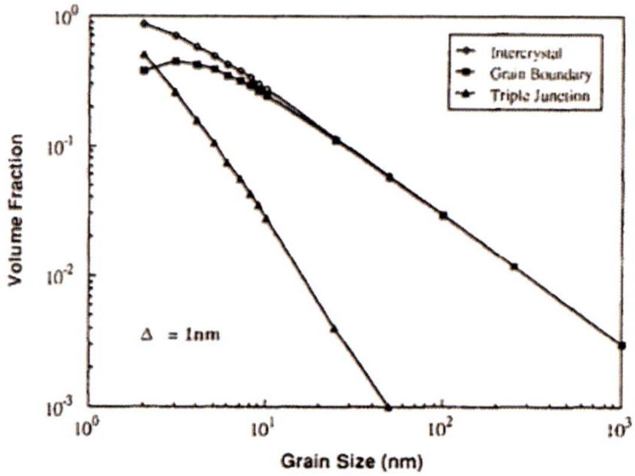

Fig. 6.1 Dependence of the volume fraction of various intercrystalline regions on the crystallite size for a boundary region thickness of $\Delta = 1$ nm. The total volume fraction of the intercrystalline region is the sum of those of the grain boundaries and the triple junctions. Reprinted from Ref [15]. Copyright (1990), with permission from Elsevier

affect material properties. Although the synthesis of nanocrystalline materials is often termed as "grain boundary engineering", this notation is by far too ambitious since one seldom can predetermine the exact boundary structure when nanocrystalline materials are produced. Instead, at most the description of the grain boundary is possible, but this is also difficult to carry out.

An important factor in the grain boundary structure is its thickness. Although this is a relatively arbitrary parameter, we can take it as the thickness of the zone where atoms are situated in positions where no species are expected to be found if either of the lattices of the neighbouring crystals is extended. For various other definitions of the thickness of the boundary region (related to either local crystallinity, local stress or positional disorder) and their relation to diffraction-based phase analysis methods, see [16].

The relative arrangement of the neighbouring crystals is often described with the ΣN parameter. Here, N is an integer that indicates the rotational correspondence of the facing crystal planes at the grain boundary. N means that every Nth of the atoms can be found in identical positions as if the neighbouring crystal were continued. The smallest value of N is 3 for relatively well-matching interfaces and it can be also above 100 for strongly misaligned neighbouring crystals. The field where the grain alignment is of high importance is the mechanical properties and deformation of nanocrystalline materials.

The grain boundary structure of many electrodeposited nanocrystalline metals was found to be similar to those of the same composition produced with other methods. At least, no distinctive feature has been identified yet for electroplated nanocrystalline materials.

Chemical aspects of the grain boundary formation: impurity accumulation. In some physical methods of materials processing (like high-pressure torsion), grain boundaries are formed as a result of the fragmentation of existing crystals, and hence, such processes have no significant chemical aspect. In contrast, electrodeposition is notorious of being influenced by all chemical species present at the surface of the newly forming phase. As it was described in connection with the additive effect, adsorbed chemical species (both molecules and ions) take part in the acceleration of the nucleation, which finally leads to grain refinement. Since the additives (or their decomposition products) may remain at the surface as adsorbate at a longer time scale than the grain growth, they can still be present at the grain boundaries in the electrodeposited nanocrystalline materials. There are only a few studies dealing with the impurity effect in the stabilization of the nanocrystalline deposits. Since the absolute concentration of the impurities is usually as low as a few tens or at most a few hundreds of ppm, a conventional EDS analysis cannot reveal the relationship between the impurity concentration and the grain size. Instead, analysis methods applied in metallurgy for the quantification of non-metallic elements can be taken advantage of for the detection of additive residues. However, the location of the impurities can be revealed by a TEM analysis only.

Establishment of the grain size. The most common methods for the assessment of the grain size, d, are application of the Scherrer equation for the full width at half maximum of a peak seen in an X-ray diffractogram recorded in the Bragg–Brentano geometry:

$$d = K\lambda/\beta \cos \Theta \tag{6.2}$$

where λ is the wavelength of the diffracted radiation, Θ is the Bragg angle, β is the line broadening (full width at half maxima, corrected with the instrumental line width). The dimensionless shape factor, K, is usually taken as 0.9.

However, the peak widening may have various reasons in addition to the grain refinement. For instance, internal stress also has a contribution to XRD peak widening. If the nanocrystalline deposit is weakly structured, the Williamson–Hall method can be used for the grain size assessment. For materials composed of cubic crystals, the full line profile analysis can be applied, even though this method requires a high-sensitivity XRD instrument, which is by far not common even in advanced physical laboratories. X-ray diffraction yields a grain size assessment in which the neighbouring grains separated by a small-angle grain boundary are taken as different crystals. This is because such boundaries break the coherence of the scattering. In contrast, electron backscattered diffraction (EBSD) performed on well-polished deposit surfaces is not sensitive to small-angle grain boundaries; hence, its result is not fully comparable to the grain size assessment of XRD. Finally, TEM yields a reliable basis for grain size assessment and statistics with the same constraint as EBSD concerning the inability to identify the low-angle boundaries. When the line profile analysis and TEM-based grain size assessment were used in parallel for electroplated metals, the results indicate that d_{TEM} is larger than d_{XRD}, as expected due to the neglect of the low-angle boundaries in the former method [17–19].

Mechanical properties of nanocrystalline materials. A key feature of the nanocrystalline materials is the hardness enhancement as the grain size decreases. This can be characterized with the Hall–Petch relationship:

$$H = H_0 + Kd^{-1/2} \tag{6.3}$$

H being the hardness, H_0 is the hardness of the same material with very large grains, and K is an empirical proportionality factor.

In parallel to the improvement of hardness, nanocrystallinity gives rise to increased yield strength in accord with a similar proportionality law. The explanation of both lies in the behaviour of dislocations. Deformations take place with the motion of the dislocations and their annihilation at the grain boundary while the crystal itself is deformed by one atomic size along a slipping plane. As the grain size decreases, the length scale span by a dislocation motion is restricted to a smaller and smaller spatial period. This gives rise to the breakdown of the Hall–Petch relationship at such a small grain size where a crystal can no longer accommodate a dislocation (about 10 nm or less). In this grain size regime, grain boundary sliding becomes the major deformation mode and the hardness decreases again. The latter is often named as inverse Hall–Petch relationship. For further details, some relevant reviews can be recommended [20, 21].

Nanocrystallinity leads to other favourable mechanical properties such as a decrease in the friction coefficient and a dramatic decrease in wear loss. Both these changes can be partly the consequence of the lower surface roughness of electroplated nanocrystalline metals as compared to their microcrystalline counterparts. While the decrease in the friction coefficient is often rated to 20–60%, the wear loss can be reduced by one or two orders of magnitude. It is straightforward that the strongly diminished wear loss can be just partly ascribed to the decrease of the friction coefficient but it is a consequence of the increased hardness, too. The same trend will be seen later for electrodeposited nanocomposites (see Chap. 7) where the incorporation of small foreign particles leads to grain refinement.

Annealing: relaxation and crystallization. Nanocrystalline materials are thermodynamically unstable, and hence, they tend to recrystallize upon annealing. However, annealing impacts the defect structure of nanocrystalline materials in various ways, depending on the temperature. As a nanocrystalline specimen is heated up, the first step is the grain boundary relaxation without a significant recrystallization during which atoms tend to move towards an energetically better position without a long-range diffusion. In parallel to the relaxation, both the dislocation density and twin fault probability decrease, while grain growth takes place at a higher temperature (characteristic temperature ranges for the above-mentioned processes highly depend on the composition of the material).

While annealing necessarily leads to a grain coarsening, the originally unimodal grain size distribution of a nanocrystalline material may develop in various ways. In materials with high purity, the grain size distribution may remain unimodal with a much larger mean value than that in the original specimen. However, when chemical segregation can also take place during annealing and it creates heterogeneous zones with significantly different diffusion characteristics, the grain size distribution may become bimodal. In this case, the large-grained zones exhibit a low density of segregation and a relatively unhindered diffusion characteristic. The fraction with small-sized grains, however, exhibits a relatively large segregation density which has a kinetic limiting effect on grain growth.

Electrical resistivity. As the grain size decreases, the resistivity of a metal increases approximately with the reciprocal grain size. This is due to the electron scattering at the grain boundaries. The grain boundary resistivity contribution, ρ_{GB} [22], can be taken as an additive factor to the resistivity originating from the point-like defects in the crystals and the thermal excitation. Interestingly, the grain boundary resistivity contribution depends on the nature of the grain boundary, i.e., low-angle grain boundaries do not acts as scattering centres. This is why grain size as determined from either EBDS or TEM can be used for the calculations, while grain size obtained from XRD yields to incorrect resistivity result. The measurement of the electrical resistivity is a sensitive and simple test of the grain size of metallic elements [23]. However, it must be kept in mind that metallic elements of relatively large self-diffusion coefficient (like Cu and Ag) tend to undergo self-annealing, and their grain size can drastically grow after the synthesis of the as-received nanocrystalline specimens, which can be easily detected by resistivity measurements. In contrast to elementary metals, the measurement of the electrical resistivity is not a proper test of the grain size of alloys since the chemical disorder itself usually causes about an order of magnitude resistivity enhancement as compared to the microcrystalline form of the constituent elements which hides the impact of the grain size on the resistivity.

Magnetic properties. Many early reports on consolidated nanocrystalline magnetic materials showed a drastic decrease in saturation magnetization as the grain size was diminished. Later on, it turned out that the consolidated nanocrystalline specimens studied were not fully compact and the magnetization loss was strongly related to the contact with reactive gases (also with air). It was shown first for electrodeposited nickel [24] that the grain size dependence of the saturation magnetization is nearly negligible, which was later confirmed for properly treated consolidated nc-Ni [25]. Numerical calculation showed for nickel [26] that the disorder occurring at the grain boundaries is responsible for at most 15% loss in saturation magnetization contribution of the atoms in the boundary layer. Nowadays, the general view is that the grain size itself does not impact significantly the saturation magnetization of magnetic metals. Studies on the saturation magnetization of electrodeposited nickel as a function of grain size show a slight but systematic decrease of saturation magnetization with decreasing grain size, which may occur as a synergetic effect of the grain boundary disorder and the preferential impurity incorporation at the grain boundaries. Since the presence of alloying elements itself and the chemical disorder due to alloying has a much larger effect on the saturation magnetization, the assessment of the structure effect alone is possible for elements only.

In contrast to the saturation magnetization, coercivity, H_C, is rather sensitive to the grain size. For microcrystalline materials, coercivity is inversely proportional to the grain size ($H_C \sim d^{-1}$). However, in nanocrystalline materials, the grain size is often smaller that the exchange correlation length. This is the limit where the local magnetocrystalline anisotropy of the grains with statistical orientation distribution becomes negligible. In this case, the local magnetic anisotropy is averaged out and the remagnetization process is determined by an effective anisotropy constant that is proportional to d^6. This means that nanocrystalline materials often show ultra-small coercivity as compared to their microcrystalline counterparts. Even though the transition from the $H_C \sim d^{-1}$ regime to the $H_C \sim d^6$ regime is well accepted as nanocrystallinity occurs [27], it is scarce that the entire crystallite size range can be mapped for one single material. For electrodeposited nanocrystalline materials, the impurities accumulated at the grain boundaries may serve as pinning centres that block the domain wall motion and may distort the behaviour as compared to theoretical predictions.

Corrosion rate of nanocrystalline metals. As it is agreed in various reviews of the corrosion of nanocrystalline metals [28–31], the crystallite size has a great impact on the corrosion rate. As it was claimed above in connection with the thermal stability, the nanocrystalline state is thermodynamically unstable, which may suggest that such materials can dissolve in an aggressive medium at a higher rate than a similar material with larger grain size. Although this is true, the higher inclination for dissolution does not necessarily leads to an increase in corrosion rate. Due to the increased driving force for dissolution, the key parameter is the fast formation of the passive layer as compared to microcrystalline materials. Hence, the corrosion rate can be significantly reduced for nanocrystalline metals as compared to their microcrystalline counterparts due to the rapid passivation and the uniformity of the passive layer. In contrast, when no passivity can take place or the corrosion rate estimation is performed with a high-current method, the dissolution rate of a nanocrystalline material is usually larger than that of the corresponding coarse-grained metal.

In either of the above cases, nanocrystalline materials exhibit an advantage that is related to the *uniformity* of corrosion. Due to the nanocrystalline structure, the possibility to local corrosion attack that often leads to a fatal corrosion-related material failure is very limited, and the corrosion of a nanocrystalline metal is rather of a laterally even rate without a development of deep cracks. This feature lends a high potential to nanocrystalline coatings in corrosion protection.

6.2 Nanocrystalline Deposits of Metallic Elements

6.2.1 Noble Metals: Au, Ag and Pd

The electrolyte solutions used for electrodeposition of nanocrystalline gold include a commercial sulphite-type solution with Au(III) [32] and an Au(I) solution prepared by the reduction of $HAuCl_4$ with 3-merkapto-1-propanesulphonate, resulting in $[Au(–S–C_3H_8–SO_3)_2]^-$ species [33, 34]. All the three above-mentioned studies demonstrate the synergetic effect of the simultaneous application of pulse plating and additives in the reduction of the grain size by a factor of 2–3. The smallest grain size achieved was around 16 nm. Figure 6.2 illustrates the trends related to the pulse current density and the application of various additives. An interesting feature of the preparation of nanocrystalline gold was that the increase in temperature had an unexpected anomalous impact on the grain size. Although the temperature increase was expected to lead to grain coarsening because of the increase in both the precursor transport rate and the surface diffusion of the adatoms, it resulted in a grain refinement [34]. Such behaviour can occur when the precursor concentration is small, the pulse current density exceeds the pulse-limiting current density, which means that the deposition process is accompanied by a side reaction. If the rate of the side reaction increases with temperature more abruptly than the deposition rate, this is well possible due to the decrease in the current efficiency. However, the description of the experiments was not sufficient to clarify the reason of the apparent anomaly. The grain growth kinetics of electrodeposited gold was studied in detail but no other properties were investigated.

Nanocrystalline silver can easily be deposited with d.c. current from complex solutions containing either a high concentration of iodide salts [35] or a silver cyanide electrolyte with commercial additives [36, 37]. The initial crystallite size of the silver deposits was around 25 nm [36, 37]. However, it was observed in all cases that the grain size changes spontaneously after the sample preparation and the as-received crystallite size increases drastically, i.e., nanocrystalline Ag samples undergo a self-annealing process at room temperature. The stabilization of the crystallite size may last as long as a week.

The self-annealing of nanocrystalline silver could be successfully studied by monitoring the electrode potential of the deposit [35]. Since the crystallite size is

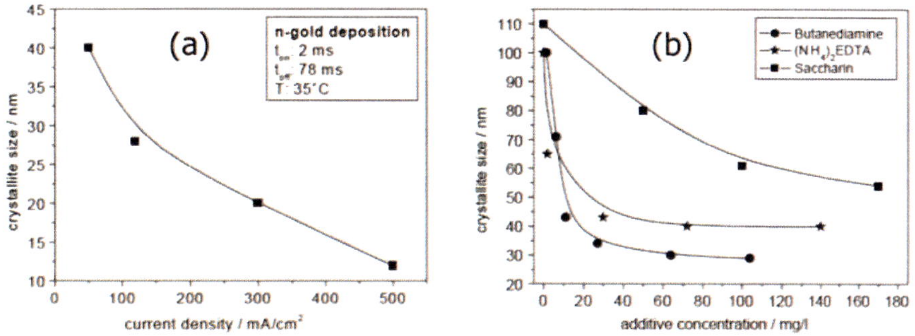

Fig. 6.2 Impact of the experimental parameters on the grain size of electrodeposited gold. **a** Grain size as a function of the pulse current density. **b** Grain size achieved with the application of various additives as a function of additive concentration at fixed deposition parameters. Reproduced from [32]. Copyright (2003), with permission from Elsevier

small, there is an electrode potential deviation from that of microcrystalline reference in the same solution. The order of magnitude of this deviation is between 6 and 10 mV, which is in accord with the theory for the grain size given. The analysis of the time evolution of the electrode potential indicated that a parabolic rate law can be applied for the self-annealing process of nanocrystalline silver.

When the self-annealing process is studied with X-ray diffractometry, not only the change in grain size can be inferred but the texture evolution is also obtained [36, 37]. It was revealed that the original texture substantially changes with time. While the as-received deposits were of <100> texture, a strong <111> texture develops within a week. The grain growth characteristics of the crystals of different orientation are also diverse. While the size of the grains with 100 orientation increases quite modestly (from about 15 nm to the double of the initial value), the 25 nm 111-oriented crystals grow up to a size of 200–240 nm (see Fig. 6.3). It is clear that the application of nanocrystalline silver is much limited by the self-annealing phenomenon, and whichever kind of application is aimed at, it should allow a spontaneous and uncontrolled change in both the crystal size and the dominant grain orientation. Note that the above-mentioned studies used the Scherrer method for the grain size assessment, and hence, the relaxation of the structural defects may have some contribution to the narrowing of the XRD peaks.

A few studies are available on nanocrystalline palladium [38–41]. Compact nanocrystalline Pd films can be deposited from a variety of solutions, containing either non-complexed, complexed or chelated Pd^{2+} species [38]. By using pulse plating, the grain size can be decreased to as low as 10 nm. Concerning the pulse parameters, a small on-time (1 ms) proved to be ideal for obtaining nanocrystalline deposits, while the process was relatively insensitive to the value of the off-time up to 80 ms. In contrast to the grain size, the surface morphology changed from globular to needle-like, even if the crystallite sizes were comparable.

Since the hydrogen solubility in Pd is quite large, nanocrystalline palladium obtained great attention concerning its hydrogen storage and diffusion properties

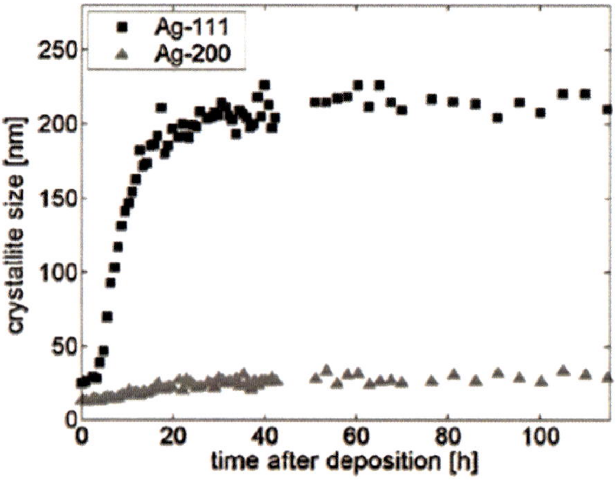

Fig. 6.3 Structural evolution of a nanocrystalline silver deposit during the self-annealing process. Reproduced from [37]. Copyright (2010), with permission from Elsevier

[39, 40]. Beside its practical importance, it is worthwhile of paying attention to this field because it is a prominent example for the difference between the properties of bulk and nanocrystalline materials. The first and most obvious observation is that hydrogen permeation through nanocrystalline palladium foils takes place at a much higher rate than in conventional polycrystalline Pd [39]. From these experiments, a 30-fold increase in hydrogen diffusivity can be concluded for nanocrystalline Pd. While it can be speculated from the permeation experiment that the greatly enhanced diffusivity is due to the high grain boundary density in the nanocrystalline specimens, this conclusion obtained direct evidence from a nuclear magnetic resonance study [40]. This revealed that the majority of dissolved hydrogen can be found at the grain boundaries where the packing of the Pd atoms is less dense than in perfect crystals. Due to the high portion of the grain boundary zones, the hydrogen solubility in the so-called α Pd–H phase increased. In contrary, the critical hydrogen concentration of the β Pd–H phase formation decreased, hence contributing to the narrowing of the two-phase regime.

Although palladium is a popular hydrogen-transfer catalyst, bulk nanocrystalline Pd is seldom used for this purpose since nanocrystals (as separate entities) are preferred for providing a high surface area. However, nanocrystalline Pd was also tested as catalyst for the oxygen reduction reaction [41]. The study of electrodeposited films with a grain size range from 13 to 30 nm showed that the most effective catalyst behaviour can be expected from the layers of highest roughness, and other parameters are of secondary importance.

6.2.2 Copper

A basic work in this field was published in 1996 by Natter and Hempelmann [42] who studied many aspects of the electrodeposition of nanocrystalline copper. In this work, several trends whose validity is much beyond the deposition of copper were demonstrated: (i) The application of pulse plating in combination with a suitable additive in the bath lead to a significant grain refinement. (ii) If the current efficiency is close to unity, the increase in temperature has an adverse effect, resulting in grain coarsening. Results related to the above-mentioned trends can be seen in Fig. 6.4.

The smallest grain size achieved by Natter and Hempelmann was smaller than 10 nm. A surprising finding was that in the presence of citric acid in an equimolar concentration with Cu^{2+}, the increase in pH lead to microcrystalline deposits, while a pH of as low as 1.5 gave the smallest grain size where a weak complex can be formed from the metal ion and the chelating ligand.

Electrodeposition of nanocrystalline copper is a relatively simple process that can be carried out with d.c. current, too [43–50]. Reports on works performed with traditional acidic copper plating baths with about 0.5 mol dm^{-3} $CuSO_4$ concentration are controversial. While some authors claim that such baths, even when additives are present, lead to polycrystalline deposits [46], others could successfully obtain nanocrystalline deposits even in the absence of additives [43]. The decisive factor may

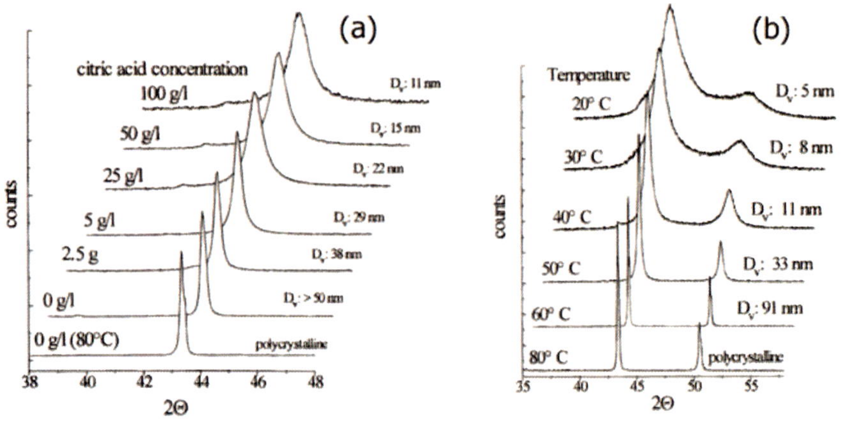

Fig. 6.4 Impact of the deposition conditions on the grain size of Cu. **a** The 111 reflection of the X-ray diffractogram for samples prepared with various additive concentrations. Samples were deposited at 40 °C and the bottom curve for a sample prepared at 40 °C is shown for comparison. **b** 111 and 200 reflections of the diffractograms for samples prepared at different temperatures with 50 g dm^{-3} citric acid concentration. In both charts, diffractograms are shifted for the sake of clarity. Reprinted with permission from [42]. Copyright (1996), American Chemical Society

be the cathodic current density whose low (10 mA cm^{-2}) and high (40–60 mA cm^{-2}) value favour microcrystalline and nanocrystalline deposits, respectively. The general conclusion is that the increase in current density is a key factor for the grain size reduction [44], although an increase in the hydrogen content of the deposit can be a collateral consequence [49]. Baths rich in additives are also suitable for d.c. plating of nanocrystalline copper [50], and the grain size of the resulting samples can be as low as 20 nm. While the grain size of nc-Cu d.c.-plated from conventional acidic sulphate solutions range from 60 to 100 nm, pulse plating is suitable to further decrease the grain size for all bath types [42, 46, 51]. In general, a long enough off-time (or, in other words, a small duty cycle) is required to fully exploit the capability of pulse plating. Additional waveforms are not superior to classical pulse plating [44].

Electrodeposition is a feasible technique to obtain thick enough Cu layers that are suitable for tensile test experiment in the self-supporting form. For such purposes, deposition of 0.25 mm [49] to 1.5 mm thick nc-Cu specimens [51, 52] is customary. Besides the usual hardness increment due to nanocrystallinity (that also leads to a diminished friction coefficient and a relatively small wear rate [46, 48]), an abnormal strain rate behaviour was observed [52, 53]. The anomaly of the strain rate means that the faster the deformation, the larger the tensile stress, although the opposite behaviour is general. This was explained with a deformation mechanism that differs from that in polycrystalline materials.

Some nc-Cu deposits were reported to be prone to self-annealing [45, 47]. Since these studies were performed with nc-Cu samples obtained from commercial baths containing proprietary components, the inclination for self-annealing cannot be directly related to the preparation conditions. The parallel study of the electrical resistivity revealed a thickness dependence of the self-annealing kinetics that was not reported for nc-Ag samples. Namely, no self-annealing was found for samples up to a thickness of 0.9 μm, above which the rate of self-annealing increased with sample thickness. Another difference as compared to nc-Ag samples was that the grain size of crystals of various orientation, 111 and 200, all increased to the 150–200 nm level upon self-annealing. In contrast to the self-annealing studies, the recrystallization of other nc-Cu samples was achieved with annealing only [51]. The latter study indicated that both the relaxation of microstrains and the grain growth has an onset temperature of about 120 °C, and the two processes take place in parallel. Parabolic grain growth kinetics was found for all annealing temperatures studied.

6.2.3 *Nickel*

Nickel was a key material in the study of electrodeposited metals. Research related to nc-Ni started well in the twentieth century [54–64], and the first short review article of the topic was published in 2000 [65]. The explanation for the role of Ni in the research on nanocrystallinity was the relative ease of Ni deposition in nanocrystalline form, the stability of the deposit, ferromagnetism of Ni and the importance of Ni in corrosion protection.

The big majority of the works related to electrodeposited nc-Ni was based on the Watts-type bath (containing nickel sulphate, nickel chloride and boric acid as major components). Baths based on nickel sulphamate also proved to be suitable [66–73], and a few studies were focused on the properties of other baths [17–19, 60]. Although there were some specific differences found between the deposits, all bath types proved to be suitable in achievement of nanocrystallinity, and the grain size was related rather to other aspects of the deposition process than the major bath components. Grain size between 12 and 200 nm could be easily achieved with any basic bath type. For decreasing the grain size below 10 nm, deep eutectic solvents had to be used [74, 75], even though this grain size regime does not offer a further property improvement and the deep eutectic solvents are considered to be very costly as compared to aqueous baths.

Additives are often used for grain refinement with any major bath type, and the variation in the additive concentration is an easy tool of grain size tuning. Saccharin is a traditional and generally used grain refinement agent in nc-Ni deposition [17–19, 62, 64, 72, 76–84], and it often leads to a tenfold or even higher degree of grain refinement (depending also on other experimental conditions). It is applied up to 20 g dm^{-3} concentration, although the maximal grain refinement efficiency is achieved at the 3 g dm^{-3} concentration (see [62, 82, 83] and references cited therein). The impact of saccharin is partly based on its slow electrochemical reduction during which sulphur atoms incorporate into the metallic deposit. Concerning the molecular-level explanation of the impact of saccharin, a study performed with additive of similar molecular structure can be recommended [85], while an attempt for a quantitative description of the additive effect based on an adsorption model was offered by Rashidi and Amadeh [82]. Other additives applied are buthanol [86], formic acid [59] and chemically non-specified proprietary additives [66].

Since the sulphur and carbon concentrations fundamentally impact the grain size of the nc-Ni deposits, their effect can be rated only if an accurate concentration measurement is performed for the electroplated specimen [62, 64, 70, 72, 87–93]. The concentration of both C and S may vary between 200 and 2200 ppm; nevertheless, the impact of sulphur is by far more important for essentially all sample properties (grain size, hardness, tensile stress and annealing behaviour). The accurate measurement of sulphur is important for translating the technical parameter of the deposition (i.e., saccharin concentration) to an inherent property of nc-Ni (i.e., the sulphur content). The two common representations of the dependence of the grain size on the saccharin addition to the bath are shown in Fig. 6.5. Additionally, the application of other sulphur-containing compounds like nickel sulphamate also lead to sulphur incorporation.

Various studies show that although the deposition mode is not crucial in achieving a grain size well below 100 nm, the fine-tuning of the deposition mode may help minimizing the grain size. Deposition with constant current is often sufficient for obtaining nanocrystalline deposits [17–19, 58, 60, 62, 66, 67, 80–84, 94, 95]. There is a general agreement in the literature that the grain size decreases with increasing current density until a limiting value is achieved, which is specific for the deposition conditions. Exceptions from this rule can also be found. Interestingly, the majority of such results refer to deposit series produced from sulphamate electrolytes [66, 71, 96]. Also, when the current efficiency of the Ni deposition process is small, the grain size can have an inverse dependence on the current density [60, 97], as it was also mentioned for Au plating baths. A theoretical model for the current density dependence of the grain size was also developed for the deposition of nc-Ni [95], although it was not yet confirmed for other metals.

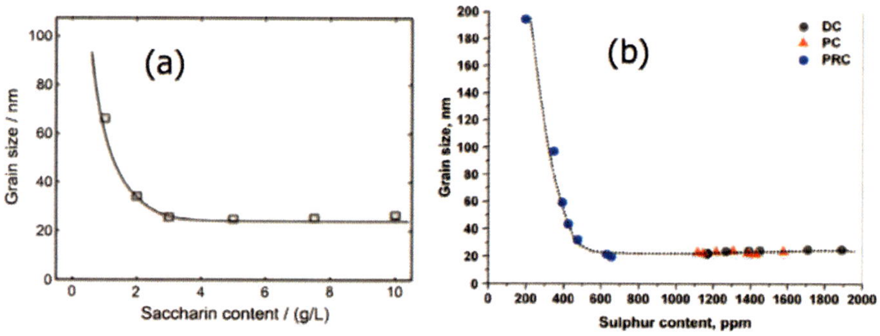

Fig. 6.5 Typical representation of the effect of sulphur incorporation into nickel deposits on the grain size. **a** Grain size as a function of the saccharin content of the bath [83]. **b** Grain size of the Ni deposit as a function of the S content of the deposit [93]. Reprinted from Refs. [83] and [93], respectively. Copyright (2010, 2016), with permission from Elsevier

Pulse plating is an adequate tool for reducing the grain size of nc-Ni below the level achievable with d.c. plating [59, 61, 64, 78, 88, 90, 92]. Studies concerning the impact of the pulse parameters show [59, 93] that a sufficiently small duty cycle is required for the grain size reduction. While the theory of pulse plating defines a practical limit for the pulse length where capacitive damping is not substantial (with $t_{ON} > \sim 0.5$ ms), several studies used ultranarrow pulses with a length of 10–200 μs [68, 77, 98, 99]. The preparation of nc-Ni with these parameters was also successful, although the background of the process is not fully understood. Some current density values applied in pulse plating of nc-Ni are also beyond the usual working range of electrodeposition, ranging up to a few A cm^{-2}, and these attempts are either related [68, 99] or unrelated [64, 71, 100, 101] to the application of ultranarrow pulses. The properties of the deposits obtained in the above-mentioned works are in line with other studies, but the impact of the application of pulses with ultrahigh current density on the current efficiency has never been mentioned. Preparation of nc-Ni with potential pulses was also exemplified [86].

The structure of the deposit is an essential issue in nearly all works dealing with nc-Ni electrodeposits. While the determination of the grain size with XRD is indispensable, the comparison of the XRD results with TEM observation is also common. Data concerning the comparison of the substrate-side and the solution-side grain sizes show that sometimes a grain refinement takes place as the deposition proceeds [63, 66]. However, in some other works, a uniform structure was found throughout the nanocrystalline deposits [17, 18, 64], and the mechanical properties measured for both sides of the deposits were in accord with the structural data.

Annealing of nc-Ni always leads to grain coarsening, hence offering a tool to produce materials with a desired grain size. However, annealing of electrodeposited nc-Ni is a complex phenomenon that may involve either segregation or abnormal grain growth. Segregation takes place when non-metallic impurities such as C and S are homogeneously distributed in the as-received specimen, and annealing leads to their accumulation at the grain boundary [69, 89, 90, 92]. This segregation has a strong impact of the mechanical properties of annealed samples. Abnormal growth [19, 76, 88, 96] is also strongly related to the grain boundary segregation. In such cases, a few grains can grow only, but the growth of many other grains is significantly hindered due to the segregation of the impurities to the grain boundary, leading to a bimodal grain size distribution.

Abundant literature data are available for the physical properties of electrodeposited nc-Ni. The study of the mechanical properties of nc-Ni has long been a core topic of the field [63, 69, 71, 72, 99, 102], including also hardness [54, 55, 81], tensile properties [87, 90, 91, 94] and wear behaviour [98, 103]. The validity of the Hall–Petch relationship is particularly often studied, for which a representative result is shown in Fig. 6.6. A few works dealt with the electrical resistivity [23, 57, 59, 60] and magnetism [56, 58, 104] of nc-Ni.

Concerning the chemical properties of nc-Ni, corrosion resistance is a key issue. It was found that the corrosion resistance of the nc-Ni specimen increase with decreasing grain size in either alkaline [84, 100, 105] or neutral chloride [84, 101] media. For sulphuric acid solutions, there is no agreement in the literature because

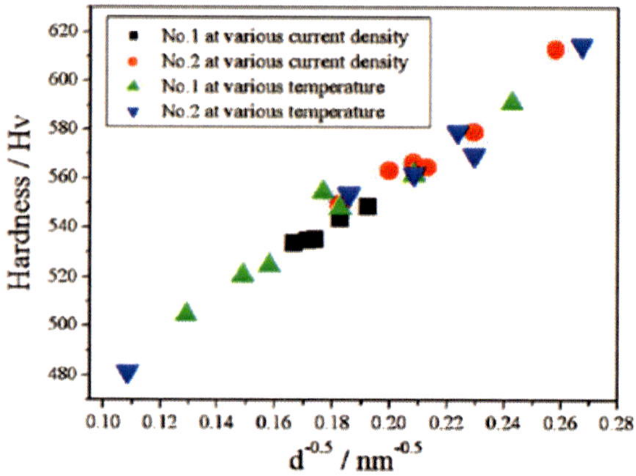

Fig. 6.6 Hall–Petch relationship for electrodeposited Ni samples produced from Watts-type baths of various saccharin concentrations (3 and 5 g dm^{-3} for No. 1 and 2, respectively) and with various deposition conditions. Reprinted from [81]. Copyright (2009), with permission from Elsevier

the decrease in grain size was rated as being either beneficial [78] or harmful [84] for the corrosion resistance. The good corrosion resistance of nc-Ni was associated with its grain boundary structure [106].

6.2.4 *Miscellaneous Transition Metals*

The electrodeposition of nanocrystalline cobalt [98, 107–110] is similar to nickel, so much that they are often mentioned in comparative studies. The generally applied electrodeposition method is pulse plating from a Watts-type bath, occasionally containing also saccharin as stress reliever and grain refiner. The resulting mean grain size falls in the 12–20 nm range. The mean grain size decreases with the increase in the pulse current density [98]. As it is common for Co electrodeposited from low-pH solutions, the nc-Co samples were reported to be of the hcp structure in all above-mentioned studies. (From solutions of higher pH, fcc-Co is occasionally obtained that exhibits nearly the same atomic distances and whose free energy differs from hcp-Co marginally. The fcc structure is favourable for Co at higher temperature.)

During the annealing of nc-Co, a relaxation process takes place around 500 K (with a wide exothermic wave in the DSC curve) followed with a recrystallization at around 625 K (narrow peak) [107, 109]. The recrystallization is complete before the hcp–fcc transformation temperature (695 K) is achieved. The thermal behaviour of electrodeposited nc-Co is influenced by the sulphur content that originates from the S-containing additive in the bath (saccharin). The higher the sulphur content, the larger the activation energy of the grain growth and the higher the peak temperature related to recrystallization in the DSC curves [109]. This behaviour was attributed to sulphur accumulation at the grain boundaries in the early stage of recrystallization that impacts the further diffusion processes at the grain boundaries.

Hardness, yield strength and ultimate tensile strength of nc-Co were larger than that of bulk microcrystalline cobalt by a factor of 2–3 [108]. However, the friction coefficient shows an adverse behaviour than that found generally for nanocrystalline materials [98]. Although the friction coefficient of nc-Co was found to be much smaller than that of nc-Ni (0.2–0.25 vs. 0.6–0.7), it slightly increased with decreasing grain size. Nevertheless, the wear loss varied with grain size in the normal way in the sense that a smaller grain size resulted in a smaller wear loss.

The corrosion behaviour of Co follows the general trend found for nanocrystalline metals. In passivating media, the passive current is significantly reduced as compared to microcrystalline Co. However, when active dissolution takes place with no passivation, the nc-Co is more active that microcrystalline Co. This is illustrated in Fig. 6.7.

In contrast to nc-Co and nc-Ni plating, electrodeposition of nc-Fe is a hard task due to the larger corrosion rate of Fe. When nc-Fe was plated from aqueous solution, pulse plating proved to be the feasible technique [111–114], while d.c. plating was also suitable from non-aqueous media [115]. For aqueous solution, either chloride-rich solutions were applied [111, 114] or chloride-free but citrate-containing solutions prepared with $(NH_4)_2Fe(SO_4)_2$ [112, 113]. Results concerning the grain size are rather controversial. While a 45 nm grain size was reported for samples prepared from an additive-free chloride-rich bath [111], a similar bath with high additive concentration led to grains in a size range of 0.22–20 μm [114]. The grain size of specimen deposited from citrate-containing solutions was 16 nm [112], but in another study applying essentially the same solution, the grain size was not reported [113]. For a non-aqueous solution prepared with ethylene glycol [115], the deposition resulted in a grain size of about 27 nm by applying both Fe(II) and Fe(III) precursor salts. The corrosion behaviour of nc-Fe was found to be in line with other nanocrystalline metals in the sense that an enhanced corrosion resistance was found for passivating

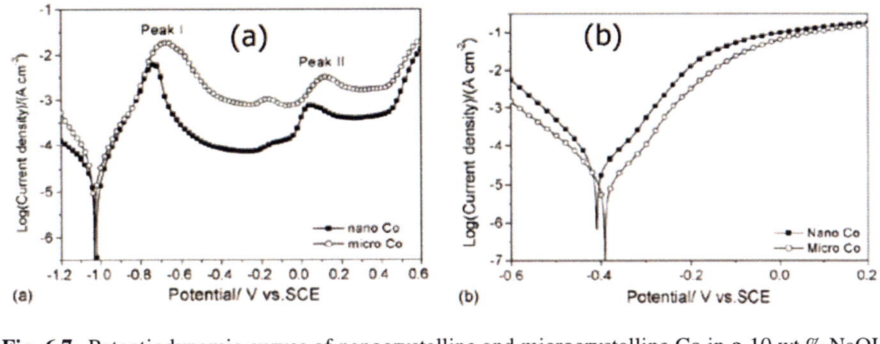

Fig. 6.7 Potentiodynamic curves of nanocrystalline and microcrystalline Co in **a** 10 wt.% NaOH and **b** 10 wt.% HCl solutions, indicating the improved and reduced corrosion resistance of nc-Co as compared to microcrystalline one in passivating and non-passivating media, respectively. Reprinted from [110]. Copyright (2007), with permission from Elsevier

media [113] but no beneficial effect of nanocrystallinity could be identified for non-passivating media [115].

Electrodeposition of nanocrystalline zinc was aimed at mainly because of the hope for improving the corrosion behaviour of Zn coatings. The exchange current density of zinc is higher than that of the iron group metals, and hence, the application of additives was expected to have relatively little impact on the grain size of Zn. Nevertheless, the application of a number of additives proved to be effective in reducing the grain size of electrodeposited zinc also under d.c.plating conditions. With d.c.plating, nanocrystallinity of Zn could be achieved with the application of either a suitable additive or a combination of several additives in either acidic [116] or alkaline [117] solution. The grain size obtained was around 45 and 22 nm, respectively. If pulse plating was applied, the application of additive was still indispensable for achieving a grain refinement [118–122], but the grain size achieved was not always smaller than that achieved with d.c. plating (20–60 nm). Nanocrystallinity was found to lead reduced surface roughness and improved corrosion resistance of Zn deposits.

6.3 Electrodeposited Nanocrystalline Alloys

Formation of essentially any electrodeposited alloy in nanocrystalline form is a relatively easy process. The works dealing with nanocrystalline alloys are so numerous that only a few representative examples can be shortly discussed here with a limited set of representative examples, in particular, studies in which nanocrystallinity is of high importance. For other examples, the reader is advised to make a literature search for the specific system.

6.3.1 Ni–Cu Alloys

Although Ni and Cu are both of fcc phase with ~2% lattice mismatch and they exhibit a wide range of miscibility, their codeposition with a single phase is hindered and often leads to two-phase deposits when a non-complexed bath is used [123]. Hence, the application of a complexing agent is essential. Since the reversible potentials of the Cu^{2+}/Cu and Ni^{2+}/Ni systems are quite much apart, the appropriate complexing agent makes a stronger complex with Cu^{2+} than with Ni^{2+}, hence promoting their intermixing in the solid phase. Traditionally, this complexing agent is citric acid [124–133], but a few other carboxylic acids have also been tested [134]. The baths used for Cu–Ni codeposition always contain Ni^{2+} in a much larger concentration than Cu^{2+}, corresponding to the normal codeposition character of the system. The metal ion ratio ranged from 2 to 50. The complexing agent is applied in at least an equimolar concentration with the copper salt (for citric acid, the concentration was at least 0.2 mol dm^{-3}). The pH of the bath was in two distinct ranges, 4–5 for acidic and

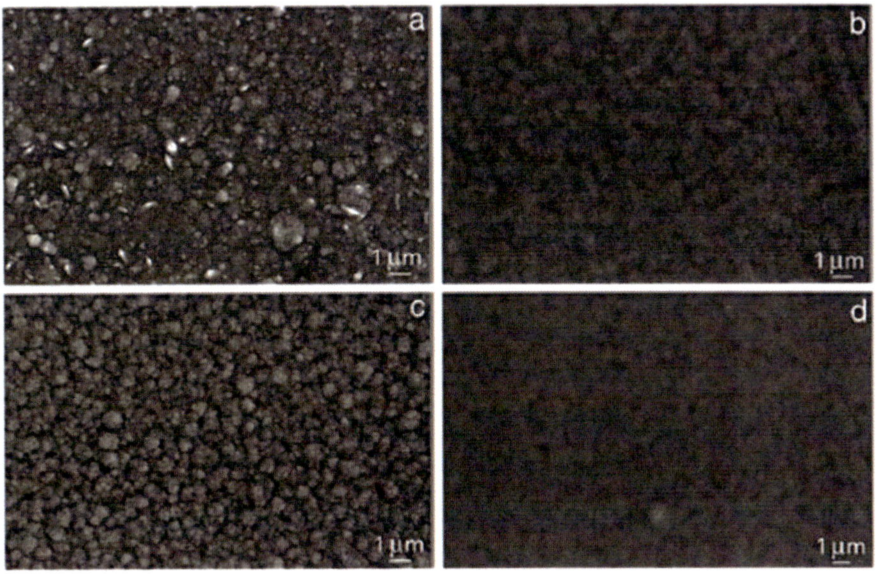

Fig. 6.8 SEM images of Cu–Ni films deposited with current densities of **(a, b)** −10 and **(c, d)** −40 mA cm^{-2} from **(a, c)** saccharine-free bath and **(b, d)** saccharine-containing bath. Nickel content of the deposits is **(a, b)** 45–47 at.% and **(c, d)** 86–87 at.%. Reprinted from [129]. Copyright (2011), with permission from Elsevier

9 for alkaline baths. The current efficiency is between 0.75 and 0.95. The available grain size was 10–30 nm depending on the conditions applied.

As it is expected for the normal codeposition, the increase in current density leads to a decrease of the Cu content of the deposit, and the same is true for the pulse current density in the case of pulse plating. The application of both pulse plating as deposition mode and saccharin as additive led to a decrease in the tendency of dendrite formation, hence helping elaborate a process suitable for practical applications. A representative image for the suppression of the surface roughness evolution upon the application of saccharin as bath additive is shown in Fig. 6.8. Nanocrystallinity of Ni–Cu alloys was reported to improve the mechanical properties and wear resistance while maintaining the good corrosion resistance of these alloys.

6.3.2 Mutual Alloys of the Iron Group Metals

A large number of iron-group alloys are important in various fields depending on their constituents and composition-dependent properties, and there is much common in the trends governing their codeposition. The so-called anomalous codeposition mode (see also Chap. 2.12.2) means that the less noble element in any of the possible metal pairs is deposited preferentially as opposed to the other metal. Therefore, the order of concentration of the metal ions in the solution has to follow a reverse trend. This is why the concentration ratio of Ni^{2+} and Fe^{2+} in the plating bath can be as high as 800, but this extreme ratio still allows the deposition of Ni–Fe alloys with a few per cent of iron. Due to the weaker deposition preference of Co against Ni, their concentration ratio is usually rather moderate ($c(Ni^{2+})/c\ (Co^{2+})$: 1–50). The mutual deposition preference is retained in ternary baths, too, in which the order of component concentrations follows the opposite trend than the deposition preference ($c\ (Ni^{2+}) > c\ (Co^{2+}) > c\ (Fe^{2+})$). In either of the baths, the increase in the current density leads to the deposit enrichment with respect to the less preferred metal.

Concerning nc-Fe–Ni alloys [135–144], their synthesis often aims at the so-called Permalloy composition (~19 wt.% Fe) at which the magnetostriction is zero and hence, the minimum of the coercivity is expected. Although the early studies of the electrodeposition of Fe–Ni alloys also dealt with this composition [145, 146], the study of their grain size and mechanical properties started much later [135–137]. It turned out that Fe–Ni alloys with grain size less than 10 nm can be obtained relatively easily with d.c. plating and by applying some suitable additive, in particular, saccharin. The modified Watts bath is the most common bath type for Fe–Ni deposition, but the deposit composition shows little variation if either all-sulphate or all-chloride baths are applied [140]. Studies dealing with the thermal stability of nanocrystalline Permalloy showed [138, 139, 143] that the grain growth starts at about 500 K, i.e., nearly at the same temperature as for Ni, but the growth rate is smaller, due to the decelerating effect of the chemical disorder. The grain coarsening upon annealing also leads to a texture change [138]. Although the improvement of the mechanical properties of nc-Fe–Ni alloys has been well documented [135–137, 139, 141, 142], the major issue is the magnetic behaviour of these alloys [138, 140, 144]. In contrast to the composition, the bath components have an immense impact on the magnetic properties of the deposits [140]. Compared with magnetron-sputtered samples of the same composition and grain size, electroplated samples exhibit much smaller coercivity than the counterparts prepared with physical deposition methods [144].

Partly similar trends have been evidenced for the properties of nc-Ni–Co deposits [140, 147–156] that for nc-Fe–Ni alloys. Pulse plating was customary for obtaining nc-Ni–Co deposits. Due to the anomalous nature of the deposition process, the increase in current density led to a decrease of Co content. In the phase composition of Ni–Co alloys, a phase transition from fcc to hcp is expected as the cobalt content increases. However, for nc-Ni–Co alloys, strict evidence for this transition was seldom found, and the fcc diffraction lines were dominant even for 75 at.% Co concentration. The complication of the phase analysis stems from the overlap of the major hcp diffraction lines with some fcc lines, and the rest of the diffraction lines belonging to the hcp crystals are very weak. The impact of saccharin on the deposit stress is more complex than for Ni. While the tensile-to-compressive stress is unambiguous for pure Ni as saccharin is applied as bath additive, the same stress reversal can be achieved up to a certain Co content only above which it does not take place. Although the thermal stability of nc-Ni–Co deposits is better than that of Ni, the grain growth at high temperature is influenced by the sulphur content the same way as for Ni (sulphur segregation at the grain boundaries, possibility of anomalous grain growth).

The systematic treatment of the nc-Fe–Co–Ni system [140, 157–161] is rather difficult due to the interplay of the variation of the composition and the grain size, and the separation of the two variables is often controversial. From structural point of view, various phase transitions take place in this ternary system (fcc, bcc and hcp for high Ni, Fe and Co mole fractions, respectively). However, the compositions associated with the phase transitions cannot be determined unambiguously, and the practical phase transition ranges strongly depend on the deposition conditions [162]. The importance of the Fe–Co–Ni system stems from the optimization opportunity of the magnetic properties. While the saturation magnetization is unambiguously determined by the deposit composition and hence, is insensitive to the bath type used, the coercivity strongly varies with the deposition conditions [140]. The minimum coercivity of electrodeposited nc-Fe–Co–Ni alloys is well below 10 Oe [140, 157, 158]. It was commonly found that the minimum of the coercivity occurs for deposits that are composed of a phase mixture in the fcc–bcc transition zone, which is the indirect consequence of the grain size minimum. It is common that the grain size minimum occurs in the phase mixture composition range because the neighbouring crystals with incompatible structure mutually impede the growth of each other.

Although zinc does not belong to the iron group, it is reasonable to discuss the deposition of nc-Ni–Zn alloys in this chapter since the anomalous nature of the codeposition of Ni and Zn. Although Zn is the metal with high deposition preference, the bath composition for the Ni–Zn deposition differs from the rest of the system mentioned above in the sense that the Zn^{2+} concentration is usually larger than that of Ni^{2+}. Due to the phase formation properties of the Ni–Zn system, alloys with Ni content less than 30 at.% can be deposited with three different crystal structures belonging to this composition range. Higher Ni content can be achieved with a pulse-reverse deposition mode, but the danger of the formation of a porous deposit is high [163]. Various baths have been tested for electrodeposition of nc-Ni–Zn alloys [163–166], and the trends found for the deposit properties varied much with the bath type.

6.3.3 Alloys of Iron Group Metals with Molybdenum or Tungsten

The induced codeposition mechanism is a common feature of the deposition of iron group metals with oxoanion-forming metals of the fifth and sixth line of the periodic table (Mo and W, respectively).This means that the iron group metals can be deposited alone, but pure Mo and W (and also Re) cannot be obtained by electrodeposition. However, when two metals from the two groups are deposited together, an alloy can be formed with continuously varying composition from the pure iron group metal to a certain composition which depends on the system studied. From chemical point of view, a suitable complexing agent is a must for achieving the codeposition. In the overwhelming majority of the relevant studies, citric acid (or its sodium or ammonium salt) is applied. The mechanistic explanation of the codeposition is based on the mixed citrate complex of the Ig^{2+} and $TrO_4{}^{n-}$ ions (Ig: iron group metal, Tr: oxoanion-forming transition metal). Without the complex facilitating the electron transfer, the reduction of the oxoanions stops at an intermediate level and a non-conducting oxide is formed, usually with a composition near to the TrO_2 formula.

From the viewpoint of the deposition process itself, plating of any binary Ig–Tr alloy fits to the following scheme: (i) If an Ig^{2+}-containing solution is enriched in the Tr compound, the Tr mole fraction increases in the deposit. (ii) As the Tr mole fraction in the deposit increases, the current efficiency drops. (iii) The impact of the increase in current density strongly depends on the deposition conditions. For baths relatively rich for the iron group metal ions, the increase in current density often leads to a larger Tr content of the deposit, while an inverse relationship is reported for baths rich in the transition metal oxoanion. (iv) The pH of the baths used may vary in a wide range. For metal-rich baths (in which $c_{\mathrm{Citrate}} < c_{\mathrm{Ig}}$), the pH is between 2 and 6, while for citrate-rich baths, the pH range is extended to 10. The pH itself is not a determining factor for the deposition process but has to be considered with all other plating conditions. (v) In parallel to the increase of the Tr content of the deposit, the grain size decreases drastically, and the deposit structure becomes amorphous (while the composition related to the structural turnover is system-specific). (vi) The variation in temperature was often found to be controversial since the increase in temperature can lead to a grain refinement, in contrast to the general trend found for many other alloys. Such irregularities are due to the variation of the deposit composition with temperature where the temperature elevation leads to a Tr enrichment and hence, a concomitant structural transition. (vi) Additives are necessary to improve the wetting of the surface (especially at low current efficiency when bubble formation is intense) and to avoid the formation of cracks in the deposits.

The grain refinement with the increase in the Tr content of the alloy make this alloy family an ideal candidate for structural studies, and the number of relevant studies amounts to several hundreds. The reason for the easy achievement of the grain size range below 10 nm probably lies in the fact that the self-diffusion coefficient of the alloy components is relatively small, and the metastable structure being formed cannot be rearranged at the temperature of the deposit formation. A representative collection of some relevant papers is listed in Table 6.1.

The outstanding feature of the Ig-Tr alloys is that the grain size range of <10 nm can be achieved fairly easily. In this range, the Hall–Petch relationship is inversed since the deformation mechanism is no longer based on dislocation slipping, and hence, hardness decreases as grains are further refined. This makes the Ig–Tr alloy group an ideal candidate to study these grain size dependent mechanical properties with samples of nearly identical composition. Another peculiar property of the Ig–Tr alloy group is that the fine-grained structure can be retained as the alloys are annealed. It is common that the grain growth is hindered by the segregation of the transition metal element and the grain structure is composed of crystals smaller than 50 nm even upon a heat treatment at 1000 K.

Table 6.1 Literature references for works on electrodeposited Ig-Tr alloys dealing with the aspects of nanocrystallinity

Constituent elements	References
Ni–Mo	[167–179]
Ni–W	[180–193]
Co–Mo	[194]
Co–W	[195, 196]
Fe–W	[181, 188]
Ni–Fe–W	[197–199]

6.3.4 *Alloys of Iron Group Metals with Palladium and Platinum*

The major motivation of the pursuits concerning the electrodeposition of the iron group metal alloys with platinum metals (Pl) is the magnetic properties of the equilibrium phase of these alloys. The Ig–Pl alloys are usually obtained as relatively disordered nanocrystalline deposits whose coercivity is small. Annealing results in a recrystallization (giving rise to the occurrence of the so-called $L1_0$ phase for Co–Pt and Fe–Pt alloys), which is accompanied with a grain growth and, importantly, a dramatic increase of coercivity, leading to coercive fields larger than 1.5 kOe. Beside the magnetic properties, hydrogen absorption and electrocatalytic properties were in the focus of research of nc-Ig–Pl alloys. A summary of the works in which nanocrystallinity of the alloys is emphasized can be seen in Table 6.2.

The major problem related to the codeposition of the iron group metals with platinum metals is the large difference in the onset potential of their deposition. In spite of this potential difference, codeposition from non-complexed acidic solutions occasionally proved to be a successful strategy [204, 206], while the application of a variety of complexing agents at pH > 5 is also common [202, 203, 205, 207, 211–217]. For solutions containing NH_4^+ ions or a chelating agent, the pH variation also results in a transition from a non-complexed bath (acidic media) to a complexed bath as the pH is varied [201, 209].

The concentration of the metal ions in the solution is relatively small. The concentration of the Ig metal ions can be as high as 0.1 mol dm^{-3} only exceptionally. Rather,

Table 6.2 Literature references for works on electrodeposited Ig-Pl alloys dealing with the aspects of nanocrystallinity

Constituent elements	References
Fe–Pd	[200–203]
Fe–Pt	[204–208]
Co–Pd	[209, 210]
Co–Pt	[211–215]
Ni–Pd	[216, 217]
Ni–Pt	[218]
Co–Fe–Pt	[207]

as the concentration of the platinum metal ions is typically a few mmol dm^{-3} only, the Ig ion concentration is only somewhat larger than this level. With the appropriate choice of the metal ion ratio and the bath components, a wide range of the deposit composition can be scanned by plating the alloys at different potentials (or, with different current densities) from one single solution [202–205, 212, 217], while other bath type lead to nearly invariant composition as the deposition potential is varied [213, 214]. For the latter case, it seems to be likely that the plating process is near to the pure mass transport control at the deposition potentials tested, which is due to the unusually small metal ion concentrations.

Regardless of the solution type, the purity of the deposits is a crucial problem. Since the platinum group metals are good electrocatalysts of the hydrogen evolution, the current efficiency of the deposition process is relatively small (10–60%), and the concomitant alkalination of the solution near the cathode leads to significant oxygen incorporation into the deposits [213]. The oxygen incorporation level is often the largest in the potential range of the formation of near-equimolar alloys [204]. The hydrogen evolution and the relatively small precursor ion concentration often lead to the development of a high roughness and porosity [207, 209, 215].

From the structural point of view, no general trends can be extracted from the available literature data. What can be outlined quite uniformly is that the increase in either the deposition current density or the overvoltage of the deposition leads to a grain refinement. The available grain size is between 5 and 20 nm. However, the impact of the pH on the grain size is controversial since grain refinement can be favoured by either the increase [201] or the decrease [209] in pH.

6.3.5 Alloys of 4d Transition Metals with Metalloid Element(s)

The direct electrodeposition of the nanocrystalline alloys of transition metals with either phosphorus or boron is the modified version of the electroplating of the amorphous alloys containing the same elements. The major difference is that the concentration of the precursor compound responsible for the metalloid content of the alloy is smaller than in the case of the production of the amorphous alloy, hence achieving a lower level of metalloid incorporation. For instance, the Ni–P alloys are nanocrystalline with a phosphorus content higher than about 4 at.%, the 10–16 at.% P range can be characterized with a mixed nanocrystalline–amorphous structure, and fully amorphous materials are obtained with electroplating at P concentrations larger than about 16 at.% [219, 220]. In studies dealing with electroplated nanocrystalline alloys with metalloid elements, the metalloid concentration is usually less than the half of the lower limit of the crystalline–amorphous transition zone.

Several baths are analogous to those developed for plating of amorphous coatings. Ni–P [221–223] and Co–P [224–226] baths contain H_3PO_3 as phosphorus source, while Fe–P baths are based on NaH_2PO_2 [227, 228]. Although NaH_2PO_2 would also be suitable to deposit Ni–P and Co–P coatings, the above choice of the precursor compounds offer a better bath stability. For boron inclusion, trimethylamine borane [229] can be used similarly to the baths developed for Ni–B amorphous alloys, but borohydride type compounds like $Na_2B_{10}H_{10}$ also proved to be feasible [230]. Unlike (Ni,Co)–(P,B) alloys that can be produced in an either an amorphous or in nanocrystalline state by melt quenching (at either sufficiently high or moderate quenching rate, respectively), alloys with carbon as the single metalloid elements are not available with this method. However, electroplating offers an opportunity to obtain nanocrystalline or even amorphous Fe–C [231] and Cr–C [232–234] alloys where the carbon content comes from various organic additives of the bath (e.g., formic acid and citric acid for Cr and Fe, respectively.)

The common structural feature of all the above-mentioned deposits is that they can be obtained by electroplating in a single-phase form in which the metalloid element is present entirely in solid solution. This means that new diffraction lines as a result of the metalloid incorporation do not arise but the usual diffraction lines of the host metal are broadened significantly. This is shown in a very spectacular graph in Fig. 6.9 in which the inverse Hall–Petch relationship is also shown for grain size smaller than 10 nm. When annealing is applied, the crystallization process takes place in various ways depending on the metalloid content. At low metalloid content (e.g., for Co–P containing 1.1 at.% P [235]) the primary process is the phosphorous segregation at the grain boundary. However, at larger metalloid concentrations [222, 224, 236], the formation of the equilibrium phases (Ni_3P, Ni_3B and Co_2P) takes place.

It is to be emphasized that should the electrodeposition of metal–metalloid alloys seem to be an easy and simple process, one often encounters difficulties that may

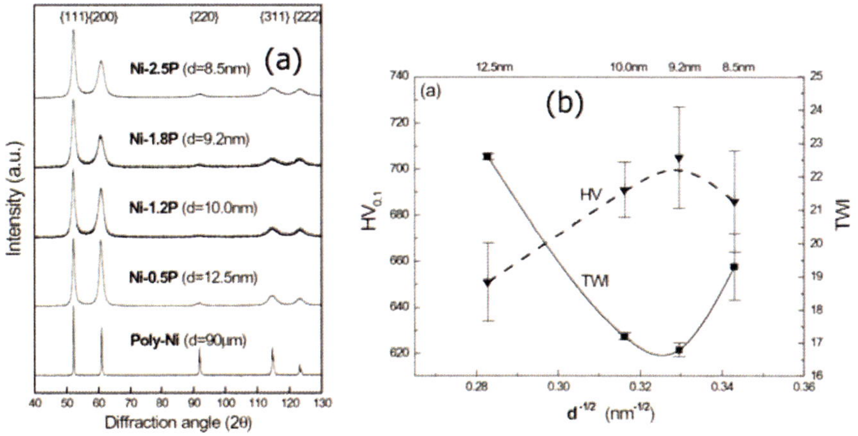

Fig. 6.9 **a** Broadening of the diffraction lines of Ni–P deposits as a result of the increase in the phosphorous content and the concomitant decrease in grain size. **b** Hardness (HV) and Taber wear index (TWI) for the same Ni–P alloys as a function of the inverse square root of the grain size. Reprinted from [223]. Copyright (2003), with permission from Elsevier

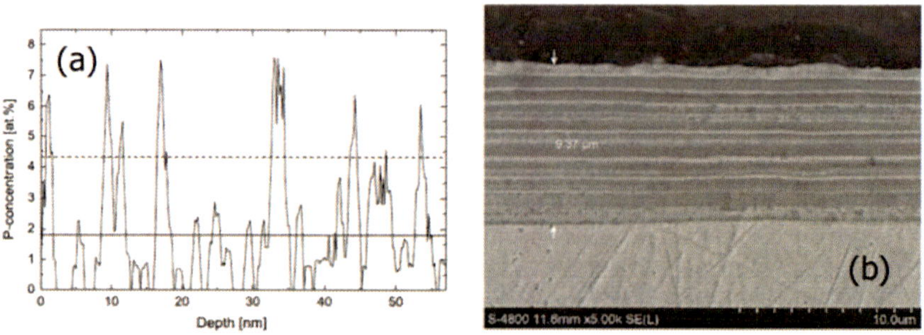

Fig. 6.10 Representations of the spontaneous composition fluctuation occurring in the deposition of metal–phosphorous deposits. **a** Composition depth profile of a Co–P deposit with 1.1 at.% P concentration. Reprinted from [235]. Copyright (2005), with permission from Elsevier. **b** Cross-sectional SEM image of a Fe–P deposit containing 13 at.% P. Reprinted from [228] with permission from Taylor & Francis Ltd., https://www.tandfonline.com

remain hidden but strongly influence the deposit properties. One of these difficulties is the strong fluctuation of the metalloid concentration in the deposit that is not revealed in simple routine diffraction measurement but can be established with composition depth profiling. Such data are now available for various deposits such as Ni–P [237], Co–P [235, 236] and Fe–P [228]. Two examples for the spontaneous composition fluctuation are shown in Fig. 6.10. The reason for this phenomenon is likely to be the oscillation of the hydrogen evolution rate during the plating process, which impacts the surface pH and the ratio of the deposition rate of the alloy components.

The major motivation of the electrodeposition of (Ni,Co)–(P,B) alloys is the improvement of the mechanical properties. Solution hardening in the as-received state combined with precipitation-hardening upon annealing offer better mechanical properties (hardness and wear resistance) than the nanocrystalline form of the pure metals. The hardness gain in the annealed nanocrystalline metal–metalloid deposits is often fourfold as compared to the pure parent metal. The improvement of the corrosion resistance of the deposit is also of importance [225, 228, 230]. The metalloid incorporation always leads to an increase of the corrosion rate, although it depends on the specific system whether the corrosion resistance of the nanocrystalline deposit can achieve that of the amorphous form with higher metalloid concentration.

References

1. Sutton AP, Balluffi NW (1995) Interfaces in crystalline materials. Clarendon Press, Oxford
2. Birringer R, Gleiter H, Klein HP, Marquardt P (1984) Phys Lett A 102:365–369
3. Gleiter H (1989) Prog Mat Sci 33:223–315
4. Erb U, Palumbo G, Szpunar B, Aust KT (1997) Nanostruct Mater 9:261–270
5. Shaw LL (2000) Min Metals Mater Soc 52:41–45
6. Erb U (1995) Nanostruct Mater 6:533–538
7. Robertson A, Erb U, Palumbo G (1999) Nanostruct Mater 12:1035–1040

8. Erb U, Aust KT, Palumbo G (2002) Electrodeposited nanocrystalline materials. In: Koch CC (ed), Nanostructures materials: processing, properties and potential applications. Noyer Publications—William Andrew Publishing, Norwich, pp 179–222
9. Erb U, Aust KT, Palumbo G (2007) Electrodeposited nanocrystalline metals, alloys and composites. In: Koch CC (ed) Nanostructured materials: processing, properties and applications, 2nd edn. William Andrew Inc., Norwich, pp 235–292
10. Koch CC (2007) J Mater Sci 42:1403–1414
11. Erb U, Palumbo G, McCrea JL (2011) Nanostructured metals and alloys. In: Whang SH (ed), Processing, microstructure, mechanical properties and applications, Woodhead Publishing, Chapter 5: The processing of bulk nanocrystalline metals and alloys by electrodeposition, pp 118–151
12. Budevski E, Staikov G, Lorentz WJ (1996) Electrochemical Phase Formation and Growth. An Introduction to the Initial Stages of Metal Deposition, Chapter 4. VCH, Weinheim–New York–Basel–Cambridge–Tokyo
13. Puippe JC, Leaman F (1986) Theory and practice of pulse plating. American Electroplaters and Surface Finishers Society, Orlando
14. Hansal WEG, Roy S (2012) Pulse Plating. Leuze Verlag KG, Bad Saulgau, Eugen G
15. Palumbo G, Thorpe SJ, Aust KT (1990) Scr Metall Mater 24:1347–1350
16. Derlet PM, Van Petegem S, Van Swygenhoven H (2005) Phys Rev B 71:024114(1–8)
17. Kolonits T, Jenei P, Tóth BG, Czigány Z, Gubicza J, Péter L, Bakonyi I (2016) J Electrochem Soc 163:D107–D114
18. Kolonits T, Jenei P, Péter L, Bakonyi I, Czigány Z, Gubicza J (2018) Surf Coat Technol 349:611–621
19. Kolonits T, Czigány Z, Péter L, Bakonyi I, Gubicza J (2019) Coatings 9:644
20. Hahn EN, Meyers MA (2015) Mater Sci Eng A 646:101–134
21. Ovid'ko IA, Valiev RZ, Zhu YT (2018) Prog Mat Sci 94:462–540
22. Bakonyi I, Isnaini VA, Kolonits T, Czigány Z, Gubicza J, Varga LK, Tóth-Kádár E, Pogány L, Péter L, Ebert H (2019) Philos Mag 99:1139–1162
23. McCrea JL, Aust KT, Palumbo G, Erb U (2000) Mat Res Soc Symp Proc 581:461–466
24. Aus MJ, Szpunar B, El-Sherik DM, Erb AM, Palumbo G, Aust KT (1992) Scr Metall Mater 27:1639–1643
25. Kisker H, Gessmann T, Würschum R, Kronmüller H, Schaefer HE (1995) Nanostruct Mater 6:925–928
26. Szpunar B, Erb U, Palumbo G, Aust KT, Lewis LJ (1996) Phys Rev B 53:5547–5556
27. Suck JB (2007) Metallic nanocrystals and their dynamical properties. In: Gemming S, Schreiber M, Suck JB (eds), Materials for tomorrow. Theory, experiments and modelling. Springer, Berlin, pp 147–190
28. Kim SH, Erb U, Aust KT, Gonzalez F, Palumbo G (2004) Plat Surf Finish 91:68–70
29. Nik Masdek NR, Alfantazi A (2010) ECS Trans 28:249–260
30. Liu L, Li Y, Wang F (2010) J Mater Sci Technol 26:1–14
31. Ralston KD, Birbilis N, Davies CHJ (2010) Scripta Mater 63:1201–1204
32. Natter H, Hempelmann R (2003) Electrochim Acta 49:51–61
33. Yevtushenko O, Natter H, Hempelmann R (2006) Thin Solid Films 515:353–356
34. Yevtushenko O, Natter H, Hempelmann R (2007) J Solid State Electrochem 11:138–143
35. Villain S, Knauth P, Schwitzgebel G (1997) J Phys Chem B 101:7452–7454
36. Hansen K, Pantleon K (2008) Scripta Mater 58:96–98
37. Pantleon K, Somers MAJ (2010) Mater Sci Eng A 528:65–71
38. Natter H, Krajewski T, Hempelmann R (1996) Ber Bunsenges Phys Chem 100:55–64
39. Natter H, Wettmann B, Heisel B, Hempelmann R (1997) J Alloy Compd 253–254:84–86
40. Sibirtsev DS, Skripov AV, Natter N, Hempelmann R (1988) Solid State Comm 108:583–586
41. Hamo E, Raviv A, Rosen BA (2019) Catalysts 9:566(1–10)
42. Natter H, Hempelmann R (1996) J Phys Chem 100:19525–19532
43. Seah CH, Mridha S, Chan LH (1999) J Mater Proc Technol 89–90:432–436
44. Ibañez A, Fatás E (2005) Surf Coat Technol 191:7–16

45. Pantleon K, Somers MAJ (2006) J Appl Phys 100:114319(1–7).
46. Tao S, Li DY (2006) Nanotechnol 17:65–78
47. Pantleon K, Somers MAJ (2006) Scripta Mater 55:283–286
48. Han Z, Lu L, Lu K (2006) Tribol Lett 21:47–52
49. Shen CJ, Zhu ZW, Zhu D (2017) Mater Sci Eng A 704:238–245
50. Seakr R (2017) Trans Nonferrous Met Soc China 27:1423–1430
51. Lu L, Tao NR, Wang LB, Ding BZ, Lu K (2001) J Appl Phys 89:6408–6414
52. Jiang Y, Hu J, Jiang Z, Lian J, Wen C (2018) Mater Sci Eng A 712:341–349
53. Lu L, Li SX, Lu K (2001) Scripta Mater 45:1163–1169
54. Hughes GD, Smith SD, Pande CS, Johnson HR, Armstrong RW (1986) Scripta Metall 20:93–97
55. El-Sherik AM, Erb U, Palumbo G, Aust KT (1992) Scripta Metall 27:1185–1188
56. Aus MJ, Szpunar B, El-Sherik AM, Erb U, Palumbo G, Aust KT (1992) Scripta Metall 27:1639–1643
57. Aus MJ, Szpunar B, Erb U, El-Sherik AM, Palumbo G, Aust KT (1994) J Appl Phys 75:3632–3634
58. Cziráki Á, Tonkovics Z, Gerőcs I, Fogarassy B, Groma I, Tóth-Kádár E, Tarnóczi T, Bakonyi I (1994) Mater Sci Eng A 179–180:531–535
59. Tóth-Kádár E, Bakonyi I, Pogány L, Cziráki Á (1996) Surf Coat Technol 88:57–65
60. Bakonyi I, Tóth-Kádár E, Pogány L, Cziráki Á, Gerőcs I, Varga-Josepovits K, Arnold B, Wetzig K (1996) Surf Coat Technol 78:124–136
61. El-Sherik AM, Erb U, Page J (1996) Surf Coat Technol 88:70–78
62. Natter H, Schmelzer M, Hempelmann R (1988) J Mater Res 13:1186–1197
63. Ebrahimi F, Bourne GR, Kelly MS, Matthews TE (1999) Nanostruct Mater 11:343–350
64. El-Sherik AM, Erb U (1995) J. Mater Sci 30:5743–5749
65. Shriram S, Mohan S, Renganathan NG, Venkatachalam R (2000) Trans IMF 78:194–197
66. Ebrahimi F, Ahmed Z (2003) J Appl Electrochem 33:733–739
67. Ebrahimi F, Ahmed Z (2003) Mater Charact 49:373–379
68. Qu NS, Zhu D, Chan KC, Lei WN (2003) Surf Coat Technol 168:123–128
69. Zhang XF, Fujita T, Pan D, Yu JS, Sakurai T, Chen MW (2010) Mater Sci Eng A 527:2297–2304
70. Savall C, Godon A, Creus J, Feaugas X (2012) Surf Coat Technol 206:4394–4402
71. Ibáñez A, Escudero-Cid R, Ocón P, Fatás E (2012) Surf Coat Technol 212:94–100
72. Matsui I, Uesugi T, Takigawa Y, Higashi K (2013) Acta Mater 61:3360–3369
73. Schüler K, Philippi B, Weinmann M, Marx VM, Vehoff H (2013) Acta Mater 61:3945–3955
74. Gu CD, You YH, Yu YL, Qu SX, Tu JP (2011) Surf Coat Technol 205:4928–4933
75. Danilov FI, Protsenko VS, Kityk AA, Shaiderov DA, Vasil'eva EA, Pramod Kumar U, Kennady CJ (2017) Prot Met Phys Chem S 53:1131–1138
76. Abraham M, Holdway P, Thuvander M, Cerezo A, Smith GDW (2002) Surf Eng 18:151–156
77. Lei WN, Zhu D, Qu NS (2002) Trans IMF 80:205–209
78. Mishra R, Balasubramaniam R (2004) Corros Sci 46:3019–3029
79. El-Sherik AM, Shirokoff J, Erb U (2005) J Alloy Compd 389:140–143
80. Moti E, Shariat MH, Bahrololoom ME (2008) J Appl Electrochem 38:605–612
81. Kang JX, Zhao WZ, Zhang GF (2009) Surf Coat Technol 203:1815–1818
82. Rashidi AM, Amadeh A (2009) Surf Coat Technol 204:353–358
83. Rashidi AM, Amadeh A (2010) J Mater Sci Technol 26:82–86
84. Qin LY, Lian JS, Jiang Q (2010) Trans Nonferrous Met Soc China 20:82–89
85. Hassan Zadeh Shirazi SM, Bahrololoom ME, Shariat MH (2016) Surf Eng Appl Electrochem 52:434–442
86. Daugherty RE, Zumbach MM, Golden TD (2017) J Appl Electrochem 47:467–477
87. Dalla Torre F, Van Swygenhoven H, Victoria M (2002) Acta Mater 50:3957–3970
88. Hibbard GD, McCrea JL, Palumbo G, Aust KT, Erb U (2002) Scripta Mater 47:83–87
89. Wang YM, Cheng S, Wei QM, Ma E, Nieh TG, Hamza A (2004) Scripta Mater 51:1023–1028
90. Yin WM, Whang SH, Mirshams RA (2005) Acta Mater 53:383–392

91. Gu C, Lian J, Jiang Z, Jiang Q (2006) Scripta Mater 54:579–584
92. Prasad MJNV, Chokshi AH (2010) Acta Mater 58:5724–5736
93. Wasekar NP, Haridoss P, Seshadri SK, Sundararajan G (2016) Surf Coat Technol 291:130–140
94. Yim TH, Yoon SC, Kim HS (2007) Mater Sci Eng A 449–451:836–840
95. Rashidi AM, Amadeh A (2008) Surf Coat Technol 202:3772–3776
96. Darnbrough JE, Flewitt PEJ (2014) Acta Mater 79:421–433
97. Bakony I, Tóth-Kádár E, Tarnóczi T, Varga LK, Cziráki Á, Gerőcs I, Fogarassy B (1993) Nanostruct Mater 3:155–161
98. Wang L, Gao Y, Xu T, Xue Q (2006) Mater Chem Phys 99:96–103
99. Shen YF, Xue WY, Wang YD, Liu ZY, Zuo L (2008) Surf Coat Technol 202:5140–5145
100. Meng G, Shao Y, Zhang T, Zhang Y, Wang F (2008) Electrochim Acta 53:5923–5926
101. Najafi Sayar P, Bahrololoom ME (2009) J Appl Electrochem 39:2489–2496
102. Shen X, Gu C, Lian J, Jiang Q, Jiang Z, Qin L (2010) J Appl Phys 108:054319(1–7)
103. Jeong DH, Gonzalez F, Palumbo G, Aust KT, Erb U (2001) Scripta Mater 44:493–499
104. Avramovic-Cingara G, Zweck J, Giallonardo JD, Palumbo G, Erb U (2018) J Appl Phys 123:234301(1–9)
105. Wang L, Zhang J, Gao Y, Xue Q, Hua L, Xu T (2006) Scripta Mater 55:657–660
106. Roy I, Yang HW, Dinh L, Lund I, Earthman JC, Mohamed FA (2008) Scripta Mater 59:305–308
107. Hibbard G, Aust KT, Palumbo G, Erb U (2001) Scripta Mater 44:513–518
108. Karimpoor AA, Erb U, Aust KT, Palumbo G (2003) Scripta Mater 49:651–656
109. Hibbard GD, Aust KT, Erb U (2006) Acta Mater 54:2501–2510
110. Wang L, Lin Y, Zeng Z, Liu W, Xue Q, Hu L, Zhang J (2007) Electrochim Acta 52:4342–4350
111. Trudeau ML (1999) Nanostruct Mater 12:55–60
112. Natter H, Schmelzer M, Löffler MS, Krill CE, Fitch A, Hempelmann R (2000) J Phys Chem B 104:2467–2476
113. Afshari V, Dehghanian C (2009) Corros Sci 51:1844–1849
114. Vijay Kumar D, Prasad MJNV (2018) Surf Coat Technol 342:121–128
115. Panzeri G, Accogli A, Gibertini E, Rinaldi C, Nobili L, Magagnin L (2018) Electrochim Acta 271:576–581
116. Li M, Luo S, Qian Y, Zhang W, Jiang L, Shen J (2007) J Electrochem Soc 154:D567–D571
117. Muralidhara HB, Arthoba Naik Y (2008) Surf Coat Technol 202:3403–3412
118. Youssef KMS, Koch CC, Fedkiw PS (2004) Corros Sci 46:51–64
119. Youssef KMS, Koch CC, Fedkiw PS (2004) J Electrochem Soc 151:C103–C111
120. Li MC, Jiang LL, Zhang WQ, Qian YH, Luo SZ, Shen JN (2007) J Solid State Electrochem 11:549–553
121. Meng G, Zhang L, Shao Y, Zhang T, Wang F (2009) Corros Sci 51:1685–1689
122. Chandrasekar MS, Shanmugasigamani, Pushpavanam M (2010) Mater Chem Phys 124:516–528
123. Sarac U, Celalettin Baykul M (2013) J Alloy Compd 552:195–201
124. Ghosh SK, Grover AK, Totlani MK (2002) Trans IMF 80:56–59
125. GhoshSK, Dey GK, Dusane RO, Grover AK (2006) J Alloy Compd 426:235–243.
126. Baskaran I, Sankara Narayanan TSN, Stephen A (2006) Mater Lett 60:1990–1995
127. Cui X, Chen W (2008) J Electrochem Soc 155:K133–K139
128. Agarwal M, Kumar V, Malladi SRK, Balasubramaniam R, Balani K (2010) JOM 62:88–92
129. Pellicer E, Varea A, Pané S, Sivaraman KM, Nelson BJ, Suriñach S, Baró MD, Sort J (2011) Surf Coat Technol 205:5285–5293
130. Calleja P, Esteve J, Cojocaru P, Magagnin L, Vallés E, Gómez E (2012) Electrochim Acta 62:381–389
131. Srinivas P, Hamann S, Wambach M, Ludwig A, Deya SR (2014) J Electrochem Soc 161:D504–D509
132. Goranova D, Avdeev G, Rashkov R (2014) Surf Coat Technol 240:204–210
133. Pellicer E, Varea A, Pané S, Nelson BJ, Menéndez E, Estrader M, Suriñach S, Baró MD Nogués J, Sort J (2010) Adv Funct Mater 20:983–991

134. Kamel MM, Anwer ZM, Abdel-Salam IT, Ibrahim IS (2014) Surf Interface Anal 46:442–448
135. Cheung C, Palumbo G, Erb U (1994) Scripta Metal Mater 31:735–740
136. Cheung C, Djuanda F, Erb U, Palumbo G (1995) Nanostruct Mater 5:513–523
137. Czerwinski F, Li H, Megret M, Szpunar JA, Clark DG, Erb U (1997) Scripta Mater 37:1967–1972
138. Czerwinski F, Szpunar JA, Erb U (2000) J Mater Sci Mater Electron 11:243–251
139. Li HQ, Ebrahimi F (2003) Acta Mater 51:3905–3913
140. Kim D, Park DY, Yoo BY, Sumodjo PTA, Myung NV (2003) Electrochim Acta 48:819–830
141. Li H, Ebrahimi F (2003) Mater Sci Eng A 347:93–101
142. Li H, Ebrahimi F (2006) Acta Mater 54:2877–2886
143. Ebrahimi F, Li H (2006) Scripta Mater 55:263–266
144. Seet HL, Li XP, Yi JB, Ooi WY, Lee KS (2008) J Alloy Compd 449:284–287
145. Cockett GH, Spencer-Timms ES (1961) J Electrochem Soc 108:906–908
146. Doyle WD (1967) J Appl Phys 38:1441–1442
147. Park DY, Song RY, Ko JM, Yoo BY, Myung NV (2005) Electrochem Solid State Lett 8:C23–C25
148. Wu BYC, Ferreira PJ, Schuh CA (2005) Metall Mater Trans A 36:1927–1936
149. Park DY, Park KS, Ko JM, Cho DH, Lim SH, Kim WY, Yoo BY, Myung NV (2006) J Electrochem Soc 153:C814–C821
150. Hibbard GD, Aust KT, Erb U (2006) Mater Sci Eng A 433:195–202
151. Wang GF, Chan KC, Zhang KF (2006) Scripta Mater 54:765–770
152. Li Y, Jiang H, Huang W, Tian H (2008) Appl Surf Sci 254:6865–6869
153. Hassani S, Raeissi K, Golozar MA (2008) J Appl Electrochem 38:689–694
154. Li J, Liu J, Jin M, Jin X (2013) J Alloy Compd 577:S151–S154
155. Qin L, Lian J, Wang G, Jiang Q (2010) J Mater Res 25:401–405
156. Tóth BG, Péter L, Révész Á, Pádár J, Bakonyi I (2010) Eur Phys J B 75:167–177
157. du Marchie van Voorthuysen EH, ten Broek FT, Chechenin NG, Boerma DO (2003) J Magn Magn Mater 266:251–257
158. Zhang Y, Ivey DG (2004) Chem Mater 16:1189–1194
159. Myung NV, Nobe K (2001) J Electrochem Soc 148:C136–C144
160. Tabakovich I, Inturi V, Riemer S (2002) J Electrochem Soc 149:C18–C22
161. Liu X, Zangari G, Shamsuzzoha M (2003) J Electrochem Soc 150:C159–C168
162. Péter L, Zuzek Rozman K, Sturm S (2018) J Electrochem Soc 165:D384–D392
163. Boonyongmaneerat Y, Saenapitak S, Saengkiettiyut K (2009) J Alloy Compd 487:479–482
164. Li GY, Lian JS, Niu LY, Jiang ZH (2005) Surf Coat Technol 191:59–67
165. Rizwan R, Mehmood M, Imran M, Ahmad J, Aslam M, Akhter JI (2007) Mater Trans 48:1558–1565
166. Mosavat SH, Bahrololoom ME, Shariat MH (2011) Appl Surf Sci 257:8311–8316
167. Chassaing E, Portail N, Levy AF, Wang G (2004) J Appl Electrochem 34:1085–1091
168. Hashimoto K, Sasaki T, Meguro S, Asami K (2004) Mater Sci Eng A 375–377:942–945
169. Donten M, Cesiulis H, Stojek Z (2005) Electrochim Acta 50:1405–1412
170. Hu J, Zheng XG, Shi YN, Lu K (2017) J Electrochem Soc 164:D348–D353
171. Hu J, Shi YN, Lu K (2018) Scripta Mater 154:182–185
172. Allahyarzadeh MH, Roozbehani B, Ashrafi A, Shadizadeh SR (2011) Surf Coat Technol 206:137–142
173. Beltowska-Lehman E, Bigos A, Indyka P, Kot M (2012) Surf Coat Technol 211:67–71
174. Ohgai T, Tanaka Y, Washio R (2013) J Solid State Electrochem 17:743–750
175. Yancheshmeh HS, Ghorbani M (2014) Surf Coat Technol 238:158–164
176. Huang PC, Hou KH, Sheu HH, Ger MD, Wang GL (2014) Surf Coat Technol 258:639–645
177. Allahyarzadeh MH, Ashrafi A, Golgoon A, Roozbehani B (2016) Mater Chem Phys 175:215–222
178. Kapoor G, Péter L, Fekete É, Lábár JL, Gubicza J (2018) Mater Charact 145:563–572
179. Kapoor G, Péter L, Fekete É, Lábár JL, Gubicza J (2018) J Alloy Compd 796:307–313
180. Yamasaki T (2001) Scripta Mater 44:1497–1502

181. Nasua T, Sakurai M, Kamiyama T, Usuki T, Uemura O, Tokumitsu K, Yamasaki T (2004) Mater Sci Eng A 375–377:163–170
182. Yamasaki T, Schloβmacher P, Ehrlich K, Ogino Y (1998) Nanostruct Mater 10:375–388
183. Iwasaki H, Higashi K, Nieh TG (2004) Scripta Mater 50:395–399
184. Somekawa H, Nieh TG, Higashi K (2004) Scripta Mater 50:1361–1365
185. Hosokawa H, Matsumoto H, Hakamada M, Mabuchi M (2006) J Mater Sci 41:8372–8376
186. Sriraman KR, Ganesh Sundara Raman S, Seshadri SK (2006) Mater Sci Eng A 418:303–311
187. Giga A, Kimoto Y, Takigawa Y, Higashi K (2006) Scripta Mater 55:143–146
188. Sriraman KR, Ganesh Sundara Raman S, Seshadri SK (2007) Mater Sci Eng A 460–461:39–45
189. Kimoto Y, Wakayama S, Fujii A, Takigawa Y, Higashi K (2007) Mater Trans 48:1483–1491
190. Haseeb ASMA, Albers U, Bade K (2008) Wear 264:106–112
191. Hou KH, Chang YF, Chang SM, Chang CH (2010) Thin Solid Films 518:7535–7540
192. Chianpairot A, Lothongkum G, Schuh CA, Boonyongmaneerat Y (2011) Corros Sci 53:1066–1071
193. Indyka P, Beltowska-Lehman E, Tarkowski L, Bigos A, García-Lecina E (2014) J Alloy Compd 590:75–79
194. Pellicer E, Gómez E, Vallés E (2006) Surf Coat Technol 201:2351–2357
195. Su FH, Huang P (2012) Mater Chem Phys 134:350–359
196. Weston DP, Gill SPA, Fay M, Harris SJ, Yap GN, Zhang D, Dinsdale K (2013) Surf Coat Technol 236:75–83
197. Ahmad J, Asami K, Takeuchi A, Louzguine DV, Inoue A (2003) Mater Trans 44:1942–1947
198. Sriraman KR, Ganesh Sundara Raman S, Seshadri SK (2006) Mater Sci Technol 22:14–20
199. Ćirović N, Spasojević P, Ribić-Zelenović L, Mašković P, Spasojević M (2015) Science Sinter 47:347–365
200. Bryden KJ, Ying JY (1997) Nanostruct Mater 9:485–488
201. Bryden KJ, Ying JY (1998) J Electrochem Soc 145:3339–3346
202. Takata FM, Pattanaik G, Soffa WA, Sumodjo PTA, Zangari G (2008) Electrochem Commun 10:568–571
203. Hernández SC, Yoo BY, Stefanescu E, Khizroev S, Myung NV (2008) Electrochim Acta 53:5621–5627
204. Leistner K, Oswald S, Thomas J, Fähler S, Schlörb H, Schultz L (2006) Electrochim Acta 52:194–199
205. Liang D, Mallett JJ, Zangari G (2010) Electrochim Acta 55:8100–8104
206. Rhen FMF, Coey JMD (2010) J Magn Magn Mater 322:1572–1575
207. Teh SH, Yaacob II (2011) Mater Res Innov 15:196–201
208. Mallett JJ, Svedberg EB, Sayan S, Shapiro AJ, Wielunski L, Madey TE, Egelhoff WF, Moffat TP (2004) Electrochem Solid State Lett 7:C121–C124
209. Noce RD, Barelli N, Sumodjo PTA, Cornejo BD, Benedetti AV (2006) J Magn Magn Mater 306:199–203
210. Noce RD, Barelli N, Marques RFC, Sumodjo PTA, Benedetti AV (2007) Surf Coat Technol 202:107–113
211. Wang F, Hosoiri K, Doi S, Okamoto N, Kuzushima T, Totsuka T, Watanabe T (2004) Electrochem Commun 6:1149–1152
212. Eagleton TS, Mallet J, Cheng X, Wang J, Chien CL, Searson PC (2005) J Electrochem Soc 152:C27–C31
213. Žužek Rožman K, Krause A, Leistner K, Fähler S, Schultz L, Schlörb H (2007) J Magn Magn Mater 314:116–121
214. Žužek Rožman K, Kovač J, McGuiness PJ, Samardžija Z, Markoli B, Kobe S (2010) Thin Solid Films 518:1751–1755
215. The SH, Johan MR, Yaacob II (2014) Mater Res Innov 18:S6–286–S6–290
216. Suresh Kumar K, Haridoss P, Seshadri SK (2008) Surf Coat Technol 202:1764–1770
217. Mech K, Chopart JP, Wróbel M, Zabinski P, Szaciłowski K (2017) J Electrochem Soc 164:D613–D620

218. Mallett JJ, Svedberg EB, Bonevich JE, Shapiro AJ, Egelhoff WF, Moffat TP (2008) J Electrochem Soc 155:D1–D9
219. McMahon G, Erb U (1989) J Mater Sci Lett 8:865–868
220. McMahon G, Erb U (1989) Microstruct Sci 17:447–457
221. Kobayashi S, Kamata A, Watanabe T (2009) Scr Mater 61:1032–1035
222. Kobayashi S, Kashikura Y (2003) Mater Sci Eng A 358:76–83
223. Jeong DH, Erb U, Aust KT, Palumbo G (2003) Scripta Mater 48:1067–1072
224. Sheikholeslam MA, Enayati MH, Raeissi K (2008) Mater Lett 62:3629–3631
225. Sheikholeslam MA, Raeissi K, Enayati MH (2010) Trans IMF 88:324–329
226. Kosta I, Vallés E, Gómez E, Sarret M, Müller C (2011) Mater Lett 65:2849–2851
227. Mikó A, Hempelmann R, Lakatos-Varsányi M, Kálmán E (2006) Electrochem Solid State Lett 9:C126–C130
228. Kovalska N, Hansal WEG, Tsyntsaru N, Cesiulis H, Gebert A, Kautek W (2019) Trans IMF 97:89–94
229. Lee KH, Chang D, Kwon SC (2005) Electrochim Acta 50:4538–4543
230. Bekish YN, Poznyak SK, Tsybulskaya LS, Gaevskaya TV (2010) Electrochim Acta 55:2223–2231
231. Müller T, Grimwood J, Bachmaier A, Pippan R (2018) Metals 8:363(1–13)
232. Przeniosło R, Wagner J, Natter H, Hempelmann R, Wagner W (2001) J Alloy Compd 328:259–263
233. Hoshino S, Laitinen HA, Hoflund GB (1986) J Electrochem Soc 133:681–685
234. Tsai RY, Wu ST (1990) J Electrochem Soc 137:3057–3060
235. Choi P, da Silva M, Klement U, Al-Kassab T, Kirchheim R (2005) Acta Mater 53:4473–4481
236. da Silva M, Wille C, Klement U, Choi P, Al-Kassab T (2007) Mater Sci Eng A 445–446:31–39
237. Nee CC, Weil R (1985) Surf Technol 25:7–15

Chapter 7
Composites

7.1 Composite Preparation by Codeposition of Metals

7.1.1 Principles of the Direct Codeposition of Composites and Their Precursor Alloys

The first group of granular materials discussed is the metallic composites in which both the matrix and the incorporated entities are formed during the deposition process (i.e., there are no pre-existing particles that incorporate into the deposit). The basic requirement for the electrodeposition of such a metallic composite is that the constituents must exhibit a nearly complete immiscibility under equilibrium conditions. Since the equilibrium properties of an alloy are not necessarily indicative of the phase(s) formed during the electroplating process, one encounters a bifurcation, depending on whether the deposit shows the equilibrium properties of the metal pair or a metastable alloy can form.

The first case is when the mixing enthalpy of the metals exhibits a large positive value and their structural properties are quite dissimilar, i.e., they have either different crystal structures or the lattice plane distances differ significantly if they are of the same crystalline form. If these conditions prevail, the metals usually crystallize independently of each other without forming any alloy, and the deposit is the so-called mechanical mixture. For many of the baths used for the deposition of mechanical mixtures, the deposition potentials of the constituents can be observed in the voltammograms in the same manner as it was shown for electrodeposited multilayers (see Fig. 5.3). Since the crystallization processes are fully unrelated to each other, the deposition potentials of the constituents are either the same as in the absence of the other reactant or that of the less noble metal may become more negative due to the nucleation barrier. However, no difference can be observed in the dissolution characteristics of the phases in a mechanical mixture as compared to their elemental forms. The codeposition mode of the mechanical mixtures is exclusively the regular one due to the absence of any interference of the deposition processes apart from the spatial

L. Péter, *Electrochemical Methods of Nanostructure Preparation*,
Monographs in Electrochemistry, https://doi.org/10.1007/978-3-030-69117-2_7

hindrance of the crystallization that reduces the grain size. The X-ray diffraction pattern of mechanical mixtures shows the diffraction lines of both constituents in the entire range of mean composition that is available with electrodeposition. Since no impurity incorporation takes place, the position of the diffraction lines is the same as for the pure elements. Many examples for such systems can be found in Watanabe's work [1] (e.g., Sn–Zn, Cu–Pb, Cd–Zn, Cd–Sn, Ag–Cu, Ag–Co) and elsewhere (e.g., Co–Pb [2]). Electrodeposited mechanical mixtures can be nanogranular materials in their as-received state. In exceptional cases when the crystallite size of both metals is smaller than about 10 nm, a single diffraction peak can only be observed [3].

In other cases, the electrodeposition process results in a metastable alloy. This is the intermediate of the granular material that can be obtained with an annealing treatment. The formation of a metastable solid solution-type alloy is possible when the crystalline forms of the constituents are structurally related to each other, e.g., they have the same crystal structure with nearly the same lattice constant, even though the equilibrium miscibility is nearly zero. The Cu–Co system is a prominent example of a metal pair that is prone to produce metastable alloys. Cu and Co both can form fcc crystals (although for Co it is stable only above 422 °C), but the Cu impurity easily induces its formation also at room temperature. The lattice mismatch of pure Cu and Co is about 2%.

It should be noted that the formation of either mechanical mixtures or metastable alloys can be regulated to some extent with the bath chemistry. As it was discussed in Chap. 2.11, the difference in the deposition potentials of the constituents affects their ability for alloy formation. While a high difference in the deposition potentials favours the formation of segregated crystals, nearly equal deposition potentials are beneficial for the formation of an alloy. Even if an alloy cannot form, a decrease in the difference in the deposition potentials leads to a reduction of the grain size, hence improving the nanogranular nature of the deposit. A change of the deposition potentials can be achieved by using a suitable complexing agent. The obvious goal is a large shift of the deposition potential of the MN metal to the negative direction, while the deposition potential of the LN metal should not be affected in the ideal case (or can decrease to a much smaller extent than that of the MN metal).

As an example of the impact of the complexing agent on the deposition of a mechanical mixture, a systematic study on the Cu–Ag system can be given here [4]. In the absence of a complexing agent, the growth of the Cu–Ag mixture leads to a dendritic coating, in agreement with the general trend that the deposition of metals with high exchange current densities leads to dendrite formation. This is particularly true for mixtures of metals with a large difference in the deposition potentials, where the MN metal is deposited with a diffusion-limited rate in the potential regime of the LN metal codeposition. In the absence of complexing agents, the lattice distances of Cu and Ag appear separately in the diffractograms. The deposit is already nanogranular under such deposition conditions since the immiscible metals mutually hinder the growth of each other, leading to a 10–30 nm grain size. Upon the application of a complexing agent (which was thiourea in the study cited above), the grain size decreased further down to 5 nm. However, the main effect is much beyond the grain size reduction itself since the deposit shows a single phase as if an alloy were formed with continuously varying composition. The collateral advantages of the application of the complexing agent include the diminished overall deposition rate and the decrease in the difference in the deposition potentials of the constituent. The latter change is important in the design of the alloy composition because the composition vs. deposition potential function changes much less abruptly than in the absence of the complexing agent.

It is common that pulse electrodeposition is used to obtain granular materials or their precursor metastable alloys. The reason for applying pulse plating is not to achieve layered deposits, in contrast to the cases presented in Chap. 5.4, and the pulse lengths are mostly not even appropriate for this purpose. The verification of pulse plating can be one of the following arguments. First, if the LN metal is codeposited with MN one, the deposition of the MN metal takes place with diffusion-limited rate, which often leads to dendritic growth. In order to avoid the dendrite formation, it is worth of operating the bath at a current density lower than the diffusion-limited current density of the MN metal in the low-current pulse. This can lead to a deposit with a relatively small roughness. The current density is to be increased only for the short periods when the LN metal is deposited, but this usually leads to a negligible change in the surface roughness. The pulse lengths can be easily chosen to deposit submonolayer quantities during each pulse, hence obtaining a homogeneous alloy also in the composition range where the constituent metals might segregate. This approach was elaborated for Cu–Ni codeposition and called by the authors as "precision electrodeposition" [5].

Secondly, the application of pulses can be useful if the nucleation barrier of the LN metal onto the MN one is large. This makes it necessary to apply a large negative potential to deposit the LN metal at all; however, the long-term maintenance of this large negative potential can lead to a continuous growth of the LN metal at a high rate. This leads to that the available mole fraction range is not favourable since a matrix composed of the MN metal with only a small amount of the LN metal becomes impossible. The start of the codeposition of the LN metal turns the system immediately from MN rich to LN rich. Hence, the application of current or potential pulses makes it possible to tune the deposit composition with a larger precision than the application of constant potential or current density.

Many granular deposits were investigated because of their magnetic properties. In such cases, the small magnetic particles have to be accommodated in a non-magnetic matrix. If the particle size is at most a few tens of nanometres, such systems have special magnetic properties and also show GMR (for the definition and the background related to GMR, see Chap. 5.4.6). The background of the GMR in granular systems is essentially the same as for layered materials. The major difference is that in granular materials, the random distribution of the particles does not allow the occurrence of an oscillatory magnetic coupling as a function of the distance of the magnetic entities since it always has a statistical distribution. The two major types of composites with non-magnetic matrix and magnetic particles will be summarized in the following chapters, and then other metallic composites will be discussed.

7.1.2 Cu(Co) Alloys

Granular Cu(Co) alloys were essentially all deposited with sodium citrate as bath additive. The works on d.c.-plated Cu–Cu alloys can be divided into two distinct groups from the viewpoint of the structure of the deposits. If either no sodium citrate was added to the bath (a few samples reported in [6]) or the sodium citrate concentration was less than the total metal ion concentration [7, 8], the hcp phase of the cobalt always appeared beside the fcc Cu (or, beside a Cu-rich alloy that might contain traces of Co). However, in all cases when the concentration of the citrate ions was large enough [6, 9–12], alloys with fcc structure could be produced without the indication of the formation of the hcp cobalt phase. As deducted from the XRD pattern, the purely fcc alloys obtained appear to be homogeneous; at least, the major XRD peak can be fitted with one single function (mostly Lorentzian) and the mean lattice distance varies systematically with the Co content as expected from the difference in the atomic volume of the Cu and Co atoms. For the correct elucidation of the result, one has to take into account that the sensitivity of the XRD method is usually claimed to be about 2 vol.%, and even this limit can be achieved only if one has a mixture of well-crystallized distinct phases. For a mixture of crystals with continuously varying composition, the sensitivity of XRD is much worse.

In contrast, the magnetization and magnetoresistance properties of the d.c.-plated deposits tell a different information on the structure and component distribution of the Co–Cu alloys. Even the alloys found to be "homogeneous" with XRD show either a significant saturation magnetization [6, 9], a measurable GMR [10] or both [12] in their as-received state, although the mean composition would not allow it if the sample were truly homogeneous. Therefore, it is obvious that the XRD patterns obtained show a mean lattice distance for the distribution of which the width of the XRD peak is somewhat indicative. However, the local fluctuation of the concentration can be rather high and the Co-rich grains can grow large enough to become magnetic already in the as-received state.

The solution composition used for the deposition of Cu(Co) granular alloys is similar to those used for multilayer deposition in the sense that the Co^{2+} concentration is significantly larger than that of the Cu^{2+} concentration. Concerning the deposit composition as a function of bath operation parameters, the influence of essentially all parameters can be taken as the natural consequence of the principles described in Chap. 2. These are as follows [11]: (i) The increase in the current density (in excess of the Cu diffusion-limited current density) leads to larger Co mole fraction in the deposit. (ii) If the bath temperature is increased under otherwise unchanged deposition conditions, the Co mole fraction in the deposit decreases, which is due to the enhanced Cu deposition rate. (iii) The increase in Co^{2+} concentration results in Co enrichment in the deposit. This latter trend does not originate from electrochemical principles but is the result of the competition between the growth of Co-rich and Cu-rich zones.

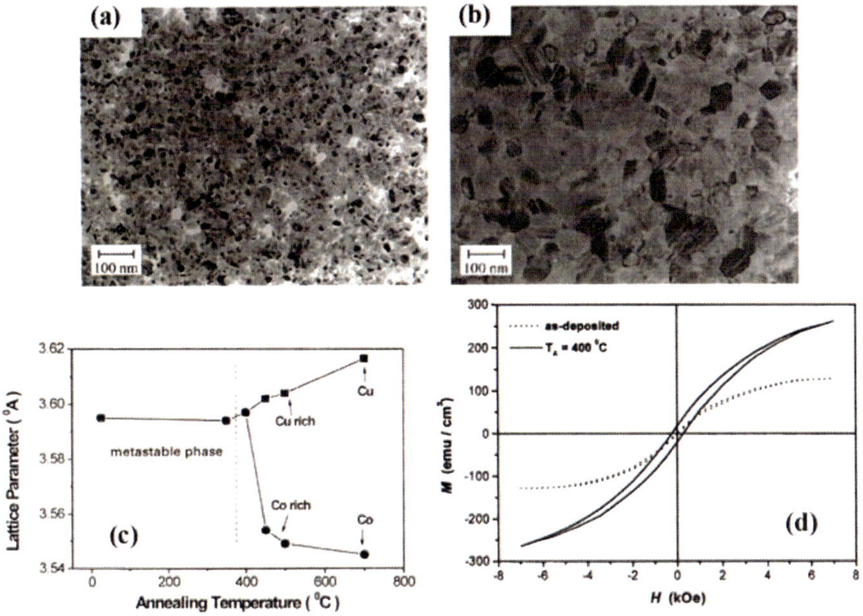

Fig. 7.1 Properties of granular $Cu_{75}Co_{25}$ deposits. TEM image of the grain size structure in the **a** as-received state and **b** after annealing at 400 °C, indicating the grain coarsening. **c** Variation of the lattice parameters as a function of the annealing temperature (observation of the phase separation). **d** Magnetization curves in the as-received and annealed states. Reprinted from [10]. Copyright (2000), with permission from Elsevier

Annealing is a straightforward method to improve the properties of the granular Cu–Co deposits. When annealing is carried out, various changes occur in the deposit properties [8–13] for which a summary is shown in Fig. 7.1.

(i) Co atoms segregate from the metastable Cu–Co alloy. New lines appear in the X-ray diffraction pattern that correspond to the pure Co phase. In parallel, the diffraction line of the Cu-rich phase is shifted and returns to the position characteristic of pure Cu. (ii) A grain coarsening takes place. (iii) The resistivity of the sample decreases. This is because the weighted average of the resistivity of all phases and that of the grain boundaries is less than the resistivity originating from the disorder in the metastable Cu–Co crystals. (iv) The saturation magnetization increases upon annealing. The reason of the increase in magnetization is that Co atoms embedded in a non-magnetic matrix have no contribution to the total magnetization in the as-received state. As a result of annealing, the segregation of Co grains enhances the number of Co atoms that form grains with nonzero total magnetization, should they belong to either SPM or FM particles. (v) The magnetoresistance ratio first increases, then above a certain temperature decreases. The explanation of the initial increase of the GMR is at least twofold, including the resistivity decrease of the matrix around the magnetic particles as well as the increase in the spin-dependent electron scattering intensity due to the increase in both the number and the volume of the magnetic particles. The decrease in GMR at annealing temperatures above 450 °C is due to the formation of larger ferromagnetic Co grains instead of very small particles with superparamagnetic properties.

The application of pulse plating is also widespread for producing granular Cu–Co deposits [6, 14–16]. Pulse-plated deposits show a smaller surface roughness than their d.c.-plated counterparts. The morphological features of the pulse-plated deposits are much more sensitive to the application of bath additives than in the case of the d.c. plating. The reason for the enhanced additive sensitivity probably lies in that the impact of the additives on the electrodeposition itself is less strong than on the exchange reaction during the off-time. Concerning the composition, it was found that the increase in the off-time leads to the decrease in the Co content of the deposit, which is an obvious consequence of the occurrence of the spontaneous exchange between Co and Cu^{2+}. Depending on the choice of the pulse parameters, pulse-plated deposits can exhibit a wide range of grain structure. Granular materials without homogeneous columns or layers can be achieved if the on-time is short enough so that the grains formed during the time of a single pulse do not coalesce. It is to be noted that two-pulse plated Cu–Cu alloys show somewhat different morphological properties. A clear layered-to-granular structural transition can be observed if either the current density in the high-current pulse is decreased [17] or the Cu^{2+} concentration in the bath is enhanced while leaving other deposition conditions unchanged [18]. However, there is always a Cu layer between the alloyed parts of the two-pulse plated deposit when the deposition conditions are set in accord with the principles discussed for multilayer deposits (see Chap. 5).

7.1.3 Ag(Co) Alloys

The Ag–Co pair is a typical example for immiscible metals. In contrast to the Cu–Co system, the large difference in the natural nearest-neighbour lattice plane distances of these two elements (14%) makes it impossible to match the two lattices to each other with little strain. This renders the Ag–Co mixtures to be nanocrystalline due to the mechanical stress that cannot be relaxed with a small deformation for large crystals. Beside the large lattice mismatch, the formation of metastable mixtures from Ag and Co is also much hindered. Therefore, the codeposition of Ag and Co leads to the mechanical mixture of the pure phases of the constituents.

Due to the weak adherence between the Co and Ag crystals in a composite deposit, the electroplating process often leads to porous deposits with insufficient mechanical cohesion. The incompact character of the deposit verifies the versatile pursuits for obtaining deposits of good quality. The most frequently mentioned bath contains sodium chloride as supporting electrolyte whose concentration ranged from 2.4 M [19, 20] to 3.5 M [21–23]. In these baths, the Ag^{+} concentration is 1–2 mM, and the Co^{2+} concentration varies from one work to another. Concerning the bath composition, other works mention metal sulphates with sodium citrate [24], sodium gluconate and thiourea as complexing agents beside metal perchlorates [25, 26], sodium thiosulphate as complexing agent [27] and a solution containing the cyanide complex of silver, potassium pyrophosphate and oxalic acid [28]. Similarly to many other structurally inconsistent deposits, it was shown that pulse plating with optimized conditions reduces the surface roughness of the deposits [25].

In the majority of studies, the X-ray diffractograms show separate lines for Ag and Co [22, 24, 28]. Nevertheless, it is difficult to perform a correct phase analysis especially for low-cobalt deposits for two reasons. First, the scattering intensity of Ag is much stronger than that of Co; hence, the Ag lines dominate the diffractogram. Secondly, the peak positions for fcc and hcp cobalt are very close to each other, which makes them nearly indistinguishable if the peak intensity is small and the peak position is uncertain. Although the formation of pure Ag and Co crystals was the general experience, one work mentioned the possibility of the formation of a metastable $CoAg_3$ phase [26]. This new metastable phase showed hcp structure, similar to the $CeAg_3$ and $GaAg_3$ intermetallic phases.

The crystallite size of the Ag(Co) deposits as obtained from X-ray diffractograms mostly refers to the matrix. The average grain size of Ag was reported to be less than 10 nm [20]. The estimate from the XRD for the Co grain size was 7 nm for a relatively Co-rich deposit [20] but TEM offers a more accurate method to assess the size of the randomly distributed Co particles. This was found to be 1–2 nm [22]. This diameter range was well confirmed by the measurement of the so-called blocking temperature of the paramagnetic Co grains, which yielded a mean Co grain diameter of 3 nm [19]. The simultaneous application of the TEM and magnetization-based methods revealed that the diameter of Co grains increases with the Co concentration in the deposit [20]. Although the grain size may vary depending on the deposition conditions, it seems to be a general trend that the size of the Co particles is smaller than the crystallite size of the Ag matrix.

For the majority of the Ag(Co) deposits, magnetoresistance was measured. The assembly of segregated nanoscale magnetic (mostly SPM) particles in the NM matrix gives rise to a significant GMR effect. Both the magnetization and magnetoresistance curves are of SPM character. The GMR as a function of the Co content of the deposit shows a maximum. This is general for NM(SPM) materials as the increase of GMR in the range of small concentration is due to the growing contribution of the electron scattering of magnetic origin, while at high concentration, the percolation of the magnetic entities cancels the independence of the local magnetizations. The Co concentration with maximum room-temperature GMR depends on the deposition conditions (like the current density) and falls in the interval of 20–50 at.% Co [20–24].

Annealing of Ag(Co) deposits leads exclusively to a grain growth [29], which is accompanied with the decrease in GMR [21, 24]. The latter is explained with the decrease in the number of independent magnetic particles and the concomitantly diminished spin-dependent electron scattering contribution.

7.1.4 Cu(Ag) Alloys

Cu(Ag) composites gained much attention due to their prospective application in the microelectronic industry as oxidation- and electromigration-resistant interconnect material in which the favourably small resistivity of Cu can be retained. The target range of the composition is < ~15 wt.% Ag. Cu–Ag deposits with a silver content larger than 10% usually show segregation, which can be accompanied by the formation of a porous deposit [30]. Even if the deposit of high silver content is compact, the X-ray diffractograms indicate [31] that the copper lattice cannot accommodate the large excess of silver and the two metals are segregated.

Numerous bath types for obtaining Cu-rich Cu(Ag) alloys were proposed for industrial application. Acidic baths were mostly formulated from the usual copper sulphate bath containing sulphuric acid by adding $AgNO_3$ and used either without [32] or with [33–35] some additives. Another acidic bath is based on methanesulphonic acid and methanesulphonate salts of both Ag and Cu [31]. One family of the alkaline bath was based on cyanide compounds such as KCN, CuCN and $K[Ag(CN)_2]$ [36–38]. A cyanide-free alkaline bath with $K_4P_2O_7$ and KI has also been elaborated [39, 40]. Regardless of the anion, the complexing agent and the additive used in these baths, their common feature is that Ag^+ is discharged at more positive potential than the Cu-containing species. The standard potential difference of the Ag^+/Ag and Cu^{z+}/Cu ($z = 1$ or 2) systems is large enough so that even the complexation of the cations does not modify the order of their reduction. Since the silver content of the deposit is usually quite small, the Ag^+ reduction process takes place at the diffusion-limited rate. For the sake of easy regulation of the Cu:Ag ratio in the deposit, the Ag^+ concentration in the bath is significantly smaller (0.1–50 mM) than that of the couprous or coupric species, and the mole fraction of Ag in the deposit can be estimated from the ratio of the silver diffusion-limited current density to the total current density.

The structural features of Cu(Ag) deposits can be easily elucidated by taking into account the atomic volumes. As Ag atoms are larger than Cu atoms, the lattice is expanded as the Ag is incorporated, and the lattice parameter measured increases with the Ag content in the deposit [31]. This impact lasts to the Ag mole fraction at which the segregation of Ag occurs, which manifests itself by the emergence of the diffraction lines characteristic of pure Ag. Interestingly, as Ag forms a segregated phase, the elimination of the nucleation barrier leads to a loss of the Ag content of the metastable Cu matrix whose lattice parameter approaches again to that of pure Cu [31, 36]. For the as-deposited samples, the Vegard law was found to be valid up to an Ag concentration where a separate Ag phase is observed [4, 40].

Concerning the grain size of Cu(Ag) deposits, it has been shown [32] that the Ag content itself has no major impact on the grain size of the deposit when the primary variable is the Ag^+ concentration of the bath and the current density is constant. However, the grain size decreases with increasing the current density [32, 33]. The latter trend agrees well with those discussed in Chap. 2.11.

Pure Cu deposits are notorious of self-annealing also at room temperature that leads to an increase in the grain size, and, consequently, the decrease in the specific resistivity due to the diminished accumulated volume fraction of the grain boundary zones. Silver doping of copper deposits can be used for the elimination of this self-annealing process by reducing the grain boundary diffusivity [33]. The inhibition of the self-annealing of Cu can be observed at as low as 1.2 at.% Ag content [32] and above.

Annealing at higher temperature is a core question in the relevant studies. An increase in grain size was reported in all annealing-related studies. As the temperature dependence of the resistivity of the as-deposited samples shows, annealing up to 200 °C results in a small relative resistivity decrease due to an impurity redistribution process. A further temperature increase up to 500 °C is required for silver precipitation and grain growth of the Cu matrix [35]. Nevertheless, one short annealing cycle even up to 500 °C does not lead to the complete segregation of silver. The presence of a residual Ag portion in the Cu crystals can be concluded from both diffraction [33] and resistivity [36] data.

The functionality of the Cu(Ag) deposits is closely related to the element distribution in the as-received and annealed deposits. The composition depth profile of the as-deposited samples shows two Ag peaks: one at the substrate/deposit boundary and another one at the external surface of the sample [33, 36]. Upon annealing, the bulk Ag concentration decreases while the surface peak becomes more intense. This indicated that, beside the segregation of Ag to the grain boundary, a significant diffusion towards the external surface also takes place [36, 38]. The latter trend explains why Ag-containing Cu deposits are much more oxidation resistant than pure Cu coatings. These features of the electroplated samples are in good agreement with those of sputtered metastable Cu(Ag) alloys [41]. The segregation trend of sulphur in Cu(Ag) alloys is completely different from that of silver. If sulphur is present in the deposit as a result of the decomposition of a bath additive [42], it accumulates at the substrate/deposit interface upon annealing.

7.1.5 *Miscellaneous Composites Obtained with Metal Codeposition*

Similarly to the binary systems discussed above, Ag and Ni also make an immiscible metal pair whose codeposition leads to various structures. The motivation of the codeposition of Ag with Ni stemmed from the successful preparation of their metastable alloys by physical deposition techniques. The common experience was that some complexing agent was necessary for the formation of a compact deposit, even though the complexing agent applied had little influence on the difference of the onset potential of the deposition of the two metals. From electrochemical point of view, the Ag deposition took place in the background of Ni deposition as the current density increased, which was accompanied by a decrease of the Ag content of the deposit [43]. This is entirely in accord with the regular codeposition mode. Baths with sodium citrate [43, 44] resulted in nanogranular deposits with a very low level of the intermixing of the components (Ni could dissolve in Ag to a little extent but not vice versa [44]). When the sample preparation was carried out by using various other complexing agents (thiourea and sodium gluconate) and with pulsed current, the X-ray diffractograms indicated the formation of a metastable Ni–Ag phase beside the nearly pure phases of the components [45]. The diffraction lines of the metastable Ag–Ni phase did not disappear after a 1-h-long heat treatment at a temperature as high as 300 °C. The atomic-level intermixing of the constituents was confirmed by the magnetization value of the mixed deposit. In the 80–92 at.% Ni content range, the saturation magnetization was only 1–5 emu/g, as compared to the 47 emu/g obtained in the same study for a pure Ni deposit from the Ag-free bath (while the room-temperature magnetization value of the metallurgically processed pure Ni is around 56–57 emu/g).

Another frequently studied system is Cu–Fe, mostly because of the interest in its magnetic properties. There is little overlap between the electrochemical data of the baths used for Cu–Fe codeposition. A composition diagram was published only for a citrate bath [46], which indicated that the dominant component of the deposit suppresses the codeposition of the other component. This is an indirect evidence for the tendency of phase separation. The current density dependence of the deposit composition corresponds to the regular codeposition, i.e., the Fe content increases with the increase in the current density for a particular bath composition [47, 48], although complexation of the cations can weaken this trend [48]. Oxygen incorporation into the deposits is also mentioned [48]. Concerning the structure of Cu–Fe deposits, the phase separation is obvious from the appearance of the XRD line systems characteristic for either bcc Fe or fcc Cu [46–50], at least in the range of the nearly equimolar deposit composition. Nevertheless, the accurate structural analysis leads to some controversy. While some studies indicate that the XRD peak positions are somewhat shifted relative to the pure form of the constituents [46, 49] which can be related to a slight intermixing of the components in the as-deposited state, other studies do not confirm this observation [48, 50] but some scattering of the lattice parameters can be seen within the standard error of the measurements. Surprisingly, a completely opposite conclusion was drawn from Mössbauer spectroscopic measurement of Fe–Cu alloy powders electroplated from an additive-free solution [51]. The analysis of the Mössbauer spectra resulted in nearest-neighbour atomic distribution characteristic of solid solutions up to 45 wt.% Fe content. Where the data for the crystallite size is available [48], it can be established that the as-obtained deposits are nanocrystalline with about 8–70 nm grain size. The magnetization data are also quite diverse. When the saturation magnetization as normalized to the Fe content is lower than that expected from a simple dilution law, an alloying can be concluded [46]. In contrast, when the reduced magnetization as a function of the temperature is not a function of the composition, a complete segregation is assumed to take place [50]. The number of the available studies is insufficient to find the correlation between the behaviour of the samples and the electrodeposition parameters.

Finally, it is worth of mentioning a study in which cathodic electrodeposition and anodization were combined in a reverse pulse deposition method to obtain a *metal–metal oxide* composite [52]. The bath contained Ag, W and Co, which resulted in the formation of a deposit containing Ag and W during the forwards (cathodic) current pulse. The tungsten content conserved the deposit during the anodic pulse by preventing its dissolution when the cobalt was deposited by the oxidation of the dissolved Co^{2+} ions, leading to the formation of a cobalt oxide whose composition was not defined. The typical deposit composition under optimized deposition conditions was about 3 wt% O, 6 wt% Co, 80 wt% Ag and 2 wt% W. The grain size of the metal was less than 100 nm for the above composition, while a further grain refinement to 20 nm was observed as the W content was increased up to 14 wt.%. The pulse parameters were optimized to obtain a dispersion-type deposit and to avoid a multilayer formation. The cobalt oxide in the deposit served as a dispersed component that offered self-lubrication properties for the coating.

7.2 Composite Deposition from Particle Suspensions

7.2.1 Preliminary Remarks on the Importance of Composite Plating

In Sect. 7.2, we deal with the cases when the particles to be incorporated into the coatings are present in the bath. This is often termed as either suspension plating or dispersion plating. Since the forces that lead to the particle incorporation are essentially not gravitational, suspension plating takes place at cathodes of nearly arbitrary position, and the cathode orientation effect is usually quite small. Composites can be synthesized also by sediment plating on horizontal and upward-facing cathodes if particles are collected by sedimentation before or during the plating process. Here, the driving force of the incorporation is related to gravity, and the interaction between the growing cathode and the particles is of secondary importance. Accumulation of particles on a conducting surface can be triggered also by electrophoresis, which forms a sediment-like quasi-continuous template. The latter two cases will be dealt with in Chap. 11 dealing with templated method, restricting here only to cases when the codeposited particles are suspended in the bulk solution until the moment of their incorporation to the solid deposit. Nevertheless, the term *metal matrix composite* (MMC) is widely used for all kinds of such phase mixtures regardless of the preparation method.

The significance of electroplating of composites is obvious if we consider that the resulting coatings cannot be synthesized in the bulk form by using a molten metal mixed with nanoparticles to be incorporated. Beside the density difference of the metals and the filling particles, the wetting of the particles is weak by the melt, and the two components can also react at elevated temperatures. The large curvature of the nanoparticle surface also prevents their wetting by molten metals.

Therefore, electrodeposition is rendered to be a unique method for the synthesis of such materials.

The practice of suspension plating is notorious of being an experiment-intensive field, despite various theories have been elaborated for the incorporation ratio of the particles and for the current density dependence of the deposit composition. The reason for the complications stems from the fact that there is no alternative way to measure the codeposition-related thermodynamic and kinetic parameters incorporated into the models but the evaluation of the suspension plating experiments themselves. Thus, although the embedded physico-chemical terms can be calculated from a series of suspension plating experiments, they cannot be used for the optimization which is hence rendered to be an experimental task. It appears that there is a long way yet until the gap between the elementary physico-chemical parameters of a plating system and the codeposition yield of the particles can be bridged.

7.2.2 *Theories of Stability of Suspensions and Their Coagulation*

Although it is seldom emphasized in the works dealing with peculiar cases of suspension plating, the particles to be incorporated must form a stable suspension. The hindrance of the coagulation stems from two conditions that are not independent of each other. First, the particles have to be charged so that an electrostatic force prevents their approach to each other. The electrostatic properties of the particle surface are characterized with the so-called ζ-potential. Suspensions are usually stable if $|\zeta| >$ 30 mV.

The theory on the stability of suspensions and their coagulation dates back to the mid 1900s. The name of the theory by which it is commonly known, DLVO, stems from the initials of the authors of two important works (Derjaguin and Landau [53] and Verwey and Overbeek [54]). On the recent progress of the DLVO theory, various review works are available [55–57]. Below, a semi-quantitative treatment will be offered only, and interested readers are referred to the above-mentioned works.

The basic concept of the DLVO theory is that the force between the particles (that are taken spherical and identical in both diameter and surface charge density in the first approximation) originates from two terms, the van der Waals force and the electrostatic force. While the former is always attractive, the latter is repulsive. In the DLVO theory, the total particle charge is taken as a parameter that is invariant as the particles approach each other; nevertheless, the charge redistribution on the surface of the particles is indeed possible, although the models usually neglect it.

Since the expressions of the two forces (van der Waals and electrostatic) depend on different powers of the interparticle distance, their sum results in the occurrence of various ranges of stability as a function of the distance between the particles. Instead of the forces themselves, the total energy of the particle pairs is usually displayed as a function of the separation distance, as it is shown in Fig. 7.2a. The minimum of

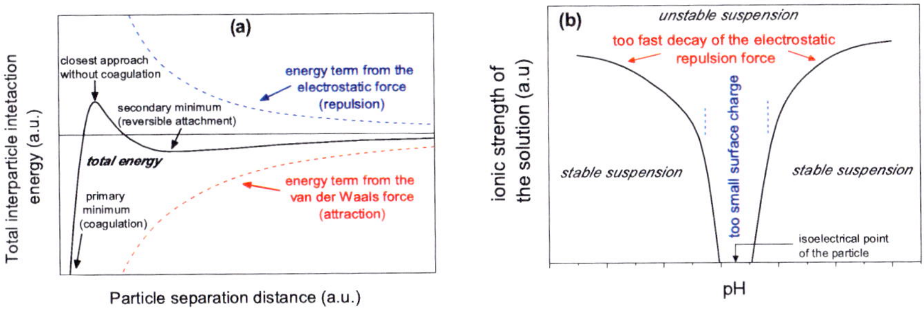

Fig. 7.2 **a** Energy diagram of the particle–particle interaction as a function of the separation distance for a relatively large surface charge and small solution concentration where the secondary minimum of the curve may occur. **b** Schematic stability diagram of suspensions as a function of pH and ionic strength of the solution.

the curve indicates an equilibrium position with stable interparticle distance, while the maximum refers to the closest approach without irreversible coagulation. The attachment of the particles in the distance of the secondary minimum usually does not lead to irreversible coagulation but the resuspension of the particles is well possible by either stirring or application of ultrasonic agitation. The coagulation with interparticle distance of the primary minimum (i.e., with a direct contact of the particle surfaces) is always irreversible.

As it was discussed in Chap. 2.5, the near-surface solution layer of a charged surface is always enriched in counterions and depleted with respect of co-ions. Since the decay of the concentration vs. distance functions of the counter- and co-ions is strongly influenced by the total ionic strength of the solution, the electrical repulsion of the particles is sensitive to the total solution concentration. The distance over which the concentration difference (or, similarly, the electrical potential difference) decays to its $1/e$ value is called the Debye length and it is proportional to the $c^{-1/2}$. The higher the ionic strength, the more abruptly the repulsion curve decays as a function of the interparticle distance. The consequence of this behaviour is that for solutions of high ionic strength (that industrial plating solutions always are), the total energy vs. distance function becomes monotonous. Because of the small Debye length, both the secondary minimum and the maximum disappear. Hence, the coagulation cannot be prevented merely by forces of electrostatic origin. The latter fact verifies the application of stabilizing agents (tensides) in suspension plating experiments.

Another factor that influences the stability of the suspensions is the charge of the particles. While the solution concentration scales the electrostatic repulsion curves along the abscissa (i.e., it determines the Debye length), the total charge of the particle determines a proportionality factor that scales the repulsion curve along the ordinate. In contrast, the van der Waals interaction strength does not depend on the solution concentrations in the first approximation. Therefore, highly charged particles are preferred to keep a suspension stable. The most common method to change the surface charge of a particle is to tune the pH far away from the electroneutrality point of the particles. The typical regime of the suspension stability as a function of pH and ionic strength of the bath is visualized Fig. 7.2b. It clearly indicates that the larger the ionic strength of a solution, the wider is the instability range in the pH scale.

Since the plating baths are usually optimized to a very narrow pH range, it is practically never a real option in the plating industry to tune the bath pH for making a suspension stable. Instead, other types of countermeasures are applied for preventing the coagulation of the particles, which will be detailed below in Sect. 7.2.5.

7.2.3 *Theories of Particle Incorporation During Electroplating*

Codeposition of particles from suspensions is possible for both positively and negatively charged particles, which means that the force acting between the particle and the metal plated is not electrostatic in nature; nevertheless, the binding mode is practically never known. Secondly, the stability of the suspensions is usually maintained by applying an appropriate surfactant which, if it is ionic, can itself provide the sufficient surface charge. Besides, the adsorbed tenside layer prevents the formation of the interaction of the core material of the particles. The preparation of stable suspensions is often described as a lengthy process with continuous stirring or ultrasonication. The bath agitation is necessary also during the plating process for the sake of both the maintenance of the suspension stability and the sufficient mechanical transport of the particles towards the growing surface.

The historical line of the models related to particle codeposition during electroplating has been summarized in a number of works [58–64]. Here, the essential details are given below with referring the interested readers to the original publications. An overview of the incorporation process is presented in Fig. 7.3.

Early notions on the incorporation of the suspended particles were based on a simple mechanical entrapment theory. This was later exceeded by claiming that the residence time of the particles at the surface of the growing metal has to be longer than during a free motion [65], i.e., an adsorption step is necessary for the immobilization at the surface prior to the incorporation itself. The pivotal point of the quantitative description of the particle codeposition was Guglielmi's work [66] in which a two-step adsorption model was proposed. The two stages of adsorption are characterized by the relevant relative surface and coverages σ and θ:

$$\sigma = S_{\mathrm{L}}/S \text{ and } \theta = S_{\mathrm{S}}/S \tag{7.1}$$

where the indices L and S stand for loosely and strongly adsorbed particles, respectively, and S represents the surface area. The loose adsorption corresponds to an approach to the metal surface in which the electrical double layers of the contacting surfaces are yet conserved and is often termed as physisorption. In the strongly adsorbed state, the electrical double layers disappear at the contact surface. In

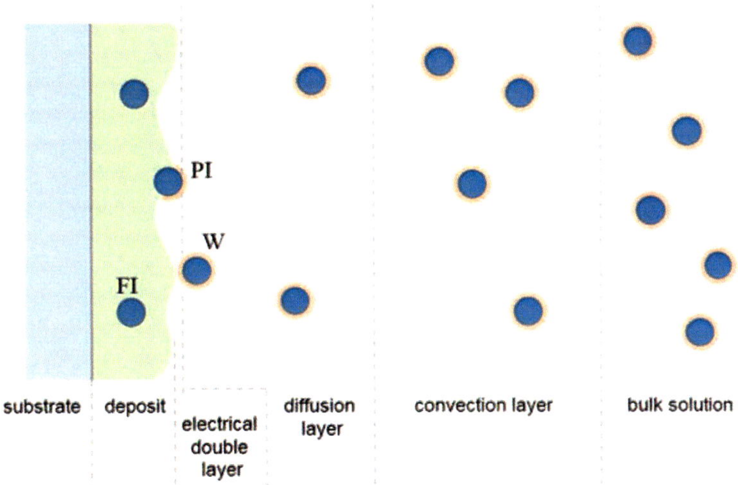

Fig. 7.3 Stages during the particle incorporation during electrodeposition. The thickness ratio of the solution layers and the particle size are highly distorted, but the number of particles shown in each layer indicates the concentration gradient of the particles. Notations refer to weakly attached (W), partially incorporated (PI) and fully incorporated (FI) particles. Redrawn after Refs. [62, 64, 67]

Guglielmi's model, the loose adsorption is taken as a reversible process while the strong adsorption is irreversible. The reversible loose adsorption is regarded as a preliminary equilibrium where a Langmuir-type rate law is applied, leading to the following relative surface coverage of the loosely attached particles:

$$\sigma = \frac{kc}{1 + kc}(1 - \theta) \tag{7.2}$$

The theorem needed for the evaluation of the particle incorporation rate is to assume that the rate of strong adsorption depends exponentially on the overvoltage. The above approximation implies that the strong adsorption step is essentially electrophoretic. This leads to

$$\frac{\mathrm{d}V_{\mathrm{P}}}{\mathrm{d}t} = \frac{kc}{1 + kc}(1 - \theta)v_0 \exp(B\eta) \tag{7.3}$$

where V_{P} is the volume of the particles already incorporated into the coating. Equation 7.3 can later be combined with the Faraday law and the Tafel-type rate law of the electrodeposition process. The final equation is

$$\frac{c}{\alpha} = \frac{M i_0}{zF\rho v_0}\left(\frac{1}{k} + c\right)\exp[(A - B)\eta] \tag{7.4}$$

in which α is the volume fraction of the particles in the deposit and i_0 is the exchange current density of the metal being deposited. The molar weight M and the density ρ also refer to the metal. From the discussion of Eq. 7.4, it can be seen that the volume fraction of the particles in the deposit increases with the particle concentration and is expected to achieve saturation. Equation 7.4 implies the dependence of the incorporation rate on the electrode potential, and, indirectly, on the current density, too. The sign in the exponential term, (i.e., the value of A–B) determines whether the increase in the current density promotes or hinders the particle codeposition. In spite of the huge leap in the electrolytic codeposition theory, Guglielmi's approach cannot reveal explicitly the dependence of the codeposition rate on many experimental parameters (e.g., pH, temperature, hydrodynamic conditions); rather, the key parameters can only be determined from the deposit analysis. Another shortcoming of Guglielmi's theory is that the metal deposition is treated as being totally analogous to the particle-free case, which is not right.

The next important step in the mathematical description of the particle codeposition was made by Celis, Roos and Buelens [67]. They took into account five consecutive subprocesses of the particle codeposition:

1. Formation of an ionic cloud around the particles including also the electroactive metal ions that will later make the coating;
2. Transport of the particles from the bulk solution to the electrode side of the hydrodynamic boundary layer by forced convection;
3. Transport of the particles through the diffusion layer by diffusion;
4. Weak adsorption of the particles with the retention of their ionic cloud;
5. Strong adsorption of the particles, loss of the ionic cloud and the reduction of the ions previously adsorbed, leading to irreversible particle incorporation.

The first step may occur independently of the deposition process and hence, its kinetic parameters are not included into the improved model either. The last two steps are nearly analogous to those in Guglielmi's model with the extension that the reduction of the ions carried by the particles is also considered (though treated the same way as the reduction of the solvated ions). The incorporation ratio of the particles, w, is expressed first as weight per cent with a simple stoichiometric ratio:

$$w = \frac{W_P N_P P}{\frac{M i}{zF} + W_P N_P P} \tag{7.5}$$

In the above equation, W_P is the weight of a single particle, N_P is the number of particles crossing the diffusion layer around the working electrode per unit of time and surface area, while P is a probability factor of the particle codeposition. The P factor was assumed to be related to the ratio of the reduced (k) vs. the total (K) number of electroactive metal ions adsorbed on a particle (where obviously $k \leq K$). Hence, the particle incorporation was approached in a totally different manner than in the Guglielmi model. The particle incorporation can take place if the reduction ratio of the adsorbed ions achieves a large enough proportion. Still, the K parameter is not available from an independent measurement. The probability is described with a binomial expression:

$$P_{(k/K,j)} = \sum_{z=k}^{K} C_z^K \left(1 - p_j\right)^{K-z} p_j^z \tag{7.6}$$

The probability factor related to the reduction of one particular ion at the current density j, p_j, can be calculated by taking into account the transport process through a Nernstian diffusion layer of thickness δ. The appropriate substitution leads to

$$p_j = \frac{j}{zFc_M\delta + \frac{j\delta^2}{2D_M} + j} \tag{7.7}$$

which shows that the ion reduction probability is related to both the transport conditions (agitation, through δ) and the current density j. The M index indicates that the relevant quantities, concentration and diffusion coefficient refer to the metal being deposited. Equation 7.7 implies that a non-monotonous incorporation ratio can be expected as a function of the current density. Another assumption included in the probability-based model was that the ions in the solution and those attached to the particles move in the diffusion layer at different rates, and hence, a differentiation is necessary when the deposit rate is no longer charge transfer- (or activation-) controlled but mass transport controlled. This effect was taken into account, though somewhat arbitrarily, in the following manner:

$$N_P = N_M \frac{c_P^*}{c_M^*} \left(\frac{j_{TR}}{j}\right)^a \tag{7.8}$$

where the free parameters (j_{TR}, a) are to be adjusted to the experimental result. The quantity j_{TR} was meant to denote the current density at which the activation control changes to diffusion control, and the exponent a differs from zero at $j > j_{TR}$ only (although no well-defined $a(j)$ function could be given).

Further models included various other physico-chemical parameters. Fransaer, Celis and Roos [68] attempted a complete force field analysis of the particles approaching the electrode surface and based their calculation to the particle trajectory, taking into account also shear forces. Hwang and Hwang improved the Guglielmi model by taking into account the reduction of protons adsorbed on the particle surfaces, hence distinguishing current density regimes in which either the proton reduction or the metal ion reduction coupled with the layer growth is the dominant driving force of the particle codeposition. The latter approach opened a way to extend the codeposition theories to non-noble metals.

7.2.4 *Experimental Observation of the Metal Growth During Particle Codeposition*

The *in situ* detection of the evolution of the shape of the metal growing around a particle being incorporated is by far not a straightforward task. A very smart way of indirect detection of the growing fronts of the metal matrix was developed by Stappers and Fransaer [69]. Although their study was performed with particles of a diameter of several micrometre, it revealed many important aspects of the field. The method of the study was that the particles were sedimented on the bottom of the electrochemical cell that served as the cathode. The metal deposited was a Fe–Ni alloy whose composition was modulated by applying two current pulses of different current densities. Hence, the composition contours in the SEM image of the cross-sectionally cut deposits revealed the temporal evolution of the process.

The observation of the incorporation of glass particles revealed that the particle immobilization started much after the start of the deposition, although the particle touched the cathode at the beginning of the process. The interaction between the glass particle and the metal was relatively weak as compared to that occurring in typical suspension plating systems. For this reason, the glass particle was raised physically by the deposit growing underneath. The cross-sectional image also showed that the particle was not totally fixed when the metal grew already around the particle but the upward motion of the particle could continue during the incorporation process. This is why a cavity formed under the particle. The curvature of the metal layers of dissimilar composition in the close vicinity of the bottom side of the glass particle indicated a preferential metal growth near the triple phase junction even in the phase when the particle was not yet fixed.

A pronounced difference was observed in the incorporation manner, depending on the nature of the particles used. In the first case, the particle being incorporated was non-conducting and hydrophobic, which also means that the surface concentration of the adsorbed metal ions is probably small. The cross-sectional figure showed that the growth profile was uniform from the start of the incorporation until the half-covered state of the particle since the composition contour lines were parallel to the substrate. However, the overgrowth of the metal can be seen after the deposit reached the half of the particle, and the top part could be covered only when the metal layer far from the particle was much thicker than the particle diameter. This means that, even though the particle was hydrophobic, the particle–metal interaction was even weaker than the particle–solution interaction. In the second case, the metal growth around a graphite particle was studied. Although the nucleation of metals is much hindered on carbonaceous materials, the coverage of the graphite particle with the growing metal preceded the rate of the metal deposition far from the particle significantly. As the contour lines showed, the graphite particle was covered with the growing metal nearly instantaneously.

The relative thickness of the same metal layer on various spots near the particle accounts for the current density distribution. It was a common experience that, regardless of the material of the particle being incorporated, the metal deposition was promoted near the bottom of the particle where the particle was fixed at the beginning of its incorporation into the growing coating.

7.2.5 Key Experimental Parameters in Suspension Plating

Effect of the surfactants and other solution components on the stability of the dispersions. The effect of the surfactants on the stability of suspensions as well as on the codeposition characteristics of the suspended particle is one of the most studied aspects of dispersion plating. Especially in the case of particles made of intrinsically hydrophobic materials (like PTFE, MoS_2), the application of surfactants is indispensable. In general, cationic surfactants are considered to be a better option than anionic surfactants since the electrophoresis of the positively charged particles helps their transport towards the cathode surface. However, since electrophoresis is by far not the sole means of the particle transport, the application of anionic surfactants is often feasible, too.

The study of Walsh et al. [70] is one of the most detailed works about the impact of various surfactants on one particular metal (particle) codeposition system (which was Ni(SiC) in their case). This work ranged to 24 surfactants of various types, including cationic, anionic and neutral ones. The conclusion of the work tells that "no simple correlation was found between surfactant type or structure and the particle content of the deposits or the performance of coatings, showing the complexity of surfactant action in this system" [70]. Since the adsorption of the surfactant on either a specific metal surface or on a particular type of particle depends on the chemical nature of all components involved, the surfactant studied show a great variety in works dealing with various metal/particle systems (e.g., Zn(SiC) [71], Ni(SiC) [72], Cu(MWCNT) [73], Ni–Zn(MWCNT) [74], etc.). While some trials are highly based on preliminary experience on the deposition of the same metal without nanoparticles, the chemical similarity of the particles and the surfactant may also be of importance. This trend prevails for cases like suspension of hydrophobic fluorocarbon particles with surfactants having high fluorine content [75] or stabilizing carbon nanotubes in hydrophobic ionic liquids [76].

It is to be noted that complexing agents and additives may also act as surfactant that adsorbs on the surface of the nanoparticles, although they are used primarily for the sake of optimizing the metal deposition. This was well shown for carbamide in the $Ni(Al_2O_3)$ deposition process [77] and sodium glutamate for the $Cu(Al_2O_3)$ system [78]. For the latter case, a detailed analysis was made for the adsorption of Cu-containing complexes on the nanoparticles by anchoring the complex through various functional groups of the glutamate ion on the surface of the alumina particle. The modification of pH can also fundamentally change the surface state of the particles and their coverage with metal ions, especially for oxide particles where the surface oxygen atoms can bind either hydrogen or metal ions, depending on the acidity of the medium.

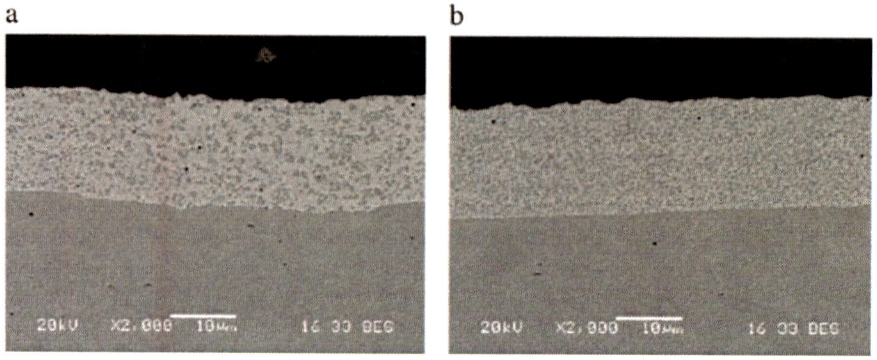

Fig. 7.4 Comparison of the cross-section of $Ni(Al_2O_3)$ electrodeposited composites obtained from mechanically stirred (**a**) and ultrasonically agitated (**b**) baths under otherwise identical deposition conditions. Bottom part of both images: mild steel substrate, light strip in the middle of the image: deposit. Reprinted from [81]. Copyright (2012), with permission from Elsevier

Maintenance of the dispersion by electrolyte flow, stirring and/or ultrasonication. Beside the application of an appropriate surfactant, the maintenance of the desired dispersity of the nanoparticles without any agglomeration can be provided also by an appropriate solution agitation. While some means of solution stirring (like with a magnetic stirrer) is indispensable and is mentioned in the majority of relevant studies, ultrasonication often yields deposits with more even particle distribution. A comparison of the cross-section of deposits obtained from mechanically stirred and ultrasonicated baths is shown in Fig. 7.4. The application of a jet electrodeposition system can eliminate the necessity of any other means of stirring [79, 80], although this setup is not suitable for an arbitrary substrate shape.

Ultrasonication was reported to improve both the dispersity of the nanoparticles and their incorporation ratio for various systems (e.g., $Ni(Al_2O_3)$ [81, 82], Ni(SiC) [82], Ni(Ti) [83], $Ni(WS_2)$ [84, 85], $Cu(SiO_2)$[86]). The combination of mechanical stirring with ultrasonic impact can also be superior of either of the agitation modes applied alone [87]. Nevertheless, for a correct assessment of the changes caused by the nanoparticles themselves, composite coatings must be compared with pure metals deposited under the same sonication conditions. This is because ultrasonication itself may improve the deposit quality, leading mostly to finer grains, higher hardness and smaller surface roughness [84].

Although ultrasonication seems to be a generally applicable tool for improving the properties of composite coatings, its application is by far not as well-defined as some other means of the hydrodynamic control. Beside the ultrasound frequency and power, the distance of the sonication head from the cathode is also a crucial parameter of the experimental setup. Even if all parameters are correctly reported, the transferability of the system parameters is not as evident as, e.g., the rotation rate of a rotating disc/cylinder electrode. The necessary ultrasonic power may also depend on various system parameters like the quality of the particles to be incorporated. For instance, the chemical modification of the surface of carbon nanotubes impacts the optimal ultrasonic power needed to achieve a deposit with the desired properties [88].

Particle size distribution of the dispersion. The common pursuit is that researchers use monodisperse particle samples for the electrodeposition of composites. The relative width of the size distribution of the particles is usually quite large and it may depend on the preparation mode. Samples of ground particle usually exhibit larger distribution width than those produced by precipitation and sol-gel methods where auxiliary components can be used to regulate the upper limit of the particle size. It is customary that 95% of the particles falls in the ±25% range of mean particle size (as calculated by the number of the particles).

In special cases, it may occur that particles with bimodal size distribution are used purposefully [81, 89]. In these cases, the particle fractions of larger and smaller sizes can be associated with different advantageous new properties of the deposit.

Polarization behaviour of the dispersion coating systems. Cyclic voltammetry is an essential element of the tool set of electrochemistry; hence, it is often applied to characterize the plating systems used to obtain dispersion coating. The comparison of the voltammograms obtained in both the absence and presence of suspended particles led to interesting experience. Both current increase and decrease at a particular potential may occur, and the direction of the current change cannot be directly correlated with the electrical conductivity of the particles.

For large and elongated conducting particles like carbon nanotubes, it is not surprising at all that the addition of nanoparticles to the plating bath can increase the cathodic current at a particular potential [76, 90–92]. The reason is the enhancement of the electrode surface area, which results in an enhanced current density despite the nucleation of metals on carbonaceous materials is usually much hindered. The current density enhancement for CNT-containing bath can be as high as 100%. In contrast, codeposition of small isotropic metallic particles may show either a similar or an opposite trend. For the deposition of Sn–Cu(Ag), the addition of the Ag particles to the metal plating bath resulted in an enhanced current density [93]. However, for a Zn–Cu(Ag) deposition system, the reduction of the current was found upon the addition of the nanoparticles [94]. Ag nanoparticles were thought to hinder both the Cu and Zn–Cu deposition, leading to smaller currents. The explanation of the inhibition of the deposition is likely the nucleation barrier effect of the Ag particles for Cu and Zn that exhibit crystal structure with very dissimilar nearest neighbour atomic distances.

Concerning non-conducting particles, the diminished current density can be elucidated on the basis of the reduction of the active surface area [95]. However, several examples can be found for the opposite trend, too; namely, a moderate [71, 87] or even twofold [96] increase of the current density can be seen upon the addition of the nanoparticles to the metal plating bath. Since this experience is highly collateral beside the properties of the deposits, no systematic study can be found for correlating the current increase with other bath properties. If we recall the codeposition model conditions setup by Fransaer et al. [68], it is apparent that adsorbed metal ions are also considered to be reactive at the cathode surface.

Plating mode: pulse and reverse pulse plating of composites. Existing theories on particle codeposition in the course of metal growth all refer to the steady-state d.c. deposition. These theories anticipate that the volumetric ratio of the particles in the coating is limited, and the codeposition ratio achieves saturation as the colloid concentration in the bath grows. From an experimental point of view, such saturation is often observed, or even a decline of the particle codeposition ratio can be seen if the particle concentration in the bath is too high. In order to overcome these limits, the reverse pulse codeposition method was suggested by Podlaha and Landolt [97]. Their systematic study of the impact of duty cycle (η) on the process revealed that the proper setup of the deposition parameters can provide a significant particle codeposition ratio increment. In the example studied, the apparent deposition efficiency showed a plateau down to $\eta \approx 0.3$, while the codeposition ratio increased steadily in the interval of $0.8 < \eta < 0.15$. This study indicates that the particle incorporation together with the metal deposition is not fully inverted as the metal dissolution was triggered by the anodic current, but the particle removal has a hysteresis. Hence, the application of the reverse pulse method opens a convenient way to increase the particle incorporation ratio.

It is to be noted that the reasons of the application of a reverse pulse deposition mode are fundamentally different for metal and composite plating. For metals, the pulse with anodic current is for surface shaping by dissolving undesired protrusions, hence leading to smoother deposit than with d.c. plating. In order to achieve a good selectivity of the dissolution of elevated spots of the deposit (peaks and dendrites), the anodic current is typically larger than the cathodic current, although the anodic pulse length is small enough to keep the charge balance in favour of deposition. For the reverse pulse plating of the composites, the dissolution has no role in the surface shaping but it merely serves to get rid of a part of the metal that is not needed to hold the particles and to create new surface sites for the adsorption of the particles in the next cathodic pulse. Therefore, the anodic pulse in reverse pulse composite plating is often of lower current density than the cathodic one (see, e.g., [98]). This current density ratio is also favourable for the retention of the already incorporated particles since the dissolution is mostly restricted to areas relatively far from the particles. Obviously, the above discussion is valid for inert (non-conducting) particles that themselves do not dissolve. For composites containing metallic particles, reverse pulse plating can be applied when the metal particles are either passive or relatively noble as compared to the matrix-forming metal.

From the viewpoint of the balance between deposition and dissolution, composite plating with cyclic voltammetry can be considered as being analogous to reverse pulse plating, provided that the potential limits allow the regulation of the desired balance between the cathodic and anodic processes [99].

Instead of reverse pulse plating, simple pulse plating (with zero current during the off-time) also proved to be appropriate for composite plating. This approach creates a synergy of composite plating and pulse plating since both lead to grain refinement. Besides, the off-time allows the replenishment of the particles available for incorporation, just like that of the metal ion concentration near the cathode.

Examples for the application of pulse plating for composite deposition are numerous [75, 83, 100–112] but no general trend can be established concerning the application of pulses on the particle incorporation ratio. Due to the large variety of the choice of the parameters of pulse plating, the comparison is difficult, especially since the bath and particle types also vary from one work to another. The application of pulse plating itself was reported to increase the particle incorporation ratio for a number of systems (Co–Ni(SiO_2) [102], Ag(PTFE) [75], Cu(SiC) [105], Ni(SiC) [113], Ni–P(SiC) [110]). However, some other pulse-plated deposits exhibited lower particle content than their d.c.-plated counterparts (Ni(Ti) [83], Ni(SiC) [108]). The application of pulsed current can be used to reduce the porosity of the resulting composite [109, 112]. Concerning the pulse frequency, the results are even more diverse than the trend on the incorporation ratio. Studies show cases of no impact of the pulse frequency on the particle incorporation ratio [83], adverse effects of micro- and nanosized particle of the same composition [103], advantage of either the low [104] or high [111] pulse frequency for the particle incorporation or existence of an optimal intermediate pulse frequency [105, 107].

An interesting collateral effect of the pulse plating is the possibility that it may lead to multilayer-like deposits at small pulse frequencies, although there is no change in the composition of the electroplated *metal* itself. A representative image of such a coating can be seen in Fig. 7.5. The reason for the occurrence of such a structure is the fluctuation of the codeposition ratio of the suspended particle within a particular current pulse. There is a chance to observe the modulation of the codeposition ratio when the metal layer thickness produced in a single pulse is larger than the size of the particles being incorporated; otherwise, the spontaneous modulation of the codeposition rate within a particular pulse remains hidden.

Effect of the external magnetic field on dispersion plating. It was found that the application of a static homogeneous external magnetic field enhances the particles incorporation into the deposit even if the particles are diamagnetic like Al_2O_3 [114, 115]. In both relatively small (0.7 T) and high (8 T) magnetic fields, the Lorentz force was identified as being effective if the magnetic field was parallel to the cathode surface (and hence, perpendicular to the current density). The magnetohydrodynamic (MHD) effect of the Lorentz force acted as a stirring mode, which reduced

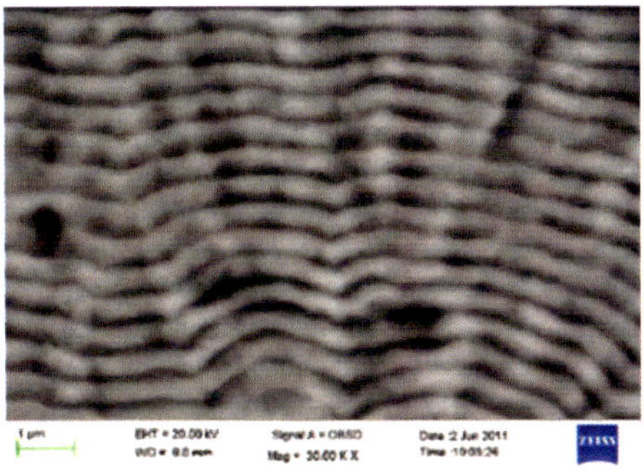

Fig. 7.5 Cross-sectional SEM image of a Ni(SiC) coating produced by pulse plating with 0.01 Hz frequency, showing a lamellar deposit structure. Reprinted from [107]. Copyright (2019), with permission from Elsevier

the diffusion layer thickness of the particles and also contributed to the stability of the suspension. The transport rate was increased so much that an eightfold enhancement in the particle incorporation rate was detected [114]. At high magnetic field, an additional agitation-enhancing effect was identified. Since the current lines are distorted around the partly incorporated particles and follow the surface tangent of the particles, the external magnetic field perpendicular to the electrode also has a Lorentz force contribution [115]. Clearly, this effect is localized to the close vicinity of the surface of the partly incorporated particles; nevertheless, it has a same effect on the reduction of the diffusion layer thickness as the solution stirring. The convection effect around the particles as exerted by the perpendicular-to-cathode magnetic field is similar to that what was found to be responsible for the modification of bubble attachment on and removal from the cathode during either electroplating [116, 117] or hydrogen evolution [118]. A similar analogy is true for the increase of the deposition rate in the early phase of metal electrocrystallization on a smooth substrate under magnetic field effect [119]. The latter was named as μ-MHD effect, regarding the size of the freshly nucleated metal particles.

7.2.6 Comparison of the Codeposition of Micrometric and Nanometric Particles

The comparison of the codeposition of micro- and nanoparticles with metals is an inherently uncertain field. The reason for this uncertainty is that the composition of the particles is given with the bulk parameters. However, the synthesis of the particles of different size often includes non-identical steps that have an immense influence on the features of the particle *surface*, i.e., the part of the particle that determines the adsorption capability and the interaction of the particles with the metal being plated. Therefore, the "size effect" of the codeposited particles may leave intrinsic parameters hidden that could be directly related to the codeposition behaviour. All discussion given below should be regarded with this reservation. Nevertheless, the investigation of the size effect in composite plating is a broad field [102–105, 107, 108, 120–126] that makes it possible to set up some trends.

A general rule for particle codeposition with metals is that the incorporating foreign body disturbs the crystallization of the metal. This leads to a grain refinement which is often accompanied with the loss of the deposit texture [122], simply because the more frequent the nucleation events are, the least the preferred texture of the freely growing deposit can manifest itself. This trend is valid for both micro- and nanoparticles, although the effect is usually stronger for the latter.

The incorporation site of the micro- and nanoparticles may vary. While microparticles are always embedded between the crystals of the metal matrix, nanoparticles can adopt another form of inclusion. Namely, due to their small size, they can be embedded into a single metal crystal or accommodate themselves at the twinning planes [82, 127]. Nevertheless, the preferred incorporation site of the nanoparticles is also a boundary between the crystals of the metal matrix.

The incorporation ratio of nanoparticles as rated by their weight (or volume) per cent often shows smaller values than that of microparticles (with otherwise identical deposition conditions, including the weight-based concentration of the particles in the solution) [103, 121]. However, if we scrutinize the numerical density of the incorporated particles, nanosized particles are much more abundant in the composite deposits, which leads to greater effect on the structure. Electroplated composites with nanoparticles are also associated with better dispersion and lower surface roughness than those obtained with microparticles.

The experience is that nearly all deposit properties investigated for composites improve when nanoparticles used for plating instead of microparticles. These features can be the microhardness [102, 124, 126], friction coefficient [102, 125] and resistance to either wear/abrasion [102, 103, 108, 122, 125], corrosion [103–105, 124, 125] or oxidation [120].

7.2.7 *Grain Size and Hardness of Granular Coatings*

Electrodeposited composites are often discussed rather with respect to their properties and functionality than their composition. The main reason for this type of treatment is that the beneficial impact of the inert particles present in a metal matrix largely stems from the structural change in the metal matrix caused by the particles instead of their chemical nature. Therefore, it is not surprising that very similar trends can be obtained for a large variety concerning the chemical nature of either the metal matrix or the particles.

The major cause of the change in the mechanical properties of the dispersion coating as compared to their purely metallic counterparts is the *grain refinement*. This is merely an effect of the disturbance of the crystal growth by the particles being present. A representative image on the grain structure of pure metallic and particle-loaded deposits can be seen in Fig. 7.6. As mentioned above, this is unrelated to the composition of the particles; moreover, even the shape of the particles is of no matter. Although the spherical particles are typical in the nanoparticle codeposition studies, essentially the same grain refinement can be achieved with nanorods [128, 129] and nanotubes [90].

The hardness change due to the presence of particles in a metal is called *dispersion hardening*. The presence of particles modifies the deformation mechanism of the metal grains by limiting the dislocation movement distance. This leads to the situation that, together with the grain refinement, MMCs are harder than the matrix metal itself.

Grain refinement-related mechanical improvement can be equally observed for hard ceramic particles (like diamond [131, 132], Al_2O_3 [77, 133–135], AlN [136], Cr_2O_3 [96], CeO_2 [128, 137–139], SiC [124, 129, 140–142], Si_3N_4 [101], SiO_2 [102], TiO_2 [143, 144], TiB_2 [145]) and also for metal particles (Al [130], Cr [146] and Ti [83]). Even soft particles can lead to hardening when their concentration in

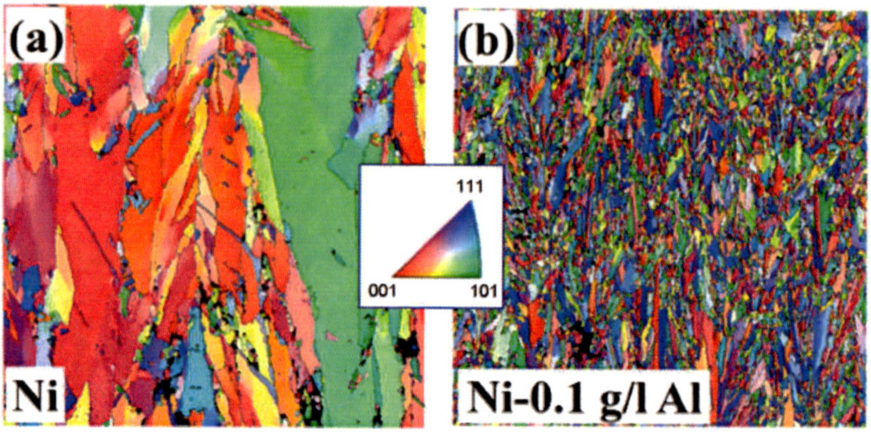

Fig. 7.6 EBSD images of the cross-section of Ni coatings plated without (**a**) and with (**b**) aluminium nanoparticles. Reprinted from [130]. Copyright (2018), with permission from Elsevier

the coating is sufficiently small so that the particle hardness itself does not impact the overall hardness but the dispersion hardening remains the decisive effect (PTFE [100] and WS_2 [84]). For a high load of soft particles, however, they can dominate the hardness of the coating and dispersion hardening is suppressed [90, 147].

It is a strong evidence for the physical nature of the particle-induced hardening that the maximum hardness is often found for the same sample in a series that exhibits the minimal grain size among the specimens studied. This trend was also observed for a great variety of codeposited particles [124, 137, 141, 142, 148].

For sake of completeness, we have to note that dispersion hardening may work also without a grain size effect. Among electroplated dispersion coatings, an example for this phenomenon is the Ni–W(ZrO_2) composite [149]. The grain size of the metal matrix was small, around 10 nm, whereas the particle size fell in the 40–50 nm range. The extremely small grain size is characteristic for (Ni,Co,Fe)–(Mo,W) alloys in general that tend to be X-ray amorphous at high refractory metal content. Although the grain size was not reduced due to the presence of the ceramic particles larger than the grains themselves, the impeded dislocation movement had a sufficient contribution to hardening.

An interesting feature of electroplated composite is their hardness evolution with annealing. For crystalline deposits (like Ni–Fe(SiC) [150]), the presence of the particles moderates the hardness loss upon annealing, which can be attributed to the hindrance of the atomic movement and hence, the retention of the relatively small grain size also at higher temperature. Concerning electrodeposited composites with amorphous matrix in the as-received state, the hardness may even increase with annealing, which is a very unusual behaviour. This was found for Ni–P(SiC) [110, 151], Ni–P(WC) [152] and Ni–P(MWCNT) [153] deposits. The reason of the hardness increase with annealing is the formation of crystallites from the amorphous matrix that themselves are harder than the original metallic matrix. The same annealing-induced hardening is known for the particle-free electrodeposited amorphous Ni–P matrices in which the number of steps in the thermal relaxation/crystallization depends on whether the actual sample contains more or less phosphorous than the eutectic composition [154–157]. The advantage of the presence of particles in such samples is that they impose a strong limitation on the crystallite growth that renders the annealed samples nanocrystalline for a larger temperature range than in the absence of dispersed particles.

7.2.8 *Influence of the Incorporated Particles on the Wear Damage and Friction of the Coatings*

The friction coefficient and the wear damage of the composite surfaces are somewhat interrelated properties, although their variation trends may be versatile (where "variation" is always meant in comparison with the surface of the particle-free deposit). Due to the diversity of the origin of their change, both the increase and the decrease in the friction coefficient may lead to diminished wear damage.

When the material of the filling particles ensures a lubrication effect upon the scratch of the layer, both the friction coefficient of the surface and the wear damage decreases as a result of the particle incorporation. In order to achieve this effect, the material of the particles has to be easy to smear out along the surface. This is the case for fluoropolymers (PTFE [75, 100, 158, 159]) and particles that exhibit layered structure where the adhesion of the layers to each other is relatively weak (WS_2 [85, 160], MoS_2 [84, 161], graphite-type carbonaceous materials [162]). The lubricating film formed from the sacrifice of the filling particles leads to the reduction of the depth of the wear scratch on the deposit surface, and, additionally, the renovation of the lubrication effect by the consumption of the composite deposits provides a temporal stability of the low-friction state. While the friction coefficient of the metals used in these studies (Ni, Ni–P etc.) is typically >0.8 in the absence of particles, a much lower value can be achieved when the particles are present. Under optimal circumstances, the friction coefficient is reduced to about 0.06–0.2 with PTFE filling particles [75, 100, 159]. Among the layered chalcogenides, MoS_2 particles resulted in the smallest friction coefficient of ~0.05 [161], while coating with both WS_2 and carbon particles exhibited a friction coefficient of ~0.2 [160, 162, 163]. An effect similar to self-lubrication was found when carbon nanotubes were codeposited with the metal matrix [153]. Here, the friction coefficient of the composite was initially larger than that of the pure metal; however, the order of the friction coefficient of the pure and composite coating was inverted as the cycle number in the scratch test increased. A friction coefficient as low as ~0.12 could be achieved with the nanotube filling, which was the third of that of the metal matrix. This behaviour was explained with the wear-induced re-alignment of the out-of-surface segments of the randomly incorporated nanotubes.

In the cases of the hard particles, the improved wear behaviour is not related to any lubrication effect caused by the redistribution of the particle material. Although the decrease of the friction coefficient as a result of the particle incorporation was detected for a variety of coatings and particles [128, 144, 164–166], the diminished wear loss is rather related to the increase of the hardness of the coating [167]. A significant decrease in the wear factor can be seen even for cases when the particle incorporation leads to an increase in the friction coefficient [108]. It was the common experience that the abrasion-induced weight loss has the minimum where the hardness exhibits the maximum [81, 122, 144, 164, 166, 168–170], and similar coincidence can be seen for the maximum hardness and the minimum of the friction coefficient [159]. The latter fact clearly indicates that, beside the surface effect of the particles themselves, the structural features of the deposit also have a fundamental role in the wear behaviour of the composite coatings.

7.2.9 Hydrophobic Dispersion Coatings

Hydrophobicity is a typical feature of a composite in which not only the presence of the particles matters (like it was shown for hardening), but the physico-chemical properties of the dispersed particles are of high importance. Namely, the hydrophobic nature of a composite surface stems from the hydrophobicity of the particles incorporated; therefore, it cannot be achieved with any arbitrary type of particle but with purposefully chosen ones only. In this respect, hydrophobic dispersion coatings differ from dispersion-hardened ones in which the particle-induced structural features had the major role in the functionality of the coatings. More details on the theory of hydrophobicity will be given in Chap. 8 where roughness-induced hydrophobicity will also be dealt with.

Although the hydrophobic properties of the composites strongly stem from the hydrophobic properties of the incorporated particles themselves, the particle concentration at which superhydrophobicity is set in varies much with the quality of the particles. For PTFE dispersions, a particle content around 70 vol.% was reported to lead to a sufficiently high water contact angle [147, 158]. Since PTFE is a soft material, the hardness of these coatings is much lower than that of the metal matrix, even though it is nanograined due to the particle incorporation. A great advantage of the PTFE filling in the metal matrix is that the hydrophobicity of the surface can be retained after sufficient wear damage since the self-lubricating PTFE maintains the hydrophobic nature of the surface. This is because the PTFE content of the coating is smeared out along the surface upon wear, hence increasing the PTFE-coated areas as compared to the undamaged surface.

Concerning the hydrophobic properties of MMCs with hard particles, CeO_2-filled composites were also of high particle content, ranging between 55 and 97 vol.% [139]. Much lower particle content was enough to achieve superhydrophobicity with ITO and WS_2 particles (about 1.8 [171] and 3.5 [160] w.%, respectively). In the hydrophobic dispersion coatings with low-concentration hard particles, the change in the deposit surface roughness was a significant factor of the enhancement of the water contact angle beside the hydrophobic properties of the particles themselves. For further superhydrophobic deposits, see also Chap. 8.2.

7.2.10 Suspension Plating with Magnetic Particles

In this chapter, we deal with composite deposits in which the particles incorporated into the coating are magnetic and the motivation of the experiments does stem from the magnetic properties of the system. The process is often named as magnetic composite electroplating (MCE). The application of an external magnetic field applied during the deposition often plays an important role in the magnetization behaviour of the resulting systems. Since the suspended magnetic particles interact with each other, both the application of a sufficient surfactant and the agitation of the bath are of high importance to solubilize the particles and to prevent their accumulation. The uniform particle dispersion is often achieved with a long stirring and/or ultrasonic agitation prior to the electrodeposition.

The incorporation of *magnetic particles into a non-magnetic matrix* can give rise to the occurrence of permanent magnetic properties in the case when the metal matrix itself if not ferromagnetic. For suspension plating of Ni particles with a Zn matrix on a magnetic steel substrate, it was found [172] that the matrix–particle magnetic interaction is predominant while the particle–particle interaction is not. As a consequence, the local concentration of the Ni particles in the coating was larger in the near-substrate zone by a factor of 4 than in the bulk deposit. In accord with the importance of the substrate–particle magnetic interaction, the incorporation ratio increased with the remanence magnetization of the Ni particles. For the codeposition of magnetite with Cu metal as matrix [173], the particle distribution was found to be even along the depth of the coating, although an external magnetic field was necessary to achieve significant particle incorporation. For the deposition of the $Cu(Fe_3O_4)$ system, a relatively low temperature and a moderate stirring rate was necessary. The deposition was successful in the mass transport-limited regime of Cu deposition, which means that the negative surface charge of the substrate may have a role in the magnetite particle adhesion to the Cu matrix.

If the *metal matrix itself is ferromagnetic*, its material is also important for the resulting magnetic properties of the composite. While Ni is a popular material as a carrier of the magnetic particles due to the simplicity of its deposition [174–176], Ni alloys with either Co [177–181] or Fe [182] can improve the magnetic properties. For optimization, matrices as complex as Co–Ni–P [183–185] or Co–Ni–Mn–P also often occur [174, 186, 187], which can be rationalized by the magnetic properties of the substrate. What concerns the codeposited magnetic particles, barium ferrite ($BaFe_{12}O_{19}$) [174, 177, 178, 180, 181, 183–187] or barium-strontium ferrite ($Ba_{0.2}Sr_{0.8}Fe_{12}O_{19}$) [179] are the most common since they exhibit both high magnetization and high coercivity (3.2–3.8 kOe). In such cases, the coercivity of the deposit is meant to be larger than the matrix alone, while the saturation magnetization usually also increases. Other options include magnetite [175, 176, 182] and cobalt [175]. For the case of magnetite incorporation, the coercivity is of marginal importance but the increase in magnetization is the goal of the deposit optimization. Composites like Ni–Co–P($BaFe_{12}O_{19}$) belong to the family of the so-called soft/hard composites in which the hard magnetic particles enhance the remanence of the matrix by pinning the matrix magnetization.

The concept of the application of the homogeneous external magnetic field in the MCE process for enhancing the coercivity is as follows. The orientation of the particles as magnetic dipoles to the direction of the external field leads to a deposit in which magnetic particles are "frozen" with a fixed magnetization direction. This direction defines the magnetization direction of the entire deposit since the hard-magnetic particles interact with each other and also with the matrix when it is magnetic. This aligning mechanism works for particles that are large enough to have a fixed magnetization direction. In contrast, composites with small enough nanoparticles exhibiting superparamagnetism have little effect on the magnetization of the composite deposit [181].

However, the impact of the external magnetic field proved to be much more complicated than representing a simple orientation effect for the particles. The application of a permanent magnet behind the substrate generates an inhomogeneous magnetic field that attracts the particles, whose accumulation may hinder the onset of the deposition process. In contrast, after the start of the plating process, the same attraction force speeds up the particle incorporation [178]. A similar field gradient effect was also demonstrated with an electromagnet if the field was perpendicular to the substrate and with a field gradient pointing from the bulk solution towards the substrate surface [175]. Interestingly, the field parallel to the substrate may lead to the decrease of the particle incorporation, too. This was explained by the Lorentz force that is comparable to the adhesion force between the small particle and the substrate. Since the Lorentz force moves the particles parallel to the surface, it can inhibit the particle incorporation by breaking the relatively weak particle/substrate interaction in the first stage of its immobilization [175].

7.2.11 Role of the Incorporation of Inert Particles in the Corrosion and Oxidation Behaviour of Metals

In the MMCs with inert particles, the change in the corrosion rate of the metal matrix as a result of the particle incorporation is mostly of physical origin, and hence, is unrelated to the chemical properties of the particles themselves. In this respect, the impact of the particle on the corrosion rate is partly similar to the phenomenon of dispersion hardening. In general, the corrosion rate of MMCs is substantially smaller than that of the metal matrix deposited without the filling particles. As for any other effect of physical origin, this was also exemplified for a large variety of codeposited particle such as Al_2O_3 [164, 165, 188], AlN [136], SiC [82, 87, 103, 104, 124, 129,

140, 141, 164, 189], SiO_2 [86, 102, 190], Si_3N_4 [101], CeO_2 [128, 138, 169], TiO_2 [143], ZrO_2 [191], TiN [79], diamond [192] and PTFE [100]. As an evidence for the physical origin of the diminished corrosion rate, it can be brought up that the particle effect works for essentially all metal matrices employed in these studies (Ni, Ni–Co, Cu, Zn etc.).

The incorporation of the inert particles usually leads to the ennoblement of the deposits, i.e., the corrosion potential is more positive than that of the metal matrix plated without particles. The estimation of the corrosion rate of the composite coatings is based on two major methods; namely, the potentiodynamic curves recorded with a small scan rate and the impedance spectra recorded at the corrosion potential. Although the Tafel-representation of the potentiodynamic curves offers an opportunity for the quantitative estimation of the corrosion rate, the linear regions usually cannot be observed clearly; hence, the evaluation of the measurements becomes somewhat arbitrary. It can also be seen that the corrosion rates tabulated appear to be overestimated as compared to the Tafel plots of the potentiodynamic curves. The electrochemical impedance (EIS) spectra make it possible to read the estimated charge transfer resistance which cannot be *a priori* associated with the corrosion rate; rather, this parameter can be used to express the estimated ratio of the corrosion rates of the samples. Whichever method is used, the decrease in corrosion rate as a result of the composite formation is reported to fall between a factor of 2–20.

The origin of the diminished corrosion rate is multifold: (i) The particle incorporation leads to grain refinement, which itself can decrease the corrosion rate of the metal matrix. The grain refinement is often coupled with a more compact deposit structure with lower roughness. (ii) The particles cover a part of the composite surface. Since the particles are inert, the particle-covered areas essentially do not corrode. This is a simple physical exclusion effect. (iii) The particles impede the penetration of the corrosion cracks in the composite.

While the surface coverage by the particles is usually rated to be the least important among the effects listed above, the hindrance of the corrosion crack penetration is strongly related to the advantage of the application of nanoparticles as opposed to microparticles. Since the numerical density of the nanoparticles can be much larger than that of the microparticles, their application can drastically enhance the corrosion inhibition efficiency because the growing corrosion cavities run into a blocking inert particle within a very short distance, which locally stops the corrosion cavity penetration. The importance of the grain size effect is evidenced by the observations where the minimum in the grain size in a deposit series can be related to the minimum of the corrosion rate [79, 124, 141, 189].

Exceptionally, the increase of the corrosion rate as a result of inert particle incorporation can also be observed [193]. In this case, the metal matrix of the Ni–W(Al_2O_3) coating has fundamentally different properties than the blank ones; namely, the particle incorporation leads to the crystallization of the amorphous matrix. The impact of the particle-induced crystallization overwrites all other advantageous effects of the particles incorporation concerning the corrosion rate of the composite.

When the particles in the MMCs are electrically conducting ones, the change in the corrosion rate can also be associated with the so-called local cell effects; or, in other words, we encounter a mixed electrode where the parts of the heterogeneous surface mutually polarize each other. This is the case for the incorporation of various forms of carbon nanotubes [74, 194]. Especially for Ni-based coatings, this polarization leads to a better conservation of passivity. For Cr(C) coatings made with carbon nanotubes [195], an improved passivity was observed due to the accelerated formation of the protecting Cr_2O_3 layer that adsorbed strongly on the nanotube-covered part of the composite coating. Here, the modification of the mechanism of the passive layer formation was a key factor of the decrease of the corrosion rate.

The incorporated non-conducting particles may also modify the oxidation behaviour of the metal matrix. This effect was shown for Ni(SiC) coatings [196]. During the oxidation in air at 1000 °C, the formation of SiO_2 was a competing process with the formation of the NiO scale, and the incorporated particles worked as scavengers of the oxygen diffusing into the depth of the NiO surface layer. Hence, the presence of the particles at the grain boundaries had a decelerating impact on the oxidation of the metal matrix itself. This effect is clearly of chemical nature and may work only for particles that themselves can be oxidized.

7.2.12 Metal–Metal Composites with Miscellaneous Applications

Electrodeposited composites with dispersed metal particles deserve special attention because the chemical nature of the particles plays a much more important role in the occurrence of the resulting functionality than for any other type of dispersion coatings. The metal–metal composites will be discussed below with attention to the special functionality gained by the particle incorporation.

Various composites were synthesized with Ag particles to improve the antibacterial feature of the metal surface. The principle of the antibacterial effect is that Ag^+ ions exert a strong bactericide effect. For achieving this effect, the coating must corrode by emitting Ag^+ ions into the contacting media. In aerated conditions, the corrosion takes place at the desired rate that is sufficient for impeding the bacteria colonization. At the same time, the Ag^+ concentration far from the surface is usually negligibly small so that Ag-containing bio-fouling surfaces are not considered to be a major heavy metal source due to the small overall amount of Ag^+ released. The concept of Ag particle codeposition is based on the retention of the protecting ability of a conventional coating without modifying the chemical background of codeposition by doping with an antiseptic agent, also in the case where alloy codeposition would not be possible due to the lack of an appropriate bath.

The coatings tested so far are all composed of metals used as typical coating materials in various applications: Zn [197], Cu–Zn [94], Cr [198] and Ni [199]. The silver nanoparticle concentration in the bath was between 0.5 and 10 g/litre. For achieving a bacteria growth inhibition efficiency of 90 to 100%, an Ag particle loading of 1 wt.% was sufficient. Although a slight change in the coating structure can usually be seen due to the Ag incorporation, it does not alter the functionality of the major metallic component of the coating; moreover, the Ag content leads to the ennoblement of Ni coatings and the reduced corrosion rate is retained also after the exposure to the bacteria colony [199]. Similar results were obtained for Zn/Ag coatings where the bio-fouling effect was not studied [200]. Here, the optimum Ag content was reported to be around 2 vol.%.

Various other goals led to the idea that composite coatings with Ag particles may exhibit beneficial properties. The enhanced catalytic activity of Ag-doped Ni–P coating was taken advantage of in the hydrogen evolution reaction in alkaline media [201]. With an Ag content of as low as 4 wt.%, a drastic decrease in the overvoltage of the hydrogen evolution reaction could be achieved. The onset potential of the hydrogen evolution was 340 mV less negative for the Ag-containing coating as observed in the potentiodynamic curves, and the hydrogen evolution rate at a constant potential was nearly doubled as compared to the Ag-free Ni–P coatings. (Similar results were obtained for Ni(TiO_2) coatings [202], although the gain both in the onset potential and in the hydrogen evolution rate was smaller than for the Ag-doped Ni–P deposits.)

Ag particle incorporation was also used as a methodical simplification of obtaining near-eutectic Sn–Ag–Cu alloys [93]. The bath optimization of the ternary Sn–Ag–Cu deposition is very cumbersome, partly due to the large difference in the onset potential of the reduction of the components. To overcome this difficulty, the Ag particles were added to a relatively simple Sn–Cu bath. The desired metal content of 0.7 wt.% (Cu) and 3.0–3.9 wt.% (Ag) could be achieved by the appropriate choice of Ag particle and Cu^{2+} ion concentrations and by tuning the current density.

In contrast to the diversity of the application opportunity of coatings with dispersed Ag particles, the motivation of plating Ni and Ni-containing alloys together with Al and/or Cr particles was nearly exclusively the optimization of the oxidation behaviour of the coatings obtained. The advantage of the nanostructured composite materials over the alloys of identical bulk composition stems from the difference of the oxide layer formation on the surface [203]. Upon the high-temperature oxidation of the nanostructured composites, a scale composed of mainly Al_2O_3 or Cr_2O_3 is produced instead of NiO whose growth is suppressed. The Al and Cr particles also serve as nucleation centres for their pure oxide scale, which is missing for the bulk alloys of the same composition. Also, the nanoparticles and oxide-forming centres are readily available at the surface of the composite coatings, while the diffusion of metal forming the protective oxide to the sample surface is needed for the alloys. There is a difference also in the oxide composition, leading to the reduction of the formation of a mixed oxide containing the matrix component when the composite is annealed [204].

For both Al and Cr particle loading, there is a threshold particle content above which the retardation effect of the matrix oxidation can be seen. This limit was found for Ni-based composites to be around 28 and 11 wt.% per cent for the two metals, respectively [205]. The beneficial effect of the codeposited particles is size-dependent. For nanoparticles, the blocking of the matrix oxidation is effective at much lower particle concentration (by wt.%) than for microparticles [120]. An interdiffusion annealing treatment at intermediate temperature (500 °C) was found to improve the oxidation resistance at high temperature (900 to 1000 °C) [206]. The advantage of the interdiffusion treatment stems from the formation of intermetallic alloys whose grains withstand to intergranular cracking.

The improvement of the oxidation resistance related to the Al or Cr particle content of the composite was investigated for Ni–Co [207] and Ni–Cu [204] alloyed matrices. Alloying Ni with Co increases the incorporation rate of the aluminium particles because the better wetting behaviour with Co. However, Co is less resistance to both oxidation and corrosion than Ni, which means that its content in the matrix was suggested to be 28 wt.% as a maximum [207]. In the case of the Ni–Cu alloys, the chromium particles effectively hindered the in-depth oxidation to several tens of micrometre. As the chromium loading of the composite increased, the mixed $CuO–NiO–Cu_2O$ layer (i.e., that produced on the particle-free specimens) could not form but a continuous chromium (III) oxide layer with a thickness of less than 5 μm formed during the oxidation treatment.

7.2.13 Combination of Various Particles in Electroplated Dispersion Coatings

The application of different kinds of particles in the electrodeposition of dispersion coatings may stem from various pursuits. When particle types of similar properties are used simultaneously, the particle load of the deposit is usually larger than that achieved with one sort of particles. An example for this case is the codeposition of Al_2O_3 and SiO_2 together with a Ni coating [167]. Both sorts of inert ceramic particles incorporate into the coating in a smaller volume ratio that in the absence of the other particle, however, the *cumulated particle content* obtained from the bath with mixed particles is larger than in the case when either of them was present alone. Consequently, the hardness and wear behaviour is better with the coating containing the mixture of the ceramic particles than that containing one particle type only. The same improvement in the deposit properties can be seen when the two sorts of nanoparticles belong to different material families, i.e., when ceramic and metallic particles are applied together [89].

The codeposition of various kinds of particles offers an opportunity to give different functionalities to a coating by doping it with particles of dissimilar properties. This pursuit can be seen when a PTFE-doped (and consequently softened) deposit is plated in the presence of TiO_2 particles that improves the wear resistance of the coating, maintaining at the same time the low friction caused by the PTFE content [159]. The application of nanoparticles of various functionality has certainly not been fully exploited so far and various other particle combinations are yet to be tested.

7.2.14 Suspension Plating in Anodic Processes

The suspension plating in an anodic process is similar to the inversion of its cathodic counterpart. The particles suspended have to be inert both in the starting solution and under the anodization conditions after the incorporation into the coating. The precursor metal ions are present in the solution in an intermediate oxidation state in the form of an ionic salt that is well soluble. In contrast, the product of the oxidation of the precursor metal ion must lead to an essentially insoluble oxide that is stable in contact with the solution. The latter condition is not fulfilled automatically since the anodization process nearly always takes place in an aqueous solution at considerably more positive potential than the stability regime of water; hence, the oxygen evolution as a result of water decomposition is preferred to be hindered on the surface of the resulting oxide. The suspended particles incorporate the same way as in cathodic processes involving metal deposition. When some of the theories developed for cathodic particle incorporation was applied to anodic processes, it was found that the evaluation of the incorporation ratio as a function of the particle concentration is possible on the basis of the Guglielmi model [208, 209]. The significant adsorption of the metal ions (Pb^{2+}) on the surface of various particles [210] indicates that the approach of Celis, Roos and Buelens may potentially be applied for anodic processes, too.

The big majority of the relevant works is based on the formation of PbO_2 from Pb^{2+}-containing solutions. The starting solution can be either acidic [208, 209] or alkaline [208, 209]. Acidic solutions are usually more concentrated, containing Pb^{2+} species up to about 1 M, while alkaline baths are somewhat more dilute with a typical concentration of 0.1 M. Regardless of the nature of the particles, the d.c.-plated PbO_2-based composites have a fine-grained structure as opposed to the coarse-grained columnar structure of their particle-free counterpart, and the texture of the PbO_2 matrix is strongly decreased as a result of the composite formation. Figure 7.7 shows a typical series of SEM images on PbO_2 deposits containing colloidal particles. The increased active surface area of the composite PbO_2 electrodes is widely considered as a key factor for the enhancement of the activity of theses electrodes as compared to pure PbO_2.

Pulse plating was seldom reported for the anodic codeposition of colloidal particles. When this was tested, the experience was that the decrease in duty cycles and the increase of the pulse frequency is favourable for the enhancement of the concentration of incorporated particles [211, 212]. The application of reverse pulse electrodeposition can further enhance the incorporation ratio of the nanoparticles, indicating that the cathodic pulse does not lead to the release of particles, but can generate adsorption sites for more particles to be incorporated [213]. The literature background does not yield unambiguous information about the effect of solution agitation on the particle incorporation ratio. Apparently, the codeposition of fairly large particles with $d > 100$ nm on downward-facing electrodes can be stimulated with solution agitation or electrode rotation [208, 209], the stable suspension of very small particles with $d \approx 5$ nm stirring is rather counterproductive. However, there is

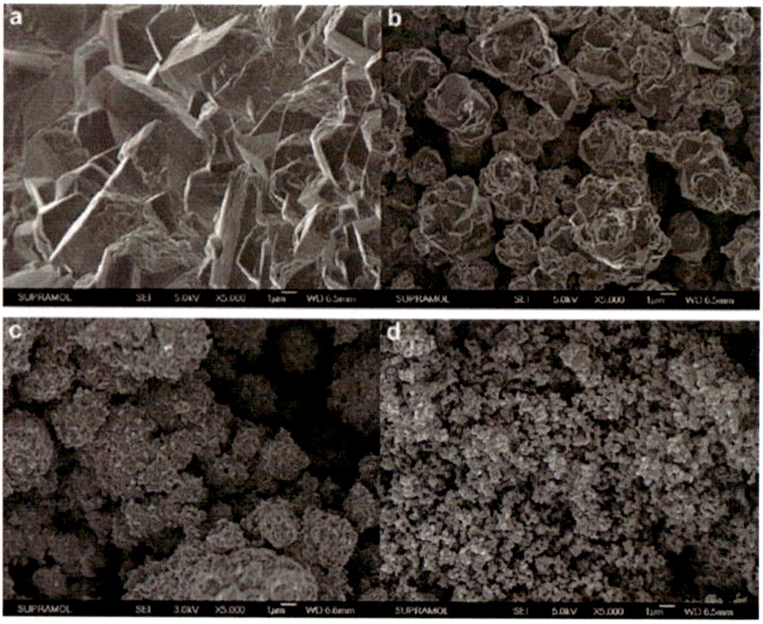

Fig. 7.7 Series of SEM images on the surface of PbO_2 deposits containing Co_3O_4 particles of 6–10 nm diameter in various weight ratios: **a** 0%, **b** 1.4%, **c** 17.0%, **d** 27.5%. Reprinted from [214]. Copyright (2011), with permission from Elsevier

an agreement on that the increase in the concentration of the colloidal particles leads to a saturation concerning the incorporation ratio.

The nature of the codeposited particles is usually of oxide type and is applied as modifiers of the electrochemical properties of the resulting PbO_2-based coating. Co_3O_4 particles can be used for decreasing the overvoltage of the oxygen evolution on the PbO_2-based electrodes, hence making it suitable for an efficient anode in water electrolysis processes [209, 214]. Not only does the incorporation of Co_3O_4 particles reduces the overvoltage of oxygen evolution by a few hundreds of millivolts but the Tafel slope of the process is also markedly affected, namely, decreased [209]. This means that the polarization of $PbO_2(Co_3O_4)$ electrodes is significantly smaller at high oxygen evolution rate that pure PbO_2 electrodes, and the better performance is accompanied with an extension of the electrode life time. The impact of the RuO_2 particle incorporation is similar to that of Co_3O_4 particles, although the change in the overvoltage of oxygen evolution reaction is smaller [215]. In contrast, the impact of either ZrO_2 [210, 211, 216, 217] or CeO_2 [212, 213, 218] nanoparticle addition to the PbO_2 deposits is the increase of the overvoltage of the oxygen evolution, which makes these electrodes suitable for anodic degradation of pollutants with a higher current efficiency than additive-free PbO_2 coatings. Manganese-containing nanoparticles modify the PbO_2 matrix so that they render the resulting coatings suitable for either oxygen evolution electrocalalyst [219] or supercapacitor applications [220].

The reason of the addition of TiO_2 particles to PbO_2 deposits was originally the enhancement of the photoelectrochemical activity of the resulting oxide electrode. However, it turned out that the TiO_2 addition leads to further advantages like the improvement of the electrode life time [221] and the enhancement of the electrode activity also without illumination [222]. From preparative point of view, the TiO_2 particles increased the rate of the PbO_2 formation in the kinetic potential regime [223], which was explained by the effect of the hydroxide radicals forming on the incorporated particles that act as nucleation centres for the PbO_2 deposition.

The anodic deposition of MnO_2-based composite coatings is based on similar principles as that of PbO_2. Briefly, the solution contains a manganese salt (usually manganese(II) nitrate in 0.1 M concentration) and the suspended particles such as Co_3O_4 [224] or WoO_3 [225]. The addition of the appropriate oxide nanoparticles results in better supercapacitive properties of the MnO_2-based coatings in terms of both the capacitance values themselves and the capacity retention with electrode cycling.

For sake of completeness, it is to be mentioned that anodic processes are capable of producing various materials other than oxides. In this example, nanoparticle-doped $BiVO_4$ coating was produced with anodic current from a solution containing Bi^{3+} and VO^{2+} ions [226]. The incorporated WoO_3 nanoparticles improved the behaviour of the resulting coating in the photoelectrochemical water oxidation process.

References

1. Watanabe T (2004) Nano-plating. Elsevier, Oxford
2. Neuróhr K, Dégi J, Pogány L, Bakonyi I, Ungvári D, Vad K, Hakl J, Révész Á, Péter L (2012) J Alloy Compd 545:111–121
3. Michaelsen C (1995) Philos Mag a 72:813–828
4. Liang D, Shao W, Zangari G (2016) J Electrochem Soc 163:D40–D48
5. Kazeminezhad I, Blythe HJ, Schwarzacher W (2001) Appl Phys Lett 78:1014–1016
6. de Paula FR, Oestereich AL, Zago LA, Spada ER (2016) J Mater Sci: Mater Electron 27:9734–9739
7. Cohen-Hyams T, Kaplan WD, Aurbach D, Cohen YS, Yahalom J (2003) J Electrochem Soc 150:C28–C35
8. Cohen-Hyams T, Plitzko JM, Hetherington CJD, Hutchinson JL, Yahalom J, Kaplan WD (2004) J Mater Sci 39:5701–5709
9. Ge S, Li H, Li C, Xi L, Li W, Chi J (2000) J Phys Condens Matter 12:5905–5916
10. Pattanaik GR, Kashyap SC, Pandya DK (2000) J Magn Magn Mater 219:309–316
11. Pattanaik GR, Pandya DK, Kashyap SC (2002) J Electrochem Soc 149:C363–C369
12. Bran J, Jean M, Lardé R, Sauvage X, Le Breton JM, Morin-Grognet S, Pautrat A (2014) J Alloy Compd 596:118–124
13. Pattanaik GR, Pandya DK, Kashyap SC (2003) Thin Solid Films 433:247–251
14. Kelly JJ, Bradley PE, Landolt D (2000) J Electrochem Soc 147:2975–2980
15. Kelly JJ, Kern P, Landolt D (2000) J Electrochem Soc 147:3725–3729
16. Kelly JJ, Cantoni M, Landolt D (2001) J Electrochem Soc 148:C620–C626
17. Rafaja D, Schimpf C, Schucknecht T, Klemm V, Péter L, Bakonyi I (2011) Acta Mater 59:2992–3001
18. Liu QX, Péter L, Pádár J, Bakonyi I (2005) J Electrochem Soc 152:C316–C323

19. Kenane S, Chainet E, Nguyen B, Kadri A, Benbrahim N, Voiron J (2002) Electrochem Commun 4:167–170
20. Kenane S, Voiron J, Benbrahim N, Chainet E, Robaut F (2006) J Magn Magn Mater 297:99–106
21. Garcia-Torres J, Vallés E, Gómez E (2010) J Magn Magn Mater 322:3186–3191
22. Garcia-Torres J, Vallés E, Gómez E (2010) J Phys Chem C 114:9146–9152
23. Garcia-Torres J, Vallés E, Gómez E (2011) Mater Lett 65:1865–1867
24. Zaman H, Ikeda S, Ueda Y (1997) IEEE Trans Mag 33:3517–3519
25. Gomez E, Garcia-Torres J, Valles E (2008) J Electroanal Chem 615:213–221
26. Garía-Torres J, Gómez E, Alcobe X, Vallés E (2009) Cryst Growth Des 9:1671–1676
27. Garcia-Torres J, Gómez E, Vallés E (2010) Mat Chem Phys 122:463–469
28. Nineva S, Dobrovolska T, Krastev I (2011) J Appl Electrochem 41:1397–1406
29. Dacuna B, Mira J, Blanco MC, López-Quintela MA, Rivas J (1999) J Magn Magn Mater 203:123–125
30. Ignatova K, Stoykova D (2011) Bulg Chem Commun 43:48–53
31. Chiang CH, Lin CC, Hu CC (2018) J Electrochem Soc 165:D550–D556
32. Strehle S, Menzel S, Wendrock H, Acker J, Wetzig K (2003) Microelectron Eng 70:506–511
33. Strehle S, Menzel S, Wetzig K, Bartha JW (2011) Thin Solid Films 519:3522–3529
34. Volov I, Swanson E, O'Brien B, Novak SW, van den Boom R, Dunn K, West AC (2012) J Electrochem Soc 159:D677–D683
35. Strehle S, Bartha JW, Wetzig K (2009) Thin Solid Films 517:3320–3325
36. Kim MJ, Lee HJ, Yong SH, Kwon OJ, Kim SK, Kim JJ (2012) J Electrochem Soc 159:D253–D259
37. Kim MJ, Yong SH, Ko HS, Lim T, Park KJ, Kwon OJ, Kim JJ (2012) J Electrochem Soc 159:D656–D658
38. Kim MJ, Park KJ, Lim T, Kwon OJ, Kim JJ (2013) J Electrochem Soc 160:D3126–D3133
39. Bernasconi R, Nobili L, Magagnin L (2014) ECS Trans 58:53–60
40. Bernasconi R, Hart JL, Lang AC, Magagnin L, Nobili L, Taheri ML (2017) Electrochim Acta 251:475–481
41. Ko YK, Jang JH, Lee S, Yang HJ, Lee WH, Lee JG, Reucroft PJ (2003) J Mater Sci Mater Electron 14:103–108
42. Strehle S, Reiche R, Hoffmann V, Acker J, Gemming T, Wetzig K (2007) Microchim Acta 156:167–172
43. Eom H, Jeon B, Kim D, Yoo B (2010) Mater Trans 51:1842–1846
44. Schneider M, Krause A, Ruhnow M (2002) J Mat Sci Lett 21:795–797
45. Santhi K, Karthick SN, Kim HJ, Nidhin M, Narayanan V, Stephen A (2012) Appl Surf Sci 258:3126–3132
46. Ueda J, Kikuchi N (1993) Jpn J Appl Phys 32:1779–1792
47. Sarac U, Baykul C (2012) J Mater Sci Technol 28:1004–1009
48. Dislaki E, Sort J, Pellicer E (2017) Electrochim Acta 231:739–748
49. Williams JM, Blythe HJ, Fedosyuk VM (1996) J Magn Magn Mater 155:355–357
50. Noce RD, Gomes ODM, de Magalhães SD, Wolf W, Guimarães RB, de Castro AC, Pires MJM, Macedo WAA, Givord D, Barthem VMTS (2009) J Appl Phys 106:093907(1–6)
51. Roy MK, Verma HC (2004) J Magn Magn Mater 270:186–193
52. Dadvand N, Dadvand M (2014) J Electrochem Soc 161:D730–D735
53. Derjaguin B, Landau LD (1941) Acta Phys Chim URRS 14:633–662
54. Verwey EJW, Overbeek JTG (1948) Theory of stability of lyophobic colloids. Elsevier, Amsterdam
55. Hidalgo-Álvarez R, Martín A, Fernández A, Bastos D, Martínez F, de las Nieves FJ (1996) Adv Coll Interf Sci 67:1–118
56. Adamczyk Z, Werofiski P (1999) Adv Coll Interf Sci 83:137–226
57. Liang Y, Hilal N, Langston P, Starov V (2007) Adv Coll Interf Sci 134–135:151–166

58. Gomes A, Pereira I, Fernández B, Pereiro R (2011). Electrodeposition of metal matrix nanocomposites: improvement of the chemical characterization techniques. In: Reddy B (ed) Advances in nanocomposites—synthesis, characterization and industrial applications. InTech, Available from: https://www.intechopen.com/books/advances-in-nanocompositessynthesis-characterization-and-industrial-applications/electrodeposition-of-metal-matrix-nanocompositesimprovement-of-the-chemical-characterization-techni
59. Haseeb ASMA (2017) Nano-/Microcomposites by Electrodeposition. In: Kar KK (ed), Composite Materials. Springer-Verlag, Berlin – Heidelberg, pp. 169–198.
60. Walsh FC, Ponce de Leon C (2014) Trans IMF 92:83–98
61. Tseluikin VN (2016) Prot Met Phys Chem S 52:254–266
62. Lekka M (2018) Electrochemical deposition of composite coatings. In: Wandelt K (ed), Encyclopedia of interfacial chemistry: surface science and electrochemistry, Elsevier, vol 5, pp 54–67
63. Hovestad A, Janssen LJJ (1995) J Appl Electrochem 25:519–527
64. Low CTJ, Wills RGA, Walsh FC (2006) Surf Coat Tech 201:371–383
65. Brandes EA, Goldthorpe D (1967) Metallurgia 76:195–198
66. Guglielmi N (1972) J Electrochem Soc 119:1009–1012
67. Celis JP, Roos JR, Buelens C (1987) J Electrochem Soc 134:1402–1408
68. Fransaer J, Celis JP, Roos JR (1992) J Electrochem Soc 139:413–425
69. Stappers L, Fransaer J (2006) J Electrochem Soc 153:C472–C482
70. Walsh FC, Low CTJ, Bello JO (2015) Trans IMF 93:147–156
71. Kazimierczak H, Szymkiewicz K, Gileadi E, Eliaz N (2019) Coatings 9:93(1–18)
72. Eroglu D, Vilinska A, Somasundaran P, West AC (2013) Electrochim Acta 113:628–634
73. Arai S, Saito T, Endo M (2010) J Electrochem Soc 157:D127–D134
74. Sudhakar R, Venkatesha VT (2013) Ind Eng Chem Res 52:6422–6429
75. Sieh R, Le H (2016) Coatings 6:31(1–14)
76. Martis P, Dilimon VS, Delhalle J, Mekhalif Z (2010) Electrochim Acta 55:5407–5410
77. Mirzamohammadi S, Khorsand H, Aliofkhazraei M, Shtansky DV (2018) Tribol Intern 117:68–77
78. Bengoa LN, Pary P, Egli WA (2016) J Electrochem Soc 163:D780–D786
79. Cui W, Song R, Zhang Y, Wang P (2018) Jap J Appl Phys 57:125001(1–6)
80. Ning D, Zhang A, Wu H (2019) Surf Eng 35:1048–1054
81. García-Lecina E, García-Urrutia I, Díez JA, Morgiel J, Indyka P (2012) Surf Coat Tech 206:2998–3005
82. Zanella C, Lekka M, Rossi S, Deflorian F (2011) Corros Rev 29:253–260
83. Lekka M, Offoiach R, Revelant F, Fedrizzi L (2019) Trans IMF 97:73–81
84. Tudela I, Zhang Y, Pal M, Kerr I, Cobley AJ (2015) Surf Coat Tech 276:89–105
85. García-Lecina E, García-Urrutia I, Díez JA, Fornell J, Pellicer E, Sort J (2013) Electrochim Acta 114:859–867
86. Bengoa LN, Ispas A, Bengoa JF, Bund A, Egli WA (2019) J Electrochem Soc 166:D244–D251
87. Bahadormanesh B, Dolati A, Ahmadi MR (2011) J Alloy Compd 509:9406–9412
88. Carpenter CR, Shipway PH, Zhu Y, Weston DP (2011) Surf Coat Tech 205:4832–4837
89. Masoudi M, Hashim M, Kamari HM, Salit MS (2013) Appl Nanosci 3:357–362
90. Arai S, Saito T, Endo M (2010) J Electrochem Soc 157:D147–D153
91. Qin X, Liu J, Wang F, Ji J (2011) J Electroanal Chem 651:233–236
92. Giannopoulos F, Chronopoulou N, Bai J, Zhao H, Pantelis DI, Pavlatou EA, Karantonis A (2016) Electrochim Acta 207:76–86
93. Han C, Liu Q, Ivey DG (2009) Mater Sci Eng B 164:172–179
94. Silva-Ichante M, Reyes-Vidal Y, Bácame-Valenzuela FJ, Ballesteros JC, Arciga E, Ţălu S, Méndez-Albores A, Trejo G (2018) J Electroanal Chem 823:328–334
95. Vaezi MR, Sadrnezhaad SK, Nikzad L (2008) Coll Surf A 315:176–182
96. Chaudhari AK, Singh VB (2017) J Appl Electrochem 47:1009–1021
97. Podlaha EJ, Landolt D (1997) J Electrochem Soc 144:L200–L202
98. Arai S, Suwa Y, Endo M (2011) J Electrochem Soc 158:D49–D53

99. Yang YJ (2012) Bull Mater Sci 35:513–517
100. Sangeetha S, Paruthimal Kalaignan G, Tennis Anthuvan J (2015) Appl Surf Sci 359:412–419
101. Kasturibai S, Paruthimal Kalaignan G (2014) Bull Mater Sci 37:721–728
102. Ghazanlou SI, Farhood AHS, Hosouli S, Ahmadiyeh S, Rasooli A (2018) Mater Manufact Proc 33:1067–1079
103. Zanella C, Lekka M, Bonora PL (2009) J Appl Electrochem 39:31–38
104. Lekka M, Lanzutti A, Zanella C, Zendron G, Fedrizzi L, Bonora PL (2011) Pure Appl Chem 83:295–308
105. Lekka M, Zendron G, Zanella C, Lanzutti A, Fedrizzi L, Bonora PL (2011) Surf Coat Tech 205:3438–3447
106. Dheeraj PR, Patra A, Sengupta S, Das S, Das K (2017) J Alloy Compd 729:1093–1107
107. Lanzutti A, Lekka M, de Leitenburg C, Fedrizzi C (2019) Tribol Int 132:50–61
108. Gyftou P, Stroumbouli M, Pavlatou EA, Asimidis P, Spyrellis N (2005) Electrochim Acta 50:4544–4550
109. Stroumbouli M, Gyftou P, Pavlatou EA, Spyrellis N (2005) Surf Coat Tech 195:325–332
110. Zoikis-Karathanasis A, Pavlatou EA, Spyrellis N (2010) J Alloy Compd 494:396–403
111. Spanou S, Pavlatou EA (2010) J Appl Electrochem 40:1325–1336
112. Hansal WEG, Sandulache G, Mann R, Leisner P (2013) Electrochim Acta 114:851–858
113. Gyftou P, Pavlatou EA, Spyrellis N (2008) Appl Surf Sci 254:5910–5916
114. Peipmann R, Thomas J, Bund A (2007) Electrochim Acta 52:5808–5814
115. Feng Q, Li T, Zhang Z, Zhang J, Liu M, Jin J (2007) Surf Coat Tech 201:6247–6252
116. Koza JA, Uhlemann M, Gebert A, Schultz L (2008) Electrochim Acta 53:5344–5353
117. Koza JA, Mühlenhoff S, Uhlemann M, Eckert K, Gebert A, Schultz L (2009) Electrochem Commun 11:425–429
118. Liu H, Pan L, Huang H, Qin Q, Li P, Wen J (2015) J Electroanal Chem 754:22–29
119. Koza JA, Mogi I, Tschulik K, Uhlemann M, Mickel C, Gebert A, Schultz L (2010) Electrochim Acta 55:6533–6541
120. Zhou Y, Peng X, Wang F (2005) Oxid Met 64:169–183
121. Offoiach R, Lekka M, Johanns KE, Fedrizzi L (2019) Surf Eng 35:135–143
122. Lekka M, Lanzutti A, Casagrande A, de Leitenburg C, Bonora PL, Fedrizzi L (2012) Surf Coat Tech 206:3658–3665
123. Kazimierczak H, Szymkiewicz K, Bobrowski P, Swiatek Z, Rogal L, Gileadi E, Eliaz N (2018) J Electrochem Soc 165:D774–D782
124. Salari Mehr M, Akbari A, Damerchi E (2019) J Alloy Compd 782:477–487
125. Ghazanlou SI, Farhood AHS, Ahmadiyeh S, Ziyaei E, Rasooli A, Hosseinpour S (2019) Metall Mater Trans A 50:1922–1935
126. Pavlatou EA, Stroumbouli M, Gyftou P, Spyrellis N (2006) J Appl Electrochem 36:385–394
127. Rashkov S, Atanassov N (1980) J Appl Electrochem 10:535–541
128. Shanmugasamy S, Balakrishnan K, Subasri A, Ramalingam S, Subramania A (2018) J Rare Earths 36:1319–1325
129. Abdel Aal A, El-Sheikh SM, Ahmed YMZ (2009) Mater Res Bull 44:151–159
130. Chhangani S, Prasad MJNV (2018) Mater Charact 136:247–256
131. Das MK, Li R, Qin J, Zhang X, Das K, Thueploy A, Limpanart S, Boonyongmaneerat Y, Ma M, Liu R (2017) Surf Coat Tech 309:337–343
132. Liu M, Liu H, Wang D, Liu B, Shi Y, Li F, Gong Y, Li L, Li L, Zhang W (2019) Materials 12:1105(1–19)
133. Jung A, Natter H, Hempelmann R, Lach E (2009) J Mater Sci 44:2725–2735
134. Beltowska-Lehman E, Góral A, Indyka P (2011) Arch Metall Mater 56:919–931
135. Góral A (2017) Surf Coat Tech 319:23–32
136. Waware US, Hamouda AMS, Wasekar NP (2018) Surf Coat Tech 337:335–341
137. Sen R, Das S, Das K (2011) Mater Charact 62:257–262
138. Ramalingam S, Muralidharan VS, Subramania A (2013) Surf Eng 29:511–515
139. Tam J, Palumbo G, Erb U, Azimi G (2017) Adv Mater Interfaces 1700850(1–11)
140. Yao Y, Yao S, Zhang L, Wang H (2007) Mater Lett 61:67–70

141. da Costa Pereira RF, Dantas de Oliveira ES, Santos Vieira MR, Gomes de Andrade Lima MA, Urtiga Filho SL (2017) Mater Res 20(Suppl. 2):221–230
142. Mirsaeed-Ghazi SM, Allahkaram SR, Molaei A (2019) Tribol Int 134:221–231
143. Abdel Aal A (2008) Mater Sci Eng A 474:181–187
144. Yousefi E, Sharafi S, Irannejad A (2018) J Alloy Compd 753:308–319
145. Gyawali G, Tripathi K, Joshi B, Lee SW (2017) J Alloy Compd 721:757–763
146. Zhou Y, Zhang G, Zhang H (2010) Trans Nonferrous Met Soc China 20:104–109
147. Iacovetta D, Tam J, Erb U (2015) Surf Coat Tech 279:134–141
148. Carpenter CR, Shipway PH, Zhu Y (2011) Surf Coat Tech 205:5059–5063
149. Beltowska-Lehman E, Indyka P, Bigos A, Szczerba MJ, Kot M (2016) J Electroanal Chem 775:27–36
150. Nagayama T, Yamamoto T, Nakamura T, Fujiwara Y (2017) Surf Coat Tech 322:70–75
151. Ahmadkhaniha D, Eriksson F, Leisner P, Zanella C (2018) J Alloy Compd 769:1080–1087
152. Zoikis-Karathanasis A, Pavlatou EA, Spyrellis N (2009) Electrochim Acta 54:2563–2570
153. Suzuki Y, Arai S, Endo M (2010) J Electrochem Soc 157:D50–D53
154. Bakonyi I, Cziraki A, Nagy I, Hosso M (1986) Z Metallkde 77:425–432
155. Berkh O, Zahavi J (1996) Corr Rev 14:323–341
156. Huang HC, Chung ST, Pan SY, Tsai WT, Lin CS (2010) Surf Coat Tech 205:2097–2103
157. Lelevic A, Walsh FC (2019) Surf Coat Tech 369:198–220
158. Tam J, Jiao Z, Lau JCF, Erb U (2017) Wear 374–375:1–4
159. Ying LX, Wu K, Li DY, Wu CX, Fu Z (2019) Wear 426–427:258–264
160. He Y, Sun WT, Wang SC, Reed PAS, Walsh FC (2017) Electrochim Acta 245:872–882
161. He Y, Wang SC, Walsh FC, Chiu YL, Reed PAS (2016) Surf Coat Tech 307:926–934
162. Suzuki Y, Arai S, Endo M (2010) Appl Surf Sci 256:6914–6917
163. Tudela I, Cobley AJ, Zhang Y (2019) Friction 7:169–180
164. Akhtar K, Khan ZU, Gul M, Zubair N, Shah SSA (2018) J Mater Eng Perf 27:2827–2837
165. Nicolenco A, Mulone A, Imaz N, Tsyntsaru N, Sort J, Pellicer E, Klement U, Cesiulis H, García-Lecina E (2019) Frontiers Chem 7:241(1–15)
166. Qu NS, Hu XY, Qian WH, Zhu ZW (2014) Surf Eng 30:159–164
167. Masoudi M, Hashim M, Kamari HM (2014) Appl Nanosci 4:649–656
168. Li B, Li D, Mei T, Xia W, Zhang W (2019) J Alloy Compd 777:1234–1244
169. Qu NS, Zhu D, Chan KC (2006) Scr Mat 54:1421–1425
170. Li B, Zhang W, Li D, Huan Y, Dong J (2018) Appl Surf Sci 458:305–318
171. Sivasakthi P, Ramesh Bapu GNK, Chandrasekaran M, Sreejakumari SS (2016) RSC Adv 6:44766–44773
172. Tacken RA, Jiskooot P, Janssen LJJ (1996) J Appl Electrochem 26:129–134
173. Roldan A, Gómez E, Pané S, Vallés E (2007) J Appl Electrochem 37:575–582
174. Guan S, Nelson BJ, Vollmers K (2004) J Electrochem Soc 151:C545–C549
175. Thiemig D, Kubeil C, Gräf CP, Bund A (2009) Thin Solid Films 517:1636–1644
176. Wang R, Wan Y, He F, Qi Y, You W, Luo H (2012) Appl Surf Sci 258:3007–3011
177. Pané S, Gómez E, García-Amorós J, Velasco D, Vallés E (2007) J Electroanal Chem 604:41–47
178. Pané S, Gómez E, Vallés E (2008) J Electroanal Chem 615:117–123
179. Cojocaru P, Spreafico M, Gomez E, Vallés E, Magagnin L (2010) Surf Coat Tech 205:195–199
180. Cojocaru P, Magagnin L, Gomez E, Vallés E (2011) Mater Lett 65:2765–2768
181. Cojocaru P, Pahari S, Magagnin L, Dietrich D, Liebig A, Lampke T (2011) Z Phys Chem 225:351–361
182. Fazli S, Bahrololoom ME (2016) Metal Mater Trans A 47:4316–4324
183. Guan S, Nelson BJ (2006) J Microelectromech Sys 15:330–337
184. El-Sayed AH, Hemeda OM, Tawfik A, Hamad MA (2016) J Magn Magn Mater 402:105–109
185. El-Sayed AH, Hemeda OM, Tawfik A, Hamad MA (2017) Phase Trans 90:325–334
186. Qu NS, Jiao F (2012) Trans IMF 90:92–97
187. Qu NS, Jiao F (2013) Trans IMF 91:88–95
188. Tian BR, Cheng YF (2007) Electrochim Acta 53:511–517
189. Bakhit B (2015) Surf Coat Tech 275:324–331

190. Alipour K, Nasirpouri F (2019) J Electrochem Soc 166:D1–D9
191. Akhtar K, Hira U, Khalid H, Zubair N (2019) J Alloy Compd 772:15–24
192. Sajjadnejad M, Omidvar H, Javanbakht M, Mozafari A (2017) J Alloys Compd 704:809–817
193. Zhou Q, Xie W, Zhang Y,Sheng M, Hu A, Cheng X, Zhang L (2018) Surf Rev Lett 25 185001(1–9)
194. Chen XH, Chen CS, Xiao HN, Cheng FQ, Zhang G, Yi GJ (2005) Surf Coat Tech 191:351–356
195. Zeng Z, Yu Y, Zhang J (2009) J Electrochem Soc 156:C123–C126
196. Zhou Y, Ding Y (2007) Trans Nonferrous Met Soc China 17:925–928
197. Reyes-Vidal Y, Suarez-Rojas R, Ruiz C, Torres J, Tălu S, Méndez A, Trejo G (2015) Appl Surf Sci 342:34–41
198. Méndez-Albores A, González-Arellano SG, Reyes-Vidal Y, Torres J, Tălu S, Cercado B, Trejo G (2017) J Alloy Compd 710:302–311
199. Raghupathy Y, Natarajan KA, Srivastava C (2017) Surf Coat Tech 328:266–275
200. Punith Kumar MK, Srivastava C (2013) Mater Charact 85:82–91
201. Elias L, Chitharanjan Hegde A (2016) Electrochim Acta 219:377–385
202. Kullaiah R, Elias L, Chitharanjan Hegde A (2018) Int J Min Met Mater 25:472–479
203. Yang XY, Peng X, Wang FH (2012) Adv Mat Res 472–475:137–140
204. Huang Z, Peng X, Wang F (2006) Oxid Met 65:223–236
205. Peng X, Zhou Y, Zhang Y, Wang F (2004) Mat Sci Forum 461–464:409–417
206. Zhou Y, Zhang H (2011) Trans Nonferrous Met Soc China 21:322–329
207. Srivastava M, Balaraju JN, Ravisankar B, Anandan C, Grips VKW (2012) App Surf Sci 263:597–607
208. Casellato U, Cattarin S, Guerriero P, Musiani MM (1997) Chem Mater 9:960–966
209. Musiani M, Furlanetto F, Guerriero P (1997) J Electroanal Chem 440:131–138
210. Velichenko AB, Knysh VA, Luk'yanenko TV, Velichenko YA, Devilliers D (2012) Mater Chem Phys 131:686–693
211. Yao Y, Zhao M, Zhao C, Zhang H (2014) Electrochim Acta 117:453–459
212. Yao Y, Cui L, Jiao L, Chen X, Yu N, Dong H (2016) J Solid State Electrochem 20:725–731
213. Yao Y, Dong H, Jiao L, Yu N, He L (2016) J Electrochem Soc 163:D179–D184
214. Dan Y, Lu H, Liu X, Lin H, Zhao J (2011) Int J Hydr Energ 36:1949–1954
215. Musiani M, Furlanetto F, Bertoncello R (1999) J Electroanal Chem 465:160–167
216. YaoY, Zhao C, Zhu J (2012) Electrochim Acta 69:146–151
217. Cui W, Chen Z, Yu O, Zhu W, Li H, Wang H (2018) Int J Electrochem Sci 13:1400–1412
218. Yao Y, Dong H, Yu N, Chen X, Jiao L, Zhao C (2017) Russ J Electrochem 53:411–416
219. Li Y, Jiang L, Liu F, Li J, Liu Y (2014) RSC Adv 4:24020–24028
220. Dan Y, Lin H, Liu X, Lu H, Zhao J, Shi Z, Guo Y (2012) Electrochim Acta 83:175–182
221. Velichenko AB, Knysh VA, Luk'yanenko TV, Danilov FI, Devilliers D (2009) Prot Met Phys Chem S 45:327–332
222. Amadellia R, Samiolo L, Velichenko AB, Knysh VA, Luk'yanenko TV, Danilov FI (2009) Electrochim Acta 54:5239–5245
223. Velichenko AB, Amadelli R, Knysh VA, Luk'yanenko TV, Danilov FI (2009) J Electroanal Chem 632:192–196
224. Yuan C, Lin H, Lu H, Xing E, Zhang Y, Xie B (2014) RSC Adv 4:64675–64682
225. Yuan C, Lin H, Lu H, Xing E, Zhang Y, Xie B (2015) Mater Lett 148:167–170
226. Cho SK, Park HS, Lee HC, Nam KM, Bard AJ (2013) J Phys Chem C 117:23048–23056

Chapter 8
Porous Nanostructured Materials

8.1 The Dynamic Bubble Template Method

8.1.1 Overview of the Dynamic Bubble Template Method

During the electrodeposition of a compact film, side reactions are usually much undesired. In aqueous solutions, the typical side reaction during metal deposition is the evolution of hydrogen. The harmful fingerprint of the hydrogen evolution is a pit on the surface where a hydrogen bubble could stay attached for a sufficiently long time so that the deposit growth is obstructed at the bubble-covered area. Here, both the structure and, in the case of alloy deposition, the local deposit composition differ from the rest of the surface. Such imperfections are the starting points of the coating damage, originating from either wear or corrosion. Therefore, the application of both wetting agents and a sufficiently intense solution agitation are the common counter-measures against the hydrogen bubble attachment. Concerning structural integrity and durability, the hydrogen absorbed into the metal coating being deposited may cause hydrogen-induced embrittlement.

Unlike in the processes leading to smooth coatings, the dynamic bubble template method requires an intense gas evolution on the cathode. Since the most common solvent is water, the bubbles are formed of hydrogen, and the method is often called as the dynamic hydrogen bubble template (DHBT) technique. The growing metal fills up the voids between the bubbles, and the solution remaining in the space between the growing dendritic or foamy metal soon becomes depleted with respect to the precursor of the deposit being formed. Therefore, the growth proceeds near the solid–liquid–gas triple-junction zones. The dynamic nature of the bubble template methods means that the bubbles form at the surface of the cathode and then leave it. Hence, neither the size of the bubbles nor their attachment site at the surface is fixed during the process, although it is much influenced later by the morphology of the deposit itself. The scheme of the process can be seen in Fig. 8.1.

In all literature resources offering a schematic image on the DHBT method, a transition from small-bubble to large-bubble templating can be seen as the porous

L. Péter, *Electrochemical Methods of Nanostructure Preparation*,
Monographs in Electrochemistry, https://doi.org/10.1007/978-3-030-69117-2_8

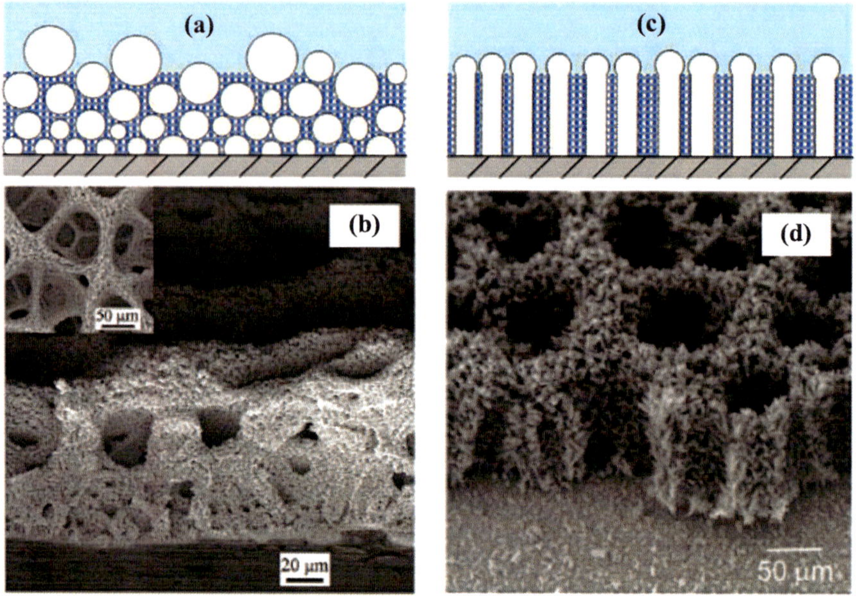

Fig. 8.1 **a** and **c** Schematic cross-sectional images of the formation of the porous deposit around the dynamic hydrogen bubble template. The pattern in the schematic figures is to indicate the secondary porosity within the wall of the bubble-templated deposit. **a** Increasing bubble size with the thickness of the porous deposit. **b** Cross-sectional SEM image of a porous copper deposit indicating the increase in the pore size with the deposit growth (the inset is the top view of the same deposit). Reprinted from [18]. Copyright (2010), with permission from Elsevier. **c** Schematic cross-sectional image for a deposit with even primary pore size along the deposit thickness. **d** Cross-sectional SEM image of a porous copper with even pore size and with dendritic structure of the pore wall. Reprinted with permission from [6]. Copyright (2004) American Chemical Society

layer grows [1–5], as shown in Fig. 8.1a. This can be verified by the increase in the primary pore size with the deposition time (i.e., with the deposit thickness) [6, 7]. The cross-sectional SEM images of the DHBT-plated porous layers, however, do not verify this model unambiguously. Where the surface of the substrate is fully covered at the early phase of the plating process, there are no pores near the substrate, and the transition form compact to porous structure naturally leads to an apparent pore size increase with thickness. The DHBT-plated porous layers are often not thick enough; i.e., the total deposit thickness in the cross-sectional images is somewhat larger than the size of the top pores but the ratio of the total thickness and the diameter of the top pore is less than an order of magnitude. Also, the magnification of the cross-sectional images published is often too small for underpinning the model of the bubble size increase with the deposit thickness [4, 8, 9]. When cross-sectional images of dendritic deposits are shown, they are rather contradictory to the notion of the pore size variation, and a pore with nearly even thickness and with an axis perpendicular to the substrate can be seen [6, 10], as indicated in Fig. 8.1c. Even if there is some change in the primary pore size (large pores), the secondary porosity (i.e., the wall structure) seems to be invariant throughout the deposit thickness [2, 11]. As it can be assessed from the image of cross-sectionally polished deposits, the even pore size is strongly related to the high dendrite length to pore wall thickness ratio [11–13], albeit all porous structures appear to be dendritic at a high enough magnification.

For finding the right balance between the gas evolution and the metal deposition, the composition of the electrolyte solution must somewhat differ from that of the typical plating baths. The concentration of the metal salts is smaller, but the acidity is higher in solutions used for the dynamic bubble template method than for deposition of smooth coatings. The acid concentration is often as high as 2 mol dm^{-3}. For transition metals whose ion can form a complex with ammonia, buffered solutions containing an ammonium salt and ammonia also proved to be feasible to prevent the chemical precipitation of metal hydroxides [11, 14–16]. The necessity of an intense bubble formation rationalizes a relatively large current density which must definitely be larger than the mass transport-limited current density of the metal deposition. While the deposition of a compact metal layer seldom can be carried out with a current density higher than −200 mA cm^{-2}, porous structures are often produced with current densities of about −3 A cm^{-2} (typical value) or exceptionally up to −5 A cm^{-2} [17]. The duration of the sample preparation procedure with a total thickness of at most a few tens of micrometers is between 5 s to 40 min, and very thick deposits cannot be prepared due to the low mechanical strength of the resulting structure.

When an electrode potential value is given for DHBT experiments, their elucidation of the experimental conditions is very difficult due to the necessity of compensating the ohmic drop between the working and reference electrode, and the reference electrode cannot be positioned close enough to the working electrode due to the vigorous hydrogen evolution. Nevertheless, the application of the ohmic drop correction during the bath optimization can yield valuable information on the kinetic background of the DHBT process [16]. The problem of the correct electrode potential referencing is even more difficult when the cell voltage is given, which often ranges to a few tens of volts. It is the pronounced opinion of the author of the present work that not only is current control simpler in this case but also provides a higher level of reproducibility. It is straightforward that stirring of the electrolyte solution influences the DHBT process to a negligible extent only due to the impact of the vigorous hydrogen evolution taking place right on the cathode surface.

Concerning the technical background of the DHBT experiments, the reactivity of the solution components with the atomic hydrogen as an intermediate of the hydrogen evolution has to be considered. The alkalization around the cathode in the presence of ammonium salts may result in ammonia production, and organic acids can also be reduced. The anode reaction is also of importance, and the penetration of the undesired anode reaction product to the vicinity of the cathode must be strictly controlled. In contrast to these requirements, many experimental works report that the porous deposits were made in a single-compartment cell with a relatively small anode–cathode distance, which was typically 2 cm.

Nearly all deposits produced with the dynamic bubble template method have a two-level hierarchical structure. The primary pores defined by the bubble size itself are not at all of the nanometer scale but can extend up to a few tens of micrometers. However, the walls surrounding the large primary pores are not compact either but exhibit another level of porosity. When high-magnification images are available for the walls of the primary pores, the interconnected dendritic nature of the wall structure can often be seen. The typical range of the secondary pore diameter can be assessed to be around or below 100 nm. Two sets of low- to high-magnification images indicating the secondary pore structure are shown in Fig. 8.2. The fine structure of the deposits formed from non-noble metals (Cu, Ni) is typically more dendritic than the noble metal foams shown in Fig. 8.2.

The roughness factor of the DHBT-plated materials is a key parameter since the porous materials are mostly used as catalysts. The measurement of the roughness

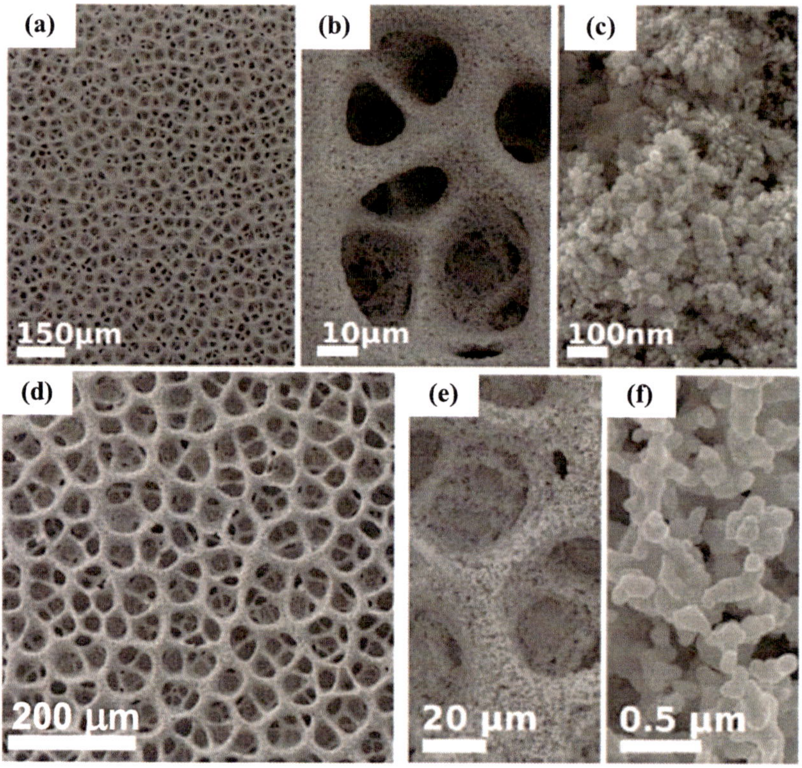

Fig. 8.2 SEM images of DHBT-plated porous structures with various magnifications, indicating the primary and secondary pore structure (from left to right). **a**, **b** and **c**: Porous Au deposit obtained from a solution containing 0.1 M $HAuCl_4$ and 2 M NH_4Cl Reprinted from [19]. Copyright (2011), with permission from Elsevier. **d**, **e** and **f**: Porous silver deposit obtained with -1 A cm^{-2} from a solution of 0.01 M Ag_2SO_4, 1.5 M KSCN and 0.5 M NH_4Cl. Reprinted from [20]. Copyright (2010), with permission from Elsevier

factor can take place with a suitable UPD process (mostly with noble metal deposits) or by double-layer charging (for any metallic deposit). The surface roughness factor commonly achieved ranges up to several hundreds.

8.1.2 Chemical Aspects of the DHBT-Plated Metals: Deposit Types and Bath Components

Copper is by far the most common material applied in DHBT studies [1, 2, 4, 6, 7, 12, 13, 18, 21–28]. The concentration of Cu^{2+} ions reported varied between 0.05–0.8 M. This concentration range is close to that applied for other relatively inert non-noble metals like Ni (0.1–0.2 M [10, 14, 16, 17, 29]), Co (0.1 M [14, 16]), Sn (0.15 M [30]) and Zn (0.61 M [31]). The concentration of noble metal cations was usually lower, such as 0.02–0.1 M [19, 32] or occasionally 0.4 M [33] for Au, 50 mM for Pt [34], 14–60 mM for Pd [3] and 10–60 mM for Ag [8, 20, 35]. Porous lead foams were obtained from solution of 0.5–20 mM Pb^{2+} concentration [36], and bismuth porous structure in the hydrogen evolution regime proved to be successfully produced from a solution of exceptionally small metal ion concentration of 1 mM [37]. Although it is not possible to set up a fully systematic trend, it is apparent that the larger is the exchange current density of the metal ion/metal system, the lower metal ion concentration is necessary. This is in good agreement with the inclination of metals with high exchange current density to develop dendritic deposits at current densities near to the diffusion-limited deposition regime.

If an alloy is deposited in one single step with the DHBT method, the concentration ranges are somewhat similar to those applied for elemental metallic porous structures. Noble metal alloys (like Au–Pt [38], Pt–Pd [39, 40] and Pd–Au [41]) are produced from solutions of small metal ion concentrations ($c < 16$ mM), and the concentration of the precursor compound for the minor component of the noble metal alloys is often less than 1 mM. For alloys composed of metals with significantly different deposition potentials, some bath recipes follow the trend established for bulk alloy deposition that the concentration of the LN metal ions is considerably larger than that of the MN metal ions. Such examples can be seen in work in Au–Cu [42], Ni–Ag [43] and Cu–Ni [9, 44, 45] alloy deposition with the DHBT method.

In several DHBT-related studies on alloy deposits [11, 46, 47], the data published are not detailed enough to establish a *concentration ratio—deposit composition* relationship. In works where the concentration ratio of the precursor compounds was systematically varied [15, 48, 49], it can be seen that the mole fractions of the components in the deposit are a monotonous function of their precursor ion concentration in the solution, similarly to smooth deposits. Where detailed data are available, it can be seen that the codeposition trends established for smooth alloys cannot be always transferred to the DHBT method. For instance, the composition diagram of the porous DHBT-plated Ni–Co deposit showed that the anomalous codeposition character of the Ni–Co pair is effective at small relative Co^{2+} concentration only, but the composition ratio of the constituents is similar to normal codeposition at high relative Ni^{2+} concentrations [15].

The application of bath additives proved to be a suitable strategy to tune the pore size of the deposits. Surface-active components that stabilize the small bubbles and prevent their coalescence lead to a reduction of the pore size. This is why acetic acid is a popular component of the DHBT baths, being an acid source and a bubble stabilizer at the same time [6]. Classical polymeric additives (e.g., polyethylene glycol [18]), originally applied as brighteners for smooth coatings, play a similar role in the DHBT method. A pronounced difference compared to the smooth coating is that the blocking of the dendritic growth is undesired in the DHBT method; hence, application of grain-refining additives (whose preferred adsorption site is the tip of the growing crystals) is not necessary.

Interestingly, electrolyte components like sulphate salts were also found to be effective in the hindrance of the bubble coalescence [4]. The role of the ammonium salts is controversial since the liberation of NH_3 is expected due to the alkalization of the solution around the growing foam, and ammonia can act as a complexing agent for the metal ions, hence changing the deposition mechanism [10, 22, 24]. While ammonium ions were found to decrease the hydrogen evolution rate during the deposition of porous Cu [22], an adverse effect was reported for Ag deposition [20, 35]. For preparing a porous Ni layer, ammonium chloride was found to be indispensable [10, 29]. Ammonium chloride was applied in most of the baths used for various porous deposits since the first study was published on a DHBT process [17]. Therefore, the impact of ammonium ions as growth modifier and their buffering capability are difficult to separate from the effect of halogenide ions. It was occasionally mentioned that the maximum applicable current density from the viewpoint of the mechanical stability of the deposit is strongly interrelated to the NH_4Cl concentration in the solution [17].

Halogenide ions are identified as promoters for the deposition, most likely due to the bridging effect at the surface and the facilitation of the electron transfer reaction between the metal and the adsorbed intermediate [6, 22]. The increase in the pore size with halogenide ion concentration was also shown for bath not containing NH_4Cl as supporting electrolyte [4, 18]. Another additive, 3-mercapto-1-propanesulphonic acid was found to lead to smooth Cu pore walls by decreasing the secondary porosity [22] through a similar effect as brighteners and levelling agents work for planar deposits. As it was stressed in Sect. 2.11, the impact of additives is a highly empirical field where an *a priori* approach seldom works.

While the DHBT method was quite well characterized concerning the role of the bath components, the role of the other deposition conditions received less attention. The available information in the literature does not make it possible to establish any trend concerning the current efficiency of the process since this parameter is practically never given. It was demonstrated for porous Cu that both the deposit grain size and the primary pore size decrease with increasing current density, but the increase in temperature has an adverse effect [26]. The application of pulsed potential was found to be suitable for tuning the morphological properties of porous Cu deposits, even though the period of the potential modulation applied (3–20 ms) was much smaller than the bubble lifetime [12]. The reverse pulse method resulted in a non-monotonous dependence of the primary pore size on the reverse pulse current density, exhibiting a minimum pore size at an intermediate current density of the anodic pulse [13]. These results, especially the pore size variation during the reverse pulse method, have not been completely elucidated.

8.1.3 Miscellaneous Other Materials and Methods Related to the Dynamic Bubble Template Approach

Although water is the typical medium for the DHBT method, the variety of the solvents can be extended to non-aqueous molecular liquids in which the hydrogen evolution takes place at very negative potentials, hence allowing the deposition of metals whose deposition is not accessible in aqueous systems. It has been demonstrated [50] that Fe–Ce intermetallic compounds with porous structure can be deposited from a dimethyl sulphoxide–urea plating bath with potential scans ranging to the hydrogen evolution regime (−2.8 V vs. SCE).

An interesting as well as economic method was shown for the application of DHBT-plated non-noble metal foams to transform them into noble metal-coated porous structures [51–53]. In the first step of the process, an inexpensive non-noble metal foam was prepared with the standard DHBT method, and then it was exposed to a solution containing a noble metal salt of small concentration. The displacement of the LN metal at its rest potential (i.e., without any electrochemical control) led to a conformally coated foam with the displacing noble metal. With this technique, the pore structure can be shaped with an inexpensive metal (e.g., Cu), and pore sizes occasionally not available with a direct deposition of Pd or Ag can be achieved. The unused portion of the expensive noble metal salt can be simply washed out from the porous structure after the completion of the displacement reaction. The average noble metal content of the displacement-modified porous structure was shown to be only about 2 at.% [52], which makes it possible to drive the process efficiently, achieving a high-surface-area noble metal coating on an inexpensive dendritic porous scaffold.

The feasibility of a second electrodeposition step producing a coating on the already formed porous metals was also realized for Ni [54] and Cu [23] deposition onto porous Cu. The composition of the second plating bath was similar to those applied in the deposition of conventional coatings, not to those applied for the DHBT method. Surprisingly, the deposition of the additional layer led to a conformal coating, which means that the deep layers of the pore system were coated similarly as the top of the sample. Therefore, no significant electric field effect was experienced that would be expected to result in a preferential deposition on the top of the samples. However, the second electrodeposition step leads to the disappearance of the secondary porosity of the samples.

Some metal foams obtained with the DHBT process can be oxidized electrochemically, hence obtaining a porous metal oxide structure. This metal-to-oxide transformation was demonstrated for electrodeposited porous Pb deposits [55], where the oxidation was performed also with an electrochemical step after changing the Pb bath to a Na_2SO_4 solution. The oxidation process was shown to take place through an initially formed $PbSO_4$ surface layer with a typical progressive nucleation and growth mechanism. During the electrochemical oxidation process, the dual porosity of the samples was retained, but the diameter of the nanowires increased, primarily due to the difference in the densities of Pb and PbO_2. The porous deposit was not stable in air and underwent a colour change due to the formation of Pb_3O_4 by the synproportionation of the structure-forming Pb and the PbO_2 surface coating. In spite of the ill-defined oxide composition, the oxide-coated Pb porous structures could be used for galvanic exchange processes to deposit oxides that are not accessible for direct deposition due to their small electrical conductivity. For example, a Co_3O_4 surface layer on the Pb/Pb_3O_4 dendrites was obtained by a spontaneous oxidation of dissolved Co^{2+} ions ($Pb_3O_4 + 3Co^{2+} \leftrightarrows Co_3O_4 + 3\ Pb^{2+}$). The resulting heterogeneous porous structure showed an outstanding activity for electrochemical oxygen evolution reaction.

The significance of the above mentioned three-step process can be understood if we compare the oxidation of porous Pb with the direct oxygen-templated deposition of PbO_2 from Pb^{2+} solution. As it was reported in several studies [56–59], the bubble-assisted direct anodic deposition of PbO_2 leads to porous structures but no second-level nanoscale porosity occurs in these deposits. Nevertheless, these deposits are also suitable for galvanic exchange processes with metal ions (like Mn^{2+}, Co^{2+} and Sn^{2+} [59]) to obtain oxides whose direct deposition in a nanoporous self-supporting form is yet to solve.

Since the dynamic bubble template method involves gas evolution, it is logical that the change in the system pressure may modify the pattern forming with the assistance of bubbles. The opportunity of changing the external pressure was exploited in the usual process of copper foam deposition with the DHBT method [60]. Since the change of the standard electrode potential of the redox systems (H^+/H_2 and Cu^{2+}/Cu) involved in the DHBT process is negligible with the change of the pressure, the *volume* of the gas evolved could be modified by keeping the *amount* of materials produced constant. The variation of the pressure from 0.02 to 0.8 MPa resulted in the decrease in the primary pore diameter from 70 to about 35 μm. The decrease in the primary pore size was explained with the reduction of the frequency of the bubble coalescence as the bubble size itself decreased due to the elevated pressure. In contrast to the primary pore size, the secondary porosity did not change to an appreciable level, and the copper dendrites formed at all pressures applied appeared very much alike.

8.2 Superhydrophobic Porous Surfaces by Electrodeposition

8.2.1 General Aspects of Morphology-Based Hydrophobicity

The preparation of hydrophobic surfaces gained an immense attention in the last decade [61, 62]. Deposits for which the hydrophobic behaviour is due to the presence of co-deposited hydrophobic particles were already discussed in Chap. 7. Here, the cases will be dealt with when the hydrophobic behaviour is strongly related to the porosity and high surface roughness of the deposits.

When the hydrophobic behaviour is related to the surface porosity, the wetting state of the surface has to be considered. The possible wetting states of surfaces with various porosities are depicted in Fig. 8.3. The surface tension of a liquid on the surface of specific composition determines the wetting angle at the smooth surface. However, at very high surface roughness, the trapped air between the submicroscopic columns of the rough surface may prevent the wetting. This behaviour, often called the "lotus leaf effect", determines the goal of the production of liquid-repellent surfaces: the achievement of fractal-like surfaces with high dendrite-branching probability and hierarchical porosity at various scales. Due to the similarity in the hierarchical porosity, such coatings are often called biomimetic, even though the morphology of the coating is completely different from those observed in the living nature.

The quantitative description of the wetting behaviour of fractal-like surfaces can be found elsewhere [63]. Below a short summary is given on the relationship of the wetting angles when the ratio of the *liquid droplet—solid* and *liquid droplet—entrapped gas* surface areas is taken into account. According to Cassie's law [64], the cosine of the contact angle of a liquid droplet on a heterogeneous surface with two components (the so-called Cassie angle, Θ_C) follows the equation:

$$\cos\Theta_C = f_1\cos\Theta_1 + f_2\cos\Theta_2 \tag{8.1}$$

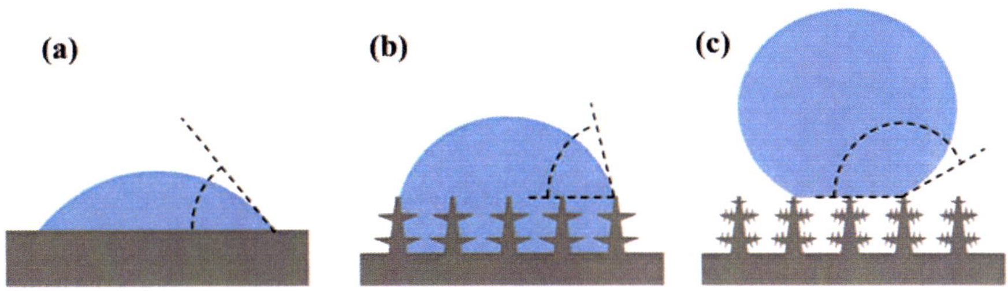

Fig. 8.3 Wetting state of surfaces with various porosities with the indication of the wetting angles. **a** Flat substrate (Young state); **b** Rough surface, Wenzel state; and **c** Rough surface, Cassie–Baxter state

In Eq. 8.1, f_i means the surface area fraction of the component i (defined as $f_i = A_i/\Sigma A_i$ where $f_1 + f_2 = 1$ is obvious for a two-component substrate) and Θ_i is the corresponding wetting angle of the liquid on this component. If one of the components is a gas, the contact angle has to be taken as 180°, and the thus obtained Cassie–Baxter angle (Θ_{CB}) is obtained as [65]

$$\cos \Theta_{CB} = f_1 \cos \Theta_1 - f_2 = f_1(\cos \Theta_1 + 1) - 1. \tag{8.2}$$

As Θ_{CB} increases, the sliding angle (i.e., the tilting angle of the substrate at which the droplet rolls down along the hydrophobic surface) decreases. The Cassie–Baxter angle and the sliding angle are larger than 150° and lower than 5°, respectively, for superhydrophobic surfaces.

The hydrophobic character of a surface with fractal-like hierarchical porosity is related to a number of other properties. Among others, anti-icing surface properties stem from the same surface morphology as the hydrophobic character. The self-cleaning nature of a surface can also be due to the high porosity when dust particles stay at the top part of the surface and their penetration between the dendrite- or column-like surface elements is hindered; therefore, they can be washed away from the surface without the application of either any tenside or a mechanical treatment. The small corrosion rate of surfaces with morphology-related hydrophobic character is also related to the small wetted area. The reduced inclination to corrosion is explained with the small effective contact area between the corrosive liquid and the porous surface, as shown in Fig. 8.3c. The corrosion aspects of the superhydrophobicity have been discussed in detail in various topical reviews [66, 67]. However, the above described wetting model implies that the wear or corrosion damage of the surface may lead to an irreversible degradation when the fractal-like surface morphology is destroyed.

The literature of the electrodeposition of superhydrophobic coatings has been nicely reviewed recently by Tan, Palumbo and Erb [68]. Here, their categorization is followed partly, except for suspension plating systems that were discussed in the previous chapter. Beside the works to be listed below, it is worthwhile of noting that several works cited in Sect. 8.1 mention the superhydrophobic nature of DHBT-plated samples. The morphological origin of the superhydrophobicity is the same for both sample groups.

8.2.2 Electroplated Metallic Coatings with Hydrophobic Properties

As it can be seen from Fig. 8.3, the surface morphology of the roughness-related superhydrophobic coatings is completely the opposite as that for smooth deposits. This offers some guidelines for the working regime in connection with the Winand diagram (see Fig. 2.19). On the one hand, highly dendritic coatings can be obtained by choosing a slightly inhibited bath only and operating it close to or beyond the mass transport-controlled current density range. A commonly applied technique is that a smooth base layer is plated, and then the current density is increased in order to obtain a dendritic (porous) adlayer. On the other hand, it is straightforward that additives used to inhibit the dendrite-like growth of metal crystals are rendered to be useless if the goal is the production of a surface layer with high roughness. This does not mean that the metallic coatings can be plated in additive-free solutions only, but at least the choice of the appropriate additives is different from other bath types. Concerning the inhibition of the deposition process, additives should block the growth of the side of the growing metal crystal but be inactive concerning the top of the crystals, allowing the side branching at the same time. The suitable additives can be chosen in a very heuristic manner since the elemental physico-chemical properties of the additives yield no easy-to-use guidelines concerning the deposit morphology regulation. The morphological features of purely metallic superhydrophobic electrodeposits are illustrated in Fig. 8.4, and a short literature summary concerning their material is given in Table 8.1.

Fig. 8.4 SEM images of electrodeposited metals with hydrophobic properties. **a** Ni [69]. **b** Ni–Co [70]. **c** Sn [71], **d** Ni–Cu-P [72]. Reprinted from [69–72], respectively. Copyright (2010) (**a** and **b**), (2013) (**c**) and (2018) (**d**), with permission from Elsevier

Table 8.1 Summary of electrodeposited porous metals exhibiting superhydrophobic properties and their deposition conditions

Material	Key experimental details	References
Ni	$NiCl_2$ 1 M, H_3BO_3 0.5 M, ethylenediamine dihydrochloride 1.5 M, $T = 60$ °C, $j = 10$–70 mA cm^{-2}	[69, 73–75]
	$Ni(NH_2SO_3)_2$ 1.5 M,·$CaCl_2$ 0–2.4 M $T = 40$–80 °C, $j = 10$–40 mA cm^{-2}	[76]
	$NiSO_4$ 0.15 M, trisodium citrate 0.28 M, pH = 8, T = 60 °C. $j = 50$ mA cm^{-2}	[75]
Co	$CoCl_2$ 0.1 M, Na_2SO_4 0.1 M, $j = 1$–30 mA cm^{-2} (with potential control), $T = 5$–25 °C	[77]
	$CoCl_2$ 0.42 M, H_3BO_3 0.56 M, unspecified crystal modifier, $T = 60$ °C	[78]
Cu	$CuSO_4$ 1 M, H_2SO_4 0.5 M	[79]
Sn	$SnCl_2$ 0.1 M, H_2SO_4 1.5 M, $E = -1.1 \ldots -1.9$ V versus SMSE	[71]
Zn	2 wt.% $ZnSO_4$ solution, $T = 35$ °C	[80]
Cu–Ni	$Ni(NH_2SO_3)_2$ 1 M, ·H_3BO_3 0.26 M, $CuSO_4$ 0–40 mM	[81]
Ni–Co	$NiCl_2$ 1 M, $CoCl_2$ 0.17 M, H_3BO_3 0.5 M, ethylenediamine dihydrochloride 1.5 M, $T = 60$ °C, $j = 20$–50 mA cm^{-2}	[70, 82]
Cu–Zn	$CuSO_4$ 0.08 M, $ZnSO_4$ 0.08 M, sodium potassium tartarate 0.48 M, NaOH 1.25 M, $j = 20$ mA cm^{-2}	[83]
Ni–Cu–P	$NiSO_4$ 0.125 M, $CuSO_4$ 5 mM, NaH_2PO_2 26 mM, citric acid 50 mM, sodium dodecyl sulphate 0.12 g L^{-1}, Na_2SO_4 0.5 M, $j = 50$–200 mA cm^{-2}	[72, 84]

In some cases, the dendritic structure was achieved by applying a high current with a significant hydrogen evolution rate [71]. In this case, however, no bubble templating was achieved, and the high overpotential leading to hydrogen evolution was a tool to achieve a mass transport limited deposition rate and the generation of the morphological instability. The variation in the current density (or the electrode potential) can lead to a transition from dendritic to DHBT-like surface morphology. This is the reason why hydrophobicity is often mentioned for DHBT-plated porous structures, too.

The contact angle of a water droplet on a rough surface is not necessarily constant in time. It was shown quantitatively for essentially all Ni-containing porous surfaces [70, 72–74, 82, 85, 86] that the contact angle increases drastically upon storage in air, hence turning the surface properties from hydrophilic to hydrophobic. When such observations were accompanied with a surface analysis, an increase of the hydrocarbon-related XPS carbon peak with time was observed [70, 72, 82]. Plasma cleaning of the porous surface eliminated the hydrophobicity but it returned again after air exposure [72]. The accumulation of airborne organic adsorbates was concluded from the temporal change of the infrared spectra of porous surfaces [85]. Therefore, one can speculate that the hydrophilic-to-hydrophobic transition with time was due to the condensation of apolar impurities on the porous surface that changes the surface energy. The oxide formation on the surface was also named as a possible reason for the hydrophobic properties observed, which was verified by either the electrochemical corrosion behaviour of the deposits [78] or the change in oxide-related XPS peaks of the metal atoms [70, 82]. The oxidation of Co was more significant with the increase in storage time than that of Ni. Electrodeposited porous Zn [80] and Sn [71] surfaces are oxide-covered already in the as-received state without ageing, and time dependence of the hydrophobicity was not mentioned for these materials.

8.2.3 *Post-deposition Impregnation of Deposits with High Surface Roughness*

The time dependence of the contact angle of as-received metal deposits with large surface roughness verifies the necessity of a reproducible production of hydrophobic surfaces so that the spontaneous adsorption of impurities should not be relied on. This is why several methods were developed for the surface treatment of porous deposits that result in a reproducible surface state, many of them exhibiting a better layer adherence than the spontaneous adsorption of airborne hydrocarbons. Electroplating of porous surface layers with a high surface roughness is not a strong necessity, although much desired. The general experience is that the post-deposition treatment yields the better results, the larger porosity the electroplated surface exhibits. Therefore, the principles concerning the electrodeposition of the surface layers are the same as for layers without a post-deposition treatment, and the deposition conditions are often the same. When a metal of technical importance is treated with a hydrophobic cover layer, the process often involves several steps [87–90], in accord with the forthcoming sequence: (i) Deposition of the first compact protecting layer, either by electrochemical or electroless manner. The key factor for the deposition of this layer is the compatibility with the technical metal substrate (like an Mg or Al alloy) and the good adherence; (ii) Deposition of a second layer with large porosity; (iii) Post-deposition treatment of the surface.

The post-deposition treatment can happen by simply immersing the workpiece coated with a layer of high surface area into a solution of polymer (like polypropylene [91] or vulcanized polymer [92]). Then, the solvent of polymer solution is evaporated and the protecting hydrophobic organic coating is dried onto the porous surface. In this case, the protecting layer is attached to the metal surface with weak adhesion force only, but the insolubility of the polymer in aqueous solution provides the desired hydrophobicity. If the post-deposition treatment is carried out with plasma-polymerization of fluorinated hydrocarbons [76, 93, 94], coatings with a thickness of about 10 nm provide a sufficiently hydrophobic character. The details of the plasma polymerization are not given in detail; therefore, it is not clear by what kind of chemical force the coatings are attached to the metal surface. Another surface treatment process based on the reaction with a gas is the carbon fibre growth from acetylene gas [95]. In this process, the carbon fibres are typically 1 μm thick; i.e., much thicker than the dendrite of the porous metal layer.

An ionically bonded hydrophobic protecting layer can be produced on the deposit surface when it is immersed into the solution of an organic (mostly stearic) acid. Mostly, ethanol is the solvent in this process (or, occasionally, methanol or acetone), and the monovalent organic acid has a long apolar chain. It is not clear if the bond to the surface is produced by the reduction of the acidic hydrogen atom of the acid ($Me + 2R\text{–}COOH \leftrightarrows Me(R\text{–}COO)_2 + H_2$) or the reaction of the acid with the already existing thin surface oxide ($MeO + 2\ R\text{–}COOH \leftrightarrows Me(R\text{–}COO)_2 + H_2O$). The latter reaction is more likely when the electroplated porous metal was oxidized prior to the organic acid treatment [96, 97]. Whichever reaction pathway is valid, however, the result is that the apolar chains are ordered and point away from the surface, hence inhibiting the penetration of a corrosive hydrophilic liquid to the metal surface. It was generally assumed, although not evidenced directly, that the immersion and spontaneous adsorption method leads to a monomolecular coverage. In order to enhance the hydrophobic character of the organic layer, highly fluorinated organic acids were also tested as adsorbates and yielded good results [89, 96–98].

The acid chemisorption method was shown to work for a variety of electroplated porous layers composed of either metallic elements such as Zn [98–101], Ni [87, 89, 90, 102], Co [88], Cr [103] and Cu [104–106], or alloys such as Cu–Zn [96, 97]. An aluminium nanoparticle layer attached to steel surface by electrophoretic deposition was also suitable for the same treatment [107]. The surface modification of metals with organic acids works in the same manner in the case when the coating contains codeposited non-metallic particles [108, 109]. The immersion time applied to accomplish the formation of the hydrophobic coatings varies from minutes to a few days. Electrochemical tests of the thus obtained specimens in laboratory environment indicate a reduction of the corrosion rate by a factor between 2 and 30. The stability of the self-assembled fatty acid layer is sufficient at $pH > 3$, but acidic solutions lead to a loss of the hydrophobic character, indicating the degradation of the chemical bond between the adlayer and the surface. The removal of the organic acids by heat treatment also leads to the loss of hydrophobicity, which can be recovered by a repeated immersion treatment into the organic acid solution [78, 99].

Attachment of thiols to porous metal surfaces can take place under similar conditions than the chemisorption of organic acids. However, thiols are known to form self-assembled monolayers and the bond between the thiol group and the metal surface is rather covalent. The surface modification with thiols was elaborated for Zn [110] and Cu [111–113] porous layers. When alkyl trialkoxy silane compounds with a long alkyl chain are used for achieving the hydrophobic character [114, 115], the bond to the surface is assumed to form with the condensation of the surface –OH groups of the metal. The chemical bond to the surface with siloxanes is assumed to be the strongest among the surface layer types mentioned because of the triple anchoring of the R–Si(O–)$_3$ group to the metal surface.

8.2.4 *Hydrophobic Porous Salt Films Prepared with Electrochemical Methods*

The methods of the impregnation of porous deposits mean some achievement as compared to the spontaneous formation of hydrophobicity. Nevertheless, the impregnation methods yield fairly thin surface layers, and thicker intrinsically hydrophobic layers are desired for many applications. Hydrophobic salt films synthesized with electrochemical methods comply with this criterion.

Salt films with superhydrophobic properties can be deposited with both cathodic and anodic electrodeposition. Since the hydrophobic nature of the weakly soluble (or practically insoluble) salt films partly stems form the anion having a long apolar chain, the porosity achieved is less critical than for stand-alone porous metallic coatings. Nevertheless, the majority of the salt films exhibit a significant porosity, though with a primary grain size falling often to the micrometer range. Within the fairly large grains, a secondary porosity level can often be seen, which is illustrated in Fig. 8.5.

The principle of the *cathodic salt film deposition* is as follows. The solution is based on an organic solvent, mostly ethanol, and it contains the fairly dilute solution (10–50 mmol L^{-1}) of a simple and well-soluble salt of the metal which will later form the coating. This salt can be $MnCl_2 \cdot 4H_2O$ [118*, 119], $NiCl_2 \cdot 6H_2O$ [120], $CeCl_3 \cdot 7H_2O$ [121–123], $Ce(NO_3)_3 \cdot 6H_2O$ [116*, 124–128], $LaCl_3 \cdot 6H_2O$ [117, 129], $La(NO_3)_3$ [130]*, $Ca(NO_3)_2$ [131]* or $Zr(NO_3)_4$ [132*, 133*] (*refers to works where the water content of the crystalline form of the metal salt was not specified;

Fig. 8.5 SEM images of hydrophobic porous salt film coatings prepared by cathodic electrodeposition. **a** Cerium(III) stearate. Reprinted from [116]. Copyright (2017), with permission from Elsevier. **b** Lanthanum(III) myristate. Reproduced from [117] with terms of Creative Commons Attribution License, version 4.0

for the rest, hydrated salts were used). Although solutes with water in the crystalline form were often used, the enhanced water content of the solution hence applied did not seem to be an obstacle of the salt formation. The dissolution of the counter electrode has never been mentioned even if an aggressive chloride salt was applied and the counter electrode was made of the same material as the cathode (copper or stainless steel).

Another solution component is a fatty acid ($H_3C–(CH_2)_n–COOH$) with a long apolar chain ($10 < n < 16$) whose concentration is typically a few times larger than that of the metal salt. The sufficient solubility of the fatty acids in ethanol verifies the choice of the solvent as opposed to water and so does the prevention of the formation of metal hydroxides instead of the fatty acid salts. For achieving a hydrophobic surface that is oleophobic at the same time, the application of fluorinated organic acids is required [123]. When the metal is quite prone to be deposited (like Ni), a mixed metal/metal salt film can be obtained [120].

Supporting electrolyte is not applied in most of the studies published. When the metal to be protected with the hydrophobic salt film is connected as cathode in the electrochemical cell, the hydrogen evolution is the dominant reaction at this electrode, which causes the alkalization of the cathode vicinity. This leads to the release of the protons from the acidic groups and hence, an electrochemically induced precipitation takes place at the cathode without any change in the oxidation number of the cation. The overall process time varies between 1 and 60 min. Due to the fairly inert solution, coatings can be produced on metals that are easily corroded in aqueous solution like Mg [116, 124, 125, 127] and Al [119, 128] alloys. The great advantage of the process is that it takes place in one single step. It was revealed in several studies that the contact angle of the water droplet at the hydrophobic surface showed a maximum at some deposition time [116, 118, 120, 121], and the maximum of the contact angle could be sufficiently correlated to the finest pore structure in the SEM images. However, when a nanogranular coating was obtained, the achievement of the full coverage led to saturation in the contact angle [125, 127, 130, 133], occasionally with a slight decline only for thick deposit coatings.

Although the major processes in the salt film formation can be understood, the role of a number of factors is unclear. The salt film formation process is usually driven with a large cell voltage (5–60 V) in a two-electrode cell; therefore, we cannot speak about electrode potential since the potential distribution within the cell is not known. In many cases, a voltage optimum was found. Based on the few reports where the current density values were given, we can say that it can vary from about 0.5 mA cm^{-2} [131] to a few mA cm^{-2} [118]. The resulting current density may be a function of the substrate applied, too [118]. There is no data in the literature about the Faradaic efficiency of the cathode process. The salt formation can be effectively induced by a constant voltage, too, but the application of voltage pulsing often leads to deposits with finer dendritic structures [117, 129].

It is specified in the majority of the studies that the electrode separation was fixed, and a distance of 2 cm is given in the majority of the studies. This raises the concern that the reaction product of one electrode can reach the other electrode, which is a huge factor of uncertainty when cations of varying oxidation state are present [118, 125]. For these reasons, the salt film produced at the cathode may contain higher-valency metal ions. Although some studies draw the attention to the possible formation of a mixed-valency product [127], the XPS data published do not indicate that higher-valency cations might be also present in the salt films produced [119, 122, 127, 128]. It was shown only for the electrolysis performed in aqueous $Ce(NO_3)_3$ solution that the cathodic process leads to the formation of a CeO_2 coating [134] (Although an oxidation process taking place during the cathodic deposition is unusual, it occurs also during the cathodic deposition of ruthenium oxide, and the oxidation process is due to the instability of the $Me(OH)_3$ compounds in both cases.).

Anodic processes for obtaining porous hydrophobic salt film coatings are much less common than cathodic ones. If the solution contains the salt of the metal that forms the salt film in the anodic process, the salt has to contain the metal in an intermediate oxidation state so that oxidation could be possible. This happens when a $Ce(NO_3)_3$ solution is anodized to form CeO_2 [135]. The CeO_2 surface showed a time-dependent hydrophobicity and a hydrophobic–hydrophilic transition upon plasma cleaning, indicating that the hydrophobicity of the surface was due to impurity adsorption.

Another means of the anodic salt film formation is when the substrate itself is used as a conversion anode. This film formation mode was demonstrated with Zn as anode in the presence of tetradecanoic acid [136]. Many deposition conditions (ethanol used as solvent, 30 V cell voltage applied) were similar to those used in the cathodic processes. The anodic fatty acid salt coating showed intrinsic hydrophobicity, similarly to the cathodically obtained ones.

The structure of the salt film deposits was analyzed with XRD method in a few works. For the high-angle range of the diffractogram, mostly the spacing of the $-CH_2-$ units can be determined [131, 133]. The periodicity of the unit cell of the compound produced (i.e., the *metal ion—apolar chain pair—metal ion* sequence) can be seen in the low-angle range (between $2° < 2\,\Theta < 20°$). Since usually no standards are available for comparison, the exact deposit structure is not known, and only the layered nature of the deposit can be concluded from the presence of the higher-order reflections [120, 129]. The most common method of the check of the presence of the hydrophobic layer is the record of the infrared spectra from the surface and its comparison with that of the fatty acid used. In most cases, the agreement is nearly perfect, and the presence of the metal ions in the deposit just negligibly modifies the spectra.

8.2.5 *Electroplated Hydrophobic Polymers with High Surface Roughness*

Electropolymerization of organic monomers is a common method to modify metal surfaces. When redox-active polymer films are produced electrochemically, the films are mostly hydrophilic and can be oxidized and reduced reversibly at least in a narrow potential window in aqueous solution. This is valid for polymers composed of simple monomers like aniline, pyrrole, thiophene or ethylene dioxithiophene. In order to obtain elecropolymerized films in a hydrophobic form, the monomers have to be substituted with highly hydrophobic side groups. A few monomers with hydrophobic side groups are shown in Fig. 8.6.

Due to the hydrophobic character of the deposits obtained with electropolymerization, the electrochemical preparation of the hydrophobic coatings is carried out in a non-aqueous solvent, mostly acetonitrile [137–142]. The supporting electrolyte is a tetraalkylammonium salt, and the monomer concentration is of the order of 10 mM, which favours the formation of fibre-like polymer deposits instead of continuous films. When the attachment of the hydrophobic polymer to the substrate metal is not sufficient, the electropolymerization of the hydrophobic monomer is preceded by the deposition of another well-adhering thin polymer film. Morphological characteristics of a hydrophobic fibrous polymer deposit are presented in Fig. 8.7.

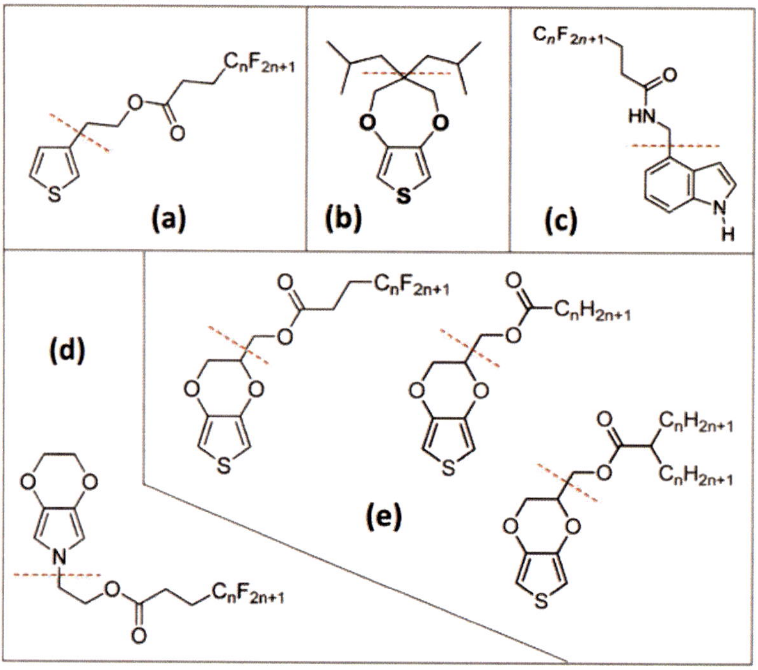

Fig. 8.6 Examples for monomers modified with apolar side groups in order to obtain electrodeposited porous hydrophobic polymers. Conventional monomers are separated from the side groups with the dashed lines. The modified monomers are **a** thiophene [137], **b** 3,4-propylenedioxythiophene [138], **c** indole [139], **d** 3,4-ethylenedioxipyrrole [137], and **e** 3,4-ethylenedioxithiophene [137, 140]

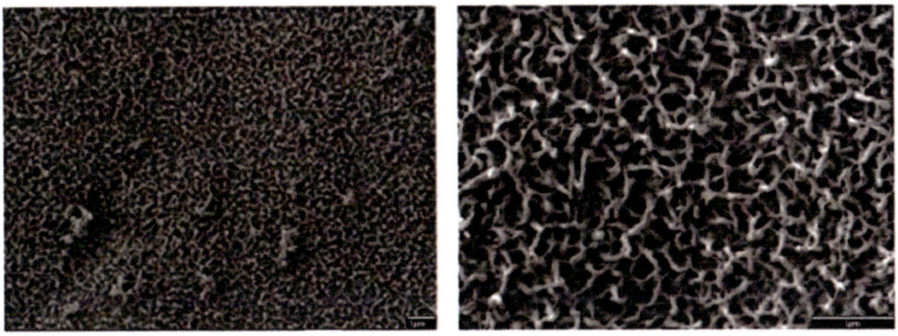

Fig. 8.7 SEM image obtained for a hydrophobic fibrous electropolymerized coating. Material: PEDOT-F4 (see Fig. 8.6e, left formula, $n = 4$, charge used for polymerization: 35 mC cm^{-2}). Reproduced from [142]. Copyright (2017) with permission from Elsevier

8.3 Porous Structures Electrodeposited from Dilute Solutions

8.3.1 Porous Metallic Deposits from Dilute Solutions

When the solution applied for electrodeposition is very dilute with respect to the metal salt to be deposited and the deposition process takes place in the mass transport-limited regime, one can expect that the metal growth soon turns into dendritic. This deposition mode is rationalized by at least two parameters. The first one is the mass transport of the solute which provides the reactant species preferably at the top of the nanocrystals. The second parameter is the electric field that provides the largest surface charge density at the top of the growing nanocrystals. While the mass transport effect of the reactant in a dilute solution always prevails, the field effect strongly depends on the supporting electrolyte concentration and, hence, on the conductivity of the plating solution. If the concentration of both the reactant and the supporting electrolyte is below 1 mM, the result is the growth of a random set of nanocrystal columns perpendicular to the surface. This process was termed as filamentary one-dimensional nanocrystal growth in an ultradilute electrolyte (FONGUE) [143, 144].

Two more parameters have to be mentioned that influence the growth of the nanocrystals. First, the anisotropy of the crystal itself determines which crystallographic axis points to the growth direction. Since one of the crystal faces usually can grow faster than the rest, the set of nanocolumns shows a uniform texture. Secondly, side branching is to be avoided. As it was shown in the previous chapters for various types of deposits, side branching of the growing column may be significant. For the elimination of side branching, a reverse pulse deposition mode was applied for nanocrystal growth from ultradilute solutions. During the oxidation period, the side branches dissolved more likely than the rest of the columns; therefore, a columnar growth could be maintained for a long time without any significant side branching.

The application of the reverse potential mode was also thought to lead to a unimodal and narrow crystal diameter distribution. The increase in the reverse pulse potential resulted in narrower crystals and higher porosity at the same time [144], which indicates that sidewise dissolution also may take place during this pulse. The addition of sulphuric acid to the bath as small-concentration supporting electrolyte was also found beneficial in avoiding the formation of side branches.

Here, the attention has to be drawn to the fact that the electrode potentials given for the reverse pulse method (either for the deposition or the dissolution pulse) have to be treated with some care. Since the conductivity of the low-concentration solutions is a few tens of $\mu S\ cm^{-1}$, the potentials ranging to ±15 V account for a huge ohmic drop, and the values given cannot be directly compared to the usual potential scale.

Metal nanocolumns thus obtained are as follows (with the solution composition): Ag (20 μM $AgNO_3$ + various supporting electrolytes [144] or 20 μM $AgNO_3$ [145]), Cu (50 μM $CuSO_4$ + 26 μM H_2SO_4 [143]), Au (650 μM $HAuCl_4$ [143]) and Ag–Cu (20 μM $AgNO_3$ + 15 μM $CuSO_4$ + 26 μM H_2SO_4 [143]). Due to the small reactant concentration, the deposition time ranges to hours (if the time is too short, a thin coating or a particle assembly can form on the surface). The cross-sectional SEM observations clearly underpin that columnar growth take place in all these systems, as shown for Cu as an example in Fig. 8.8a. A very interesting aspect of the process is that Ag–Cu columns proved to be homogeneous single crystals, too, although Ag and Cu are immiscible in equilibrium at room temperature. Moreover, the crystals were dense, which indicates that no dealloying took place during the reverse pulse. (Similar effect was found at nanoscale electroplated objects for another immiscible system, Ir–Au [146], indicating that miscibility, phase nucleation and segregation at the nanoscale significantly differ from the corresponding phenomena in bulk materials.)

It has to be noted that relatively dilute solutions combined with pulse reversing lead to different nanostructures. As it was demonstrated for a solution containing 2 mM K_2PdCl_4, 100 mM NH_3 and 10 mM of cetyltrimethyl ammonium bromide

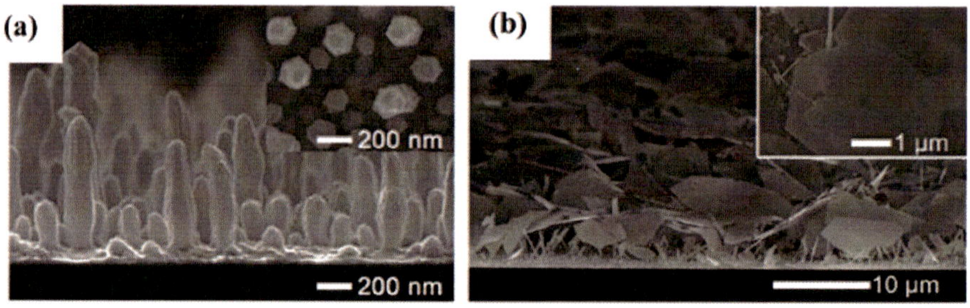

Fig. 8.8 Cross-sectional (main images) and top-view (insets) SEM images taken for sets of **a** Cu and nanocolumns plated from ultradilute ($c = 0.05$ mM) $CuSO_4$ solutions with a reverse pulse method. Reproduced from [143]. Copyright (2013) with permission from Elsevier. **b** Ag nanosheets deposited from $AgNO_3$ solution ($c = 0.2$ mM). Reproduced from [148] with terms of Creative Commons Attribution License, version 2.0

[147], stacks of interconnected Pd nanoplatelets could be synthesized with a potential scan. The result is very similar for Ag if the concentration of the silver salt is increased from the 10^{-5} M range to 10^{-3} M [148], as shown in Fig. 8.8b. Here, the detailed data of the original work reveal that a nanorod-to-platelet transformation happens as a prolonged deposition of several hours is carried out.

In contrast to the relatively noble metals, the deposition of non-noble metals is always accompanied by hydrogen evolution that necessitates a different approach to template-free nanowire deposition. As it was demonstrated for cobalt [149], successful nanorod deposition can be carried out in electrolyte solutions with orders of magnitude larger metal ion concentrations than for similar noble metal deposits (0.05 M $CoSO_4$ with 0.4 M H_3BO_3). The relatively large current density of -150 mA cm^{-2}, although much less than that used for the DHBT method, indicates an intense hydrogen evolution. It was probably due to the combined impact of the electric field, mass transport and gas evolution that the nanorod diameter distribution was less homogeneous than for noble metals. The Co nanorod system could be successfully transformed to Co_3O_4 without any appreciable morphological change by annealing in air.

8.3.2 *Non-metallic Nanocolumnar Deposits*

The most important nanostructured material obtained with electrodeposition from dilute solution is zinc oxide. It is an inexpensive n-type semiconductor with a direct band gap of 3.37 V, and its application in photoelectrochemical devices is emerging fast. The literature of this field is very rich, and a large variety of deposition conditions have been tested, whose summary can be found in a recent review article [150].

The chemical background of the formation of ZnO nanorods is similar to the coatings discussed in Sect. 8.2.4 in the sense that the charge transfer reaction does not modify the oxidation number of the metal ion. Instead, a reactant is produced in an electrochemical process which acts as a co-reactant in the precipitation of the product. The solubility of ZnO is the lowest between pH values of about 9–12, and the optimum pH range of the ZnO deposition changes slightly with temperature [151, 152]. When the synthesis of zinc oxide is carried out in aqueous media, the temperature has to be above 40 °C so that zinc oxide is produced instead of zinc hydroxide. The increase in deposition temperature strongly improves both the crystallinity and the texture of the deposits [152] and leads to a slight increase in the nanocolumn diameter. The reaction is as follows:

$$Zn^{2+} + 2OH^- \rightleftarrows ZnO + H_2O. \qquad (8.3)$$

Concerning the source of the hydroxide ions, various reactants are possible. The simplest way is to saturate the solution with oxygen by bubbling air through the solution and reducing the dissolved oxygen:

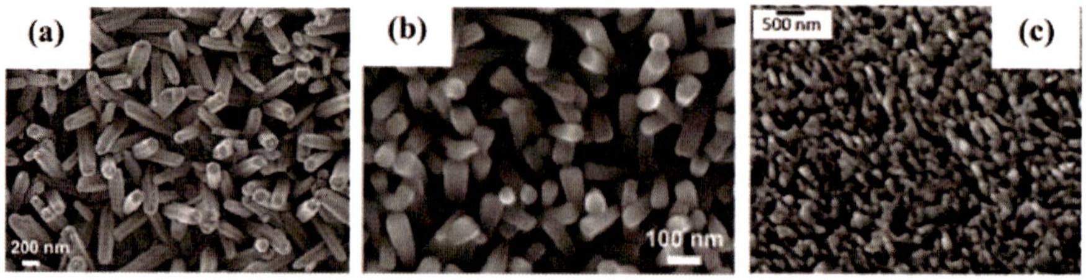

Fig. 8.9 SEM images of a few electrodeposited nanorod systems obtained from dilute solutions. **a** ZnO. Reproduced from [156]. Copyright (2007) with permission from Elsevier. **b** CuSCN. Reprinted by permission from Springer [157]. Copyright (2015). **c** CdO. Reproduced from [158]. Copyright (2011) with permission from Elsevier

$$O_2 + 2H_2O + 4e \rightleftarrows 4OH^-. \tag{8.4}$$

Another way is to use a reactant whose reduction is a proton-consuming process. Therefore, the reduction of the auxiliary solution component leads to the alkalization of the close vicinity of the cathode. The most common co-reactant is the nitrate ion, whose reduction to nitrite ions produces hydroxide ions, too:

$$NO_3^- + H_2O + 2e \rightleftarrows NO_2^- + 2OH^-. \tag{8.5}$$

If the zinc oxide production is carried out in an aprotic solvent (mostly in an ionic liquid), the reaction does not take place with the participation of hydroxide ions but the dissolved oxygen is reduced to peroxide anions that can later react with the dissolved zinc(II) ions:

$$O_2 + 2e \rightleftarrows O_2^{2-} \tag{8.6}$$

It is crucial to understand why regular columnar structure can be obtained for zinc oxide and a few other materials (see Fig. 8.9) during the precipitation-based electrodeposition process. For many other electrodeposited compounds, both the structure and the morphology of the deposit are much less regular (see essentially all materials discussed in Sect. 8.2.4). For the chemical precipitation of ZnO, a strong influence of the anion quality and the pH was established [153], but the anion type during the electrochemically induced ZnO formation never arose as an issue. The reason is the well-defined structure of the ZnO deposit and its preferential growth along one specific crystallographic axis (mostly the c-axis of the hexagonal wurtzite crystal). In the anisotropy of the crystal growth is large, the side branching becomes negligible, in contrast to the metal nanocolumns. Additionally, the relative oversaturation of the solution is the largest near the top of the growing columns. This is because (i) the metal ion concentration between the growing columns is essentially zero and increases from the growth front to the bulk solution, and (ii) the hydroxide ion concentration is the maximum near the tip of the columns and decreases towards the bulk solution. The large relative oversaturation in the presence of the growth centres makes it possible to obtain parallel columns perpendicular to the substrate surface. The production rate of the hydroxide ions is usually much larger than the growth rate allowed by the Zn^{2+} ion transport; therefore, the deposition efficiency is mostly below 50%.

Nevertheless, the morphology of the deposit changes with deposition time if no seed layer is applied. At the beginning, the near-substrate part of the coating appears to be somewhat irregular, and a morphologically stable columnar structure can be observed after a long enough deposition time only (see, e.g., [154, 155]).

The zinc salt applied is usually either $Zn(NO_3)_2$ [155, 159] or $ZnCl_2$ [156, 160–162] with nitrate ion or oxygen reduction as the hydroxide ion source, respectively. The concentration of the zinc salt varies between 0.1 and 5 mM. The minimum total Zn concentration is determined by the solubility of ZnO; therefore, the further decrease in the Zn^{2+} concentration does not lead to deposit growth. In contrast, with Zn^{2+} concentration higher than about 20 mM, the nanocolumns start to coalesce and the resulting deposits are composed of platelets instead of columns, although the deposit remains somewhat porous [155].

The supporting electrolyte can be chosen in accord with the zinc compound applied. In the case of the nitrate-based solution, the addition of another nitrate salt (e.g., $NaNO_3$ [163]) can increase the concentration of the hydroxide ion precursor, hence opening a way to tune the Zn^{2+} and OH^- concentrations independently of each other. For chloride-based media, KCl as supporting electrolyte is not a reactant but a growth-regulating agent. It improves the deposit quality in the concentration range of about 0.1 M; however, if $c(KCl) > 0.5$ M, the flat-capped ZnO nanorods adopt a pencil-like peaky shape [164].

Apart from ZnO, various other materials have been synthesized as porous columnar structures with the template-free approach by applying a relatively small solution concentration (usually 1–12 mM). Lanthanum(III) hydroxide plated from nitrate solution with NH_4Cl as additive showed a fairly regular columnar structure with unimodal column diameter distribution [154], while the application of an additive-free $La(NO_3)_3$ solution led to a less regular column structure under essentially the same deposition conditions [165]. Cadmium oxide nanocolumn assembly can be plated from oxygen-saturated dimethyl sulfoxide solution of 0.01 M $CdCl_2$ at 150 °C [158].

Finally, a few examples will be mentioned where the oxidation state of some of the structure-forming ion also changes during the nanocolumn deposition. In these cases, the reasons for the formation of nanocolumns as morphological units remain the same, but the process becomes more complicated. The first example is the nitrate-based electrochemical production coupled with the oxidation of the manganese (II) ions, leading to the deposition of Mn_3O_4 (hausmannite) nanocolumns [166]. In this process, dissolved oxygen must be present for the oxidation of a part of the Mn^{2+} ions to Mn^{3+}; otherwise the Mn_3O_4 composition could not be achieved. Hence, the nitrate ions and oxygen have a synergetic effect in the overall reaction. The second example is the formation of CuSCN nanocolumns from $CuSO_4$ solution [157]. Here, the reduction of the Cu^{2+} ions takes place without any side reaction that forms the desired compound with the SCN^- ions present. The third example is the formation of

CdSe from 0.01 M solution of both $CdSO_4$ and Na_2SeO_3 [167]. Here, the reduction of the selenate ions takes place. The CdSe columnar deposits showed more frequent branching than ZnO.

8.4 Electrochemical Dealloying

8.4.1 Background of Dealloying

Dealloying is a process in which an alloy is selectively dissolved so that the more noble (MN) component is enriched at the surface while the less noble (LN) component dissolves preferentially. Dealloying can be a simple chemical process and it can be controlled electrochemically by applying a desired electrode potential. For alloys having several components, dealloying may lead to the dissolution of either one single or several components, leaving behind either an alloy as a partly dealloyed material or a metal composed of solely the most noble element of the original alloy, respectively. Topical reviews of various aspects of dealloying are available in the literature [168–170].

Electrochemical dealloying has been known historically from essentially the beginning of the electrochemical corrosion studies, and the spongy nature of the dealloyed material was also recognized already in the 1920s for the corrosion/dissolution of several alloys (see, e.g., the references in [171]). The polarization behaviour of the alloys undergoing dealloying was also studied with a great accuracy [171, 172], and three electrode potential zones were identified. For dealloying, there must be a difference between the standard electrode potentials (E^0) of the Me^{z+}/Me systems for the constituents, and there must be a comparable difference between the onset potentials of the dissolution (E_D) of the constituent metals. While ΔE^0 is a relevant parameter if the LN metal dissolves reversibly, ΔE_D can be used only if the LN metal can be passive; hence, even the sign of the two parameters can be different (as will be shown later for the Ni–Cu system).

The relative position of the anodic parts of the polarization curves is shown in Fig. 8.10a. For dealloying, the prerequisite is that at least for a range of intermediate alloy composition, the polarization curve of the alloy must run in between the polarization curves of the pure metals. There is a monotonous dependence of the critical dealloying potential (E_{crit}) on the alloy composition (see Fig. 8.10b). The composition range in which dealloying can take place is the narrower, the smaller is ΔE^0 for the alloy constituents [172], as indicated with the trend line in Fig. 8.10c.

In line with the research on dealloying, it was recognized that atomic-scale component redistribution is necessary which involves surface diffusion. In the potential region where the surface enrichment of the MN takes place without inducing porosity, the surface diffusion was observed real-time with STM [175, 176]. The *in situ* STM study also revealed that at small overpotential, dealloying-related surface diffusion of the MN metal leads to a smooth surface. Dealloying in this potential range influences

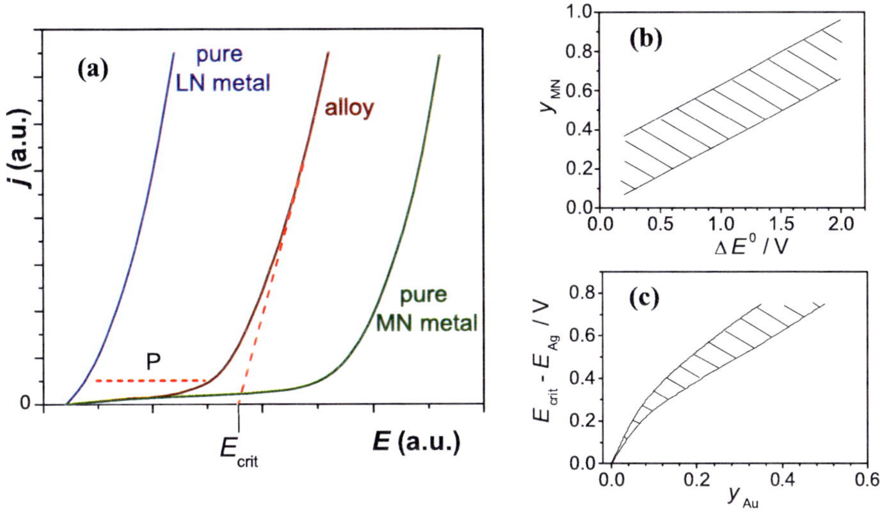

Fig. 8.10 **a** Relative position of the polarization curve of the LN and MN metals and their alloy. The extrapolation of the alloy dissolution curve shows the common determination method of the critical dealloying potential (E_{crit}), and the dotted line labelled with P stands for the quasi-passive potential interval where the surface enrichment of the MN metal does not lead to the formation of a porous surface. **b** Dependence of the mole fraction where dealloying is possible on the standard potential difference of the constituent. **c** The critical dealloying potential as referred to the dissolution potential of the LN metal as a function of the MN metal mole fraction for the Ag–Au system. The actual value of the dependent parameter in graphs **b** and **c** is a function of various experimental conditions like the solution composition. Graphs **b** and **c** were sketched after Refs. [173, 174]

only the topmost atomic layers in Au_3Cu, and smoothening happens by the coalescence of the voids left behind by the Cu atoms dissolved [177], hence leading to large terraces instead of frequent step edges. Faceting and the appearance of smooth terraces were observed for dealloying of brass, too [176]. However, upon approaching the critical potential of dealloying, atomic-scale roughness occurs [175], which may be correlated to solid-state defects that also act as initiation sites for stress corrosion cracking (SCC) assumed earlier [172]. It happens only above the critical dealloying potentials that voids caused by the dissolving LN atoms are extended into the direction perpendicular to the surface [177]. Since the dealloying process in strongly related to stress corrosion cracking, it often happens that the dealloyed material has a bimodal pore structure where the small pores originate from short-scale displacement of the atoms and large pores are formed due to the cracking of the stressed master alloy.

Surface diffusivity of the atoms of the element that later form the porous structure is the key factor in the quantitative description of the dealloying process. The surface diffusivity (D_S) can be determined from the dealloying parameters with the formula below:

$$D_S = \frac{[d(t)]^4 kT}{32\gamma t a^4}. \tag{8.7}$$

The meaning of the parameters is as follows: $d(t)$ is pore diameter at etching time t, γ is the surface energy and a is the lattice parameter (other parameters have their usual meaning). Interestingly, the mole fraction of the pore wall-forming metal is missing from Eq. 8.7, although it is known to influence the pore size of the dealloyed porous structure. Hence, we must conclude that the curvature of the surface of the pore wall and the surface diffusivity are not independent of each other. Equation 8.7 also suffers from the problem that the determination of D_S with a dealloying-independent method is cumbersome, which means that the formula has small predictive value for assessing the expected pore diameter, especially when other dealloying conditions (e.g., dealloying potential or dealloying rate) may also influence the formation of the pore structure.

Attention must be drawn to the fact that in accord with Eq. 8.7, a particular dealloying method leads to various pore structures as the dealloying process proceeds. Qualitatively, the dealloying process starts with pore nucleation, and the diffusion distance is small at low dealloying time, which defines a fine pore structure at the beginning that undergoes a coarsening upon prolonged dealloying. The quantitative treatment tells that the time dependence of the pore size is in accord with a $d \propto t^{1/4}$ proportionality, which means that the coarsening is well visible at the sample surface at small dealloying times and appears to be negligible after the pore penetration to the bulk material is more than an order of magnitude larger than the pore size itself. This can be well seen where the morphology of dealloyed materials is reported as a function of the dealloying time [178, 179]. However, Eq. 8.7 is difficult to apply for multicomponent master alloys where the leaching out of the accompanying components beside the pore wall-forming metal(s) takes place at different time scales. This can explain why both pore and ligament size reduction was occasionally observed for multicomponent master alloys for long dealloying times [180]. Here, the feature size reduction at large time scale can be explained with the dealloying of the already formed but only partly dealloyed ligaments without any new pore formation.

While the formation of a nanoscale surface pattern during dealloying was recognized long ago [181], the theoretical description of the dependence of the critical dealloying potential as a function of the properties of the alloy components appeared much later. Various parameters such as the surface free energy and its coupling with surface diffusion [182] and the local surface curvature [173] were considered. The correct definition and experimental determination of the critical dealloying potential obtained a great attention. The onset of dealloying can be considered as a kinetically influenced phase transition whose apparent starting potential depends, among others, on the sweep rate in a similar manner as the glass transition temperature depends on the heating rate [183]. This was the reason why a steady-state determination method for the critical dealloying potential was also recommended [184, 185]. The latter studies revealed that the difference in dealloying potentials determined with dynamic and steady-state methods is about 0.1 V on average. Since this deviation is little as compared to the difference between the onset of the dissolution potential of the LN metal and the dealloying potential applied in practice for essentially all alloys, the exact determination of E_{crit} is not crucial from the preparative point of view. Nowadays, the application of various computational methods to simulate dealloying phenomena is rather common [186–188].

It must be stressed that the volume fraction of the structure-forming (i.e., MN) and leaving (i.e., LN) metal in the dealloyed master alloy is not indicative of the ligament-to-pore diameter ratio in the dealloyed porous structure. This is because a volume change also takes place during dealloying, which can lead to a compression up to 8 vol.% [189]. Beside the change of the specimen shape, another discrepancy is that the purity of the dealloyed samples is limited since the slow solid-state diffusion does not allow the removal the LN component entirely. A 2–5 at.% impurity of the LN metal in the porous MN structure is typical for both chemically and electrochemically dealloyed samples. Concerning the mean ligament diameter, the general trend is that the smaller the mole fraction of the porous structure-forming (i.e., MN) metal in the master alloy, the smaller the ligament diameter and the most fragile porous structure is obtained upon the dealloying process (see, e.g., [190–194]).

8.4.2 Dealloying of Binary Alloys

Typical binary alloys processed by electrochemical dealloying and leading to nanoporous materials are listed in Table 8.2. The master alloys were synthesized with a variety of methods but mostly with classical metallurgical procedures.

Special care has to be taken when the dealloying potential is beyond the stability range of water when gas evolution may accompany the dealloying process. This usually occurs when both constituents are noble metals like Au and Ag [217]. In such cases, the solution is acidified near the dealloying front, which has an impact on the dealloying process. The pH of the solution chosen for the dealloying experiments also has a role in the entire dealloying process. In acidic solutions, the surface pH change is insignificant, while in unbuffered neutral solution it is substantial. Another phenomenon to be taken into account is that the mechanism of dealloying may change if the dealloying potential is high and is close to the oxidation of the MN metal. Namely, the dealloying process may take place with a *dissolution–redeposition* mechanism. In this case, the oxidation of the MN metal is possible strictly at the dissolution front where the LN metal leaves the alloy, but later these MN metal ions are redeposited onto the pure parent metal surface. The mode of dealloying (e.g., simple or redeposition-mediated) can be chosen by changing the dealloying potential, as it was shown for the Au–Sn system [204]. The consequence of the redeposition process on the sample morphology is that nanoparticle-like additional deposit is formed on the top of the dealloyed structure.

The pore (and ligament) size of the electrochemically dealloyed porous structure cannot be related merely to the composition of the master alloy but is a function of the dealloying conditions, including the electrolyte solution and the dealloying potential.

Table 8.2 List of binary alloys applied in dealloying studies with inclination to nanopore formation

Dissolved metal	Nanoporous structure-forming metal					
	Ni	Bi	Cu	Pd	Au	Pt
Mg			0.33–0.67 [191]			
Al			0.6–0.7 [195] 0.25 [196]	0.2 [197]		
Zr			0.8 [198] 0.62 [199]			
Mn	0.38 [200] 0.3 [201, 202]		0.15–0.45 [192]			
Zn	0.103 [203]					
Sn		0.05–0.5 [193]			0.4 [204] 0.45 [205]	
Co				0.2 [206]		
Ni					0.82 [207]	0.25 [208]
Cu	0.5–0.86 [190] 0.2–0.8 [209]			0.25 [210] 0.15–0.25 [194]	0.18–0.23 [211] 0.2–0.3 [212] 0.18–0.48 [213]	0.25–0.75 [188] 0.2–0.29 [214] 0.78–0.82 [215] 0.25 [216]
Ag					0.35 [178, 217, 218] 0.2–0.3 [185] 0.3–0.35 [219] 0.32–0.35 [220] 0.3 [221]	

The numbers mean the mole fraction of the structure-forming metal in the alloy processed. Further data on systems that are suitable for dealloying can be found in Ref. [172]

The latter is a degree of freedom that lends more versatility to the electrochemical dealloying procedure as compared to the chemical ones, yielding an opportunity of tuning the pore size. Figure 8.11 presents representative examples of the impact of the dealloying conditions on the porous structure formed. If all other conditions are identical, less positive dealloying potentials lead to smaller pore sizes [211] provided that the process takes place in a single step. The above described pore size rule may not be true if multistage dealloying takes place with some intermediate composition that has exceptional stability, as it was found for Au–Cu alloys [212].

The shape of the electrical signal used for dealloying may vary. A potential sweep instead of an abrupt potential change was found to lead to a crack-free porous structure [220], which may be associated with the stress occurring during the atomic-scale

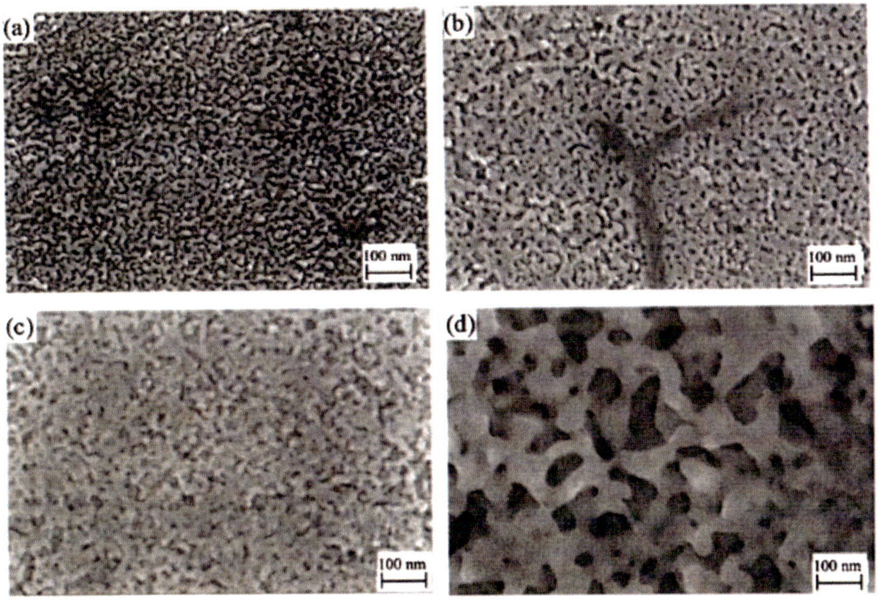

Fig. 8.11 Scanning electron micrographs showing bi-continuous Au porous structure of dealloyed $Ag_{65}Au_{35}$ alloy at various dealloying conditions. **a** 1.285 V versus NHE in 0.1 M $HClO_4$. **b** 0.95 V in 0.1 M $HClO_4$ + 0.1 M KCl. **c** 0.74 V in 0.1 M $HClO_4$ + 0.1 M KBr. **d** 0.404 V in 0.1 M $HClO_4$ + 0.1 M KI. Republished with permission of The Electrochemical Society, Inc., from Ref. [219]; permission conveyed through Copyright Clearance Center, Inc.

material redistribution. In order to keep the dealloying rate and the concomitant stress below the level that leads to the development of cracks, slow-rate galvanostatic dealloying was also applied successfully [221].

The application of pulses opens a new opportunity for the selection of the desired sample morphology. By using alternating on/off periods (which corresponds to dealloying followed by relaxation), ultrafine porous structures can be achieved, which is explained with the renucleation of the dissolution sites in each anodic pulse [178, 218]. When the dealloying step is followed with a negative pulse, the reduction of the surface and the hydrogen evolution taking place lead to a dendritic morphology. When pulse dealloying is applied to Au–Sn alloys in alkaline media, the surface morphology is similar to what is obtained for pure Au with a similar treatment but without dealloying [205].

8.4.3 Dealloying of Ternary Alloys

The number of ternary systems studied for dealloying is fewer than binary ones. From technically important structural materials, Al alloys were found to undergo dissolution, especially those containing the so-called S phase of Al_2CuMg composition [222–224]. The dealloying of these materials leads to a Cu-rich porous structure, although exact composition has not been given in any of the relevant studies. The Cu-rich porous structure can form from sputtered Al_2CuMg layers [224] or isolated grains of the same composition [222]. Nevertheless, the relatively high-rate dissolution of the intermetallic Al_2CuMg grains and the formation of a Cu-rich porous network is a major process in the corrosion of 2024-T3 Al alloys with <5 wt.% Cu and <2 wt.% Mg content [223]. The dealloying studies of Al-based structural materials have no synthetic aspect for obtaining nanostructures (although the dealloyed Cu-rich pore system fully complies such criteria), but are focused on the passivation behaviour of alloys containing the intermetallic Al_2CuMg phase and the reaction of the Al alloys during the formation of conversion coatings.

Concerning the ternary systems containing noble metals, a number of studies were published for the Ag–Au–Pt system [225–227]. The silver content of the ternary master alloys varied between 65–80 at.%, and Pt was the minority component with a concentration between 1 and 7 at.%. Pt was found to act as a stabilizing agent with additional beneficial properties like the decrease of the ligament diameter (below 10 nm), the prevention of the fragmentation of the dealloyed material, the retention of a part of the silver content of the dealloyed porous structure and the enhancement of the catalytic properties of the dealloyed material. With Pt addition to the Ag–Au system, the dealloying process could be performed in two steps with a mild heat treatment in between [225]. With a two-step dealloying, a two-level porous system could be created in which the already formed ligaments were retained and they hosted a secondary pore system with much smaller pore size than that formed in the first dealloying step. The residual silver content of the dealloyed pore system varied in a wide range depending on the dealloying condition. With an increase of the Pt content of the precursor mater alloy from 0 to 3 at.%, the residual silver content increased form 28 to 48 at.% [227], while about 11 at.% was given for the two-step dealloying process [225]. The accumulation of the Pt content of the dealloyed structure at the surface of the ligaments was found, which contributed to the enhancement of the catalytic properties of the ternary system.

8.4.4 Dealloying of Non-equilibrium Alloys

Already among the binary alloys discussed in Sect. 6.4.2, there were several metastable specimens that were produced and/or processed by a non-equilibrium method (e.g., mechanically alloyed and then spark-plasma-sintered $Pd_{20}Al_{80}$ samples [197]). In this chapter, a variety of alloys will be presented that can be manufactured with non-equilibrium methods, hence widening the master alloy composition range to be used for dealloying.

Concerning the preparation method of the metastable materials, co-sputtering is a relatively easy method that was used for binary alloys, mostly for Pt–Si precursor materials [228, 229]. The quenching of a well-homogenized binary alloy from a temperature at which the miscibility of the components is unlimited proved to be successful for Pd–Co alloys [230]. In the Pd–Co system, quenching of a homogeneous alloy is a prerequisite for the application of dealloying because the slow cooling would lead to a phase segregation, which results in the dissolution of some grains of high LN metal content instead of bulk dealloying. The big majority of the electrochemical dealloying experiments with non-equilibrium alloys uses amorphous metals obtained with the rapid solidification (also named as melt quenching) of a molten mixture [179, 180, 231–240]. The quenching rate in this sample preparation procedure often achieves 10^5–10^6 K s^{-1}, which means that the solid ribbon obtained conserves essentially the same atomic arrangements that prevail in the melt. The number of the components of such alloys can be as high as five. An increase in the number of the components leads to a decrease in the eutectic temperature of the melt and also leads to an impediment of the crystallization process upon quenching.

The components of the electrolyte solution used for electrochemical dealloying are often the same as used for equilibrium alloys. Dilute solutions of strong acids are suitable to dissolve various metals, while for aluminium dissolution from non-noble alloys, alkaline solutions or sodium chloride solutions are preferred. An important difference arises when a metalloid component is to be removed during the dealloying process. In this case, the application of HF is often indispensable, especially when only silicon has to be dissolved [228, 229].

The composition of the dealloyed non-equilibrium alloys shows a great variety, and no general trend can be outlined. From some binary alloys of noble metals, the accompanying component can be removed quantitatively during dealloying (e.g., Au–Al [231, 233] and Pd–Co [230] alloys can be transformed to pure porous Au and Pd, respectively). The dealloying of other binary alloys like $Pt_{0.1}Si_{0.9}$ proceeds with the selective dissolution of Si, but the Pt:Si atomic ratio increases to 1:1 only in accord with both EDS and XPS observations [228].

Dealloying of ternary $Mg_{90-x}Cu_xY_{10}$ alloys leads to pure Cu porous structure since there is a sufficient difference in the nobility of the constituents and the surface diffusion rate of Cu is high enough to expose the non-noble atoms to the surface [179]. For another ternary alloy family, $Pd_{80-x}Ni_xP_{20}$, dealloying led to a Pd-rich porous structure up to 89 at.% Pd even if the Ni molar percentage in the master alloy was 60%, and a measurable amount of Ni and P remained in the dealloyed material [235]. For a slightly more phosphorous-rich Pd–Ni–P alloy, $Pd_{32}Ni_{48}P_{30}$, the final Pd enrichment upon electrochemical dealloying was between 70 and 80 at%, the larger value being achieved with more positive dealloying potential [239]. An interesting dataset in the latter study shows that the release of Ni and P is not proportional during the dealloying process (see Fig. 2f in Ref. [239]). In the early stage of the dealloying process, Ni is released first, which leads to a temporary increase in the phosphorous mole fraction in the partially dealloyed material. Later, the composition changes monotonously. Although such time-resolved composition data are not available for the majority of the dealloying studies, it can be assumed that such sequential dealloying may take place in various other alloys, which is likely to be accompanied with a concomitant change of the surface composition of the intermediate-stage ligaments. Partial dealloying of ternary materials under mild electrochemical etching was also shown for an amorphous $Fe_{60}Pt_{10}B_{30}$ precursor alloy [240]. Here, B and Fe were removed selectively, but the residual Fe content was limited by the formation of an fcc Fe–Pt alloy with an approximately 2:1 molar ratio.

Selective dissolution of the multicomponent alloys leads to various compositions in term of the completeness and the selectivity of the removal of the components. Dealloying of an $Au_{30}Si_{20}Cu_{33}Ag_7Pd_{10}$ master alloy at room temperature led to a gold-rich Au–Pd porous structure with enhanced Au-to-Pd ratio as compared to the master alloy composition [232]. However, the dealloying of another alloy with similar composition, $Au_{40}Si_{20}Cu_{28}Ag_7Pd_5$, at 70 °C resulted in a complete dealloying of all four accompanying element beside the most noble one (i.e., Au) [180, 234], indicating the importance of the thermal activation.

A very interesting secondary structural transformation may happen when the pore wall-forming components of the dealloyed material are not miscible in equilibrium. This was demonstrated for an Mg–Ag–Cu–Y precursor alloy with equimolar Ag and Cu ratio (12.5%) [238]. As the dealloying process took place, the initially amorphous structure turned into a crystalline one, and in parallel, the Ag and Cu crystals started segregating. The segregation leads to a primarily Ag-rich pore wall and Cu crystals at the sample surface with much larger diameter than the feature size of the bicontinuous pore/ligament system.

The morphology of the porous dealloyed materials strongly depends on the homogeneity of the master alloy. If there are "pre-existing features" (grain boundaries, composition inhomogeneities etc.) in the sample, the etching process is likely to take place preferentially along these parts [229]. Trenches at the surface of melt-spun ribbons also influence the dealloying morphology. For amorphous alloys, the increase in the number of the components and the corresponding increase in the homogeneity of the precursor alloy have a beneficial impact on the morphology of the dealloyed porous structure.

8.4.5 Dealloying of Electrochemically Produced Layers

Dealloying procedures discussed in the previous chapters result in a porous material where the resulting structure is open at all surfaces. These materials can be used as filters or flow-through catalysts. If the porous material must form a layer, the alloy to be etched has to be grown on an existing surface, and the dealloying procedure has to take place on this surface only. Therefore, the dealloyed material is obtained as a supported porous layer that is capped at the bottom side (which means that the specimen is not permeable). These materials can be used, e.g., as porous electrodes with high surface area. The procedures based on deposition combined with dealloying have the advantage that the thickness of the porous layer can be easily regulated by an appropriate choice of the thickness of the alloyed layer.

In the simplest case, the coating to be dealloyed can be electroplated, and the resulting deposit can be dealloyed in the same solution. This procedure can be performed, e.g., for Au–Cu alloys deposited from a solution containing $KAuCl_4$, $CuSO_4$ and H_2SO_4 [241]. The suggested procedure involved one single deposition step at constant negative potential followed with one dealloying step where Cu dissolved selectively. Although the composition of the primary Au–Cu layer was not reported, it seems that the dealloying process exhibits the same features as the dealloying of metallurgically processed bulk Au–Cu alloys. The opportunity that the primary layer composition can be regulated with the composition of the solution and the deposition current density for tuning the pore size was not exploited for Au–Cu alloys but was demonstrated in a set of experiments related to Ni–Cu alloys deposited from a sulphamate bath [190]. For dealloyed Ni obtained from Ni–Cu alloys, a detailed magnetization study confirmed the formation of an isotropic nanoporous structure [190].

Here, the attention must be drawn to the fact that Ni–Cu alloys tend to segregate under some deposition conditions because the unlimited miscibility observed at high temperature is followed by segregation at lower temperature. This low-temperature miscibility gap was not confirmed directly by structural studies but was concluded from model calculations [242–244]. In accord with the theoretical calculations, some physical properties of the Ni–Cu alloys indicate that bulk Ni–Cu alloys exhibit a tendency for the formation of Ni-rich segregations. The same tendency for segregation was also confirmed in electrodeposition experiments [245–249]. The formation of Ni-rich and Cu-rich zones is suitable to obtain porous structures that are otherwise not available. As demonstrated by Chang et al. [209], dealloying of electrodeposited Ni–Cu alloys with columnar structure leads to an irregular nanochannel system with channel axes perpendicular to the substrate and with Ni wall thickness around 20 nm. As shown in Fig. 8.12, the structure of the dealloyed inhomogeneous Ni–Cu alloy is not conformal to the usual bicontinuous pore/ligament systems obtained for dealloyed homogeneous alloys.

The segregation tendency of the alloys containing Ni and Cu can be also exploited when other components are also present, as it was exemplified for Ni–Co–Cu alloys [250]. The mole fraction of the alloying element beside Ni has to be limited in this case because the release of the Co content of the alloys is accompanied with some Ni loss, too. On the one hand, this makes the process difficult to design quantitatively, and, on the other hand, may lead to a collapse of the residual Ni network.

When a pure metal layer is plated onto the surface of another one, the spontaneous diffusion of the adlayer into the substrate usually does not take place to the desired extent at room temperature. Therefore, essentially all interdiffusion-related dealloying processes require some thermal activation, either by performing the electrodeposition step itself at an elevated temperature or treating the coated substrate at high temperature *ex situ*. The interdiffusion temperature may vary between 60 and 150 °C. Electroplating at room temperature and interdiffusion at higher temperature may be followed also by a chemical etching, as it was shown for nanoporous Au and Cu plated with a Zn layer [251].

Zinc was used as a much preferred plated adlayer and etched metal in many other deposition-combined dealloying studies because its relatively large diffusion coefficient in other metals, the availability of alloying at a relatively low temperature

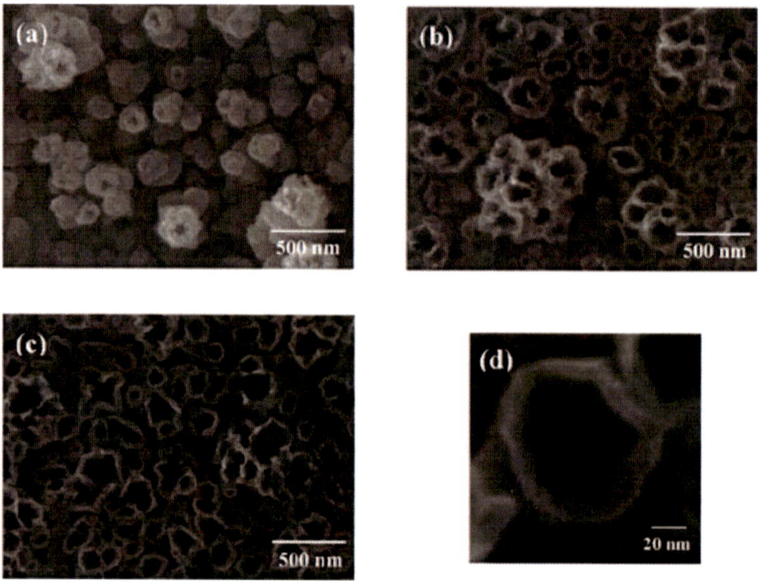

Fig. 8.12 SEM micrographs of Ni–Cu alloy films that were electrodeposited with −1 C of charge followed by selective dissolution as characterized with the following anodic charge: **a** 0.1 C, **b** 0.3 C, and **c** 0.5 C where the dealloying current decayed nearly to zero. **d** Magnification of a characteristic part of the micrograph in (**c**). Reprinted with permission from [209] Copyright (2008) American Chemical Society

and its high selectivity in dealloying processes. Non-aqueous media are typical in these cases and can be classified as four major kinds: organic molecular liquids [252–254], deep eutectic solvents [255, 256], ionic liquids [257–261] and molten salts [262, 263]. While the first three types of media was sufficient to produce Cu–Zn, Ag–Zn and Au–Zn alloys, Ni–Zn alloys required molten salts because of the smaller mobility of the Zn atom in the Ni lattice. A single deposition-and-dealloying cycle is sometimes enough for producing a porous layer, but repeated cycling proved to be more efficient. Sn proved to be an equally suitable sacrificial metal or co-reactant with Zn in deposition/dealloying processes [253]. The structures obtained with Zn alone and with Zn and Sn together are shown in Fig. 8.13.

The principles of the establishment of the desired working conditions of deposition followed by dealloying are explained with the help of Fig. 8.14. When the deposit makes a distinct layer on the top of the substrate, one single stripping peak is obtained in the cyclic voltammograms. As the interdiffusion becomes effective (as a result of, e.g., the raise of the temperature), the stripping peak splits up and a new dealloying peak arises at more positive potential than the stripping peak of the pure deposit (see curves in Fig. 8.14a). The larger portion of the deposit forms an alloy by diffusing into the substrate, the larger the ratio of the surface area of the dealloying and the stripping peaks. The ratio of the dealloying and stripping peaks increases also with the cycle number. When the chronoamperometric records are scrutinized at a constant

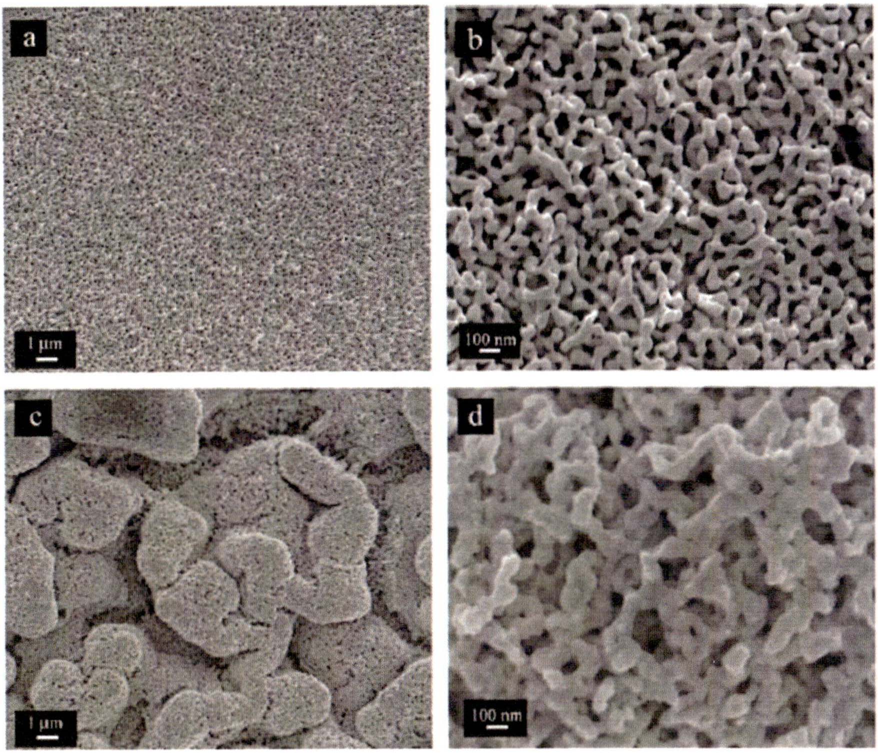

Fig. 8.13 Nanoporous gold films obtained from benzyl alcohol solution of 1.6 M $ZnCl_2$ (**a** and **b**) and 1.321 M $ZnCl_2$ + 0.264 M $SnCl_2$ (**c** and **d**) after 30 deposition–dealloying cycles at 120 °C. Reprinted from [253]. Copyright (2009), with permission from Elsevier

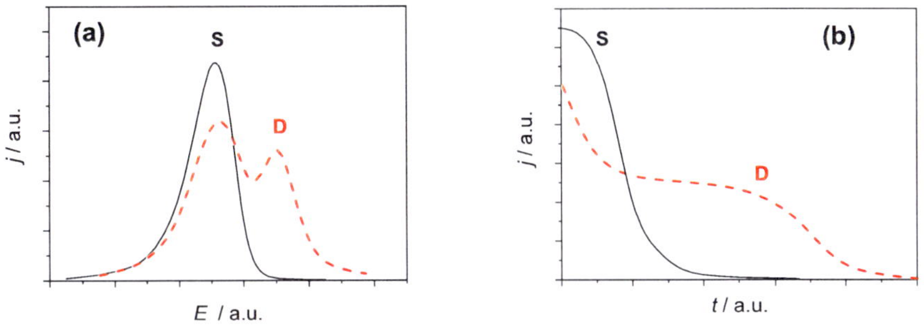

Fig. 8.14 Electrochemical measurement results for stripping and dealloying. **a** Anodic-going sweeps corresponding to stripping (S) and dealloying (D). The second peak at more positive potential than the stripping peak is responsible for dealloying. The decay of the dealloying peak is longer than that of the stripping peak due to the solid-state diffusion of the reactant. **b** Chronoamperometric curves for stripping and dealloying at a fixed electrode potential. The occurrence of a new plateau in the chronoamperometric data is associated with dealloying. The schemes are based on the measurements published in Refs. [194, 254, 258, 261]

dealloying potential (Fig. 8.14b), the increase in the alloying ratio results in a two-step curve. The first and fast step accounts for the stripping of the pure deposit layer and the second plateau-like dissolution period is related to the dealloying when the small anodic current is due to the relatively slow transport of the dealloyed metal from the bulk to the surface.

Although Zn as a dealloyed metal is by far the most popular one in the deposition-coupled dealloying studies, there are other possibilities, too. For instance, Li as a dealloyed metal can lead to high surface area aluminium [264]. This reaction differs from Zn alloying/dealloying in various aspects. The Li insertion requires a strictly oxygen- and water-free environment, and can be carried out in the solution of a Li salt in an organic carbonate, a typical medium for Li ion batteries. In contrast to the Zn insertion, Li uptake by Al takes place at room temperature.

All the processes described above involved a deposition step (at sufficiently negative electrode potentials) followed by a dealloying step (oxidation). However, the formation of porous layers is also possible with an *oxidation—reduction* sequence. This is an analogy of the deposition-combined dealloying procedure, even though it can be debated whether the final process truly has the dealloying character. Here we cite a work related to the formation of an AgCl layer on Ag sheets in an aqueous solution containing NaOH and NaCl [265]. The electrode potential used for the reduction step was the determining factor in tuning the pore size: at the more negative potential the AgCl layer was reduced, the smaller pore size was obtained. The reduced nanoporous Ag electrodes were shown to exhibit a high catalytic activity towards both oxygen reduction and oxidation of organic compounds.

8.5 Combination of Porosity-Related Electrochemical Methods

The combination of the methods described in the earlier parts of Chap. 8 makes it possible to prepare porous materials with porosity levels that are not available with one single method. Electrodeposition by using a dynamic bubble template is a bottom-up method that results in a hierarchical structure with a primary pore size above 10 μm and a secondary pore size around 100 nm. In contrast, dealloying, which is a top-down method, leads to pore sizes between 5 and 100 nm. It is expected that a combination of these methods, i.e., dealloying of alloys produced with the dynamic bubble template method, can lead to a porous material with three levels of porosity. A few studies will be presented below in which the combinatorial application of the dynamic bubble template method and dealloying was exploited.

Images recorded for dealloyed bubble-templated alloys verify the above described expectation concerning the morphology of the resulting structures, as shown in Fig. 8.15. The high-magnification images indicate that the dendrites constituting the spongy walls of the primary pores developed internal pores after dealloying.

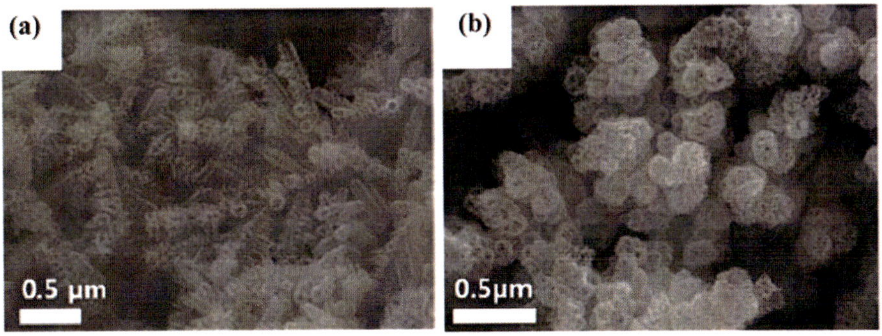

Fig. 8.15 **a** Nickel foam obtained with dealloying of a DHBT-plated Ni–Cu alloy. Reprinted by permission from Springer [45], Copyright (2009). **b** Ruthenium foam prepared with dealloying of a DHBT-plated Ru–Cu alloy specimen. Reprinted from [266]. Copyright (2013), with permission from Elsevier

Concerning the methodology of the combined fully electrochemical DHBT–dealloying process, two approaches have been reported so far. The two-step sequential method involves the deposition of the hydrogen-templated porous structure, from which the less noble component is dealloyed after a solution change. This method was applied for Ni–Cu [45], Ru–Cu [266] and Pd–Cu [267] deposits. For all these alloys, 1 M sulphuric acid could be used for the Cu dealloying process (for Ni–Cu foams, chemical dealloying with $(NH_4)_2S_2O_8$ led to similar result in terms of the filament morphology [268]). When the remaining porous structure was composed of mainly Ni and Ru, the remaining metal was partly oxidized in the dealloying process, and the target property was the supercapacity. In contrast, the porous metallic Pd was used for electrochemical sensing.

Instead of the two-step deposition–dealloying sequence, a pulse reversal system was elaborated for the deposition of porous Pt by periodical dealloying of a Pt–Cu alloy [269]. In the latter work, a cathodic pulse, typically 0.1 s long, was followed with a dealloying pulse for 0.3 s, and this cycle was repeated for several hundred times. The cathode potential during the first pulse was negative enough so that the deposition process took place in the DHBT regime, although the bubble size over the short pulse length is not comparable to those forming during the continuous polarization. The anodic pulse, however, led to dealloying only with no oxygen evolution, although the oxidation of a part of the hydrogen evolved during the first pulse might also have happened. The resulting deposit morphology was completely different from those obtained in the sequential two-pulse method. The cavities at the deposit surface corresponding to the primary pore structure of the DHBT-plated layers were completely missing. Instead, an apparently even surface of a porous material was seen with some cracks that are typical for the dealloying process (having a stress-related origin). The porosity of the deposit corresponded to the dealloying process and appeared to be even along the deposit thickness. The cross-sectional cut of the reverse pulse plated samples was columnar with no significant morphological variation along the growth direction.

References

1. Shin HC, Dong J, Liu M (2003) Adv Mater 15:1610–1614
2. Zhang H, Ye Y, Shen R, Ru C, Hu Y (2013) J Electrochem Soc 160:D441–D445
3. Yang GM, Chen X, Li J, Guo Z, Liu JH, Huang XJ (2011) Electrochim Acta 56:6771–6778
4. Zhang W, Ding C, Wang A, Zeng Y (2015) J Electrochem Soc 162:D365–D370
5. Plowman BJ, Jones LA, Bhargava SK (2015) Chem Commun 51:4331–4346
6. Shin HC, Liu M (2004) Chem Mater 16:5460–5464
7. Wang H, Wang N, Hang T, Li M (2016) Appl Surf Sci 372:7–12
8. Mattarozzi L, Cattarin S, Comisso N, Guerriero P, Musiani M, Verlato E (2016) Electrochim Acta 19:296–303
9. Wei L, Goh K, Birer Ö, Karahan HE, Chang J, Zhai S, Chena X, Chen Y (2017) Nanoscale 9:4401–4408
10. Guo J, Guo X, Zeng J, Nie L, Dong J, Peng L, Ding W (2018) Infrared Phys Techn 93:162–170
11. Rafailović LD, Gammer C, Rentenberger C, Kleber C, Whitehead AH, Gollas B, Karnthaler HP (2012) Phys Chem Chem Phys 14:972–980
12. Nikolić ND, Branković G, Maksimović VM, Pavlović MG, Popov KI (2009) J Electroanal Chem 635:111–119
13. Nikolić ND, Branković G, Maksimović VM (2011) J Electroanal Chem 661:309–316
14. Jović VD, Maksimović V, Pavlović MG, Popov KI (2006) J Solid State Electrochem 10:373–379
15. Jović VD, Jović BM, Pavlović MG, Maksimović V (2006) J Solid State Electrochem 10:959–966
16. Jović VD, Jović BM, Pavlović MG (2006) Electrochim Acta 51:5468–5477
17. Marozzi CA, Chialvo AC (2000) Electrochim Acta 45:2111–2120
18. Tan K, Tian MB, Cai Q (2010) Thin Solid Films 518:5159–5163
19. Cherevko S, Chung CH (2011) Electrochem Commun 13:16–19
20. Cherevko S, Xing X, Chung CH (2010) Electrochem Commun 12:467–470
21. Nikolić ND, Popov KI, Pavlović LJ, Pavlović MG (2007) J Solid State Electrochem 11:667–675
22. Kim JH, Kim RH, Kwon HS (2008) Electrochem Commun 10:1148–1151
23. Jiang T, Zhang S, Qiu X, Sun M, Chen L (2008) Electrochem Solid State Lett 11:D50–D52
24. Nam DH, Kim RH, Han DW, Kim JH, Kwon HS (2011) Electrochim Acta 56:9397–9405
25. Qi B, Yang H, Zhao K, Bah MM, Bo X, Guo L (2013) J Electroanal Chem 700:24–29
26. Niu J, Liu X, Xia K, Xu L, Xu Y, Fang X, Lu W (2015) Int J Electrochem Sci 10:7331–7340
27. Hoang TTH, Ma S, Gold JI, Kenis PJA, Gewirth AA (2017) ACS Catal 7:3313–3321
28. Singh H, Dheeraj PB, Singh YP, Rathore G, Bhardwaj M (2017) J Electroanal Chem 785:1–7
29. Yua X, Wang M, Wang Z, Gong X, Guo Z (2016) Appl Surf Sci 360:502–509
30. Jeun JH, Kim DH, Hong SH (2012) Sensor Actuat B—Chem 161:784–790
31. Chamoun M, Hertzberg BJ, Gupta T, Davies D, Bhadra S, Tassell BV, Erdonmez C, Steingart DA (2015) NPG Asia Mater 7:e178(1–8)
32. Huan TN, Ganesh T, Kim KS, Kim S, Han SH, Chung H (2011) Biosens Bioelectron 27:183–186
33. Plowman BJ, O'Mullane AP, Selvakannan PR, Bhargava SK (2010) Chem Commun 46:9182–9184
34. Ott A, Jones LA, Bhargava SK (2011) Electrochem Commun 13:1248–1251
35. Cherevko S, Chung CH (2010) Electrochim Acta 55:6383–6390
36. Cherevko S, Xing X, Chung CH (2011) Appl Surf Sci 257:8054–8061
37. Yang M (2011) J Mater Chem 21:3119–3124
38. Liu J, Cao L, Huang W, Li Z (2011) ACS Appl Mater Interfaces 3:3552–3558
39. Liu J, Cao L, Huang W, Li Z (2012) J Electroanal Chem 686:38–45
40. Ojani R, Hasheminejad U, Raoof JB (2014) Int J Hydrogen Energ 39:8194–8203
41. Liu J, Wang J, Kong K, Huang T, Yu A (2016) Catal Commun 73:22–26

42. Najdovski I, Selvakannan PR, Bhargava SK, O'Mullane AP (2012) Nanoscale 4:6298–6306
43. Yu X, Wang M, Wang Z, Gong X, Guo Z (2016) Electrochim Acta 211:900–910
44. Zhang J, Baró MD, Pellicer E, Sort J (2014) Nanoscale 6:12490–12499
45. Jeong MG, Zhuo K, Cherevko S, Chung CH (2012) Korean J Chem Eng 29:1802–1805
46. Cherevko S, Kulyk N, Chung CH (2012) Langmuir 28:3306–3315
47. Jeun JH, Kim WS, Hong SH (2015) Mater Lett 138:33–36
48. Mattarozzi L, Cattarin S, Comisso N, Gambirasi A, Guerriero P, Musiani M, Vázquez-Gómez L, Verlato E (2014) Electrochim Acta 140:337–344
49. Mattarozzi L, Cattarin S, Comisso N, Gambirasi A, Guerriero P, Musiani M, Vázquez-Gómez L, Verlato E (2015) J Electrochem Soc 162:D236–D241
50. Li GR, Ke QF, Zhang ZS, Dawa CR, Liu P, Liu GK, Tong YX (2007) Chem Mater 19:2283–2287
51. Verlato E, Cattarin S, Comisso N, Guerriero P, Musiani M, Vázquez-Gómez L (2010) Electrochem Commun 12:1120–1123
52. Shahbazi P, Kiani A (2011) Electrochim Acta 56:9520–9529
53. Vázquez-Gómez L, Verlato E, Cattarina S, Comisso N, Guerriero P, Musiani M (2011) Electrochim Acta 56:2237–2245
54. Herraiz-Cardona I, Ortega E, Vázquez-Gómez L, Pérez-Herranz V (2012) Int J Hydrogen Energ 37:2147–2156
55. Comisso N, Armelao L, Cattarin S, Guerriero P, Mattarozzi L, Musiani M, Rancan M, Vázquez-Gómez L, Verlato E (2018) Electrochim Acta 273:454–461
56. Comisso N, Cattarin S, Guerriero P, Mattarozzi L, Musiani M, Verlato E (2015) Electrochem Commun 60:144–147
57. Comisso N, Cattarin S, Guerriero P, Mattarozzi L, Musiani M, Verlato E (2016) Electrochem Commun 73:59–62
58. Comisso N, Cattarin S, Guerriero P, Mattarozzi L, Musiani M, Verlato E (2016) Electrochim Acta 200:259–267
59. Comisso N, Armelao L, Cattarin S, Guerriero P, Mattarozzi L, Musiani M, Rancan M, Vázquez-Gómez L, Verlato E (2017) Electrochim Acta 253:11–20
60. Pang K, Hou YC, Wu WZ, Tian SD, Sun NN (2012) Sci China Chem 55:1325–1329
61. Motlagh NV, Khani R, Rahnama S (2015) Colloid Surface A 484:528–546
62. Wen Q, Guo Z (2016) Chem Lett 45:1134–1149
63. Jain R, Pitchumani R (2017) Langmuir 33:7181–7190
64. Cassie ABD (1948) Discuss Faraday Soc 3:11–16
65. Cassie ABD, Baxter S (1944) Trans Faraday Soc 40:546–551
66. Zhang D, Wang L, Qian H, Li X (2016) J Coat Technol Res 13:11–29
67. Vazirinasab E, Jafari R, Momen G (2018) Surf Coat Tech 341:40–56
68. Tam J, Palumbo G, Erb U (2016) Materials 9:151(1–27)
69. Hang T, Hu A, Ling H, Li M, Mao D (2010) Appl Surf Sci 256:2400–2404
70. Khorsand S, Raeissi K, Ashrafizadeh F, Arenas MA (2015) Chem Eng J 273:638–646
71. Cao L, Liu J, Xu S, Xia Y, Huang W, Li Z (2013) Mater Res Bull 48:4804–4810
72. Wang J, Liu J, Neate N, Bai M, Xu F, Hussain T, Scotchford C, Hou X (2018) J Alloy Compd 765:221–228
73. Khorsand S, Raeissi K, Ashrafizadeh F (2014) Appl Surf Sci 305:498–505
74. Esmailzadeh S, Khorsand S, Raeissi K, Ashrafizadeh F (2015) Surf Coat Tech 283:337–346
75. Salehikahrizsangi P, Raeissi K, Karimzadeh F, Calabrese L, Patane S, Proverbio E (2018) Colloid Surface A 558:446–454
76. Lee JM, Jung KK, Lee SH, Ko JS (2016) Appl Surf Sci 369:163–169
77. Qiu R, Zhang D, Wang P, Zhang XL, Kang YS (2011) Electrochim Acta 58:699–706
78. Xiao H, Hu A, Hang T, Li M (2015) Appl Surf Sci 324:319–323
79. Haghdoost A, Pitchumani R (2014) Langmuir 30:4183–4191
80. You Q, Ran G, Wang C, Zhao Y, Song Q (2018) Sep Purif Technol 193:21–28
81. Lee JM, Bae KM, Jung KK, Jeong JH, Ko JS (2014) Appl Surf Sci 289:14–20
82. Khorsand S, Raeissi K, Ashrafizadeh F, Arenas MA (2015) Surf Coat Tech 276:296–304

83. Li H, Yu S (2017) New J Chem 41:5436–5444
84. Yu Q, Zeng Z, Zhao W, Li M, Wu X, Xue Q (2013) Colloid Surface A 427:1–6
85. Rahimi E, Rafsanjani-Abbasi A, Kiani-Rashid A, Jafari H, Davoodi A (2018) Colloid Surface A 547:81–94
86. Khorsand S, Raeissi K, Ashrafizadeh F, Arenas MA, Conde A (2016) Appl Surf Sci 364:349–357
87. Liu Y, Yin X, Zhang J, Yu S, Han Z, Ren L (2014) Electrochim Acta 125:395–403
88. Li W, Kang Z (2014) Surf Coat Tech 253:205–213
89. Liu Y, Li S, Wang Y, Wang H, Gao K, Han Z, Ren L (2016) J Coll Interf Sci 478:164–171
90. Song Z, Xie Z, Ding L, Zhang Y (2018) Int J Electrochem Sci 13:6190–6200
91. Zhang X, Liang J, Liu B, Peng Z (2014) Colloid Surface A 454:113–118
92. Brassard JD, Sarkar DK, Perron J, Audibert-Hayet A, Melot D (2015) J Coll Interf Sci 447:240–247
93. Lee JM, Jung KK, Ko JS (2016) Curr Appl Phys 16:261–266
94. Lee JM, Ko JS (2017) Int J Precis Eng Man 18:871–877
95. Qiu R, Zhang Q, Wang P, Jiang L, Hou J, Guo W, Zhang H (2014) Colloid Surface A 453:132–141
96. Li H, Yu S, Han X, Zhao Y (2016) Colloid Surface A 503:43–52
97. Li H, Yu S (2017) J Alloy Compd 691:195–205
98. Li H, Yu S, Han X, Zhang S, Zhao Y (2016) J Bionic Eng 13:622–630
99. Qing Y, Yang C, Yu Z, Zhang Z, Hu Q, Liu C (2016) J Electrochem Soc 163:D385–D391
100. Gao B, Poduska KM (2018) J Electrochem Soc 165:D472–D476
101. Feng Y, Chen S, Cheng YF (2018) Surf Coat Tech 340:55–65
102. Tan J, Hao J, An Z, Liu C (2017) Int J Electrochem Sci 12:40–49
103. Zhang B, Feng H, Lin F, Wang Y, Wang L, Dong Y, Li W (2016) Appl Surf Sci 378:388–396
104. Liu L, Liu W, Chen R, Li X, Xie X (2015) Chem Eng J 281:804–812
105. Zhang JL, Gu CD, Tong YY, Wang XL, Tu JP (2015) J Electrochem Soc 162:D313–D319
106. Jain R, Pitchumani R (2018) Langmuir 34:3159–3169
107. Xu Z, Jiang D, Wei Z, Chen J, Jing J (2018) Appl Surf Sci 427:253–261
108. Zhao G, Li J, Huang Y, Yang L, Ye Y, Walsh FC, Chen J, Wang S (2017) RSC Adv 7:44896–44903
109. Bagheri H, Aliofkhazraei M, Forooshani HM, Rouhaghdam AS (2018) Appl Surf Sci 436:1134–1146
110. Wang H, Zhu Y, Hu Z, Zhang X, Wu S, Wang R, Zhu Y (2016) Chem Eng J 303:37–47
111. Cao H, Gu W, Fu J, Liu Y, Chen S (2017) Appl Surf Sci 412:599–605
112. Zhang C, Li Y, Bai N, Tan C, Cai P, Li Q (2017) Mater Sci Eng B 224:117–124
113. Zhao Y, Xiao X, Ye Z, Ji Q, Xie W (2018) Appl Phys A 124:193(1–10)
114. Liang J, Li D, Wang D, Liu K, Chen L (2014) Appl Surf Sci 293:265–270
115. Zhang XF, Chen RJ, Hu JM (2016) Corros Sci 104:336–343
116. Kang Z, Li W (2017) J Ind Eng Chem 50:50–56
117. Jiang S, Guo Z, Liu G, Gyimah GK, Li X, Dong H (2017) Materials 10:1229(1–13)
118. Chen Z, Li F, Hao L, Chen A, Kong Y (2011) Appl Surf Sci 258:1395–1398
119. Zhang B, Xu W, Zhu Q, Li Y, Hou B (2018) J Coll Interf Sci 532:201–209
120. Chen Z, Hao L, Chen A, Song Q, Chen C (2012) Electrochim Acta 59:168–171
121. Liu C, Su F, Liang J, Huang P (2014) Surf Coat Tech 258:580–586
122. Liu Y, Li S, Zhang J, Liu J, Han Z, Ren L (2015) Corros Sci 94:190–196
123. Pijáková B, Klíma M, Alberti M, Buršíková V (2016) Mater Lett 184:243–247
124. Liu Q, Kang Z (2014) Mater Lett 137:210–213
125. Liu Q, Chen D, Kang Z (2015) ACS Appl Mater Interfaces 7:1859–1867
126. Zhang B, Zhao X, Li Y, Hou B (2016) RSC Adv 6:35455–35465
127. Liu Y, Xue J, Luo D, Wang H, Gong X, Han Z, Ren L (2017) J Coll Interf Sci 491:313–320
128. Zhang B, Zhu Q, Li Y, Hou B (2018) Chem Eng J 352:625–633
129. Jiang S, Guo Z, Deng Y, Dong H, Li X, Liu J (2018) Appl Surf Sci 458:603–611
130. Chen X, He Y, Fan Y, Yang Q, Li H (2016) Appl Phys A 122:1030(1–10)

131. Zhang Y, Tang S, Hu J, Lin T (2016) Corros Sci 111:334–343
132. Fan Y, He Y, Luo P, Chen X, Yu Z, Li M (2017) Mater Lett 188:115–118
133. Chen C, He Y, Fan Y, Chen X, Yang Q (2018) Surf Innov 6:106–115
134. Pedraza F, Mahadik SA, Bouchaud B (2015) Phys Chem Chem Phys 17:31750–31757
135. Nakayama K, Hiraga T, Zhua C, Tsuji E, Aoki Y, Habazaki H (2017) Appl Surf Sci 423:968–976
136. Wang P, Zhang D, Qiu R, Hou B (2011) Corros Sci 53:2080–2086
137. Darmanin T, Taffin de Givenchy E, Amigoni S, Guittard F (2010) Langmuir 26:17596–17602
138. Darmanin T, Guittard F (2014) RSC Adv 4:50401–50405
139. Ramos Chagas G, Darmanin T, Guittard F (2015) Beilstein J Nanotechnol 6:2078–2087
140. Wolfs M, Darmanin T, Guittard F (2014) Surf Coati Tech 259:594–598
141. Darmanin T, Guittard F (2015) Synthetic Met 205:58–63
142. Bruzaud J, Tarrade J, Celia E, Darmanin T, Taffin de Givenchy E, Guittard F, Herry JM, Guilbaud M, Bellon-Fontaine MN (2017) Mater Sci Eng C 73:40–47
143. Park SH, Shin HS, Kim YH, Park HM, Song JY (2013) J Alloys Compd 580:152–156
144. Park SH, Shin HS, Kim YH, Park HM, Song JY (2013) Nanoscale 5:1864–1869
145. Wang J, Gupta A, Pan S (2017) Int J Hydrogen Energ 42:8462–8474
146. Park SH, Park HM, Han SS, Han SY, Song JY (2016) RSC Adv 6:3210–3212
147. Jia F, Wong KW, Du R (2009) Electrochem Commun 11:519–521
148. Park SH, Son JG, Lee TG, Park HM, Song JY (2013) Nanoscale Res Lett 8:248(1–6)
149. Sj K, Kim EJ, Liu M, Shin HC (2016) J Electrochem Sci Technol 7:206–213
150. Skompska M, Zarębska K (2014) Electrochim Acta 127:467–488
151. Yamabi S, Imai H (2002) J Mater Chem 12:3773–3778
152. Goux A, Pauporte T, Chivot J, Lincot D (2005) Electrochim Acta 50:2239–2248
153. Oliveira APA, Hochepied JF, Grillon F, Berger MH (2003) Chem Mater 15:3202–3207
154. Zheng D, Shi J, Lu X, Wang C, Liu Z, Liang C, Liu P, Tong Y (2010) CrystEngComm 12:4066–4070
155. Lin Y, Yang J, Zhou X (2011) Appl Surf Sci 258:1491–1494
156. Elias J, Tena-Zaera R, Lévy-Clément C (2007) Thin Solid Films 515:8553–8557
157. Gan X, Liu K, Du X, Guo L, Liu H (2015) J Mater Sci 50:7866–7874
158. Henríquez R, Grez P, Muñoz E, Dalchiele EA, Marotti RE, Gómez H (2011) Thin Solid Films 520:41–46
159. Hsieh CT, Yang SY, Gu JL, Jiang RY (2012) Solid State Ionics 209–210:43–50
160. Pauporté T, Lincot D (1999) Appl Phys Lett 75:3817–3819
161. Pauporté T, Lincot D, Viana B, Pellé F (2006) Appl Phys Lett 89:233112(1–3)
162. Tamvakos D, Lepadatu S, Antohed VA, Tamvakos A, Weaver PM, Piraux L, Cain MG, Pullini D (2015) Appl Surf Sci 356:1214–1220
163. Khajavi MR, Blackwood DJ, Cabanero G, Tena-Zaera R (2012) Electrochim Acta 69:181–189
164. Ghannam H, Bazin C, Chahboun A, Turmine M (2018) CrystEngComm 20:6618–6628
165. Khosrow-pour F, Aghazadeh M, Sabourb B, Dalvand S (2013) Ceram Int 39:9491–9498
166. Yousefi T, Golikand AN, Mashhadizadeh MH, Aghazadeh M (2012) J Taiwan Inst Chem E 43:614–618
167. Zhang Z, Tian L, Jiang Y, Fu W, Yang H (2013) Mater Lett 106:266–269
168. Qiu HJ, Peng L, Li X, Xu HT, Wang Y (2015) Corros Sci 92:16–31
169. Xiao X, Si P, Magner E (2016) Bioelectrochemistry 109:117–126
170. Kunduraci M (2016) J Solid State Electrochem 20:2105–2111
171. Pickering HW, Wagner C (1967) J Electrochem Soc 114:698–706
172. Pickering HW (1983) Corros Sci 23:1107–1120
173. Erlebacher J, Sieradzki K (2003) Scripta Mater 49:991–996
174. Erlebacher J, Aziz MJ, Karma A, Dimitrov N, Sieradzki K (2001) Nature 410:450–453
175. Moffat TP, Fan FRF, Bard AJ (1991) J Electrochem Soc 138:3224–3235
176. Morales J, Esparza P, Gonzalez S, Vazquez L, Salvarezza RC, Arvia AJ (1996) Langmuir 12:500–507

177. Chen SJ, Sanz F, Ogletree DF, Hallmark VM, Devine TM, Salmeron M (1993) Surf Sci 292:289–297
178. Chen AY, Shi SS, Qiu YD, Xie XF, Ruan HH, Gu JF, Pan D (2015) Micropor Mesopor Mat 202:50–56
179. Luo X, Li R, Huang L, Zhang T (2013) Corros Sci 67:100–108
180. Paschalidou EM, Scaglione F, Gebert A, Oswald S, Rizzi P, Battezzati L (2016) J Alloy Compd 667:302–309
181. Forty AJ (1979) Nature 282:597–598
182. Martín H, Carro P, Hernández Creus A, Morales J, Fernández G, Esparza P, González S, Salvarezza RC, Arvia AJ (2000) J Phys Chem B 104:8229–8237
183. Sieradzki K, Dimitrov N, Movrin D, McCall C, Vasiljevic N, Erlebacher J (2002) J Electrochem Soc 149:B370–B377
184. Dursun A, Pugh DV, Corcoran SG (2003) Electrochem Solid-State Lett 6:B32–B34
185. Dursun A, Pugh DV, Corcoran SG (2003) J Electrochem Soc 152:B65–B72
186. Erlebacher J (2004) J Electrochem Soc 151:C614–C626
187. Eilks C, Elliott CM (2008) J Comput Phys 227:9727–9741
188. Strasser P, Koh S, Greeley J (2008) Phys Chem Chem Phys 10:3670–3683
189. Parida S, Kramer D, Volkert CA, Rösner H, Erlebacher J, Weissmüller J (2006) Phys Rev Lett 97:035504(1–4)
190. Sun L, Chien CL, Searson PC (2004) Chem Mater 16:3125–3129
191. Zhao C, Wang X, Qi Z, Ji H, Zhang Z (2010) Corros Sci 52:3962–3972
192. Zheng J, Zhang S, Liu W, Xing Y (2011) Rare Metals 30:370–374
193. Lia K, Hong L, Han D, Wang S, Xiao M, Meng Y, Bao D, Ren S (2017) Electrochem Commun 81:88–92
194. Shi S, Markmann J, Weissmüller J (2018) Electrochim Acta 285:60–69
195. Li J, Jiang H, Yu N, Xu C, Geng H (2015) Corros Sci 90:216–222
196. Song T, Yan M, Shi Z, Atrens A, Qian M (2015) Electrochim Acta 164:288–296
197. Kong Q, Lian L, Liu Y, Zhang J, Wang L, Feng W (2015) Micropor Mesopor Mat 208:152–159
198. Lu HB, Li Y, Wang FH (2006) Corros Sci 48:2106–2119
199. Lu HB, Li Y, Wang FH (2007) Scripta Mater 56:165–168
200. Dan Z, Qin F, Sugawara Y, Muto I, Hara N (2012) Intermetallics 31:157–164
201. Qiu HJ, Kang JL, Liu P, Hirata A, Fujita T, Chen MW (2014) J Power Sources 247:896–905
202. Hakamada M, Mabuchi M (2014) Procedia Engineer 81:2159–2164
203. Stein M, Owens SP, Pickering HW, Weil KG (1998) Electrochim Acta 43:223–226
204. Chen S, Chu Y, Zheng J, Li Z (2009) Electrochim Acta 54:1102–1108
205. Hu S, Huang W, Li Z (2010) Mater Lett 64:1257–1260
206. Steyskal EM, Wiednig C, Enzinger N, Würschum R (2016) Beilstein J Nanotechnol 7:1197–1201
207. Inkrott DK, Willingham RL, Balster T, Gan YX (2011) J Alloy Compd 509:L47–L51
208. Rudi S, Gan L, Cui C, Gliech M, Strasser P (2015) J Electrochem Soc 162:F403–F409
209. Chang JK, Hsu SH, Sun IW, Tsai WT (2008) J Phys Chem C 112:1371–1376
210. Li W, Ma H, Huang L, Ding Y (2011) Phys Chem Chem Phys 13:5565–5568
211. Chauvin A, Delacôte C, Boujtita M, Angleraud B, Ding J, Choi CH, Tessier PY, El Mel AA (2016) Beilstein J Nanotechnol 7:1361–1367
212. Xia J, Ambrozik S, Crane CC, Chen J, Dimitrov N (2016) J Phys Chem C 120:2299–2308
213. Xia J, Achari I, Ambrozik S, Dimitrov N (2017) Mater Res Bull 85:1–9
214. Pugh DV, Dursun A, Corcoran SG (2005) J Electrochem Soc 152:B455–B459
215. Tominaka S, Shigeto M, Nishizeko H, Osaka T (2010) Chem Commun 46:8989–8991
216. Wang D, Yu Y, Zhu J, Liu S, Muller DA, Abruña HD (2015) Nano Lett 15:1343–1348
217. Snyder J, Livi K, Erlebacher J (2008) J Electrochem Soc 155:C464–C473
218. Chen AY, Wang JW, Wang Y, Jia YQ, Gu JF, Xie XF, Pan D (2015) Electrochim Acta 153:552–558
219. Dursun A, Pugh DV, Corcoran SG (2003) J Electrochem Soc 150:B355–B360
220. Okman O, Lee D, Kysar JW (2010) Scripta Mater 63:1005–1008

221. Okman O, Kysar JW (2011) J Alloy Compd 509:6374–6381
222. Dimitrov N, Mann JA, Vukmirovic M, Sieradzki K (2000) J Electrochem Soc 147:3283–3285
223. Vukmirovic MB, Dimitrov N, Sieradzki K (2002) J Electrochem Soc 149:B428–B439
224. Yoon Y, Buchheit RG (2006) J Electrochem Soc 153:B151–B155
225. Qi Z, Vainio U, Kornowski A, Ritter M, Weller H, Jin H, Weissmüller J (2015) Adv Funct Mater 25:2530–2536
226. Snyder J, Asanithi P, Dalton AB, Erlebacher J (2008) Adv Mater 20:4883–4886
227. Vega AA, Newman RC (2016) J Appl Electrochem 46:995–1010
228. Kraehnert R, Ortel E, Paul B, Eckhardt B, Kanis M, Liuc R, Antoniou A (2015) Catal Sci Technol 5:206–216
229. Antoniou A, Bhattacharrya D, Baldwin JK, Goodwin P, Nastasi M, Picraux ST, Misra A (2009) Appl Phys Lett 95:073116(1–3)
230. Hakamada M, Mabuchi M (2009) J Alloy Compd 479:326–329
231. Zhang Q, Wang X, Qi Z, Wang Y, Zhang Z (2009) Electrochim Acta 54:6190–6198
232. Lang XY, Guo H, Chen LY, Kudo A, Yu JS, Zhang W, Inoue A, Chen MW (2010) J Phys Chem C 114:2600–2603
233. Zhang Z, Wang Y, Wang Y, Wang X, Qi Z, Jia H, Zhao C (2010) Scripta Mater 62:137–140
234. Rizzi P, Scaglione F, Battezzati L (2014) J Alloy Compd 586:S117–S120
235. Zeng YQ, Yang SC, Xiang H, Dong X, Chen L, Chen M, Inoue A, Zhang X, Jiang J (2015) Intermetallics 61:66–71
236. Scaglione F, Celegato F, Rizzi P, Battezzati L (2015) Intermetallics 66:82–87
237. Paschalidou EM, Celegato F, Scaglione F, Rizzi P, Battezzati L, Gebert A, Oswald S, Wolff U, Mihaylov L, Spassov T (2016) Acta Mater 119:177–183
238. Li R, Wu N, Liu J, Jin Y, Chen XB, Zhang T (2017) Corros Sci 119:23–32
239. Wang S, Zhang C, Li H, Liu L (2017) Intermetallics 87:6–12
240. Ou S, Ma D, Li Y, Yubuta K, Tan Z, Wang Y, Zhang W (2017) J Alloy Compd 706:215–219
241. Tominaka S (2011) J Mater Chem 21:9725–9730
242. Massalski TB (ed) (1996) Binary alloy phase diagrams, Second Edition Plus Updates on CD-ROM. ASM International, Materials Park, Ohio, USA
243. Chakrabarti DJ, Laughlin DE, Chen SW, Chang YA (1991). In Nash P (ed) Phase diagrams of binary nickel alloys. ASM International, Materials Park, OH, USA
244. Wang CP, Liu XJ, Jiang M, Ohnuma I, Kainuma R, Ishida K (2005) J Phys Chem Sol 66:256–260
245. Jafari Fesharaki M, Péter L, Schucknecht T, Rafaja D, Dégi J, Pogány L, Neuróhr K, Széles É, Nabiyouni G, Bakonyi I (2012) J Electrochem Soc 159:D162–D171
246. Tsakalakos T (1981) Scripta Metall Mater 15:255–258
247. Kazeminezhad I, Schwarzacher W (2001) J Magn Magn Mater 226:1650–1652
248. Ghosh SK, Grover AK, Dey GK, Kulkarni UD, Dusane RO, Suri AK, Banerjee S (2006) J Mater Res 21:45–61
249. Liu Z, Guo L, Chien CL, Searson PC (2008) J Electrochem Soc 155:D569–D574
250. Peecher BE, Hampton JR (2016) Int J Electrochem. Article ID 2935035(1–10)
251. Jia F, Yu C, Deng K, Zhang L (2007) J Phys Chem C 111:8424–8431
252. Jia F, Yu C, Ai Y, Zhang L (2007) Chem Mater 19:3648–3653
253. Jia F, Yu C, Zhang L (2009) Electrochem Commun 11:1944–1946
254. Dong H, Cao X (2009) J Phys Chem C 113:603–609
255. Zhang QB, Abbott AP, Yang C (2015) Phys Chem Chem Phys 17:14702–14709
256. Yang C, Zhang QB, Gao MY, Hua YX, Xu CY (2016) J Electrochem Soc 163:D469–D475
257. Yeh FH, Tai CC, Huang JF, Sun IW (2006) J Phys Chem B 110:5215–5222
258. Lin YW, Tai CC, Sun IW (2007) J Electrochem Soc 154:D316–D321
259. Huang JF, Lin BT (2009) Analyst 134:2306–2313
260. Jiang J, Zhang L, Wang X (2013) ACS Appl Mater Interfaces 5:12689–12694
261. Hsieh LY, Wang PK, Chang CC, Sun IW (2017) J Electrochem Soc 164:D752–D757
262. Katagiri A, Nakata M (2003) J Electrochem Soc 150:C585–C590
263. Fukumizu T, Kotani F, Yoshida A, Katagiri A (2006) J Electrochem Soc 153:C629–C633

264. Nishio K, Yanagishita T, Yoshida M, Hayakawa T, Masuda H (2015) Electrochem Commun 59:13–15
265. Yuan L, Jiang L, Zhang T, Wang G, Wang S, Bao X, Sun G (2016) Catal Sci Technol 6:7163–7170
266. Jeong MG, Zhuo K, Cherevko S, Kim WJ, Chung CH (2013) J Power Sources 244:806–811
267. An CY, Zhuo K, Kim WJ, Chung CH (2015) Sens Actuat B Chem 213:329–333
268. Choi WS, Chang W, Shin HC (2014) J Solid State Electrochem 18:427–433
269. Cherevko S, Kulyk N, Chung CH (2012) Nanoscale 4:568–575

Chapter 9
Electrosynthesis of Nanostructures Without a Coating Formation on Electrodes

9.1 Synthesis of Nanoparticles on Electrodes

9.1.1 Metal Nanoparticle Formation by Direct Electrochemical Reduction of Metal Ions

The common goal of electroplating is the production of a compact, well-adherent layer on the substrate material. Although various additives are used in conventional industrial plating processes to tune the properties of the deposits, these additives should never make a film that prevents the deposition of the metal atoms on the substrate surface. In contrast to the metal film formation, the cathodic nanoparticle production is based on the application of stabilizers that, on the one hand, prevent the formation of a compact deposit film on the cathode; and, on the other hand, act as stabilizer of the metal particles formed.

Here we have to recall Fig. 2.2 where the electrical double layer structure is depicted with strongly adsorbing ions that are comparable in size with both the solvent molecules and the cations that can be reduced on the electrode. The difference between this classical case and the studies discussed in this chapter is as follows: For metal particle production, the charge on the metal is negative, and the ions electrostatically attached to the electrode are bulky cations that prevent the metal cations to approach the metal surface itself, and the charge transfer is likely to take place through the fairly thick cationic film on the negatively charged metal.

In all studies published so far, tetraalkylammonium salts with different alkyl chain lengths were applied as adsorbent layer. If the alkyl groups are identical, tetrabutylammonium is the smallest cation applied [1–3], but the alkyl chain may be significantly larger, too (like the octyl group [4]). If an asymmetrical cation is used, the common solution is the application of alkyltrimethylammonium cations, most frequently with a cetyl group as the long-chained part of the cation [5–9]. Salts with hydrophobic cations having unusually large alkyl chains are applied as growth modifiers [10, 11] that result in the growth of elongated nanoparticles. Unfortunately, in the majority of

L. Péter, *Electrochemical Methods of Nanostructure Preparation*,
Monographs in Electrochemistry, https://doi.org/10.1007/978-3-030-69117-2_9

the relevant studies, the chemical name of the additive mentioned, its abbreviation and the formula given show a large discrepancy [6–9, 12], and sometimes nanocubes are obtained instead of spherical nanoparticles under conditions that are formally analogous to other studies [12]. It was mentioned that Ag^+ ions are necessary to grow rod-like Au nanoparticles [10], although the desired concentration was not defined. The change of the supporting electrolyte to a non-stabilizing conventional salt (like $NaClO_4$) leads to a layer formation instead of particle production [1].

The cathode was a platinum sheet in all studies mentioned above. The role of the cathode material has not been studied yet; therefore, it cannot be established whether the good electrocatalytic capability of platinum plays any role in the formation of nanoparticles. All metals from which nanoparticles were electrosynthesized can be easily plated onto platinum in the absence of the particle-stabilizing surfactant; therefore, it is likely that the same surfactant can make a film on the Pt cathode, too. It is mentioned in the works of one research group only [1, 2] that the metal deposition and nanoparticle formation are competing processes.

It has to be mentioned here that the cathode potential has to be limited for metal particle formation without a deposition process. First, this is because a too high cathodic overvoltage can lead to both solvent decomposition at a high rate and a reduction of the solute that destroys the deposition-preventing film. Secondly, the noble metal electrode applied for particle production may also undergo a transformation in the presence of tetraalkylammonium salt, as it is described in a number of studies [13–17]. This transformation involves an in-depth insertion of the tetraalkylammonium cations into the metal, similarly to intercalation processes. The electrochemical cation insertion into the noble metals is reversible but is accompanied with the ramification of the cathode.

All methods described in the literature combine the dissolution of the metal to be precipitated in the form of nanoparticles and the reduction of its ion. Hence, the sacrificial anode is always made of the material of the particles to be produced, even if it is a noble metal. Due to stability reason, metallic particles that can be obtained with this technique are composed of relatively stable metals such as Pd [4], Au [5–9] Ag [1, 2] and Cu [3]. Interestingly, the nature of the anion of the supporting electrolyte (i.e., complexing or non-complexing) does not seem to be important. Beside the most common bromide salts, tetraalkylammonium acetate [2] and nitrate [3] salts have also been reported as appropriate solution components that allow the dissolution reaction at the anode. What is the most unexpected experience concerning the *metal cation–anion* compatibility is that silver can readily dissolve in the presence of a bromide salt [1], despite AgBr is notorious of its weak solubility.

The solvent seems to be related to the material dissolved. If the nanoparticles are made of gold (which is not sensitive to either water or dissolved oxygen), aqueous media are sufficient [5–7], although isopropanol is also commonly used as solvent [8, 9]. Occasionally, acetone was used as co-solvent, in particular when shape-modifying additive with large hydrophobic groups has to be applied. Acetonitrile proved to be suitable for Ag nanoparticles [1, 2], while an acetonitrile/tetrahydrofuran solvent mixture was applied to synthesize Pd nanoparticles [4]. Surprisingly, no organic solvent was necessary for Cu nanoparticle synthesis [3]. The number of reports is yet insufficient to clarify the role of the solvents accurately.

Since the cell configurations are optimized for the current density and the anode–cathode distance, the critical concentration of the metal ions that allow the formation of nanoparticles at the cathode has not been reported. Another piece of information missing is the cationic form produced in the dissolution when the metal may form various cations [like Au(I) versus Au(III) and Cu(I) versus Cu(II)]. The current density is between 0.1 and 25 mA cm^{-2}, regardless whether galvanostatic of potentiostatic control was applied, and the typical interelectrode distance was 5–20 mm. It was demonstrated for Pd [4], Ag [1] and Au [6] that the diameter of the nanoparticles decreases with the increase in current density. It has to be noted that the range of the mean particle size for Pd and Ag was between 1.4–6.0 nm, the gold particles were an order of magnitude larger (although the current density intervals were comparable). Since the current density is a monotonous function of the electrode potential, it is widely accepted that the current density dependence of the particle size is related to the potential dependence of the critical nucleus size. By applying the same considerations that were already discussed in Chap. 6, one can get that

$$r_{\mathrm{CRIT}} = \frac{2M\gamma}{zF\eta\rho} \tag{9.1}$$

where γ is the specific interfacial energy (assumed to be constant) and η is overvoltage (e.g., the potential as compared to the equilibrium potential of a macroscopic piece of metal of the same material as that of the nanoparticle in the same solution). The rest of the symbols have the same meaning as applied before. The equation shows that the current density dependence of the critical particle size can be understood, and substituting realistic data into Eq. 9.1 a semi-quantitative agreement with the particle size obtained can be seen [4]. Nevertheless, the standard deviation of the particle size distribution function is usually about 1/3 part of the mean particle size as determined from TEM measurements.

The above treatment has to be completed with at least four aspects that have not been mentioned concerning the studies published so far:

- The variation in the current density means that the concentration of the precursor metal ion concentration is also a function of the current density. Therefore, the overpotential of the metal deposition process also depends on the synthesis conditions applied, which makes the complete quantitative elucidation of the process rather difficult.
- It requires an explanation why the particles formed at higher current density (and, consequently, from solution of higher concentration) cannot grow, although the supply of the precursor materials would make it possible. For explaining the stop of the particles growth, we have to assume that the desorption of the particles formed at the cathode and its solubilization with the surfactant applied is relatively fast as compared to the further growth of the nuclei that achieved the critical size.

- The classical thermodynamic explanation that considers the energy of the bulk and the surface (which is the indirect consideration of the surface curvature) treats the material as a continuum. However, the particle size is often close to the size of the atoms, which means that the validity of the continuum treatment is by far not guaranteed. It was shown for Ag nanoparticle synthesis with optical methods that Ag_4^{2+} clusters play an important role in the particle formation [2], and Cu clusters formed from about 14 Cu atoms also proved to be stable [3]. It is clear that quantization effects, concerning both the incorporation of new atoms and the transfer of one electron, also play role in the particle formation.
- The classical treatment can be applied for nearly spherical particles. However, the growth of nanoparticles with a large aspect ratio [5, 7, 10, 11] would certainly require an explanation including kinetic terms with the involvement of the surface energy of various crystal faces, which is beyond the scope of the classical thermodynamic approach.

It is noteworthy that the micelle encapsulation approach used for Ni nanoparticle synthesis in an anode process was also reported [18]. The electrochemical setup applied in the latter study was very similar to those described for cathodic particle generation processes and made it possible to obtain Ni particles with about 5 nm mean diameter. Since a two-chamber cell was not used in either of the studies mentioned hitherto, it remains a question what is the role of the anodic and cathodic processes in the nanoparticle formation.

At last, an example for a completely adverse process is to be brought up in which the primary product of the electrode reaction is an anion which is oxidized in the solution to form nanoparticles. In this method, selenium particles were dispersed in a carbon paste electrode, and the reduction of selenium resulted in Se^{2-} ions [19]. The dissolved oxygen in the solution was the oxidation agent. The oxidation of the Se^{2-} ions led to either the production of elemental selenium directly or that of Se^{4+} species, depending on the concentration ratio of the reactant, and the Se^{4+} could also react with Se^{2-} in a synproportionation reaction. Nearly spherical Se nanoparticles with 43–85 nm diameter could be obtained this way in the presence of a suitable stabilizing agent.

9.1.2 Mediated Formation of Metal Nanoparticles with Electrochemical Generation of a Reducing Agent

The mediated nanoparticle formation is based on the electrochemical formation of at least one of the reactants of the redox reaction that leads to the nanoparticle formation. The indispensable electrochemical step of the mediated nanoparticle formation is the generation of a reducing agent on the cathode that later react with the metal ions in the solution. Although it is principally possible that the solution used contain a metal salt as solute and this configuration is common in lab-scale research, practical aspects dictate that the metal ions are generated in the anode reaction in large-scale

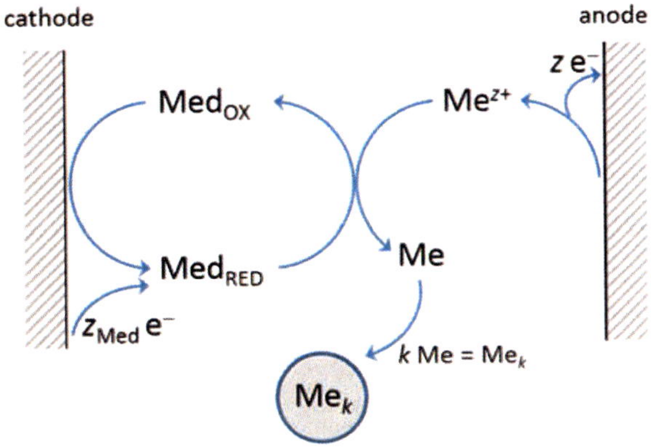

Fig. 9.1 Scheme of the mediated formation of metal nanoparticles from metal cations in the solution. If the Me^{z+} cations are present in the starting solution, too, the anode reaction can be omitted

processes. In the latter case, the reaction is nearly quantitative and no metal-rich waste is produced, which is important for green chemical processes. Another reason for the application of sacrificial anodes is that the reaction zone for nanoparticle formation can be separated from the cathode since the metal ions do not reach this electrode, hence eliminating metal deposition as a competing process on the cathode. The diagram of the mediated metal particle formation is shown in Fig. 9.1. Although the topic of electrochemically mediated metal nanoparticle production is a relatively new field with essentially all papers dated after 2014, a comprehensive review of the field is already available [20]. A contracted overview of the field will be given below.

The prerequisite of the mediated nanoparticle formation is that the reduction potential of the Med_{OX}/Med_{RED} couple has to be more negative than that of the Me^{z+}/Me couple, otherwise there is no driving force for the electron transfer reaction between Me^{z+} and Med_{RED}. Apart from the thermodynamic terms, kinetic parameters such as the overvoltage of both the metal deposition and the Med_{OX} reduction on the cathode as well as the nucleation barrier of the free metal particle in the solution determine how the reaction can proceed. If a capping agent or micelle can incorporate the metal particle being formed, the nucleation barrier decreases due to the stabilization of the product.

The tendency of the mediated nanoparticle formation can be efficiently studied with electrochemical method. Cyclic voltammetry with an appropriate pretreatment is the most commonly applied one and will also be used below to present some diagnostic criteria of nanoparticle formation in preliminary experiments. As shown in Fig. 9.2, a solution can be used for the preliminary test that contains the ions of the metal to be reduced and the oxidized form of the mediator. A solution with these reactants is stable and no reaction takes place. If a cyclic voltammetric experiment is run with a negative-going sweep first (no such curve is shown in Fig. 9.2), the first peak arising refers to the reduction of the metal ions to a metal coating (C1), and this is followed with the reduction peak(s) of the oxidized form of the mediator (C2 and so forth, corresponding to the reduction steps of the oxidized form of the mediator). This peak order corresponds to the thermodynamic requirement of a metal–mediator

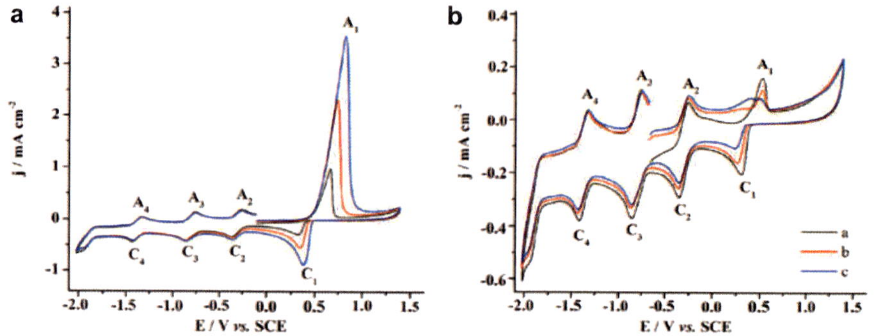

Fig. 9.2 Establishment of the mediated formation of metal particles that do not accumulate on the cathode by using cyclic voltammetry. Solvent: toluene-DMF (2:1 volume ratio), solutes: 0.1 M Bu_4NBF_4 as supporting electrolyte, 2 mM C_{60} as Med_{OX} and 1.5 mM $AgNO_3$ as metal precursor compound. The electrolysis during the pretreatment was performed at −0.10 V (**a**) and −0.66 V (**b**) neighbourhood of the cathode for a period of 5 (a, black), 60 (b, red) and 180 s (c, blue) in the increasing C1/A1 peak height order for graph A. All voltammograms were recorded at 0.1 V s^{-1} sweep rate by going to the anodic direction first. For further details, see the text. Republished with permission of The Electrochemical Society from [21]; permission conveyed through Copyright Clearance Center, Inc.

pair which is feasible for mediated nanoparticle formation. When the sweep direction is returned, the peaks appear in the reverse order, the last anodic peak corresponding to the dissolution of the metal layer (A1).

The test of the mediated nanoparticle formation in the solution is based on cyclic voltammograms recorded after holding the electrode at different potentials for various time intervals. If the pretreatment potential is selected so that the reduction of the metal ion can take place without the generation of the oxidized form of the mediator (Fig. 9.2a), the amount of the metal accumulated on the electrode increases with time, and the stripping peak is proportional to the charge passed through the system during the pretreatment. However, when the pretreatment potential is more negative than the (first) reduction potential of Med_{OX}, the dominant product is not a metal coating on the electrode. As Med_{OX} is reduced and Med_{RED} leaves the electrode surface, the reduction of the metal ions takes place in the solution before the metal cations can reach the electrode surface. Therefore, the accumulation of the metal coating on the electrode is reduced; moreover, the time dependence of the charge corresponding to the metal stripping peak becomes negligible (see Fig. 9.2b). This is because the metal ions in the close vicinity of the cathode can be reduced at the beginning of the pretreatment, but later this layer cannot grow any further because of the lack of the precursor material that is reduced rather in the solution by Med_{RED}.

Single-component metal nanoparticles produced by metal ion reduction with an electrochemically produced mediator comprise Ag [21–25], Pd [24, 26, 27], Au [24, 28–31], Pt [32] Co [33], Cu [34]. The production of bimetallic particles is possible if the reduction potentials of the precursor ions are close enough to each other so that one single mediator can be used. The synthesis of Ag–Pd nanoparticles in a wide composition range proved to be possible [35].

The mediators show a great variety, including both organic and inorganic compounds. Among organic compounds, methyl viologen [23, 27, 29, 31, 32] and its derivative, tetraviologen calix[4]resorcinol [22, 23, 26], are the most commonly applied ones, but fullerenes (either C_{60} or C_{70}) [21, 30] and anthracene [33] also appear as mediator. Concerning inorganic mediators, O_2 [24, 28] was applied first, and later Cr(III) and Co(III) triple-charged complexes such as $[Cr(bipy)_3]^{3+}$, $[Co(bipy)_3]^{3+}$ and $[Co(sep)]^{3+}$ were tested [25] (bipy = 2,2′-bipyridyl, sep = sepulchrate).

The solution has to be chosen in accord with the mediator to be used. Large organic molecules requires less polar solvents (N,N-dimethylformamide being the most common one), whereas processes mediated by metal complexes can take place in aqueous solutions.

Nanoparticles obtained with the electrochemically mediated method are mostly spherical. The diameter of the particles may vary even for the same metal depending on the synthesis conditions. Taking Ag as an example, particles obtained in a fullerene-mediated process are of 10–50 nm in diameter [21], the oxygen-mediated Ag particles have a size distribution between 20 and 100 nm [24], and inorganic Me(III) complexes as mediators result in the formation of Ag particles with mean diameter ranging between 40 and 150 nm [25]. The application of stabilizers (like polyvinyl pyrrolidone/PVP/) tentatively decreases the mean particle size [21]. The particle size observation strongly depends on the method used (SEM, TEM, DLS or AFM). An Ostwald ripening can be seen upon storage of particles for months in spite of the stabilizer present. A 60-day-long storage of PVP-stabilized Ag particles led to a doubling of the particle size as observed by DLS [24].

9.2 Metal Oxide Nanoparticles

9.2.1 General Aspects of the Electrosynthesis of Free Oxide Particles

Methods discussed in Sect. 9.2 all belong to the large family of the coagulation processes. Electrocoagulation is the method in which at least one of the reactants that later form a precipitate is produced in a heterogeneous electrochemical reaction, and the precipitate forms in a simple chemical reaction without a charge transfer step (which is a clear distinction condition from mediated reactions where charge transfer takes place between the dissolved reactants). The most common goal of the electrocoagulation methods is to produce a colloidal adsorbent with very large surface area that adsorbs pollutants that range from simple heavy metal ions to dyes in industrial waste. For industrial purposes, electrocoagulation setups involve a sacrificial anode that can be either Al or Fe for both economic and environmental affordability. On the cathode, hydrogen is evolved, which itself is not used but the alkalination of the medium provides the hydroxide ions in the system that makes colloidal precipitates with the metal cations. The precipitate is filtered out from the solution together with the pollutant that is adsorbed on the colloid surface, which means that the balance of the process is the input that includes the electrical current, the waste solution and some inexpensive metal, but the output is the purified solution and the concentrated waste.

In industrial electrocoagulation processes, both the particle size and its homogeneity are of secondary importance, while the adsorption capability of the colloidal particles is the key parameter to optimize. However, when electrocoagulation is used for synthetic processes to obtain nanoparticles, the optimization aims at the particle size and the size distribution. Since the resulting particles are often superparamagnetic, they are also named as superparamagnetic iron oxide nanoparticles (SPIONs). The forthcoming chapters will detail such processes. Since SPIONs are by far the most frequently produced materials with electrocoagulation methods, the various versions of electrocoagulation will be presented with Fe_xO_y particles as examples. The processes are partly analogous to various other coagulation methods that are summarized in [36] and those that lead to well-adherent coatings on electrode (see, e.g., [37–39]). Thereafter, processes will be just shortly mentioned for all other materials.

9.2.2 Cathodic Processes for Obtaining Iron Oxide Particles

The common feature of the "cathodic" processes of iron oxide nanoparticle preparation is that the solution used contains an iron salt already at the time of the start of the experiment. Therefore, the mean iron concentration of the bath is constant during the experiment also in the case when an iron (or steel) electrode is used as anode whose dissolution provides the replenishment of the solution with the ions that form the precipitate. Besides, the oxidation state of the dissolved Fe source can be taken as known.

Since the key reaction partner of the precipitate formation is the hydroxide ions, it is straightforward that aqueous solutions can be used [40–44], and the alkalization takes place in the neighbourhood of the cathode due to the hydrogen evolution. Even in the cases when ethanol is used as solvent, the solution must contain water in order to provide a source for hydroxide ions, originating either solely from the crystalline form of the iron salt applied [45, 46] or from the salt and the water content of the solvent together [47]. It is generally assumed that the precipitation leads to some iron hydroxide in the first step and the oxide particles are obtained after the decomposition of these particles by releasing water.

The cathodic electrosynthesis of iron oxide nanoparticles is very divergent from the viewpoint whether the formation of the final product takes place by a simple precipitation reaction (possibly combined with the release of some water) or some further oxidation/reduction reaction has to be taken into account. The list below makes an attempt to categorize the *source–product permutations* with the above viewpoints:

- *Fe_2O_3 particles obtained from an Fe(III) solution* [40]. The method described made use of $FeCl_3$ as the iron source with a concentration range of 0.01–0.1 M. The current density applied ranged from 0.15 to 2 A cm^{-2}. The product composition was Fe_2O_3, both the diffraction pattern and XPS line shape of which matching closely to maghemite. Despite the vigorous hydrogen evolution on the cathode, no Fe(II) species were found in the precipitate, which is probably due to fact that the solubility of $Fe(OH)_3$ is much smaller than that of $Fe(OH)_2$.
- *Fe_3O_4 particles from Fe(III) solution* [45, 46]. This method involves the application of an ethanol solution with $Fe(NO_3)_3{\cdot}9H_2O$ as solute and requires the very accurate adjustment of the bath operation parameters to obtain magnetite nanoparticles. The salt concentration was found to be critical which has an optimum value (~0.04 M) above which the increase in the concentration of both Fe(III) and H_2O leads to the formation of an amorphous non-magnetic precipitate instead of magnetite nanoparticles. Also, a current density larger than 10 mA cm^{-2} was required to obtain nanoparticles with a current efficiency nearly 100%. In the case when the current efficiency was close to 1, the process was formulated with the following equation [46]:

$$3Fe^{3+} + 2NO_3^- + 2H_2O + 9e \rightleftarrows Fe_3O_4 + 2NO_2^- + 2H_2 \qquad (9.2)$$

  which means that the reduction of Fe(III) to Fe(II) is a part of the particle formation process at the cathode (in other words, no Fe_2O_3 particles have to form that might later be reduced to Fe_3O_4). It was the surprising experience of the magnetite synthesis in ethanol that no additional stabilizer was necessary to obtain monodispersed particles with diameter in the 4.5–8.9 nm range.
- *Fe_3O_4 particles from Fe(II) solution* [41]. It was found that the particles formed from aqueous $Fe(ClO_4)_2$ solution in the neighbourhood of the cathode have the composition of Fe_3O_4. The key cathode process is the hydrogen evolution, and the formation of the precipitate is accompanied with the oxidation of Fe(II) species by dissolved oxygen. Since the anode used was Fe, it can be ruled out that the occurrence of Fe(III) species is the anode reaction since it leads to Fe(II) formation only. The size range of the Fe_3O_4 particles produced in the absence of any stabilizing agent was 15 ± 6 nm.
- *Fe_3O_4 particles from solution or stoichiometric Fe(II)–Fe(III) mixtures* [42–44, 47]. The methods applying Fe(II)–Fe(III) mixed solution corresponding the stoichiometry of the desired particles (1:2) makes it unnecessary to combine the coagulation process with any redox transformation of the metal ions themselves. Therefore, the key process is the hydroxide ion formation at the cathode and the forthcoming precipitation in the solution with the release of water to achieve the desired Fe_3O_4 particle composition:

$$Fe^{2+} + 2Fe^{3+} + 8OH^- \rightleftarrows Fe_3O_4 + 4H_2O \qquad (9.3)$$

The Fe(II) compound used was always $FeCl_2{\cdot}4H_2O$, while both $FeCl_3{\cdot}6H_2O$ [44] and $Fe(NO_3)_3{\cdot}9H_2O$ [42, 43, 47] were suitable Fe(III) compounds. For the reaction medium, ethanol-based [47] and aqueous [42–44] solutions were applied.

In addition to the synthesis of the nanoparticles themselves, various stabilizing agents were also tested in the above-mentioned works. In each study, a clear difference was found between the size distribution of the particles synthesized in additive-free and stabilizer-doped solution and, surprisingly, the stabilized particles were always larger. An example for the stabilizer-induced size change is shown in Fig. 9.3.

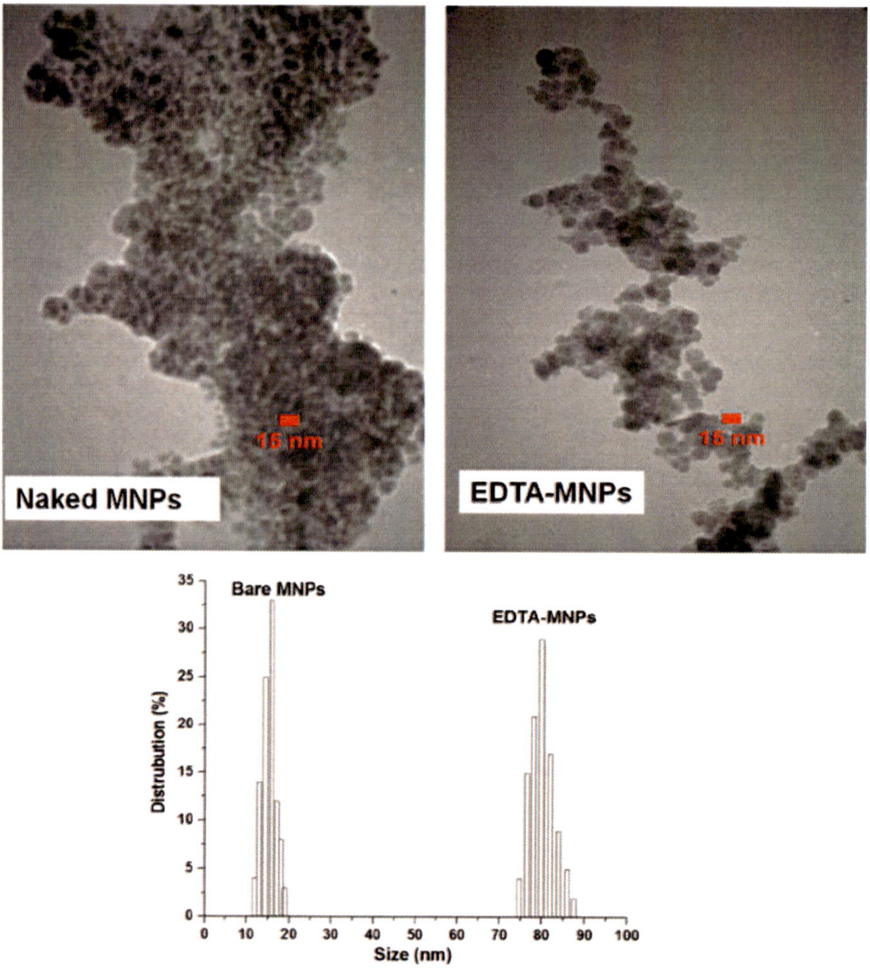

Fig. 9.3 TEM images of bare and EDTA-coated iron oxide nanoparticles (top) and the characteristic particle size distribution of the two sorts of particles (bottom). Reprinted from [42] Copyright (2017), with permission from Elsevier

The mean diameter of the non-stabilized particles varied between 16 and 19 nm, while the size increment upon the application of additives exceeded much the effect that could be explained merely by the molecular size of the stabilizing agent that may coat the particles. The mean diameters of the coated particles were 42 nm (additive: polyethylene glycol [47]), 82 nm (EDTA [42]), 52 nm (dextran [43]), 64 nm (glucose [44]) and 70 nm (dextran [44]). At the same time, the X-ray diffraction patterns revealed a mean crystallite size below 10 nm for both bare and coated particles, indicating that the particles seen with TEM or DLS may be composed of smaller grains. Nevertheless, the large particle size increment with the application of additives raises questions on the generally accepted mechanism of the particle stabilization by hindering the growth process.

9.2.3 *Iron Oxide Synthesis with Sacrificial Anodes*

In Sect. 9.2.2, systems were discussed in which the cathode process plays a key role in the nanoparticle synthesis since the metallic precursor species were present in the solution as dissolved compounds, even though in some cases a sacrificial Fe anode was applied. This chapter deals with the processes in which the only source of iron in the system is the anode reaction.

If the both the cations and the anions needed for the coagulation are provided in the anode and cathode reactions, respectively, the optimization of the setup requires much attention. The zone of the particle formation is the interelectrode space where the oversaturation of the solution is the highest. No wonder, studies dealing with the anodic synthesis of iron oxide particles all report that the fine tuning of the interelectrode distance (typically 1–2 cm) and the current density are crucial for the successful nanoparticle synthesis [48–54]. Following a similar classification to that applied for the cathodic synthesis, various pursuits can be distinguished as presented below:

- *Synthesis of Fe_2O_3 particles* [48]. The method based on a DMF–water mixture as this solvent is the only one in which Fe(III) species are formed directly in the anode process. The water content was limited to at most 5 vol.%. While the high water content may decrease the stability of the Fe(III) species formed on the anode, water itself is indispensable for the OH^- formation on the cathode. The supporting electrolyte was 0.05–0.15 M tetraoctylammonium bromide. Although it was not mentioned in the original work, the bromide ions may play a role in the solvation of Fe(III) species in the solution. Oxidation of bromide ions to bromine was not discussed. The absence of Fe(II) species in the solution was checked with a redox titration with chromate ions. Since the polarization behaviour was not reported, it cannot be assessed whether the oxidation of Fe takes place in a step-wise process with any soluble Fe(II) species that may be stable in a moderately positive potential interval. The diameter of the particles decreased as a result of the increase in current density, varying between about 3.5 and 7.5 nm. Since the formation of the particles does not take place on the electrode surface in an electrochemically assisted nucleation process, Eq. 9.1 is clearly not applicable here. The explanation given by the authors for the current density dependence of the particle size refers to the diffusion length of the ions and its relation to the thickness of the Helmholtz layer. This explanation was not elaborated quantitatively and neglects the fact that anions are assumed to accumulate in the vicinity of the anode.

- *Synthesis of* Fe_3O_4*particles* [49–54]. Although the goal of the synthesis modes discussed in this section was to obtain Fe_3O_4, the formation of either Fe_2O_3 or FeOOH as a competing reaction pathway was also evidenced [49, 51, 52, 54]. The formation of a product mixture itself indicates the complexity of the processes. Since the synthesis conditions show a great variety, no unambiguous reaction mechanism can be given. The spectrum of solutions used includes pure water [52], aqueous solutions with various supporting electrolytes and complexing agents [49, 50, 52, 54] and water–ethanol mixtures [51, 53]. The current density applied scales somewhat with the conductivity of the solution: while it was smaller than 1 mA cm^{-2} for pure water, values between 2 and 200 mA cm^{-2} were chosen for solutions containing a supporting electrolyte (the larger current density values typically belong to larger interelectrode distances). The particle size varied between 5 and 40 nm. In contrast to several experiments discussed above, the increase in the current density led to an increase in the particle size [52, 54], which is yet to be explained. Mössbauer spectra indicated that the particles obtained cannot be identified as a pure phase of either magnetite of maghemite.

Since the primary goal of the papers discussed so far was the synthesis and the characterization of nanoparticles, the systematic change of the experimental parameters did not make it possible to conclude a fully feasible particle formation mechanism. The anodic polarization curve of iron in various water–ethanol mixtures indicated [51] that the working regime of the sacrificial anode falls to the potential region where Fe(III) species are stable. This was called the "transpassive region", even though the experiment was performed in the presence of LiCl as supporting electrolyte which promotes iron dissolution instead of allowing the formation of a passive oxide layer. The possibility of the Fe(III) species in the dissolution of the sacrificial anode indicated that an additional oxidation process is not necessary for the formation mechanism of either Fe_3O_4 or Fe_2O_3 particles. Rather, the formation of oxide particle mixtures is a natural consequence of the experimental conditions.

Concerning the reaction mechanism of formation of various iron oxide particles, those works can be recommended that aimed at the scrutiny of the particle formation mechanism by systematically changing a large number of experimental conditions and performing *in situ* observations [55, 56]. The authors of these works concluded that the $Fe(OH)_2$ formation is the initial step of the oxide particle formation, and an oxidation steps follows with the participation of dissolved oxygen. This is partly contradictory to the result of the works in which oxide particle formation in deaerated solutions were observed. In spite of the large number of unresolved controversies, the occurrence of two processes could be convincingly excluded: (i) reduction of Fe(III)-containing particles by hydrogen evolved on the cathode in the formation of

Fe_3O_4; (ii) direct oxidation or reduction of the precipitated particles on either of the electrodes. The variety of the experimental conditions used makes it natural that no uniform picture can be outlined on the anodic iron oxide particle formation process.

9.2.4 Co-precipitation of Mixed Magnetic Oxide Particles Containing Iron

The co-precipitation method is based on similar processes to the iron oxide nanoparticle formation, except for that the solution contains the ions of other metallic elements. Since the solubility of $Fe(OH)_2$ and $Fe(OH)_3$ is smaller than that of the hydroxide of the other metals present, these additives do not precipitate in their simple oxide/hydroxide phase but co-precipitate with iron oxide only. The metal beside iron (that all are present in the bath as Me^{2+} cations) can be Sr [57, 58], Mn [59], Zn [59] and Co [58], either as a single additive or a part of an additive mixture to form ternary oxides [58, 59]. The element combination of choice is rationalized to obtain magnetic nanoparticles that are similar in composition to hard magnetic ferrites.

Two solution types have been elaborated for the synthesis of ferrate-like nanoparticles. The suitable pH range of the bath based on $Sr(NO_3)_2$ and $Fe(NO_3)_3$ was 1–3, while the solution containing chloride salts could be operated between pH 1 and 12. Ternary oxides were obtained from mixed chloride-nitrate baths. While a relatively large Sr^{2+} concentration could be used in these baths ($c = 0.1–1.5$ M), the concentration of other metal cations was always $\leq$60 mM. Although the formation process of the binary and ternary oxides is described in all relevant papers as if the cations other than iron entered the iron-containing precipitate after the formation of the particles, this view has not been strictly evidenced. It is unlikely that a solid-state diffusion might lead to the mixed multielement oxide particles (if so, the process should also work *ex situ*, but no such results are available). Therefore, it is the opinion of the author of this monograph that the formation of the mixed oxide nanoparticles is a simultaneous co-precipitation of the constituents. For stabilizing the particles, an organic stabilizer (aniline) was applied in a concentration of a few hundred ppm. Concerning other synthesis conditions, the temperature range was 20–80 °C and the upper limit of the current density applied was 250 mA cm^{-2}.

The metal ratio in the nanoparticles varied in a wide range. In $SrFe_wO_z$ particles, iron was always the predominant metallic component, and Sr proved to be nearly an impurity with $12 < w < 400$ [57]. In contrast, for $MnZn_iFe_jO_k$ in occurred that Mn was the majority component with $i = 0.6$ and $j = 0.4$ [59]. The comparison with the solution composition from which the $MnZn_iFe_jO_k$ particles were obtained tells that Zn has a stronger doping ability than Mn. The data available do not make it possible to establish clear trends, especially since the materials collected from the anode, from the cathode and from the solution were of dissimilar composition even in the same experiment. For all compositions reported, the particle size was in the range from 17 to 32 nm.

The structure of the nanoparticles was in accord with their composition. When iron was the majority component, the X-ray diffractograms differed from that of magnetite and maghemite to a negligible extent only. However, when Mn and Zn were the major metallic element in the particles, the diffraction lines could be found nearly at the same position as for the corresponding $MeFe_2O_4$ ferrite. Magnetic properties of the particles were similar to the corresponding ferrites, whereas some difference was found in their saturation magnetization and remanence values as compared to their bulk counterparts. Due to the small particle size, some samples were superparamagnetic.

9.3 Electrochemically Assisted Synthesis of Miscellaneous Non-metallic Nanoparticles

ZnO nanoparticles were synthesized in a similar manner as iron oxide nanoparticles. Concerning the formation of ZnO and its stability in comparison with $Zn(OH)_2$, Chap. 8.3.2 already gave abundant information. In both works reported here [60, 61], Zn^{2+} ions originated from the anode reaction of the sacrificial Zn electrode. The supporting electrolytes were totally different in the synthesis methods applied. While oxalic acid proved to be suitable in spite of its high buffering ability due to the step-wise acidic dissociation [60], tetrabutyl ammonium bromide could act as both supporting electrolyte and stabilizer [61]. The variation of the solvent revealed the flexibility of the precipitation-based ZnO synthesis method, as shown in Fig. 9.4. Pure ethanol as solvent led to a porous structure with nearly spherical grains. As the water content of the solvent was increased, the porosity as measured by the specific surface area strongly decreased, the grain size increased and the shape of the particles became more and more rhombohedral. In parallel to the morphological change, $Zn(OH)_2$ also appeared beside ZnO in the deposit when the solvent was pure water. The ZnO structure was highly conformal to the wurtzite structure with a good crystallinity as shown by the X-ray diffractograms. TEM images revealed that the grains having a diameter of a few hundred nm were made of nearly spherical grains of 25–30 nm. This subgrain structure may indicate that the particles primarily produced in the precipitation reaction correspond to the spherical subgarins and the larger particles formed as a result of the seizure of the primarily formed particles.

In the synthesis of CuS particles [62], the products of the electrode reactions are Cu^{2+} at the sacrificial Cu anode and OH^- ions at the cathode. The peculiar feature of this synthesis mode is that the hydroxide ions produced on the cathode are not a reactant but a catalyst. The solution component that provides sulphur for the formation of CuS is the thiosulfate ion whose decomposition is facilitated by two other components of the solution. One of these components is the hydroxide ion produced at the cathode, and the other is the thioglycerol which is also a solution component which is nonreactive but act as a catalyst. At relatively large thioglycerol concentration (0.1 M), the primary reaction product is a metastable form of CuS with 8–12 nm

Fig. 9.4 ZnO particles synthesized from various media by the electrochemically assisted precipitation method. SEM images: solvent is **a** ethanol, **b** water–ethanol mixture (7:3), and **c** water. **d** TEM image of a particle synthesized in the water–ethanol mixture, revealing the internal spherical subgrains of the rombohedral particle. Reprinted from [61]. Copyright (2017), with permission from Elsevier

particle diameter that undergoes a transformation with time and larger nanoparticles and nanorods are formed with 30 diameter and 100 length. The application of more dilute thioglycerol solution (0.01 M) leads directly to the spherical particles with a diameter of about 50 nm and a crystalline form corresponding to a covellite-type hexagonal phase.

References

1. Rodríguez-Sánchez L, Blanco MC, López-Quintela MA (2000) J Phys Chem B 104:9683–9688
2. Rodríguez-Sánchez ML, Rodríguez MJ, Blanco MC, Rivas J, López-Quintela MA (2005) J Phys Chem B 109:1183–1191
3. Vilar-Vidal N, Blanco MC, López-Quintela MA, Rivas J, Serra C (2010) J Phys Chem C 114:15924–15930
4. Reetz MT, Helbig W (1994) J Am Chem Soc 116:7401–7402
5. Yu YY, Chang SS, Lee CL, Wang CRC (1997) J Phys Chem B 101:6661–6664
6. Huang CJ, Chiu PH, Wang JH, Chen KL, Linn JJ, Yang CF (2006) J Electrochem Soc 153:D193–D198

7. Huang CJ, Chiu PH, Wang JH, Chen WR, Meen TH, Yang CF (2006) Nanotechnol 17:5355–5362
8. Huang CJ, Wang JH, Chiu PH, Lin JJ (2006) Chem Lett 35:30–31
9. Huang CJ, Chiu PH, Wang YH, Yang CF, Feng SW (2007) J Coll Interf Sci 306:56–65
10. Mohamed MB, Wang ZL, El-Sayed MA (1999) J Phys Chem A 103:10255–10259
11. Huang CJ, Chiu PH, Chen MD, Meen TH (2005) Jap J Appl Phys 44:L941–L944
12. Huang CJ, Chiu PH, Wang JH, Chen WR, Meen TH (2006) J Electrochem Soc 153:D129–D133
13. Cougnon C, Simonet J (2001) Electrochem Commun 3:209–214
14. Cougnon C, Simonet J (2001) J Electroanal Chem 507:226–233
15. Simonet J (2003) Electrochem Commun 5:439–444
16. Ghilane J, Guilloux-Viry M, Lagrost C, Hapiot P, Simonet J (2005) J Phys Chem B 109:14925–14931
17. Simonet J (2005) Platinum Met Rev 50:180–193
18. Pachón LD, Thathagar MB, Hartl F, Rothenberg G (2006) Phys Chem Chem Phys 8:151–157
19. Ye X, Chen L, Liu L, Bai Y (2017) Mater Lett 196:381–384
20. Yanilkin VV, Nasretdinova GR, Kokorekin VA (2018) Russ Chem Rev 87:1080–1110
21. Yanilkin VV, Fazleeva RR, Nasretdinova GR, Nastapova NV, Osin YN (2018) ECS J Solid State Sci Technol 7:M55–M62
22. Nasretdinova GR, Fazleeva RR, Mukhitova RK, Nizameev IR, Kadirov MK, Ziganshina AY, Yanilkin VV (2015) Electrochem Commun 50:69–72
23. Nasretdinova GR, Fazleeva RR, Mukhitova RK, Nizameev IR, Kadirov MK, Ziganshina AY, Yanilkin VV (2015) Russ J Electrochem 51:1029–1040
24. Yanilkin VV, Nastapova NV, Fazleeva RR, Nasretdinova GR, Sultanova ED, Ziganshina AY, Gubaidullin AT, Samigullina AI, Evtyugin VG, Vorobev VV, Osin YN (2018) Russ J Electrochem 54:265–282
25. Yanilkin VV, Fazleeva RR, Nastapova NV, Nasretdinova GR, Gubaidullin AT, Berezin NB, Osin YN (2018) Russ J Electrochem 54:650–664
26. Yanilkin VV, Nasybullina GR, Ziganshina AY, Nizamiev IR, Kadirov MK, Korshin DE, Konovalov AI (2014) Mendeleev Commun 24:108–110
27. Nasretdinova GR, Osin YN, Gubaidullin AT, Yanilkin VV (2016) J Electrochem Soc 163:G99–G106
28. Yanilkin VV, Nastapova NV, Nasretdinova GR, Fazleeva RR, Osin YN (2016) Electrochem Commun 69:36–40
29. Yanilkin VV, Nastapova NV, Nasretdinova GR, Fedorenko SV, Jilkin ME, Mustafina AR, Gubaidullina AT, Osin YN (2016) RSC Adv 6:1851–1859
30. Yanilkin VV, Nastapova NV, Nasretdinova GR, Osin YN, Gubaidullin AT (2017) ECS J Solid State Sc 6:M19–M23
31. Yanilkin VV, Nastapova NV, Nasretdinova GR, Osin YN (2017) Mendeleev Commun 27:274–277
32. Yanilkin VV, Nastapova NV, Nasretdinova GR, Fazleeva RR, Fedorenko SV, Mustafina AR, Osin YN (2017) Russ J Electrochem 53:509–521
33. Yanilkin VV, Nasretdinova GR, Osin YN, Salnikov VV (2015) Electrochim Acta 168:82–88
34. Kokorekin VA, Gamayunova AV, Yanilkin VV, Petrosyan VA (2017) Russ Chem B+ 66:2035–2043.
35. Nasretdinova GR, Fazleeva RR, Osin YN, Evtjugin VG, Gubaidullin AT, Ziganshina AY, Yanilkin VV (2018) Electrochim Acta 285:149–163
36. Ramimoghadam D, Bagheri S, Hamid SBA (2014) J Magn Magn Mater 368:207–229
37. Yousefi T, Davarkhah R, Golikand AN, Mashhadizadeh MH, Abhari A (2013) Prog Nat Sci Mater 23:51–54
38. Meng Q, Wang Z, Chai X, Weng Z, Ding R, Dong L (2016) Appl Surf Sci 368:303–308
39. Elrouby M, Abdel-Mawgoud AM, El-Rahman RA (2017) J Mol Struct 1147:84–95
40. Park H, Ayala P, Deshusses MA, Mulchandani A, Choi H, Myung NV (2008) Chem Eng J 139:208–212
41. Gopi D, Thameem Ansari M, Kavitha L (2016) Arab J Chem 9:S829–S834

42. Aghazadeh M, Karimzadeh I, Ganjali MR (2017) J Magn Magn Mater 439:312–319
43. Aghazadeh M, Karimzadeh I (2017) J Nanoanalysis 4:228–238
44. Karimzadeh I, Aghazadeh M, Ganjali MR, Norouzi P, Doroudi T, Kolivand PH (2017) Mater Lett 189:290–294
45. Marques RFC, Garcia C, Lecante P, Ribeiro SJL, Noé L, Silva NJO, Amaral VS, Millán A, Verelst M (2008) J Magn Magn Mater 320:2311–2315
46. Ibrahim M, Serrano KG, Noe L, Garcia C, Verelst M (2009) Electrochim Acta 55:155–158
47. Karimzadeh I, Dizaji HR, Aghazadeh M (2016) Mater Res Express 3:095022(1–11)
48. Pascal C, Pascal JL, Favier F, Elidrissi Moubtassim ML, Payen C (1999) Chem Mater 11:141–147
49. Franger S, Berthet P, Berthon J (2004) J Solid State Eletrochem 8:218–223
50. Cabrera L, Gutierrez S, Menendez N, Morales MP, Herrasti P (2008) Electrochim Acta 53:3436–3441
51. Starowicz M, Starowicz P, Zukrowski J, Przewoźnik J, Lemański A, Kapusta C, Banas J (2011) J Nanopart Res 13:7167–7176
52. Fajaroh F, Setyawan H, Widiyastuti W, Winardi S (2012) Adv Powder Technol 23:328–333
53. Marín T, Ortega D, Montoya P, Arnache O, Calderón J (2014) J Appl Electrochem 44:1401–1410
54. Manrique-Julio J, Machuca-Martinez F, Marriaga-Cabrales N, Pinzon-Cardenas M (2016) J Magn Magn Mater 401:81–85
55. Lozano I, Casillas N, Ponce de Léon C, Walsh FC, Herrasti P (2017) J Electrochem Soc 164:D184–D191
56. Montoya P, Marín T, Mejía S, Arnache O, Calderón JA (2017) J Electrochem Soc 164:D1056–D1065
57. Amigó R, Asenjo J, Krotenko E, Torres F, Tejada J, Brillas E (2000) Chem Mater 12:573–579
58. Asenjo J, Amigó R, Krotenko E, Torres F, Tejada J, Brillas E, Sardin G (2001) J Nanosci Nanotechno 1:441–449
59. Torres F, Amigó R, Asenjo J, Krotenko E, Tejada J, Brillas E (2000) Chem Mater 12:3060–3067
60. Anand V, Srivastava VC (2015) J Alloy Compd 636:288–292
61. Dhayagude AC, Nikam AV, Kapoor S, Joshi SS (2017) J Mol Liq 232:290–303
62. Yang YJ, Hu S (2008) J Solid State Electrochem 12:1405–1410

Part III
Nanostructured Materials Obtained by Using Structured Substrates and Other Special Electrode Arrangements

Chapter 10
Electrochemical Manufacturing Methods Based on Surface Inhomogeneities at the Nanoscale

10.1 Pulse Sequences for the Regulation of Nucleation, Growth and Post-deposition Treatment

When electrodeposition is used for producing nanoscale objects on foreign surface, distinct objects can be obtained in the cases when the deposit does not form a uniform layer at the substrate surface at the atomic scale. This means that the nucleation and growth process is mostly of the Volmer–Weber type (see also Sect. 2.10.1). In order to regulate the relative importance of the nucleation and growth rates and to achieve a sufficient areal density of active nucleation sites, the application of special pulse sequences may be necessary. A schematic overview of the possible pulses is outlined in Fig. 10.1. Concerning the selection of the pulse potential values and their interrelation with the cyclic voltammograms, a review of the topic can be recommended [1]. The pulses listed therein are not applied in each single deposition process, but some of them are related to a particular substrate type or post-deposition sample treatment procedure, which will be discussed below.

The first pulse in Fig. 10.1 is an anodic pulse (A). This step is mostly applied for pre-conditioning of the substrate. It is common when highly oriented pyrolytic graphite (HOPG) is used as substrate in aqueous solutions. It is a common observation that at highly oxidized step edges at the end of the basal plane of the graphene layers, the nucleation of metals takes place with smaller activation energy than without this anodic step. Nevertheless, it is ill-defined which kind of terminal functional groups are produced with such a treatment. It is rather assumed that the terminal groups are the same as those produced in the oxidation process of graphite, like ≡C–OH, = CO, –CHO and –COOH. Although these terminal groups can be partly reduced in a forthcoming nucleation pulse, their reduction takes place typically at a longer time scale than the metal nucleation on these active centres; therefore, the oxidative treatment can provide favourable nucleation conditions for metal deposition for a sufficiently long time period. The oxidative pre-treatment step seldom occurs for metal-on-metal deposition processes since the metallic state of the substrate is always a more favourable nucleation environment for another metal than a surface

L. Péter, *Electrochemical Methods of Nanostructure Preparation*,
Monographs in Electrochemistry, https://doi.org/10.1007/978-3-030-69117-2_10

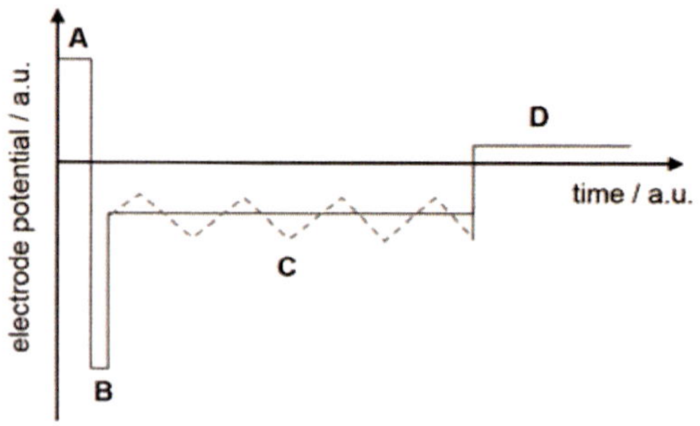

Fig. 10.1 Scheme of the possible pulses for the electrodeposition of nanocrystals and nanowires laying on an electrode surface. A: Anodic pulse for substrate conditioning. B: Nucleation pulse at high cathodic overvoltage. C: Growth period typically with negligible nucleation rate. Potential cycling is applied if compound semiconductor is deposited and periodic stripping of the excess deposit is required for achieving a desired composition (dashed line). D: Post-deposition treatment. Remember that some of the pulses, mostly A and D, are not applied in all electrochemical nucleation and growth methods. The relative length of the pulses may vary

oxide layer. When the substrate is a noble metal, the anodic pre-treatment can be applied for desorbing spontaneously adsorbed adatoms (e.g., a partial UPD layer) and to achieve an adsorbed anion layer, hence modifying the nucleation conditions for the subsequent deposition step [2].

Pulse B in Fig. 10.1 represents a so-called nucleation pulse. This pulse leads to the formation of a relatively large number of nuclei within a short time period. The large cathodic overvoltage applied in this pulse helps to overcome the nucleation barrier of the deposition on a foreign substrate, but the limited-time polarization at a high overvoltage prevents the occurrence of a significant solvent decomposition. Since the polarization in the growth pulse (C in Fig. 10.1) is rather mild, the solvent decomposition is limited to the nucleation pulse. When the pulse pair for nucleation and growth is applied (B and C in Fig. 10.1), the deposition process *as a whole* is typically instantaneous, whichever nucleation mode would be valid without the nucleation pulse. The explanation is that the nuclei already formed in a large number act as growth centres for the rest of the process, hence contributing to the decrease of the reactant concentration in the uncovered part of the substrate, which diminishes the probability of the occurrence of further nucleation events. A consequence of the pulse combination is that the size distribution of the nuclei at the surface is narrower than for progressive nucleation processes, which is a beneficial factor in the design of nanoparticle synthesis with a desired diameter range.

The more negative the applied nucleation potential is, the larger nucleation densities can be achieved. When the growth is mass transport-limited during the upcoming deposition pulse and the deposition is carried out with a fixed deposition time, the particle surface density and the particle size will be roughly inversely proportional to each other since the same amount of material is distributed among the particle assemblies composed of different numbers of particles. An example for the systematic variation of the particle size and particle density is shown in Fig. 10.2.

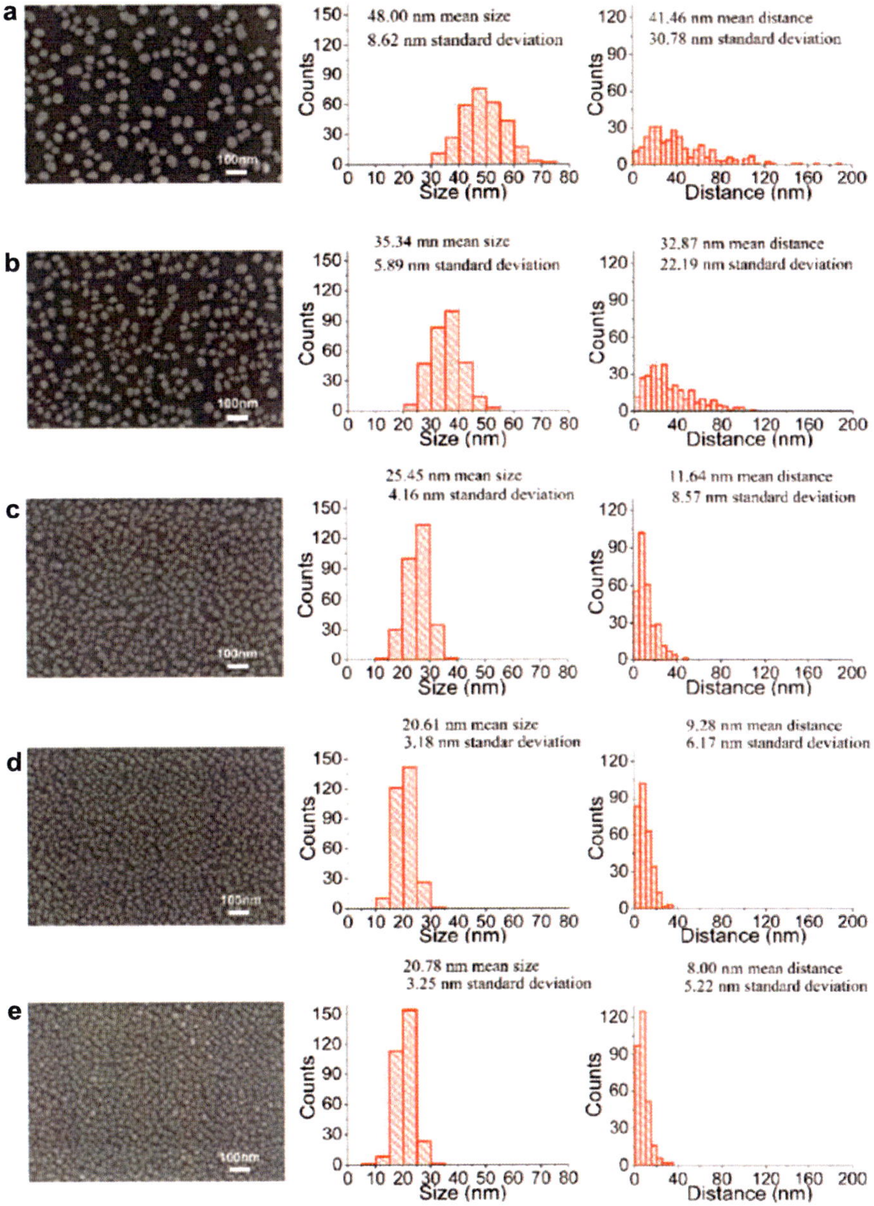

Fig. 10.2 Ag particles deposited onto ITO substrate with various nucleation potentials: **a** −0.6 V, **b** −0.8 V, **c** −1.0 V, **d** −1.2 V and **e** −1.4 V followed with a constant growth time of 3600 s with growth potential of −0.2 V (all potentials referred to SCE). The graphs show the corresponding diameter distribution (left) and interparticle distance histograms (right). Reprinted from [3]. Copyright (2011), with permission from Elsevier

Finally, pulse D in the above scheme is highly optional. It is applied when a post-treatment is indispensable, like for the electrochemical oxidation of a metal to a metal oxide.

It has to be noted that the role of the surface inhomogeneity is not always evidenced directly. The indirect evidence for the role of the importance of the site-specific nucleation is the immiscibility of the substrate material with the deposit, their lattice mismatch and the resulting particle formation. Nevertheless, nanoparticles can be produced with many substrate–deposit pairs with small enough deposition times.

10.2 Electrodeposition of Nanocrystals

10.2.1 General Aspects of the Electrochemical Nucleation of Nanoparticles

Whenever the Volmer–Weber-type nucleation and growth process takes place, nanocrystals can be deposited onto the electrode surface, provided that the deposition process is short enough and the particles formed do not coalesce. Although the role of surface defects in the nucleation process often remains hidden, there are a large number of studies for substrate–deposit pairs that reveal which cases are to be considered as inhomogeneity-influenced deposition processes. In most of the cases, the Volmer–Weber-type nucleation process involves incommensurable atomic distances of the substrate and the deposit.

The direct evidence for the influence of the surface inhomogeneity can be revealed by the study of the atomic-scale surface structure of the substrate surface (e.g., with STM). This is by far not always affordable. An indirect evidence for the role of some specific surface position in the nucleation process is that during repeated experiments with the same substrate, the deposit nuclei form on identical positions. This method can be applied for deposits only that can be removed from the surface without modifying the substrate, either by anodic dissolution or with a chemical etching method. However, in this case one cannot exclude that the nucleation itself causes some irreversible change in the substrate. Beside this uncertainty, the repeated nucleation method also requires the imaging of the surface at the nanoscale.

If the goal is the deposition of monodispersed particles on a surface, it is indispensible to make the nucleation and growth process quasi-instantaneous. This is carried out by the application of a nucleation pulse (pulse B in Fig. 10.1). As it was demonstrated for a large number of metals deposited as random nanoparticle assemblies on HOPG surface, a nucleation overvoltage of −500 mV for 5 ms and a growth overvoltage of −5 to −300 mV (depending on the metal to be deposited) can lead to a nearly monodisperse particle diameter distribution [4]. Note that no anodic activation was applied here for the HOPG substrate and the deposit formed particles on the HOPG basal plane (unlike in anodically activated processes where the preferred nucleation spots were the step edges; see Sect. 10.3).

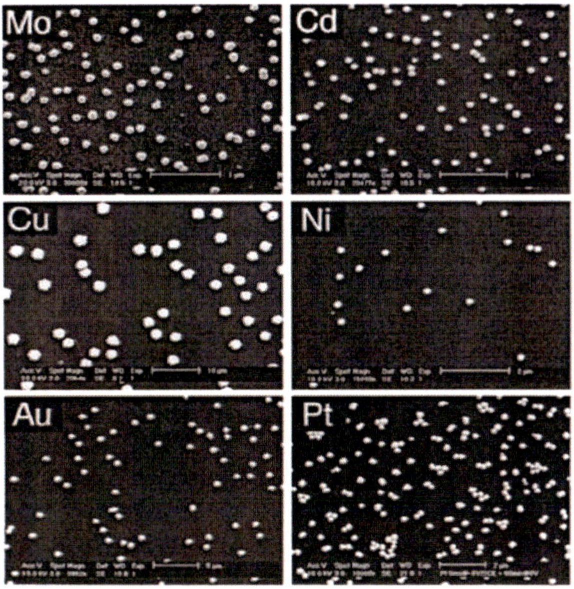

Fig. 10.3 SEM images of metal particles obtained with the slow growth mode on freshly cleaved HOPG surfaces. Reprinted from [4]. Copyright (2001), with permission from Elsevier

In [4], the growth overvoltage applied was established so that the metal deposition took place in the so-called *slow growth mode* with nearly identical current during the deposition (c.f. the method applied for nucleation studies where the deposition is *diffusion controlled* throughout the entire process, as shown in Sect. 2.10). The slow growth differs from simple kinetic control when the current is proportional to the surface area and increases with time as the particles grow. However, the constant-current deposition at small overpotential leads to an $r \propto t^{1/3}$ time dependence of the particle radius. This also means that the initial radius distribution is narrowing during the growth process, which contributes to the achievement of the monodispersity of the deposit. Scanning electron micrographs of the particle assemblies of different metals are shown in Fig. 10.3.

10.2.2 Substrate–Nanoparticles Pairs and Deposition Conditions for Nanoparticle Preparation

The field of nanoparticle deposition is very rich; therefore, a few representative examples are highlighted here only with discussing some trends in the experimental conditions of high relevance. One typical substrate for electroplating metallic nanoparticles is the group of the carbonaceous materials, including natural graphite [5], HOPG, glassy carbon [6], boron-doped diamond [7] and various forms of graphene oxide [8]. Carbonaceous materials as substrate are typical in electroanalytical studies when the surface decoration with nanoparticles is essential to improve the electrocatalytic properties. For the same reason, catalysts in fuel cells are often composed of a carbon substrate partly coated with noble metal particles. Another typical substrate is indium tin oxide (ITO) that is commonly used for depositing either metallic [3, 9–11] or semiconductor [12–14] nanoparticles. Whenever an optical study is aimed at, the application of ITO is very convenient since it is transparent and has nearly negligible interference with the light absorption or scattering properties of the particles studied. Metallic substrate, although they are less inert than either carbon-type or ITO substrates, can be used if the deposit is less noble than the substrate [11, 14].

The concentration of the metal ions in the solution is mostly in the range of 0.05–10 mM (the more noble is the metal used, the concentration is usually the smaller). The reason for the small metal ion concentration is that the probability of the nucleation is proportional, among others, to the precursor ion concentration. Therefore, the small precursor ion concentration leads to the nucleation of relatively well-separated crystals that do not coalesce shortly after the nucleation event. The deposition time may vary between a few seconds to about an hour, and obviously the amount of the deposit is determined by the charge passed that leads to deposit formation. In the case when the deposition mode does not correspond to the pulse sequence shown in Fig. 10.1, continuous potential cycling or square-wave voltammetry could be used for the particle synthesis. All methods can be used for the synthesis of either elemental or alloyed nanoparticles. For alloyed particles, the codeposition mode has little influence on the composition of the particles when the deposition takes place in the diffusion-controlled regime.

The shape control of the particles deposited is often crucial to achieve the desired catalytic properties. While the substrate itself may have a crucial role in the determination of the particle shape [8], the deposition temperature has also a great influence [15], various additives can be used to achieved different particle shapes [11, 13, 14] and the shape of the particles can also be modified with post-deposition electrochemical treatments [7].

10.3 Electrodeposition on Surfaces with Step Edges

10.3.1 General Considerations Concerning the Deposition Along Step Edges

Step edge decoration was developed by physical deposition techniques on single-crystal metal surfaces [16, 17]. The basic idea of this process is that adatoms tend to accommodate at the step edge of the single-crystal surfaces, and growth may start along the step edge as a deposited nanowire, especially at small deposition rate of atoms with high surface mobility. Step edge-decorated samples obtained with physical methods played an important role in studies of one-dimensional magnetism with non-magnetic substrates and ferromagnetic nanowire deposits at the step edge. When physical methods were applied for the sample preparation, the specimens were studied in an as-received state, and the removal of the nanowires could not be performed efficiently. Since physical deposition techniques work also with non-conducting substrates, removal was not even necessary even if the electrical transport was the property of interest.

Electrodeposition of a nanostructure along the step edge of a metal is highly non-conventional. The explanation is that the step edge of a non-noble metal surface is ill-defined, regarding that it is formed after the reduction of the oxide usually present on a metal surface. For noble metals, step edges can be well visualized and some preference of the nucleation along the step edges can be identified for a few substrate–deposit pairs like Ag(111)–Ni [18], as presented in Fig. 10.4a. Even if a deposition preference at the metal step edge could be seen, it was stressed that the mobility of the surface atoms of the metal substrate was high and also influenced by the deposit being formed. The high mobility of the step edges and terraces of a few-atom height was reported also independently of any deposition pursuit [19]; therefore, step edges on metals surfaces cannot be prepared with a fixed atomic configuration. Also, the difference in the nucleation barrier at the flat terrace surface and at the step edge is usually relatively little. Therefore, a metal surface is not ideal for a highly selective deposition along the step edge. This was also presented in Sect. 4.3.2 where a large number of examples were shown for deposition processes leading to coatings that are laterally fairly even. Also, the removal of a metallic nanowire from the step edge of a substrate is yet to be solved; hence, the application of metal substrates is rather limited.

A successful attempt for electrodeposition of a polymer onto the step edges of a reconstructed Au(111) surface was shown for the irreversible reduction of nitrobenzene and picric acid in acidic solution [20]. The explanation offered for the deposition preference at the step edge was related to the rearrangement of an ordered adlayer

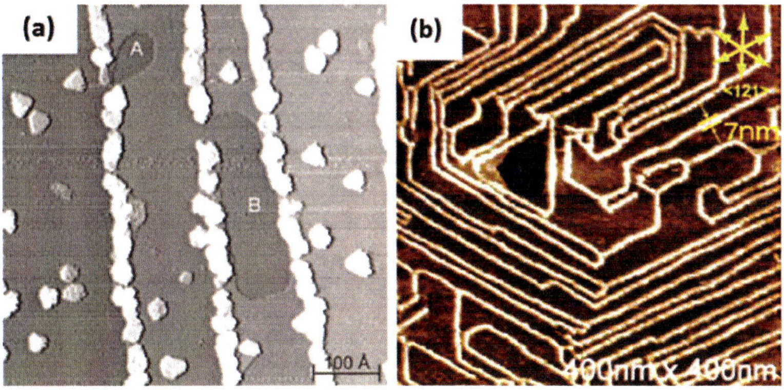

Fig. 10.4 Examples for nanowire-like structures obtained at the step edges of single-crystalline noble metal surfaces. **a** Ni on Ag(111), A and B indicating pools of one monolayer depth [18]. **b** polymer deposited with the cathodic reduction of picric acid on Au(111) along the step edges [20]. Reprinted with permission from Refs. [18, 20]. Copyright (2000) Elsevier and (2008) American Chemical Society, respectively

during the negative shift in the electrode potential from the double-layer region to the potential interval of electroreduction of the adlayer. Although the deposit along the step edge formed a labyrinth-like network (see Fig. 10.4b), it was anticipated that such polymerization processes can be used for surface nanopatterning.

10.3.2 Deposition on the Step Edges of Graphite

Among the common substrates applied in electrodeposition, carbonaceous materials and particularly HOPG play an important role. The reason for this importance is multifold. First, the interatomic distance of the carbon atoms in the graphene plane differs significantly from that of any metal lattice, which results in an incommensurable surface structure of the graphene plane and the metal deposit forming thereon. This makes the deposition of separated grains favoured over a layer formation. Secondly, carbonaceous substrates withstand to anodic treatments, which allows the oxidative post-treatment of the deposit. Furthermore, the atomic position of the surface carbon atoms can be taken as being fixed due to the covalent bond within the graphene planes, which makes unwanted intermixing of the deposit atoms with substrate ones much hindered.

It has long been known that the areal density of appropriate nucleation sites on carbon surface for metal deposition is quite small regardless of the nature of the carbon substrate. This is why various early studies on electrochemical nucleation applied carbonaceous substrates. Even before the discovery of atomic-scale imaging methods, it was speculated that the step edge of the basal plane plays a crucial role in the deposition process. Later, this was evidenced and taken advantage of in the process called electrochemical step edge decoration (ESED).

In the simplest case, the electrodeposition process from aqueous solutions leads to *metallic nanowires*. The three-pulse deposition process (steps A, B and C as seen in Fig. 10.1) seems to be typical [21]. Pulse A with a length of about 5 s ensures the appropriate chemical environment at the step edges. The nucleation pulse initiates the growth of nanoparticles, most of them being formed along the step edges. The length of this pulse varies between 5 and 100 ms, depending on the metal to be deposited and the concentration of the precursor ions. Then, the individual particles coalesce during the growth pulse lasting up to several hundreds of seconds, which lends a pearl chain-like appearance to the nanowires obtained. Examples of morphological features of various nanowires are presented in Fig. 10.5. Depending on the deposit composition, the nanowire formation can take place without either the initial oxidation pulse or the nucleation pulse [22], although this method can be taken rather as an exception. Regardless of the number of the components, the concentration of the precursor metal cations was at most a few tens of mM. The length of the nanowires is independent of the electrochemical parameters of the process but is determined by the length of the appropriate step edges of the substrate. In an ideal case, even millimeter-long nanowires can grow. Such nanowires form a class of nanoscopic objects with an extremely high aspect ratio.

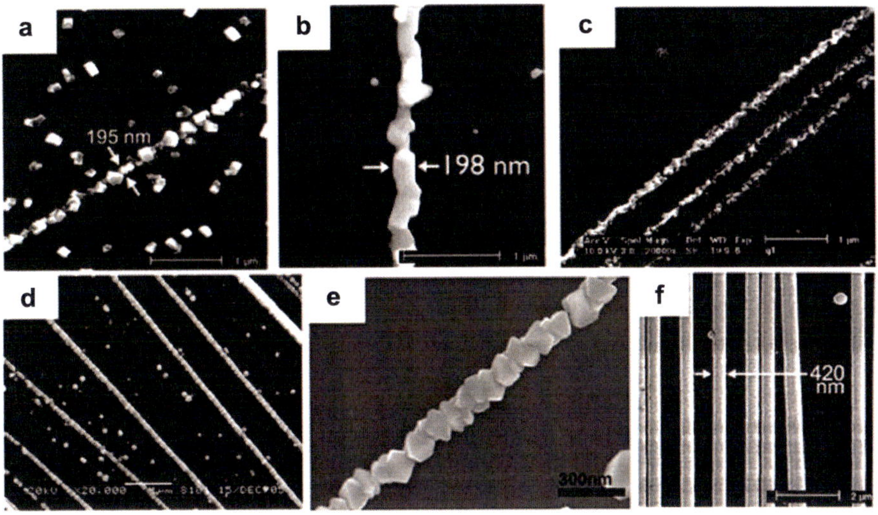

Fig. 10.5 Scanning electron microscopic images of ESED-plated nanowires of different composition: **a**—Cu [21], **b**—Cd [23], **c**—Ag [24], **d**—Pd–Ni [25], **e**—Cu_2O [26], **f**—MoO_2 [27]. For deposition conditions, refer to the original works. Reprinted with permission from the works identified next to the figures. Copyright, in the order of appearance, (2002), (2005), (2002), (2008), (2013) and (2002). With permission from the American Chemical Society (**a**, **b** and **f**), Elsevier (**c** and **e**) and Springer (**d**)

For understanding the choice of the electrode potential for nucleation and growth pulses, the schematic voltammograms in Fig. 10.6 are shown as a guide. When a potential scan is run to the negative direction, the current on HOPG is practically

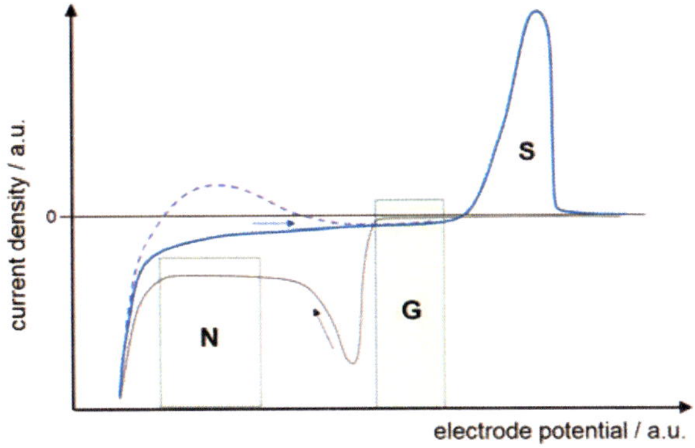

Fig. 10.6 Schematic graph of a cyclic voltammogram for metal nucleation and deposition on HOPG. Arrows indicate the scan direction. Solid and dashed lines in the anodic-going scan indicate the absence and presence of the oxidation of absorbed hydrogen. Potential intervals: N—nucleation, G—growth. S stands for the stripping peak of the deposit. Current ratios strongly depend on both the sweep rate and the cathodic limit of the sweep. Sketched after Refs. [21, 23, 24, 28, 29]

zero until the nucleation of the deposit takes place. Then, a typical voltammetric peak arises due to the combined impact of the growth and solution depletion with respect to the reactant. At high negative potential, the decomposition of the solvent takes place, which involves hydrogen evolution in aqueous solutions. In the anodic-going scan, the current depends on the quality of the deposit. For a hydrogen-absorbing metal like Pd, the hydrogen oxidation takes place and the current becomes anodic prior to reaching the onset potential of metal dissolution. This anodic peak is obviously missing for metals exhibiting no hydrogen absorption. Should the deposit absorb hydrogen or not, the negative current is maintained to potentials more positive than the onset of the nucleation-related current rise in the cathodic-going scan. This hysteresis is taken advantage of for choosing the growth potential without the influence of further nucleation events. In the potential range of nucleation-free growth, the current is much less than the diffusion-limited current.

The range of the metallic element from which ESED-synthesized nanowires were obtained directly from aqueous solutions includes Pd [21, 28, 30], Ag [21], Au [21], Cu [21] and Cd [23]. For Ag nanowire deposition, acetonitrile was also used as solvent [24]. If the nanowire is a homogeneous alloy, the composition is hardly predictable but an empirical relationship between the experimental conditions (such as the concentrations of the metal ions and the deposition potential) and the nanowire composition has to be established. All metal pairs used so far for ESED-based nanowire deposition belong to systems exhibiting regular codeposition (see Chap. 2), which has the general consequence that the increase in the deposition overvoltage results in an enrichment of the less noble component. This was well seen for Pd–Ag [22] and Pd–Ni [25, 31] nanowires. Interestingly, it was found that the potential during the nucleation pulse also has a great impact on the nanowire composition [31], and the trend concerning the nucleation potential and the composition is the same as for the deposition pulse. This raises the question whether the initial nuclei and the shell connection between them grown later are of the same composition or different. Since the nucleation density along the step edge is typically in the range of 4–15 μm^{-1}, it is obvious that a simple EDS composition analysis is unable to resolve internal inhomogeneities in a particular nanowire. This problem might be addressed with a TEM investigation, but no such study was published so far.

Binary nanowires can be obtained also by tailoring two nanowire sections with different composition; however, this requires the combination of ESED with lithographic techniques [32]. In the first step, about the half of the HOPG substrate has to be covered by lithography with a mask whose edge is perpendicular to the step edges on the substrate. Then, the step edges on the uncovered half of the substrate are decorated with nanowires. After removing the insulation applied in the first step and protecting the already deposited nanowires with another masking layer, the step edge decoration can be performed at the second half of the substrate. Hence, the two segments of the wires can be deposited sequentially by applying a protective layer for insulating the zone that is not used in a particular step.

Electrodeposition of non-noble transition metals as step edge-decorating nanowires is possible from nonaqueous environments only, in particular, from ionic liquids. Titanium nanowires were successfully produced from 1-butyl-3-methyl imidazolium bis(trifluoromethylsulfonyl) imide containing $TiCl_4$ [33]. Since the ionic liquid is non-reactive, the oxidative treatment of the HOPG substrate was not necessary, and the entire deposition process could be performed with one single potential pulse. The STM images of the edge-aligned nanowires were composed of a series of coalesced grains, similar to those obtained from aqueous solutions; therefore, this growth characteristic of metal nanowires along HOPG step edges seems to be general. It was confirmed later with the same type of ionic liquid that the oxidation pulse has either no impact on the ESED process or, in the worst case, disfavours the process reproducibility [34]. In the latter study, the deposition of Zn, Ti and Si nanowires was demonstrated, all of them having the same morphological properties as those obtained from aqueous electrolyte solutions.

It is also possible to produce *metal oxide nanowires*, typically from solutions of the oxoanions of transition metals. In these cases, the common characteristic of the reduction process of the oxoanions is that it takes place in several steps. The nanowire growth can be driven at a potential where an intermediate oxide is formed without much interference of the solvent reduction process. This usually occurs near to the onset potential of the first reduction wave of the anion.

In the pioneering study of this field [35], MoO_2 nanowires were deposited on HOPG at a fixed electrode potential. Besides using MoO_2 nanowires as they are, they can also be transformed to either metallic Mo [27, 35] or MoS_2 [36] upon annealing in a hydrogen or hydrogen sulphide atmosphere, respectively. Molybdenum forms oxoanions with the Mo(VI) oxidation state (whose speciation ranges from MoO_4^{2-} at pH > 6.5 to $Mo_7O_{24}^{6-}$ for 6.5 > pH > 2.5 and $Mo_8O_{26}^{4-}$ for pH < 2.5; the protonation degree of the latter two forms varies with pH). The reduction of the Mo(VI) species to metal from aqueous solution in the absence of another metal ion is hindered, and the process stops at intermediate oxidation state, even though the stability range of metallic Mo is just slightly more negative than the stability regime of water. Molybdenum can be reduced to metal in an induced codeposition process together with the deposition of iron group metals only. This is why pure metallic molybdenum nanowires laying along HOPG step edges could be performed in a two-step process only. It is apparent that Mo nanowires as well as their MoO_2 precursor deposits thus produced were quite smooth, in contrast to other metallic nanowires electrodeposited in one single step (see Fig. 10.5). It is possible that for the production of MoO_2, the linear density of the growth centres along the step edge is much larger than for metals, which makes the individual grains indiscernible after their coalescence. Concerning the growth site preference, it was shown [27] that nanowire growth can be expected if the overvoltage of the deposition is small. The more negative deposition potential was applied, the more MoO_x "parasitic particles" at spots other than step edges were obtained and the smaller was the probability of nanowire-wise growth along the HOPG step edges.

Another oxide produced with ESED was MnO_2 [37]. Here, $KMnO_4$ was applied as precursor material in a solution of pH = 6.5 which tends to prevent from the full reduction of the central metal anion to Mn^{2+} but stabilizes the intermediate-valency oxide. The fully crystalline MnO_2 nanowires were achieved with a heat treatment after the electrodeposition. The MnO_2 wires were of granular structure, similar to metal nanowires obtained with ESED.

In a few cases, metal oxide nanowires can be produced from solutions containing simple metal cations (i.e., not oxoanions). As it is usually done with $CuSO_4$ solutions to obtain Cu_2O nanocrystals, the ESED synthesis of Cu_2O nanowires is possible when the solution is quite dilute and unbuffered [26]. For the production of Ag_xO, deposition can be carried with anodic oxidation of Ag^+ from a nearly neutral acetate solution [38]. The composition of the deposit exhibited some uncertainty concerning the oxidation state of silver, and the result of various characterization methods was in acceptable agreement with Ag_2O, too.

As an example for the ESED with a *semiconductor* obtained by direct deposition, the formation of Bi_2Te_3 can be mentioned [39, 40]. The deposition conditions of Bi_2Te_3 are similar to those mentioned in Chap. 4 (i.e., solutions with at most a few mM precursor concentrations). An important difference as compared to other ESED processes is that the single-step deposition of Bi_2Te_3 is not possible. For this reason, the pulse sequence shown in Fig. 10.1 was slightly modified. After the anodic pre-treatment and the cathodic nucleation pulses, the deposition could be performed with running subsequent cyclic voltammograms in the potential range comprising the UPD region of both components. The anodic limit was positive enough to strip off the excess Bi and hence to achieve the desired composition. A similar process was proved to be feasible fore CdSe nanowire deposition by ESED [41]. The major difference was that the desired composition could be achieved by running the voltammograms during the deposition process in an unusually wide 1.4 V potential interval to achieve the stripping of the excess of both components. For indirect preparation of semiconductor nanowires along the surface, the reaction of Cd deposits with H_2S opens the way for CdS nanowire formation [23].

Concerning the kinetics of the nanowire growth along the step edge, it was found for essentially all deposits that they exhibit a hemicylindrical shape. Assuming that (i) the accumulated length of the step edges along which the growth takes place does not vary in time, (ii) the coalescence of the particles forming the nanowires takes place early enough during the nanowire growth (or, at least, the overlap of the diffusion field takes place shortly after their nucleation), (iii) the growth process is controlled by the diffusion of the precursor ions, the current is expected to reach a constant value. In this case, the expected radius of the nanowire can be obtained as

$$r = \left(\frac{2 I t V_{\mathrm{M}}}{\pi z F l} \right)^{1/2} \tag{10.1}$$

where I is the total current of the deposition process, t is the deposition time, V_{M} is the molar volume of the deposit, and l the total length of the step edges where the deposition process takes place (other variables have the usual meaning). The basic character of this rate law concerning both the constancy of the deposition current and the validity of the $r \propto t^{1/2}$ relationship were confirmed for essentially all metallic nanowires and also for MoO_2, although the exact compliance with Eq. 10.1 could not be evidenced since l was typically unknown. In contrast, the current decayed in time for MnO_2 deposition, which was accompanied with a lower exponent in the $r(t)$ relationship. This is probably due to the activation limitation of the MnO_2 deposition process via the $MnO_4^- + 2H_2O + 3e \leftrightarrows MnO_2 + 4OH^-$ reaction leading to a pH change in the unbuffered solution.

Not only may the kinetics of the deposition but that of the dissolution play an important role in the regulation of the diameter of nanowires obtained by ESED [42]. When a dissolution pulse with moderate anodic current was applied after the deposition of the nanowires deposited onto HOPG step edges (see step D in Fig. 10.1), the experience was that the grains along the step edge forming the nanowire remain interconnected down to a thickness at which only segregated grains can be seen during the deposition step. This opens a way for tuning the nanowire diameter in a range that is inaccessible without a dissolution pulse. However, if the dissolution conditions were not mild enough, the dissolution took place in the intergranular zone, hence leading to segregated grains. Concerning the kinetics of the slow-rate dissolution (i.e., in the nanowire regime), a linear relationship was obtained between the diameter and the anodization time. This can be also understood from a quantitative approach. It was assumed that at a moderate dissolution rate a uniform current density predominates at the surface of the hemicylindrical nanowire. Then, the relationship between the charge and wire diameter is

$$Q = -\pi r l j t \tag{10.2}$$

with the same notations as for Eq. 10.1, with a negative sign indicating that cathodic charge corresponds to growth. By expressing the volume of the object from the charge and from geometric data of the hemicylindrical wire, one obtains that

$$\frac{-\pi r l j t V_M}{z F} = \frac{\pi r^2 l}{2} \tag{10.3}$$

which leads to

$$\frac{dr}{dt} = -\frac{2 j V_M}{z F}. \tag{10.4}$$

It is also remarkable that, if the current distribution along the nanowire is indeed uniform, the wire corrugation does not change during the anodic dissolution. This is what was shown in the SEM study of thinned nanowires regardless of the composition [40, 42].

The practical application of the ESED process stems form the opportunity that the nanowire bundle deposited on the HOPG surface can be removed with a gentle lift-off procedure that does not damage the wires. The steps of this process are presented in Fig. 10.7 where the electrical connection mode of the wires on a non-conducting new substrate is also shown. The sensoric application mode depends on the material of the nanowires. Palladium nanowires can be used as hydrogen sensor [28, 30].

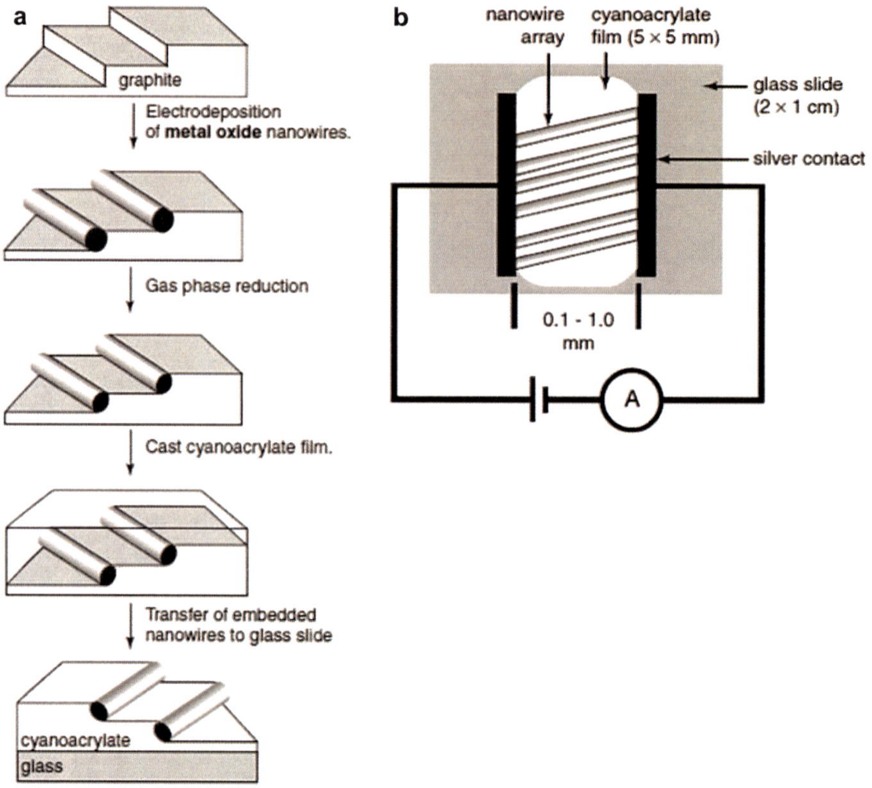

Fig. 10.7 **a** Steps of the nanowire preparation with ESED and relocation of the wires onto a non-conducting substrate. The second step in the process is ommitted if the deposit is metallic. **b** Electrical connection of the nanowire array transferred onto a non-conducting substrate when they are used for sensor with resistivity change as the response parameter. Reprinted from Ref. [29] with permission of John Wiley and Sons

The change in resistivity of the Pd nanowires is not simply because of the hydrogen absorption in the Pd lattice but also because of the expansion and contraction of the grains making the wires upon uptake and release of hydrogen, respectively. This volume change modifies significantly the interconnected cross section at the grain boundaries, which leads to larger change in resistivity that could be rationalized from the bulk resistivity change due to merely the hydrogen absorption. CdS nanowires have large photocurrent responsivity [23], while other semiconductor nanowires like Bi_2Te_3 can be used as thermoelectric materials [40]. Segmented nanowires can be used as microscopic thermocouples [32].

10.4 Electrodeposition on Surfaces with Mechanically Induced Nanoinhomogeneities

In Sect. 10.3, atomic-scale surface inhomogeneities occurring as natural features of the substrates were discussed from the viewpoint of electrochemical nanomanufacturing. In this chapter, the nanoscale surface modification methods will be presented where a tip (usually an AFM tip) is contacted to the surface. Although the primary impact of the tip is *mechanical*, either the imprint or the scratch line of the tip leads to a *chemical* modification by either thinning/removing a surface layer or changing the crystallinity of the near-surface zone due to the occurrence of dislocations and other lattice defects. Various nomenclatures can be found in the literature for such phenomena, including tip-based nanomanufacturing (TBN) or scanning probe lithography (SPL). Concerning the applications of these methods by using techniques other than electrochemistry, two recent reviews can be recommended [43, 44].

In the methods discussed in this chapter, the tip has no role related to electrochemical phenomena. The cases when a tip is electrochemically active in a nanostructure preparation method will be presented in Chap. 12 together with other techniques involving a special electrode arrangement. After a general overview, the mechanical surface modification opportunities will be listed below roughly in the order of the strength of the tip impact, which nearly coincides with an incremental order of the hardness of the surface layer modified. Examples will be mostly restricted to silicon substrate due to their importance and dominance in the literature, even though the tip-based modification of metallic surfaces by mechanical action was also studied in a few cases (see, e.g., Ref. [45]).

A thumb rule for the surface scratching and indentation methods is that the tip has to be much harder than the surface to be modified. For polymers on graphite, a normal silicon AFM cantilever is sufficient, while a diamond-coated tip is a standard choice for oxide-covered silicon surfaces. The shape of a groove depends on the surface structure, which will be detailed at the surface structure types below. However, regardless of the specific surface–tip pairs, the increase in both the load of the tip and the grooving cycle number result is deeper and wider trenches. Therefore, the tuning of the load is a tool for size adjustment, which can be performed with one single tip in a limited range.

Another common feature of the metal deposits formed on Si surface is that their nucleation and growth follows the Volmer–Weber mode. This means that metal nuclei form in the trenches similarly than at HOPG step edges; i.e., the wires on the surface are formed as a coalescence of the grains. This trend is valid for both electrochemical and electroless deposition processes. Since the deposit thickness is mostly in the range of the characteristic lateral size of the surface feature produced, the deposition time needed is fairly short, ranging from a few tenths of second to a few seconds only.

Organic molecular layers on silicon single crystals form a soft coating relative to oxide layers. Therefore, they are quite vulnerable for the grooving by AFM tips. In spite of their soft nature, they are technologically important because organic layers are resistant against HF etching and the tip-induced defects in the masking organic layer

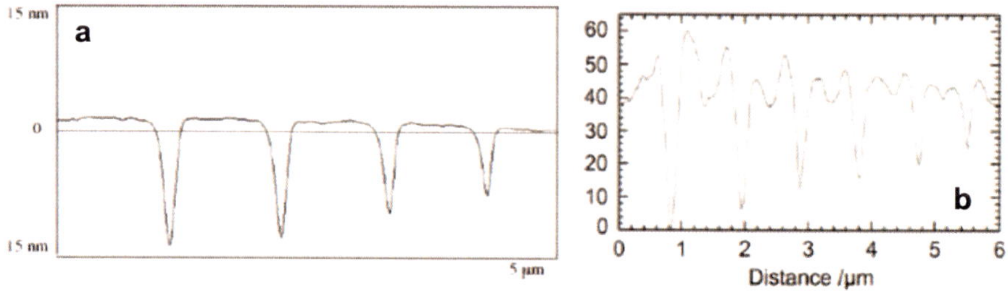

Fig. 10.8 Comparison of the cross-sectional height profiles of scratches on surfaces of different hardness. **a** 1-μm-spaced scratches produced with various loads (10–30 μN from right to left) on octadecene-coated Si surface [47]. **b** 1-μm-spaced scratches produced with various loads (40–60 μN from right to left) on Si surface coated with a 10-nm-thick oxide layer [48]. The line in graph A corresponds to the zero set-point of the AFM system, whereas in Figure B the dotted line indicates the most probable height level of the original surface. Reprinted from Refs. [47, 48]. Copyright (2006) and (2003), respectively; with permission from Elsevier

can be further treated with solutions containing HF for preventing the formation of an oxide layer. A covalently bound organic layer can be formed easily at H-terminated Si surface by reacting with either a fatty acid (like undecylenic acid [46]) or a 1-alkane (like 1-octadecene [47]). When a monomolecular organic layer is scratched, the coating can be fully removed without modifying the surface structure next to the trench; hence, dune formation next to the scratched line was not reported. This is in contrast to the scratch behaviour of the oxidized Si surface where the tip motion leads to a rearrangement of the substrate material rather than its compression or removal, which necessarily leads to a surface height enhancement next to the scratch line. A comparison of the height profile of the above-mentioned scratch types can be seen in Fig. 10.8.

Examples show that Cu deposits produced both from dilute HF + $CuSO_4$ solutions in an electroless manner [46, 47] and from H_2SO_4 + $CuSO_4$ solutions by electrodeposition [49] exhibit a very large site preference of the deposit formation along the grooved lines. (Electroless deposition is often applied as a simple immersion method [50] but it also exhibits an electrochemical mechanism due to the following reaction: $2Cu^{2+} + Si + 6HF \leftrightarrows H_2SiF_6 + 2Cu + 4H^+$). For both methods, the width of Cu deposit in the trenches can be well below 1 μm.

Various other studies dealt with the case when the Si surface was either covered with an oxide layer (native or thermally grown) or cleaned by dissolving the native oxide layer. The sequences of the sample preparation steps for both cases are presented in Fig. 10.9.

When continuous wires were deposited in deeply scratched oxide layers as a result of the coalescence of the initially formed grains, the lower limit of the width of the wires was above 100 nm, even if the scratches were narrower. It was demonstrated for both wide scratch lines and larger scratched areas that the deposition starts at the *sides* of the scratch. This results in the situation that at small deposition times (i.e., before the coalescence of the grains produced in the early phase of the electrodeposition

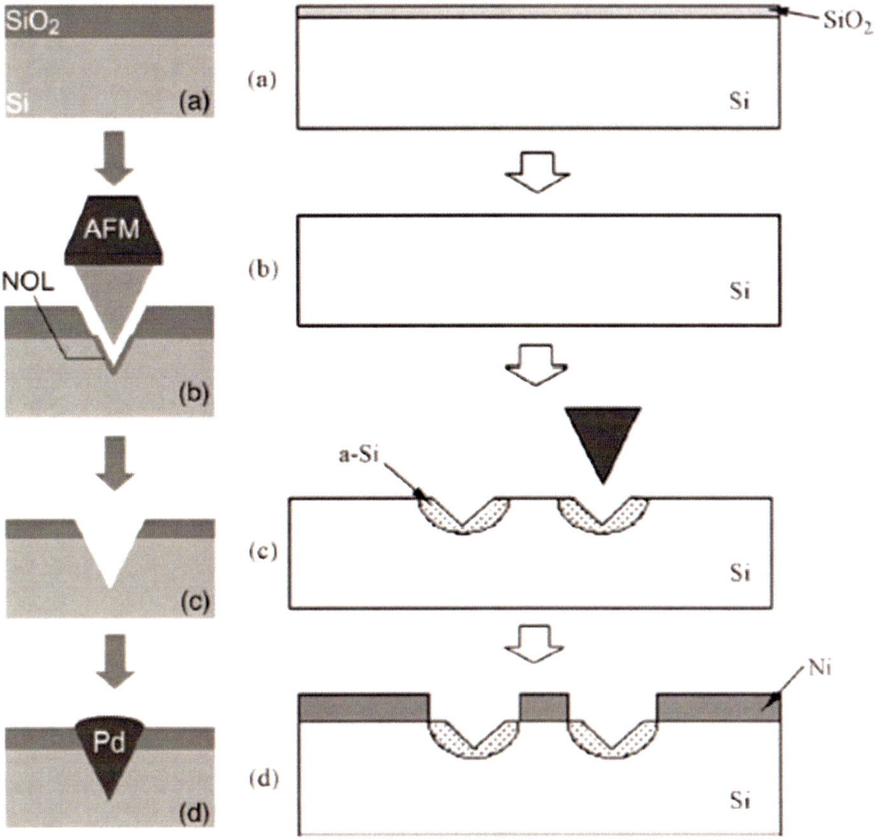

Fig. 10.9 **Left**: Steps of the activation of oxide-covered Si surface by scratching- The oxide layer is darker than the bulk Si. Si substrate with a thick oxid layer **(a)**; AFM scratching of the oxide-covered Si surface **(b)** accompanied with a native oxide formation within the trench; removal of the native oxide within the trench **(c)** and surface passivation; selective deposition within the trench **(d)** [48]. **Right**: Sample preparation by applying the scratching step. State of the Si wafer with **(a)** and after the chemical removal **(b)** of the native oxide layer; scratching the surface with a tip **(c)** after removing the surface oxide; deposition onto the structurally unmodified bare Si surface **(d)**. a-Si stands for amorphous silicon [51]. Reprinted from Refs. [48, 51]. Copyright (2003) and (2010), respectively; with permission from Elsevier

process), not one but two series of grains can be identified at both sides of the scratch. Examples for the double chains of grains and deposition preference at the edge of the scratched areas are presented in Fig. 10.10.

The formation of two parallel grain sets along the edges of the scratch raised the concern about the role of the oxide removal and the structural damage induced in Si in the overall deposition process. In spite of the studies showing uniformly the activation of the Si surface for electrodeposition by scratching, the role of the irregularities produced by the mechanical treatment seems to be controversial. A study performed with bare Si surface without even a native oxide layer showed that the scratching- and nanoindentation-based electrodeposition process can be completely

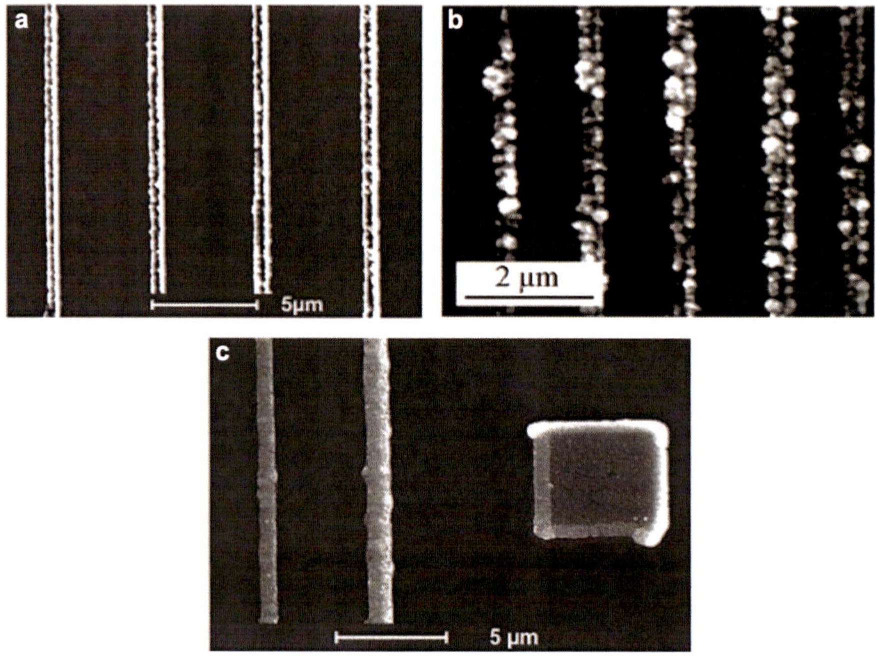

Fig. 10.10 Deposits in scratched parts of oxide-covered Si samples. **a** Cu [52], **b** Pd [48], **c** Cu in scratch lines after long enough deposition time for the occurrence of grain coalescence (left) and in a fully scrached area after the same deposition time, showing the preferred deposition along the edge of the square (right) [52]. Reprinted from Refs. [52] (**a** and **c**) and [48] (**b**) with permission from The Electrochemical Society and Elsevier, respectively. Permission conveyed through Copyright Clearance Center, Inc

inverted as compared to the SiO_2-covered surfaces [51]. In this study, the mechanical manipulation at the surface damaged the crystal structure, possibly leading to the formation of Si nanograins with a phase structure other than the diamond-like Si which is stable at ambient conditions. It was speculated that the amorphization could take place during the unloading of the tip; in particular, when it is performed at a high rate. The occurrence of amorphous silicon was evidenced with spectroscopic measurements. The force of the AFM tip used was rather high as compared to other methods, falling in the range of 200–500 μN. The explanation for the high force can be that the initial slope of the load vs. penetration function was about 40 $\mu N\ nm^{-1}$ for the bare Si surface [51] but only about 6 $\mu N\ nm^{-1}$ for the oxide-covered surface [53]. The structural irregularities thus produced were assumed to increase the resistivity locally, which led to a loss of the electrochemical activity. Therefore, the areas not impacted by the mechanical manipulation were active and damaged areas were not covered by the deposit. Nanowires at the surface could be produced by applying parallel scratches with an appropriate spacing (without allowing the overlap of the areas modified), while a nanodot systems could be achieved by scratching the surface in various directions.

10.5 Top-Down Electrochemical Synthesis of Nanosheets

10.5.1 Intercalation and Accompanying Exfoliation Processes

Electrochemical exfoliation of nanolayers, especially of graphite, belongs to the liquid-phase exfoliation methods [54]. It is a progressively expanding field of graphene production with several tens of papers appearing every year; therefore, it is not possible to strive for completeness. Reviews of various sub-topics of this field are also available [55–58].

Nanosheet-forming methods by electrochemical processes are all based on the ingress of ions or molecules into materials exhibiting a lamellar structure. In the case of delamination processes, the penetration of the species between the nanosheets mostly cannot lead to an intercalation compound of well-defined composition. In order to obtain nanosheets with monoatomic thickness, the prerequisite is that the bond strength connecting atoms within a sheet are much stronger than those acting between the planes. Therefore, an attack will primarily delaminate the layers, and bond breaking between the atoms connected within a plane will be of smaller, though not negligible, importance.

Historically, the irreversible change of graphite cathodes in the presence of teraalkylammonium salts was described already at the end of 1970s [59, 60]. The first studies published in this field dealt with the electrochemical behaviour of the resulting modified electrode. Although the elucidation of the intercalation phenomena can be taken as scientifically plausible also nowadays, no structural study was performed at the time of the above-mentioned works.

The first application of the intercalation process was the lithium uptake of graphitic carbon. This process leads to maximum lithium content at the composition of LiC_6 with various intermediate compositions, all of them having a regular equilibrium structure concerning the Li distribution. Since the electrochemical lithium intercalation and deintercalation both are reversible, it could be taken advantage of in the construction of lithium ion batteries, where graphitic carbon is the cheapest negative electrode material till nowadays.

At this point, it is worth of comparing Li-intercalated graphite with metals dissolving hydrogen. In both cases, the outer-shell s^1 electron of the intercalated (absorbed) species becomes a part of the delocalized electron system of the host lattice. Although the negative charge delocalized cannot be assigned to any location of the host lattice, the intercalated (absorbed) species are Li^+ (H^+) and they exhibit a large mobility due to their shrinkage caused by the ionization and the expansion of the lattice due to the presence of the foreign interstitial species.

The intercalation of molecular compounds was also reported in the field of graphitic materials. In particular, propylene carbonate was shown to be detrimental because it co-intercalates with Li, hence leading to electrode degradation. This process of harmful impact on battery materials was turned into a productive one by using it for nanosheet production. The example of propylene carbonate shows the highly empirical nature of the field. Although propylene carbonate exhibits co-intercalation together with Li into graphite, other organic carbonates do not, although the difference in the molecular structure is really minor. Therefore, neither theoretical predictions for the probability of intercalation nor the required electrode potential can be assessed from elementary physico-chemical principles.

The cell configuration for the electrochemical exfoliation is very simple. A two-electrode system is applied nearly exclusively. A symmetrical electrode configuration is also customary, although non-reactive electrodes (like Pt) are often used for counter electrode. Due to the two-electrode configuration, the cell voltage is the parameter reported, and the electrode potential in a classical potential scale is not defined. Since the cell voltage is typically between 1.5 and 15 V, it is obvious that the polarization limit is much beyond the stability regime of the solvent, even for an ionic liquid, and the reaction of the solvent is an important factor in the delamination process. Although the high cell voltage commonly used makes the exfoliation-based top-down synthesis method similar to those to be discussed in Chap. 13, a few aspects rationalize its discussion together with other site-preferred synthesis methods. These reasons are: (i) the occurrence of the layers already in the starting material; (ii) the difference between the intra- and interlayer bond strength, (iii) the obvious role of the already existing step edge position in the layer cleavage process, and (iv) the fact that intercalation and exfoliation processes can take place also at moderate voltages, unlike the processes to be mentioned later, and the goal of the application of the relatively high voltage is often the speeding up of the synthesis process.

It is also because of the relatively high cell voltage why functionalization reactions in parallel to the exfoliation take place in essentially all cases, and the electrochemical exfoliation leads to a product that is rather ill-defined concerning its chemical structure, especially concerning the regularity of the defect positions. Despite the chemical structure of the product of the exfoliation process is indefinite, the formation of the functional groups of the mostly apolar lamellae is an important factor of the stabilization of their suspension that prevents the fast coagulation of the exfoliated nanosheets. Since no exact electrochemical reaction is associated with either the delamination process itself or the accompanying functionalization, the faradaic efficiency of the electrochemical exfoliation process cannot be defined.

The exfoliation process can be either cathodic or anodic. The driving force of the process is the uptake of the ions into the lamellar solid structure, which is accompanied with the occurrence of a large mechanical stress. The stress-induced delamination is also possible via bubble formation between the layer of interest and its substrate, which establishes electrochemistry-based transfer procedures of nanosheets. In both cases, the atomically smooth nanosheets can exhibit a lateral dimension up to 50 μm, hence resulting in objects with extremely large aspect ratio. The aspect ratio is illustrated in Fig. 10.11 for two kinds of electrochemically exfoliated nanosheets where the height measurement method with AFM is also presented with representative step height profile functions.

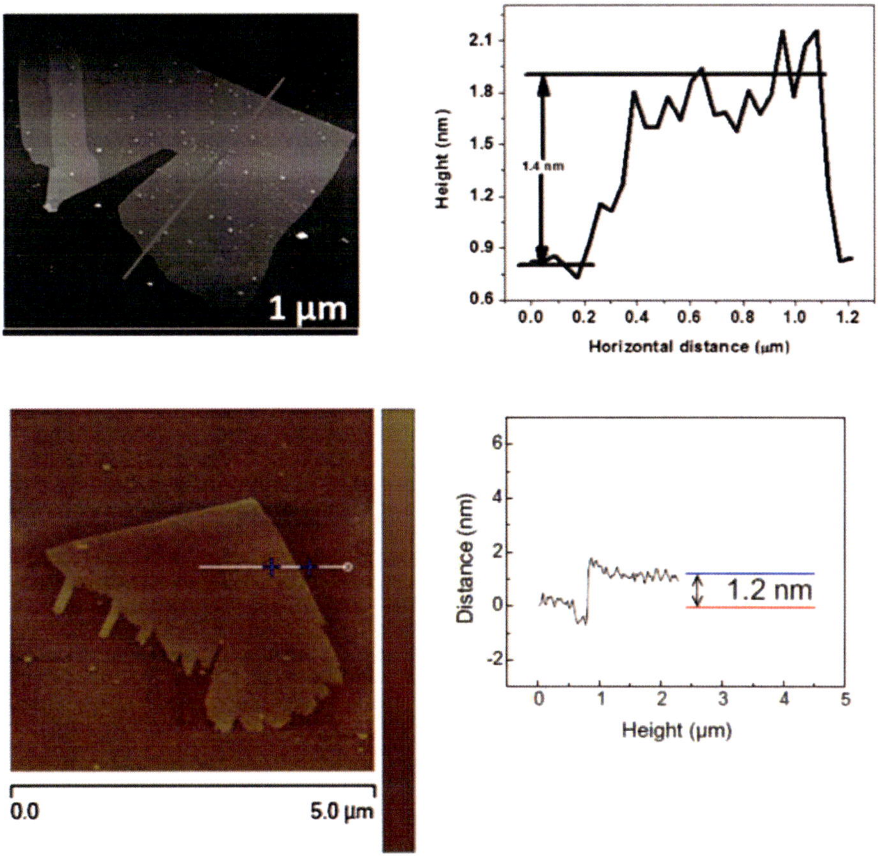

Fig. 10.11 AFM images on exfoliated sheets (left) with height profile measures in the edge zone (right). Top: Black phosphorous nanosheet exfoliated in 0.5 M Na_2SO_4 solution [61]. Bottom: Graphene trilayer exfoliated electrochemically by using 0.1 M sulphuric acid solution [62]. Reprinted form Refs. [61, 62]. Copyright (2016) and (2018); with permission from The American Chemical Society and Elsevier, respectively

10.5.2 Graphene by Electrochemical Exfoliation from Aqueous Solutions

Exfoliation of graphene starts from various carbon sources. The typical graphene exfoliation method is the anodic polarization of the carbonaceous material in a cell equipped with a Pt cathode [63–68]. Electrode potential in the classical electrochemical scale is not given, but the cell voltage is usually reported, which is typically 10 V or even below. Since commensurable electrode surface areas can be assumed and Pt is a good electrocatalyst for hydrogen evolution, it is plausible that a large part of the potential drop occurs in the close vicinity of the carbon anode surface. Pt cathode can also be replaced with carbon [69, 70] or even with stainless steel [71]. It is common that the overall process starts with a mild anodization step before achieving the final cell voltage, which serves the intercalation and swelling but no formation of by-products, gas evolution or graphene oxidation [63, 71]. Another approach is that carbon anodes are used as an electrode pair with an alternating current power supply so that both electrodes undergo intercalation and the subsequent exfoliation process [72, 73]. Harmful impact of the cathodic polarization for half of the working time on either the quality or the quantity of the exfoliation product was not reported; instead, it was assumed that the cathodic polarization reduce the oxygen content of the product [63].

Concerning the quality of the carbon source, high-purity graphite and HOPG [65] are customary; nevertheless, low-quality carbon sources are also suitable for the exfoliation. An example is the application of used carbon from non-rechargeable batteries [64]. This can be regarded as a recycling means of used material, and the low quality of the starting material seems to be tolerable since the exfoliation product has to be purified anyway (suspending in dimethyl formamide is the most common way of separation from the aqueous solution for further storage and use). Compressed carbon black also proved to be suitable for aqueous electrochemical exfoliation [74]. Since the particle size of carbon powders is not at all smaller than the typical lateral dimension of the graphene sheets produced (which is at most a few tens of micrometers), the application of powders enhances the density of the graphene plane edges in the stating material, which increase the probability of the intercalation and the subsequent delamination.

Although the application of parallel plates and rods is the most common and the simplest cell geometry reported, it was suggested that the increase of the process efficiency requires the completion of the exfoliation of partly separated sheets. This lead to the cell geometry where the carbon anode is placed to the bottom of the cell [64] or to a basket [74] from which the partly exfoliated grains cannot escape because of the sedimentation, and they remain in electrical contact with the rest of the anode. Hence, the exfoliation can take place also in the case when particles are disintegrated from the bulk of the anode. This method was named as multiple exfoliation.

The degree of the exfoliation varies from one process to another. The comparison is difficult due to the diverse nature of the exfoliation products (e.g., the number of the graphene sheets in the nanoflakes regarded as the end product). The best results are around 50–60% of the initial mass of the anode [66, 75]. The process time decreased significantly with the improvement of the exfoliation process, and high-degree exfoliation electrolysis can now be performed within minutes.

The number of graphene layers delaminating together and making the final product depends on the overall set of experimental conditions, and it is difficult to establish a trend. Since the application of the exfoliation products is by far not only the same as that of pure graphene (including also, among others, paint ingredient [76], super-capacitor materials [67, 68], printed electronic contacts [69] and electrocatalysts mostly for oxygen reduction [71, 72]), the achievement of the reproducibility of the process is often enough. Few-layer graphene sheets with a particular thickness range [64–66, 74] or flakes with a well-defined number of layers [62] all can be useful.

Since the process is highly empirical, there is a wide platform to test new and new experimental circumstances.

Concerning the compounds used in the electrolyte solution, simple inorganic salts are very common. Sulphate ion is one of the species whose incorporation and the oxidation/decomposition to SO_2 is thought to play a key role in the anodic exfoliation; therefore, various compounds containing sulphate ions can be used (dilute [62, 63] or concentrated [75] H_2SO_4, $(NH_4)_2SO_4$ [70, 74, 76, 79], alkaline metal sulphates [68, 76] or oxone [66]). A feasible pursuit in the choice of the solute is that it should be oxidized easier than graphene, hence reducing the oxygen content of the product. For this reason, hydrogen peroxide is often applied as solution component [70, 74], and sodium halides as electrolyte were also used successfully [67, 77]. Nevertheless, the exfoliation works also with various types of inorganic and organic acids [64] as well as basic solutions like NaOH [74].

One of the most important quality indicators of the exfoliation products is their content of foreign elements, mostly oxygen. The oxygen content is often given as one of the indicators of the success of the sample preparation. The oxygen weight percent of the delamination products falls between 0 and 15%. It remains mostly hidden which kinds of functional groups are produced upon the oxidation and how their ratio depend on the exfoliation circumstances. From a work in which the oxygen content of the exfoliation process was systematically studied, one can learn that an excessive amount of hydrogen peroxide leads to an increase of the oxygen content, while elevated temperature favours the exfoliation and reduces the graphene oxidation [70].

If the goal is the intentional doping of the graphene framework in parallel with the exfoliation, targeted addition of compounds containing the element to be doped proved to be a successful strategy. Examples range to mostly N and S doping. Some 6% N content can be achieved by using glycine [65] as N-doping agents, while a high nitrogen and oxygen content (about 8 and 16%, respectively) were achieved by using the reaction product of melamine and formaldehyde [71]. Parallel inclusion of S and N was achieved by using melamine and alteration of the electrode potentials [72]. Thiourea was effective in the deposition of additional sulphur without allowing the oxidation of graphene [73].

We must note that the mechanism of the anodic exfoliation is by far not as clear and well-studied as the cathodic intercalation. It can be assumed that the anodic treatment leads to the formation of oxygen-containing groups mostly at the edge of the graphene sheets [64, 78]. A severe damage of the graphene sheets can lead to the breakage of the C–C bonds far from the graphene plane edge, which can lead to the occurrence of additional locations for foreign atom incorporation. The relatively large oxygen or nitrogen content of the anodically treated graphite makes the break of the in-plane C–C bonds very likely. As an analogy of the cathodic intercalation, an alternative route can be the partial oxidation (i.e., electron release without the incorporation of foreign species) of the graphene plane in parallel to the anion incorporation. Since anion residues have not been reported for anodically exfoliated graphene, it can be assumed that the resulting adduct of the oxidized graphene sheet and the anions decomposes when electrochemical control is no longer applied.

10.5.3 Graphene by Electrochemical Exfoliation from Solutions with Non-aqueous Solvents

There is much commonality between electrochemical exfoliation form aqueous and non-aqueous systems (i.e., cell configuration, cell voltage, electrode materials, etc.). The non-aqueous solvent is often a component in a water-containing solution and serves either a co-intercalation molecular agent [79], an electrolyte when it is an ionic liquid [80, 81] or both. Solvent mixtures containing water are often used for economic reason. The application of either polar non-aqueous solvents [82, 83], pure ionic liquids [84, 85] or their mixture [86] was rationalized by the quality of the exfoliated product, although the nature of the specific ionic liquid used has a fundamental impact on the exfoliation process. Functionalization of the nanosheets often takes place during exfoliation in organic solvents; hence, the optimization of the one-step exfoliation and functionalization process is in the focus of research nowadays.

The systems containing non-aqueous solvent components(s) were partly optimized for single-electrode exfoliation, either cathodic [79] or anodic [80, 81] one. When the counter electrode is not a carbon material, the opportunity for double exfoliation cannot be revealed [80]. Double exfoliation is possible in a single- or double-compartment cell, the latter [87] yielding an opportunity to collect two kinds of exfoliation product since they may significantly differ due to the dissimilar functionalization effect of cations and anions.

Concerning the appropriateness of various ions for intercalation, only weak trends can be outlined. The clarification of the roles of individual ions is difficult when a solvent mixture is used, the components of which (in particular, water) may also take place in the intercalation and exfoliation process. In anodic exfoliation, the role of the anions seems to be very similar to the aqueous systems; hence, it is not discussed here separately. In the case of the cathodic intercalation, the complex of sodium ion with dimethyl sulfoxide (DMSO) was assumed to be responsible for the good exfoliation efficiency in DMSO-containing solutions [79]. N,N-dialkylpirrolidinium cations proved to be suitable for cathodic intercalation and exfoliation (similarly to teraalkylammonium cations [82]). For the cations containing an organic ring (like the N,N-dialkylimidasolium cation [85]), the π-π interaction between the graphene sheets and the intercalating ions decreases the energetic barrier of the intercalation, similarly to aromatic compounds applied in purely aqueous solutions [88]. Comparative studies concerning the nature of the cations in the exfoliation process are quite scarce. From a study comparing three different cations with quaternary nitrogen atoms and with the same anion, one can conclude that the increase in the cation size is preferred from the viewpoint of the exfoliation efficiency [86].

The thickness of the exfoliated sheets varies from one synthesis method to another. If the interaction between the solvent/ions used for the exfoliation and the graphene sheets is strong, the thickness measured may include an additional molecular layer on each side of the exfoliated sheet. Hence, a thickness of 1.1 nm (like the one obtained with exfoliation in 1-octyl-3-methyl-imidazolium hexafluorophosphate [81]) means the production of single but solvent-encapsulated graphene layers, although stand-alone graphene in only 0.335 nm thick. The appearance of layers with thickness much larger than 1 nm definitely means that a few layers exfoliated together [79, 82–84] (although the folding of the layers may produce artefacts when the samples are post-treated for thickness measurements either by AFM or TEM). When the dependence of the quality of the exfoliated products was studied, it was observed that higher cell voltages are preferred for the specific solution used [85].

10.5.4 Electrochemistry-Based Transfer Methods of Graphene and Graphene Oxide Nanosheets

Almost 15 years after the isolation of a single graphene sheet, in spite of the extraordinary electron mobility coupled with transparency, the adoption of the graphene technology in industry proceeds very slowly. One of the main bottlenecks using chemical vapour deposition (CVD) for graphene production is the lack of a fully reliable graphene transfer technique which would serve as an interface between the CVD process taking place on one substrate and the application demanding another substrate of different type. The delamination of graphene sheets, especially with a surface area of industrial scale, is still challenging. Among others, electrochemical techniques have been developed for delaminating single graphene sheets produced on transition metal catalysts.

The major electrochemical technique used for peeling off a CVD-produced single graphene sheet from its metal substrate is the so-called electrochemical bubbling transfer. Before the electrochemical treatment, the graphene layer is covered with a support layer [typically poly(methylmethacrylate)] in a manner that a part of the uncovered metal substrate is free and is available for contacting an electrolyte solution. Then, this workpiece is immersed in an electrochemical cell and is connected as cathode, and hydrogen generation is started. The hydrogen bubble formation delaminates the plastic cover layer from the metal together with the graphene sheet, i.e., the attack point of the stress is between the graphene sheet and the metal catalyst used for graphene growth. It is evident that the metal surface and its contact with the electrolyte solution play a key role in the bubble transfer process. This technique was reported to work with various metal substrates like Cu [89–92], Pt [93, 94] and Ni [95, 96], preventing at the same time the substrate damage and allowing its repeated application as graphene growth substrate in the CVD process. Beside the cathodic exfoliation, anodic treatment with sacrificial Cu substrate was also elaborated [97]. A simple version of the anodic exfoliation method was shown to work in the same beaker as the cathodic one, hence doubling the current efficiency of the process [98]. Cathodic exfoliation with a parallel reduction and the elimination of the majority of the oxygen-containing groups works in an analogous way for removing graphene oxide nanosheets from Au or Al substrate [99].

Although the bubbling-induced graphene delamination and transfer became widely accepted shortly after its appearance, some doubt arose concerning the delamination mechanism, and the picture on the graphene transfer process had to be refined. When Ru was the growth substrate for the CVD graphene growth, the initiation of the electrochemical delamination did not require any hydrogen bubbling but the process started in the H-UPD potential regime in KOH solution [100]. The complete delamination was achieved at highly negative potential. It seems to be likely that the negative charge of the substrate surface after the partial delamination of the supported graphene sheet and the corresponding cation ingress into the solution layer under the graphene sheet contributes to the driving force of the delamination. A similar bubbling-free transfer process was elaborated for Cu substrate [101] where oxygen reduction was found to be responsible for the delamination in a potential range where metallic Cu was stable but no hydrogen evolution could occur. Since the graphene detachment did not work in the absence of the dissolved oxygen, the oxygen reduction leading to hydroxide ions was found to be responsible for the delamination process.

A recent work published on the systematic study of the graphene delamination opportunity with various solutions claimed that the major driving force is the ion intercalation between the substrate and the graphene sheet at the applied potential, and gas evolution is at most of secondary importance [102]. The major argument for the intercalation effect is that from various solutions, the delamination takes place at potentials where essentially no steady-state current flow. Since the variety of solutions and experimental conditions tested so far is quite limited, it seems to be too early to stand up a mechanism valid for all cases, and further progress can be anticipated in this field.

10.5.5 Electrochemical Exfoliation of Various Inorganic Materials

Among the chemical elements other than carbon, phosphorous has an allotrope with atomic layer structure, i.e., black phosphorous. In contrast to graphite, the "atomic plane" of black phosphorous is not planar but plaited. Anodic exfoliation from neutral aqueous solution is possible [61, 103], just as the delamination with cathodic treatment where the application of an inert solvent is preferred [103–105]. Regardless of the media used for the cathodic exfoliation (e.g., propylene carbonate with tetrabutylammonium hexafluorophosphate [104] or dimethyl sulfoxide with tetrabutylammonium tetrafluoroborate [105]), the thickness range of the nanosheets obtained varied between 2 and 10 nm, but the diameter of the sheets achieved 10 μm. Anodic treatment led to thinner lamellae but still not single atomic layers with a mean thickness of 1.4 nm [61]. Black phosphorous nanosheets can be used from dispersion in a non-reactive solvent and applied for photosensors in thin-layer microelectronic devices.

Molybdenum disulfide, MoS_2, gained a considerable interest in electrochemical exfoliation procedures because of possible application of its nanosheets in field-effect transistors [106, 107] or in supercapacitors [108]. The thickness of nanosheets that could be produced with electrochemical exfoliation varies from a single to a trilayer, even if the preparation methods were fundamentally different [106, 108]. "Single layer" means here a formula unit containing one Mo atomic layer in the middle plane of a nanosheet with S-termination along both sides. Anodic exfoliation from aqueous solution proceeds through the uptake of sulphate and hydroxide ions which can later be oxidized to SO_2 and O_2, respectively [106, 107], similarly to the processes taking place during the anodic exfoliation of graphene. Despite the anodization up to a 10 V cell voltage, the surfaces of the S–Mo–S nanosheets were not oxidized. For cathodic exfoliation of MoS_2, the first step models the electrochemical lithium uptake process in lithium ion batteries, and the second step involves the decomposition of the lithiated bulk MoS_2 in water. This process can be extended to related materials like $MoS_{2x}Se_{2(1-x)}$ and $Mo_xW_{1-x}S_2$ [109]. The exfoliation can be performed with either a pellet containing conducting carbon black powder and polyvinylidene fluoride binder [109] or by simple pressing a binder-free pellet [106]. Electrochemical exfoliation of the MoS_2 compound family has a multifold advantage as compared to the solution-based chemical exfoliation methods: (i) The time scale of the experiment is at least an order of magnitude smaller in the case of the electrochemical exfoliation. (ii) The control of the extent of lithium uptake can be simply regulated by the charge passed during the electrolysis, while chemical modifications are rather cumbersome to control. (iii) Electrochemical lithium intercalation can be performed in ambient conditions; hence, glove box environment is not necessary. (iv) From the various phases that can be produced, electrochemical intercalation-based exfoliation leads to a larger ratio of the desired 1T-MoS_2 phase. It is interesting to note that anodic exfoliation in a water–ionic liquid mixture with dilute lithium bis(trifluoromethylsulphonyl) imide as solvent leads to nanodots instead of large surface area nanosheets [110]. The mechanistic background for such a difference in the exfoliation process is yet to be clarified.

Exfoliation from Na_2SO_4 solution was adapted to produce Bi_2X_3 nanosheets (X: Se or Te) [111]. In both calcogenides, the typical exfoliation product was a sheet composed of 5 covalently bonded atomic layers (quintuple layer) with calcogenide-terminated surfaces. In the exfoliation process, the precursor bulk material was a cathode connected against a Pt counter electrode, and the potential program applied for the cell voltage included several steps between 2 and 10 V. The structure and the composition of the nanosheets produced were confirmed by TEM, and XPS, respectively, indicating no damage as a result of the exfoliation.

10.5.6 Exfoliation of Suspended Particles with Bipolar Electrochemistry

Exfoliation methods included in the above chapters all referred to techniques in which the material composed of covalently bonded atomic planes was used as an electrode component. Hence, the electrode potential of the material to be exfoliated could be known, just as the current passed through them. Unlike in the above discussed methods, the bipolar electrochemical method applies a completely different electrochemical arrangement. As the first step of this method, a material composed of atomic planes is used for microparticle synthesis, often applying a chemical intercalation of lithium from t-butyllithium and then decomposition with water. The microparticles thus synthesized are suspended in an aqueous solution containing a supporting electrolyte (which is most often Na_2SO_4 in 0.1 M concentration). Two platinum electrodes are used for passing a current through the suspension. The cell voltage reported is around 10 V with a 2-cm interelectrode separation. Although it can be inferred from these experimental conditions that a vigorous hydrogen and oxygen evolution takes place at the cathode and anode, respectively, the electrode reaction products are neither of interest nor do they modify the chemical nature on the particles suspended. Instead, the suspended particles are further fragmented, which is attributed to electrophoretic effect. Although it does not seem to be completely understood how a relatively small voltage across the suspended micrometer-sized particles can lead to intercalation and further exfoliation, the size reduction of the particles as a result of the current passed through the solution is convincingly evidenced. The size reduction of the suspended particles takes place as a result of the current over a period of less than an hour. Therefore, an easy-to-implement, fast and scalable method is available for even the mass production of nanoparticles.

The electrophoretic exfoliation was successfully applied to various semiconductor particles such as black phosphorous [112], $MoSe_2$ [113], WS_2 [114] and MoS_2 [115]. In each case, the particle size could be reduced to below 100 nm, but the particles' shape was not fully characterized in either of the above cited studies. Nothing is known about the kinetics of the electrophoretic microparticle fragmentation since it was not studied whether the reaction time selected resulted in a saturation particle size. The particle size reduction was followed only with the absorption change of the solution since the microparticle suspension was dark in each case, but the resulting nanoparticle suspension was light or nearly transparent due to the reduction of the absorption intensity with the particle size. The nanoparticles produced were tested as components of various immunoassays.

Although the exfoliation mechanism was not completely understood for layered semiconductor particles, it seemed to be reasonable that the current passing through the solution can generate a potential difference across the particles, which induces a redistribution of the charge carriers at their surface, hence leading to electrochemical reactions. However, it is much unexpected that the bipolar exfoliation method was found to work for a typical layered insulator, the hexagonal form of boron nitride

[116]. The experimental conditions were identical as reported above for the semiconductor particles. The lateral size of the resulting flakes was 1.27 μm on the average, with a thickness distribution of 8.4 nm ± 3.3 nm.

10.6 Molten Salt Electrolysis Methods for the Synthesis of Nanostructures

10.6.1 Carbon Nanostructures Prepared from Graphitic Cathode

Although the intercalation-induced exfoliation process is the common initial step in room-temperature and high-temperature electrolytic process of the carbon nanostructure preparation, the different experimental conditions rationalize the discussion of the procedures with molten salts in a separate chapter. A wide scope of the experimental conditions and a variety of nanostructures produced with these methods have been described in a few short review papers [117–119]. Below, the common features of the processes based on molten salt electrolysis are summarized. The specific features of the carbon nanotubes will be given in Chap. 13 where their arc discharge synthesis processes are presented.

The construction of the electrolysis cells used in the molten salt processes changed little since the discovery of the carbon nanotube formation in the high-temperature intercalation processes [120]. Briefly, the container of the molten salt is a graphite pot manufactured from a rod by drilling its central part. The container serves as anode as well. The cathode rod is immersed into the molten salt from upwards to a controlled depth. This setup makes it possible to apply a current-controlled synthesis with fixed anode and cathode. Later, various modifications were made: (i) For the sake of a classical electrochemical polarization study, the cell was equipped with a quasireference electrode (a Mo wire [121] or a glassy carbon rod [122]). (ii) For the operation of the cell with alternating polarity, two identical carbon rods were immersed into the molten salt pool, and the container was no longer a part of the circuit [123]. The arrangement of the electrodes in two basic cell configurations is depicted in Fig. 10.12.

By using either of the cell types, the temperature control was maintained by an external heating unit and a thermocouple measuring the temperature of the molten salt. The flushing of the cell with a non-reactive gas (Ar) is a prerequisite for avoiding oxidation processes.

The salt applied for the electrolysis process was mostly either LiCl [120–122, 124, 125] or NaCl [122, 126–128], although comparative studies with LiBr [120] and KCl [126] are also available. The melting point of the salt used determines the temperature range applied (>615 or 810 °C for LiCl and NaCl, respectively). The anode reaction is the evolution of the halogen gas, while the cathode reaction is primarily the discharge

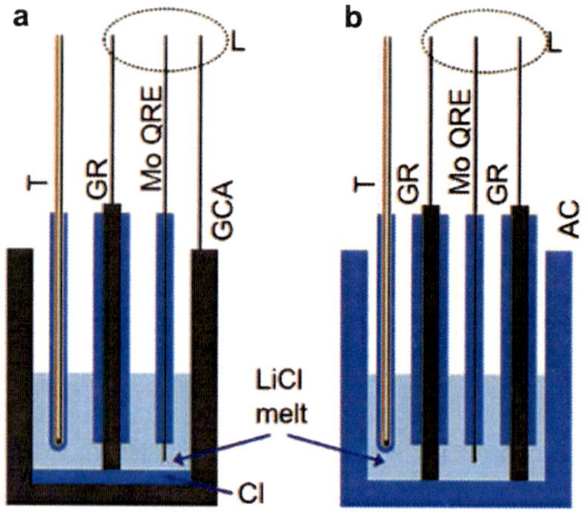

Fig. 10.12 Basic cell configurations for molten salt electrolysis. **a** Single graphite rod cathode with carbon crucible as anode; and **b** Two graphite rods used as cathode and anode alternatingly and with an insulating crucible. Notations: T: thermocouple; GR: graphite rod (electrode), Mo QRE: molybdenum quasireference electrode; CI: ceramic insulator plate; GCA: graphite crucible anode; AC: alumina crucible; L: leads to the power source. Adapted from [131]. Copyright (2011), with permission from Elsevier

of the metal ion followed by various structural and morphological changes that lead to the nanostructure formation.

As a consequence of the metal atom discharge at the cathode, its intercalation into carbon and the subsequent carbon delamination process, the erosion of the cathode was shown in all fixed-polarity experiments, while the anode remained relatively intact. As an exception, a study showed that the anode can also mechanically degrade upon long-term alumina electrolysis (1.5 years as opposed to the few-minute-long lab experiments), and its degradation product also contains carbon nanostructures analogous to those produced with the cathode-consuming setups [129].

The quality of the carbon cathode did not seem to be a crucial question during the early studies of the molten salt processes. Later, a correlation was found between the grain structure of the carbon cathode and the product distribution of the electrolysis process [123]. A mixture of tubular and spherical carbon nanostructures with a minor fraction of irregular grains were found in the electrolysis product if a graphite cathode exhibiting predominantly planar micro-sized grains were used. In contrast, only spherical carbon nanostructures could be produced from a graphite cathode with a microstructure of primarily nano-sized grains.

The necessity of the presence of the originally lamellar (graphene-like) structures in the precursor carbon cathode for producing carbon nanotubes is an indirect evidence for that the graphene sheets remain relatively intact during the electrolysis process; at least, new graphene-like sheets cannot form as a result of the electrolysis. Various other and more direct experimental findings also underpin the intercalation-induced delamination as the key step of the nanotube formation during the molten salt electrolysis technique. These results are as follows: (i) The weight of the cathode increases at low current density [122]. Achievement of a critical current density accompanied with the production of molten metal in excess to the intercalation process is required for the cathode delamination. In other works, as long as the intercalated metal can diffuse away from the surface to the bulk carbon without a supersaturation of the surface, the delamination process is negligible. (ii) By applying potential-controlled experiment, delamination of the cathode was detected at electrode potentials where the liquid metal also appeared at the surface of the cathode [121], which was preceded by the intercalation of the metal at lower overvoltages. Nevertheless, the accumulation of the liquid metal at the cathode is an inhibiting factor for the delamination process [128]. (iii) The alkali metal can be found in the cathode rod after the electrolysis process, especially in its outer layer that was in direct contact with the molten salt [128].

Quantum chemical calculations also verified the roll-up formation mechanism of carbon nanotubes from delaminated graphene sheets [130]. The most likely formation route was characterized with the surface diffusion of the alkali metal atoms toward the edge of a graphene sheet, whereas halogen atoms remain at the sites of their initial attachment at any point of the sheet. When the saturation of the sheet surface with alkali metal and halogen atoms is achieved and they start interacting with each other, the graphene sheet twists spontaneously to form a tubular structure.

The typical electrolysis time with fixed-polarity molten salt cells is around 4–5 min. The bottleneck of the process is the cathode consumption that reduces the active surface area of the cathode, which is counteracted by the degradation-induced change in the surface morphology that leads to a diminished current density. With the periodical change of polarity with identical-sized carbon rods, the electrolysis time could be increased by a factor of four [131], which was accompanied by a more homogeneous product distribution.

Concerning the electrical control of the cell, the current regulation was the early solution. The cell voltage was typically below 10 V, although this value includes all resistive contributions in the circuit (lead wires, contact resistance at the metal wire/carbon junction, resistance of the carbon electrodes and the ohmic loss within the melt). The total current was given in the range of 1–10 A, which was correctly converted to a current density unit in later works only [122]. Potential-controlled 3-electrode setups [121] and the voltage-controlled two-electrode preparative experiments appeared much later [125, 128]. The latter control mode was reported to help eliminating the cathode blocking by the molten metal (which also floats on the surface of the molten salt and can cause a short-circuit).

The floating (suspended) carbon particles were not reported to lead to a short-circuit during the limited-time experiments. After the electrolysis is finished, the melt is left to solidify in the anode crucible. The solidified salt–carbon blend is then mixed with water, and the carbon nanoparticles are extracted with a non- or slightly polar solvent. The common extraction agent is toluene [120, 126], even though a detailed study concerning the separation efficiency found [132] that ethyl acetate as separation agent has the highest yield.

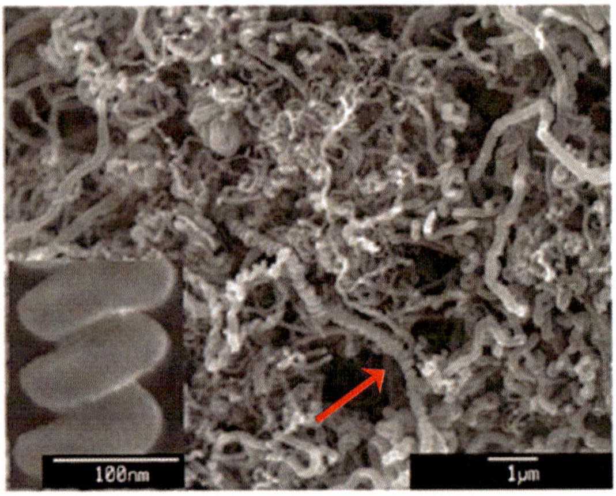

Fig. 10.13 Assemly of carbon nanotubes produced with the electrolysis method. The arow shows a part in the large image where scrolled nanotubes can be seen. A single spring-like nanotube is presented in the inset. Reprinted from [123]. Copyright (2011), with permission from Elsevier

The upscaling of the electrolytic nanotube preparation appeared to be promising [124, 133], and the 20–30% nanotube product ratio of the first attempts [120] was increased to 80% over years [134]. The disadvantage of the electrolytic method is the relatively wide product distribution, ranging from various carbon nanoparticles though nanospheres to nanotubes, while even the nanotube fraction shows a lower crystallinity as compared to other preparation methods. However, a peculiar form of nanotubes, namely the nanosprings [120], are often found is electrolysis product (see Fig. 10.13), even though their purposeful preparation was never reported. No feasible formation mechanism was suggested so far for the nanoscroll-type tubes.

10.6.2 Electrolytic Preparation of Filled Nanotubes

It was observed already in the early time of electrolytic nanotube preparation that Sn-filled multiwall carbon nanotubes can be produced with essentially the same techniques as the nanotubes themselves [135]. The key of the preparation of the filled nanotubes was the addition of the appropriate amount of $SnCl_2$ to the molten salt. An approximately 1 wt.% of $SnCl_2$ proved to be the optimal concentration [124, 135, 136] above which the formation of a continuous metallic Sn film on the cathode prevents the lithium intercalation and hence, has an adverse effect on the efficiency on the production of filled nanotubes. It was shown that Sn-filled nanotubes can be produced with the addition of Sn to the molten LiCl [136]. The encapsulating nanowires are multiwalled and have a diameter between 30 and 100 nm.

Although the elementary steps of the formation of filled nanotubes are by far not fully understood, the general view is that the molten Sn seeds are entrapped by the graphene sheets. Hence, there is neither opportunity for uncovered Sn nanowire formation, nor can the preliminary prepared nanotubes filled with the molten salt technique, but the entire complex filled nanotube structure takes shape *in situ*.

Tin is a metal with low melting point (more than 300 K below the melting point of LiCl) that forms molten droplets during the synthesis process. While this property makes tin particularly suitable for a dynamic shape transformation during the nanowire synthesis, no similar process with another metallic element has been exemplified so far.

10.6.3 Carbon Nanostructures Obtained by CO_2 Reduction in Molten Salts

Although it is not related to any surface inhomogeneity, the preparation of carbon nanostructures in inorganic molten salts by carbon dioxide reduction belongs to this chapter due to the relatively large commonality with other molten salt processes. This research field was born as the size branch of the pursuits or the electrochemical recycling of carbon dioxide to carbon. The field is reviewed below on the basis of a few recent papers [137–140] whose literature list offers rich further readings.

The salt mixtures suitable for CO_2 reduction contain at least an alkali/alkali earth metal chloride (LiCl, NaCl, KCl, $MgCl_2$, $CaCl_2$). A mixture of suitable chlorides reduces the working temperature because of the lowering of the melting point. Another typical (though not compulsory) component of the salt mixture is either a carbonate (Na_2CO_3, Li_2CO_3, $CaCO_3$) or an oxide (CaO) that forms carbonate with CO_2. A CO_2 gas stream is provided above the molten salt mixture with a pressure up to a few bars. The cell voltage is usually in the range of the 3–10 V interval, and the current is up to 2 A (depending on the size of the electrodes).

The cathode reaction is related to the reduction of the carbonate ion and can lead to a partial of fully reduced product:

$$CO_3^{2-} + 2e \rightleftarrows CO + 2O^{2-} \tag{10.5}$$

$$CO_3^{2-} + 4e \rightleftarrows C + 3O^{2-} \tag{10.6}$$

The anode reaction depends on the anode material. For a carbon anode, the recovery of the CO_2 is possible with the degradation of the anode:

$$C + 2CO_3^{2} \rightleftarrows 3CO_2 + 4e, \tag{10.7}$$

whereas the application of an inert anode leads to oxygen evolution in parallel to the formation of CO_2:

$$2CO_3^{2-} \rightleftarrows 2CO_2 + O_2 + 4e. \tag{10.8}$$

The above reactions are well evidenced with both voltammetric studies and composition measurements of the outflow gases. Nevertheless, in contrast to the thorough clarification of the elementary electrochemical processes, the driving force of the morphological arrangement of the reduced carbon atom is yet to be unveiled. In the molten salt processes, both single-wall and multiwall carbon nanotubes can be produced, occasionally filled with the electrolyte salt. Degradation products of both the reaction vessel and the electrodes (e.g., Pt or stainless steel) can be found as entrapped particles in the nanotubes. With the regulation of the temperature and the composition of the salt mixture, nanosheets and nanospheres can also be found in the product profile that is by far not homogeneous.

References

1. Penner RM (2002) J Phys Chem B 106:3339–3353
2. Guin SK, Sharma HS, Aggarwal SK (2010) Electrochim Acta 55:1245–1257
3. Bian JC, Li Z, Chen ZD, He HY, Zhang XW, Li X, Han GR (2011) Appl Surf Sci 258:1831–1835
4. Liu H, Favier F, Ng K, Zach MP, Penner RM (2001) Electrochim Acta 47:671–677
5. Cortés M, Gómez E, Vallés E (2010) Electrochem Commun 12:132–136
6. Querejeta AL, del Barrio MC, García SG (2016) J Electroanal Chem 778:98–102
7. Hernández-Lebrón Y, Cabrera CR (2017) J Electroanal Chem 793:174–183
8. Chen Z, Zhou Y, Li Y, Liu J, Zou Z (2017) Prog Nat Sci Mater Int 27:452–459
9. Plieth W, Dietz H, Anders A, Sandmann G, Meixner A, Weber M, Kneppe H (2005) Surf Sci 597:119–126
10. Ma Y, Di J, Yan X, Zhao M, Lu Z, Tu Y (2009) Biosens Bioelectron 24:1480–1483
11. Wei C, Liu Q (2017) CrystEngComm 19:3254–3262
12. Nayak J, Sahu SN, Nozaki S (2006) Appl Surf Sci 252:2867–2874
13. Sun F, Guo Y, Tian Y, Zhang J, Lv X, Li M, Zheng Y, Wang Z (2008) J Cryst Growth 310:318–323
14. Patil PB, Mali SS, Kondalkar VV, Mane RM, Patil PS, Hong CK, Bhosale PN (2015) J Electroanal Chem 758:178–190
15. Li A, Chen Y, Duan W, Wang C, Zhuo K (2017) RSC Adv 7:19694–19700
16. Jung T, Schlittler R, Gimzewski JK, Himpsel FJ (1995) Appl Phys A 61:467–474
17. Petrovykh DY, Himpsel FJ, Jung T (1998) Surf Sci 407:189–199
18. Maroun F, Morin S, Lachenwitzer A, Magnussen OM, Behm RJ (2000) Surf Sci 460:249–263
19. Hirai N, Watanabe K, Hara S (2001) Surf Sci 493:568–574
20. Wen R, Pan GB, Wan LJ (2008) J Am Chem Soc 130:12123–12127
21. Walter EC, Murray BJ, Favier F, Kaltenpoth G, Grunze M, Penner RM (2002) J Phys Chem B 106:11407–11411
22. Tang L, Yu G, Ouyang Y, Si W, Weng B (2008) Electrochim Acta 53:3305–3312
23. Li Q, Penner RM (2005) Nano Lett 5:1720–1725
24. Walter EC, Ng K, Zach MP, Penner RM, Favier F (2002) Microelectron Eng 61–62:555–561
25. Si W, Yu G, Ouyang Y, Tang L, He X, Hu B (2008) J Appl Electrochem 38:1727–1734
26. Ng SY, Ngan AHW (2013) Electrochim Acta 114:379–386
27. Zach MP, Inazu K, Ng KH, Hemminger JC, Penner RM (2002) Chem Mater 14:3206–3216
28. Walter EC, Favier F, Penner RM (2002) Anal Chem 74:1546–1553
29. Walter EC, Penner RM, Liu H, Ng KH, Zach MP, Favier F (2002) Surf Interface Anal 34:409–412
30. Favier F, Walter EC, Zach MP, Benter T, Penner RM (2001) Science 293:2227–2231
31. Xiao Y, Yu G, Yuan J, Wang J, Chen Z (2006) Electrochim Acta 51:4218–4227
32. Bourg ME, van der Veer WE, Güell AG, Penner RM (2008) Chem Mater 20:5464–5474
33. Mukhopadhyay I, Freyland W (2003) Langmuir 19:1951–1953
34. Fournier C, Favier F (2011) Electrochem Commun 13:1252–1255

35. Zach MP, Ng KH, Penner RM (2000) Science 290:2120–2123
36. Li Q, Newberg JT, Walter EC, Hemminger JC, Penner RM (2004) Nano Lett 4:277–281
37. Li Q, Olson JB, Penner RM (2004) Chem Mater 16:3402–3405
38. Murray BJ, Li Q, Newberg JT, Hemminger JC, Penner RM (2005) Chem Mater 17:6611–6618
39. Menke EJ, Li Q, Penner RM (2004) Nano Lett 4:2009–2014
40. Menke EJ, Brown MA, Li Q, Hemminger JC, Penner RM (2006) Langmuir 22:10564–10574
41. Li Q, Brown MA, Hemminger JC, Penner RM (2006) Chem Mater 18:3432–3441
42. Thompson MA, Menke EJ, Martens CC, Penner RM (2006) J Phys Chem B 110:36–41
43. Yan Y, Geng Y, Hu Z (2015) Int J Mach Tool Manu 99:1–18
44. Rosa LG, Liang J (2009) J Phys Condens Matter 21:483001
45. Nielinger M, Berenz P, Xiao X, Baltruschat H (2005) Surf Sci 597:1–10
46. Zhang Y, Balaur E, Maupai S, Djenizian T, Boukherroub R, Schmuki P (2003) Electrochem Commun 5:337–340
47. Zhang Y, Balaur E, Schmuki P (2006) Electrochim Acta 51:3674–3679
48. Santinacci L, Djenizian T, Hildebrand H, Ecoffey S, Mokdad H, Campanella T, Schmuki P (2003) Electrochim Acta 48:3123–3130
49. Santinacci L, Zhang Y, Schmuki P (2005) Surf Sci 597:11–19
50. Choi J, Chen Z, Singh RK (2003) J Electrochem Soc 150:C563–C565
51. Cecchini R, Fabrizi A, Paternoster C, Zhang W, Roventi G (2010) Electrochim Acta 55:3355–3360
52. Santinacci L, Djenizian T, Schmuki P (2001) J Electrochem Soc 148:C640–C646
53. Santinacci L, Djenizian T, Schwaller P, Suter T, Etcheberry A, Schmuki P (2008) J Phys D: Appl Phys 41:175301(1–9)
54. Cui X, Zhang C, Hao R, Hou Y (2011) Nanoscale 3:2118–2126
55. Ambrosi A, Chua CK, Latiff NM, Loo AH, Wong CHA, Eng AYS, Bonanni A, Pumera M (2016) Chem Soc Rev 45:2458–2493
56. Wu W, Zhang C, Hou S (2017) J Mater Sci 52:10649–10660
57. Paredes JI, Munuera JM (2017) J Mater Chem A 5:7228–7242
58. Gong Y, Pan C (2017) MRS Adv 2:1611–1619
59. Simonet J, Lund H (1977) J Electroanal Chem 75:719–730
60. Bernard G, Simonet J (1979) J Electroanal Chem 96:249–253
61. Erande MB, Pawar MS, Late DJ (2016) ACS Appl Mater Interfaces 8:11548–11556
62. Mir A, Shukla A (2018) Appl Surf Sci 443:157–166
63. Su CY, Lu AY, Xu Y, Chen FR, Khlobystov AN, Li LJ (2011) ACS Nano 5:2332–2339
64. Liu J, Poh CK, Zhan D, Lai L, Lim SH, Wang L, Liu X, Sahoo NG, Li C, Shen Z, Lin J (2013) Nano Energy 2:377–386
65. Yang Y, Shi W, Zhang R, Luan C, Zeng Q, Wang C, Li S, Huang Z, Liao H, Ji X (2016) Electrochim Acta 204:100–107
66. Tian S, Yang S, Huang T, Sun J, Wang H, Pu X, Tian L, He P, Ding G, Xie X (2017) Carbon 111:617–621
67. Munuera JM, Paredes JI, Villar-Rodil S, Castro-Muniz A, Martínez-Alonso A, Tascón JMD (2018) Appl Mater Today 11:246–254
68. Singh R, Tripathi CC (2018) Mater Today Proc 5:973–979
69. Chen K, Xue D (2014) J Colloid Interf Sci 436:41–46
70. Hossain ST, Wang R (2016) Electrochim Acta 216:253–260
71. Chao L, Qin Y, Liu Y, Kong Y, Chu F (2017) J Solid State Electrochem 21:1287–1295
72. Liu F, Niu F, Chen T, Han J, Liu Z, Yang W, Xu Y, Liu J (2018) Carbon 134:316–325
73. Li L, Wang M, Guo J, Cao M, Qiu H, Daia L, Yang Z (2018) J Mater Chem C 6:6257–6263
74. Sharief SA, Susantyokol RA, Alhashem M, Almheiri S (2017) J Mater Sci 52:11004–11013
75. Munuera JM, Paredes JI, Villar-Rodil S, Martínez-Alonso A (2017) Carbon 115:625–628
76. Parvez K, Wu ZS, Li R, Liu X, Graf R, Feng X, Müllen K (2014) J Am Chem Soc 136:6083–6091
77. Munuera JM, Paredes JI, Enterría M, Pagán A, Villar-Rodil S, Pereira MRF, Martins JI, Figueiredo JL, Cenis JL, Martínez-Alonso A, Tascón JMD (2017) ACS Appl Mater Interfaces 9:24085–24099

78. Liu J (2017) Electrochemical exfoliation synthesis of graphene. In: Liu J (ed) Graphene-based composites for electrochemical energy storage. Springer, Singapore
79. Zhou M, Tang J, Cheng Q, Xu G, Cui P, Qin LC (2013) Chem Phys Lett 572:61–65
80. Lu J, Yang JX, Wang J, Lim A, Wang S, Loh KP (2009) ACS Nano 3:2367–2375
81. Liu N, Luo F, Wu H, Liu Y, Zhang C, Chen J (2008) Adv Funct Mater 18:1518–1525
82. Cooper AJ, Wilson NR, Kinloch IA, Dryfe RAW (2014) Carbon 66:340–350
83. Zhang Y, Xu Y, Zhu J, Li L, Du X, Sun X (2018) Carbon 127:392–403
84. Yang Y, Lu F, Zhou Z, Song W, Chen Q, Ji X (2013) Electrochim Acta 113:9–16
85. Lei H, Tu J, Yu Z, Jiao S (2017) ACS Appl Mater Interfaces 9:36702–36707
86. Najafabadi AT, Gyenge E (2014) Carbon 71:58–69
87. Najafabadi AT, Gyenge E (2014) Carbon 84:449–459
88. Wang G, Wang B, Park J, Wang Y, Sun B, Yao J (2009) Carbon 47:3242–3246
89. Wang Y, Zheng Y, Xu X, Dubuisson E, Bao Q, Lu J, Loh KP (2011) ACS Nano 5:9927–9933
90. Zhan Z, Sun J, Liu L, Wang E, Cao Y, Lindvall N, Skoblina G, Yurgens A (2015) J Mater Chem C 3:8634–8641
91. Mafra DL, Ming T, Kong J (2015) Nanoscale 7:14807–14812
92. Wang Z, Liu Z, Monne MA, Wang S, Yua Q, Chen MJ (2016) RSC Adv 6:24865–24870
93. Gao L, Ren W, Xu H, Jin L, Wang Z, Ma T, Ma LP, Zhang Z, Fu Q, Peng LM, Bao X, Cheng HM (2012) Nature Commun 3:699(1–7).
94. Sun J, Fan X, Guo W, Liu L, Liu X, Deng J, Xu C (2015) Sensors 15:31811–31820
95. Lee J, Kim Y, Shin HJ, Lee CS, Lee D, Lee S, Moon CY, Lee SC, Kim SJ, Ji JH, Yoon HS, Jun SC (2014) ACS Appl Mater Interfaces 6:12588–12593
96. Wong CHA, Pumera M (2016) J Phys Chem C 120:4682–4690
97. Zhang D, Jin Z, Shim J, Wang X, Peng S, Wang S (2015) Chem Commun 51:2987–2990
98. Shi L, Liu Y, Yang F, Gao L, Sun J (2014) Nanotechnology 25:145704(1–7).
99. Wang X, Kholmanov I, Chou H, Ruoff RS (2015) ACS Nano 9:8737–8743
100. Koren E, Sutter E, Bliznakov S, Ivars-Barcelo F, Sutter P (2013) Appl Phys Lett 103:121602(1–5).
101. Pizzocchero F, Jessen BS, Whelan PR, Kostesha N, Lee S, Buron JD, Petrushina I, Larsen MB, Greenwood P, Cha WJ, Teo K, Jepsen PU, Hone J, Bøggild P, Booth TJ (2015) Carbon 85:397–405
102. Verguts K, Coroa J, Huyghebaert C, De Gendt S, Brems S (2018) Nanoscale 10:5515–5521
103. Ambrosi A, Sofer Z, Pumera M (2017) Angew Chem Int Edit 56:10443–10445
104. Xiao H, Zhao M, Zhang J, Ma X, Zhang J, Hu T, Tang T, Jia J, Wu H (2018) Electrochem Commun 89:10–13
105. Li J, Chen C, Liu S, Lu J, Goh WP, Fang H, Qiu Z, Tian B, Chen Z, Yao C, Liu W, Yan H, Yu Y, Wang D, Wang Y, Lin M, Su C, Lu J (2018) Chem Mater 30:2742–2749
106. Liu N, Kim P, Kim JH, Ye JH, Kim S, Lee CJ (2014) ACS Nano 8:6902–6910
107. You X, Liu N, Lee CJ, Pak JJ (2014) Mater Lett 121:31–35
108. Ejigu A, Kinloch IA, Prestat E, Dryfe RAW (2017) J Mater Chem A 5:11316–11330
109. Tan C, Zhao W, Chaturvedi A, Fei Z, Zeng Z, Chen J, Huang Y, Ercius P, Luo Z, Qi X, Chen B, Lai Z, Li B, Zhang X, Yang J, Zong Y, Jin C, Zheng H, Kloc C, Zhang H (2016) Small 12:1866–1874
110. Gopalakrishnan D, Damien D, Li B, Gullappalli H, Pillai VK, Ajayanb PM, Shaijumon MM (2015) Chem Commun 51:6293–6296
111. Ambrosi A, Sofer Z, Luxa J, Pumera M (2016) ACS Nano 10:11442–11448
112. Mayorga-Martinez CC, Latiff NM, Eng AYS, Sofer Z, Pumera M (2016) Anal Chem 88:10074–10079
113. Toh RJ, Mayorga-Martinez CC, Sofer Z, Pumera M (2016) Anal Chem 88:12204–12209
114. Leong SH, Mayorga-Martinez CC, Chia X, Luxa J, Sofer Z, Pumera M (2017) ACS Appl Mater Interfaces 9:26350–26356
115. Bouša D, Mayorga-Martinez CC, Mazánek V, Sofer Z, Boušová K, Pumera M (2018) ACS Appl Mater Interfaces 10:16861–16866
116. Wang Y, Mayorga-Martinez CC, Chia X, Sofer Z, Pumera M (2018) Nanoscale 10:7298–7303

117. Kamali AR, Fray D (2016) J Mater Sci 51:569–576
118. Rezaei A, Kamali AR (2018) Diamond Relat Mater 83:146–161
119. Kaptay G, Sytchev I, Miklósi J, Nagy P, Póczik P, Papp K, Kálmán E (2000) Electrochemical synthesis of carbon nanotubes and microtubes from molten salts. In: Berg RW, Hjuler HA (eds) Progress in molten salt chemistry, vol 1. Elsevier, Amsterdam, pp 257–262
120. Hsu WK, Terrones M, Hare JP, Terrones H, Kroto HW, Walton DRM (1996) Chem Phys Lett 262:161–166
121. Xu Q, Schwandt C, Chen GZ, Fray DJ (2002) J Electroanal Chem 530:16–22
122. Sytchev J, Kaptay G (2009) Electrochim Acta 54:6725–6731
123. Kamali AR, Schwandt C, Fray DJ (2011) Mater Charact 62:987–994
124. Dimitrov AT, Chen GZ, Kinloch IA, Fray DJ (2002) Electrochim Acta 48:91–102
125. Dimitrov AT, Tomova A, Grozdanov A, Popovski O, Paunović P (2013) J Solid State Electrochem 17:399–407
126. Chen GZ, Fan X, Luget A, Shaffer MSP, Fray DJ, Windle AH (1998) J Electroanal Chem 446:1–6
127. Bai JB, Hamon AL, Marraud A, Jouffrey B, Zymla V (2002) Chem Phys Lett 365:184–188
128. Kinloch IA, Chen GZ, Howes J, Boothroyd C, Singh C, Fray DJ, Windle AH (2003) Carbon 41:1127–1141
129. Alekseev NI, Arapov OV, Belozerov IM, Osipov YG, Semenov KN, Polovtsev SV, Charykov NA, Izotova SG (2005) Russ J Appl Chem 78:1944–1947
130. Alekseev NI, Polovtsev SV, Charykov NA (2006) Techn Phys 51:349–355
131. Schwandt C, Dimitrov AT, Fray DJ (2012) Carbon 50:1311–1315
132. Gábor T, Kármán FH, Sytchev J, Kálmán E, Kaptay G (2009) Carbon 47:1195–1198
133. Kamali AR, Fray DJ (2014) Carbon 77:835–845
134. Schwandt C, Dimitrov AT, Fray DJ (2010) J Electroanal Chem 647:150–158
135. Hsu WK, Terrones M, Terrones H, Grobert N, Kirkland AI, Hare JP, Prassides K, Townsend PD, Kroto HW, Walton DRM (1998) Chem Phys Lett 284:177–183
136. Gupta RD, Schwandt C, Fray DJ (2014) Carbon 70:142–148
137. Novoselova IA, Kuleshov SV, Volkov SV, Bykov VN (2016) Electrochim Acta 211:343–355
138. Hu L, Song Y, Ge J, Zhu J, Han Z, Jiao S (2017) J Mater Chem A 5:6219–6225
139. Jin KWM, Gakim M, Janaun JA, Liew WYH, Siambun NJ (2018) Int J Electrochem Sci 13:9771–9783
140. Abbasloo S, Ojaghi-Ilkhchi M, Mozammel M (2019) JOM 71:2103–2111

Chapter 11
Templated Systems

11.1 Definition and Classification of Template Systems

The present chapter deals with electrodeposition methods in which the shape of the deposit is determined by the geometrical constraints of a non-electroactive material exhibiting a condensed phase. This shape-regulating material will be termed hereinafter as template. The condensed phase that constitutes the template is solid in most of the cases (nanoporous templates and those obtained by the self-assembly of sub-microscopic solid particles). In a few cases, the template-forming material is not solid; this is the case for either lyotropic templates that are liquid crystalline or molecular assemblies. In either of the above-listed cases, the spatial constraint defined by the template has a much longer temporal stability than the typical deposition time of the nanoobjects into the template. In this respect, these static condensed-phase templates differ from the dynamic gas bubble templates discussed in Chap. 8. In other words, the spatial arrangement of the templates constituted from condensed phases is predefined and is mostly invariant during the deposition process.

The organization principle of the parts in this chapter is as follows. First, templates obtained from a bulk material with top-down methods will be dealt with. The most prominent example for such templates is anodized aluminum oxide (AAO); also known as porous anodic alumina, PAA) membranes with nanochannels. The synthesis of these templates is carried out also with an electrochemical method which is presented in Chap. 13 among other high-voltage methods. Therefore, the emphasis here will be on the comparison of the properties of AAO templates with other nanochannel templates like ion track-etched polymer membranes. Templates obtained with the selective dissolution of one phase of a directionally ordered multiphase system also belong to templates synthesized with the top-down approach. The top-down character of the preparation of several template types is restricted to *the last step of the process* since the synthesis of the compositionally modulated precursor materials themselves is a bottom-up process.

In the list of the template formation modes, the template group obtained through entirely bottom-up methods will be the second. In this part, an order of decreasing

L. Péter, *Electrochemical Methods of Nanostructure Preparation*,
Monographs in Electrochemistry, https://doi.org/10.1007/978-3-030-69117-2_11

feature size will be followed. The discussion will be started with the self-assembly of solid particles in which the particle size may be even in the micrometre scale. Finally, the template obtained with molecular-level self-assembly will be detailed.

The literature of the template-based synthesis methods of nanostructures is particularly rich. The number of works dealing with each topic of the present chapter ranges to thousands, and the careful selection of the relevant works is really hard. Therefore, the literature list cited in connection with each template type is meant to be representative. Neither of these lists can be taken as exclusive, and the further literature search of the reader is highly encouraged. Review works published on electrochemical applications of templates give an ample selection of additional examples [1–12].

11.2 Nanochannel Templates Obtained with Top-Down Synthesis Methods

11.2.1 Comparison of the Templates Suitable for Electrodeposition of Nanowires

The research on nanochannel templates started in 1970 when the first article was published on the electrochemical treatment of ion-irradiated and etched MICA template with nanochannels [13]. Although the instrumentation used for electrodeposition was not truly professional and a part of the observations made were abandoned later, the basic experimental setup with the nanochannel templates and their single-side metal coverage used as electrode was laid down and has been followed ever since. The deposition method based on ion track-etched MICA was improved later, making it possible to deposit nanowires with a diameter as small as 8 nm [14]. It was recognized right at the time of the introduction of the technique that it has a unique potential to produce quasi-one-dimensional materials. Although MICA templates were used in a few studies for both electrochemical [15] and electroless [16] depositions of one-dimensional nanomaterials, this template family was suppressed by anodized porous alumina and ion track-etched polymer membranes. The latter two dominate the market of nanochannel templates in electrochemical research nowadays. The distinctive feature of MICA is that, due to its crystalline nature, the cross section of the track-etched channel is rhombohedral, not circular.

The principle of the preparation of track-etched membranes is as follows. The foils are bombarded with heavy ions of energy typically larger than 10 MeV. In velocity equivalence, this means that the speed of the particles is more than 10% of the speed of the light. Although the collision cascade within a polymer material decelerates the heavy ions, their penetration depth is larger than 100 μm. During the sequential collisions, the bonds in the membrane are destroyed in a cylindrical zone of a few-nanometre radius along the heavy ion trajectory. The small molecules produced leave the membrane easily, and the ion track can be further etched chemically to fully open

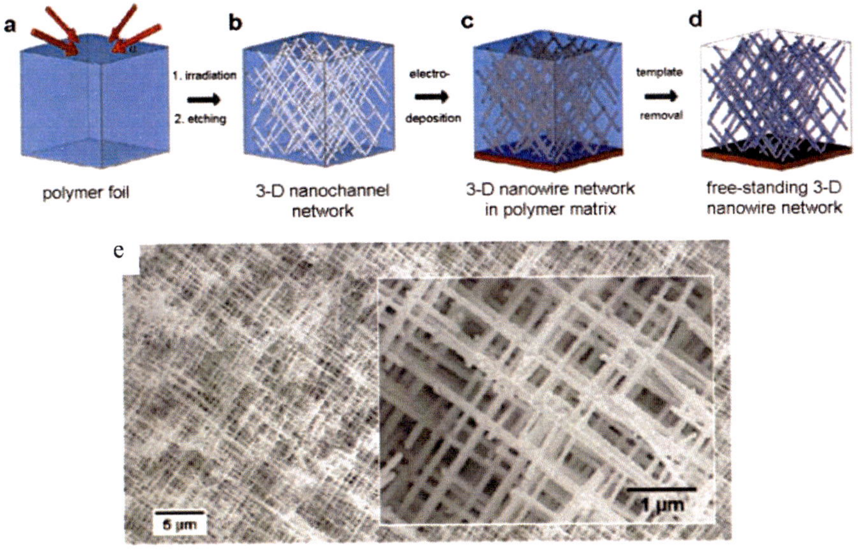

Fig. 11.1 **a** through **d** Scheme of the preparation of 3-dimensional interconnected nanowire networks in ion track-etched membranes. **e**: SEM image of a Pt network prepared in accord with the steps shown above. Reprinted with permission from [17]. Copyright (2011) American Chemical Society

and widen the pores. The pores are randomly situated within the membrane. The areal density of the pores can be regulated by the adsorbed flux up to 10^{12} ions/cm^2 where some of the etched pores tend to overlap. The pore width is determined by the etching time. The pores are not necessarily perpendicular to the sample surface but can be inclined to it. With varying the inclination of the irradiation of a particular sample, a percolating network of tilted crossing channels can be produced, Fig. 11.1.

Porous anodic alumina (PAA) can be prepared with an electrochemical process. A detailed description of this procedure is given in Chap. 13. Briefly, it is a self-organization process in which the pores are hexagonally ordered in an ideal case. Unlike for the track-etched membranes, the pore density and the native pore diameter are strongly interrelated for PAA. Since track-etched polymer and anodized alumina templates cover similar pore diameter, pore density and template thickness ranges, samples obtained with these templates are discussed in a single block from Sect. 11.2.2.

The formation of diblock copolymer and that of directionally solidified eutectic templates are discussed later in the relevant chapters. Diblock copolymer templates exhibit hexagonal ordering, similar to PAA templates, but the typical pore size and the sample thickness are much lower. The small pore size means a higher areal pore density at the same time. Directionally solidified templates exhibit pores faraway from each other due to their formation mechanism. Apart from track-etched membranes, all others contain pores perpendicular to their surface. A comparison of the nanochannel template types is given in Table 11.1.

Table 11.1 Comparison of the main properties of the nanochannel template types

	Porous anodic alumina templates	Ion track-etched membranes	Directionally solidified eutectics	Etched diblock copolymer template
Pore diameter range (nm)	12–500	10–500	200–2800	4–30
Pore density range (cm^{-2})	10^9–10^{11}	Up to 10^{10} (often much less)	6×10^5–3×10^7	5×10^{10} 5×10^{12}
Interrelation of pore diameter and interpore distance	Yes, 10% porosity can be taken as thumb rule	Independent	Varies with the composition of the eutectic system and with the cooling rate	Can be slightly tuned within the composition range of the suitable hexagonal phase
Regularity of the pore arrangement	Regular (hexagonal for properly anodized templates)	Irregular	Irregular (scarce fibres) or hexagonally ordered (dense fibrous structure)	Regular (hexagonal)
Pore diameter tuning	Diameter is defined by formation condition; widening: by chemical etching, narrowing: by ALD	Pore diameter can be tuned by the length of the chemical etching period	Slightly enhanced pore diameter upon prolonged etching	Not possible (the pore diameter is defined by the width of the column in the lyotropic phase)
Pore axis direction	Perpendicular to the membrane surface	Defined by the irradiation angle	Perpendicular to the electrode surface (as polished)	Perpendicular to the electrode surface
Typical membrane thickness (or nanochannel depth) (μm)	1–2 (evaporated films on conducting substrates) Up to 100 (free-standing Al foils)	5–50	<100	<2 (sometimes a few tens of nm only)
Electrode connection	Bulk Al by thinning the barrier layer Evaporated metal of the free-standing membrane Metal on which Al layer was anodized	Evaporated metal of the free-standing membrane	Material of the fibre in the eutectic whose dissolution leaves the channel behind	Metal on which the copolymer template was prepared

11.2.2 Deposition into Nanochannel Templates: Electrochemistry

Potential-controlled electrodeposition is used for nanochannel template filling processes in most of the studies. The advantage of the potential control is twofold. First, the control of the driving force helps avoid the onset of a side reaction when the deposition of the component of interest approaches the diffusion-controlled limit. Secondly, the chronoamperometric curves recorded during the electrodeposition are indicative of the advancement of the channel filling process, and hence, it can be used in the regulation of the pore filling ratio.

When the deposition of compact nanowires is aimed, it is important that a continuous metal layer coating is produced on one side of the membranes. This can be usually achieved with the seeding of a metal layer on the ending of the pore walls and thickening of this layer by electrodeposition until the pores are fully closed. The deposition of a well-adhering metal layer seems to be a prerequisite. If a membrane is only pressed against a metal, the growth starts with a metal layer in the gap between the metal electrode and the membrane [18].

If the deposition process leads to the formation of compact nanowires, the chronoamperometric curves can be divided into 4 sections (see Fig. 11.2 and similar graphs in other works [19–27]). After a transient and a short nucleation period, the current falls because the nanochannels become, at least partly, depleted with respect of the reactant. The gradual increase in the current density is due to the shortening of the diffusion pathway in the remaining unfilled parts of the nanochannels. The break point shown with the black arrow in Fig. 11.2 indicates the growth phase when the

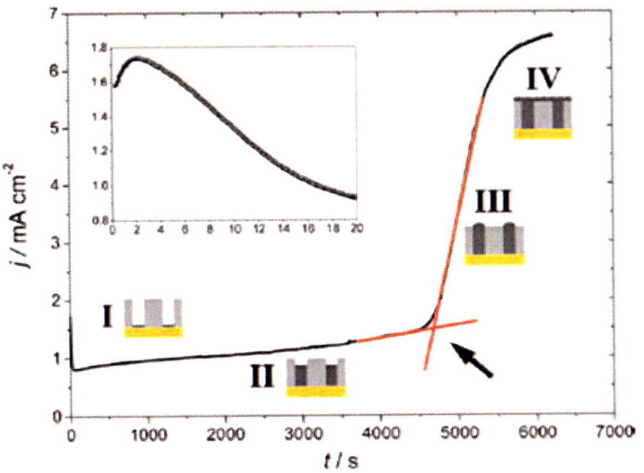

Fig. 11.2 Chronoamperometric record for the growth of homogeneous and compact nickel nanowires in a porous alumina template (*j* stands for the cathodic current density). The schemes next to the curve denote the nucleation (I), the growth period (II), the slight overgrowth of the nanowire while they are yet independent (III) and the formation of a laterally uniform film on the porous substrate (IV). Reprinted from [28]. Copyright (2011), with permission from Elsevier

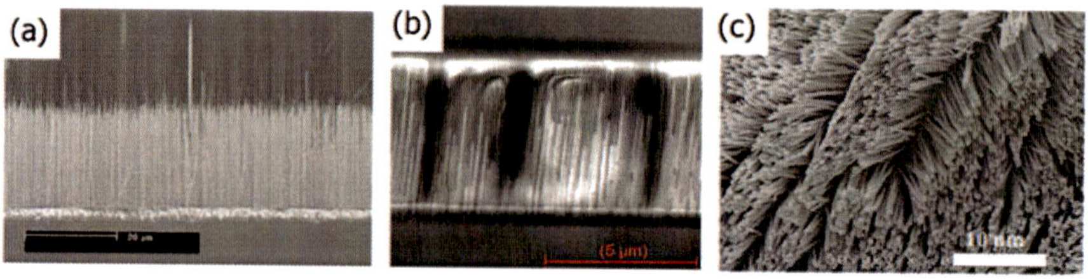

Fig. 11.3 Examples for the difference in the uniformity of pore filling as shown by cross-sectional images. **(a)** Nearly even pore filling with a few overgrown wires (Pd in PAA). Reprinted from [29]. Copyright (2008), with permission from Elsevier. **(b)** Very unevenly grown wires (Sn in PAA). Reprinted from [30]. Copyright (2019), American Chemical Society. **(c)** Typical SEM image with a "Persian carpet" pattern for a bunch of nearly evenly grown Au nanowires after the removal of the template. Reprinted from [27]. Copyright (2018), with permission from Elsevier

nanowires reach the top of the pores and the effective surface area of the nanowire assembly starts increasing. Since the individual nanowires all behave as nanoelectrodes, the spherical growth of the caps on the nanowires leads to coalescence after which a planar growth starts with a substantially higher current density.

The exact shape of the chronoamperometric track may differ substantially from one system to another. If the growth of the nanowire within the nanochannel template is uneven, the break point between the pore filling and pore end capping sections (II and III in Fig. 11.2, respectively) becomes less sharp. Example for both uneven and nearly even pore fillings is shown in Fig. 11.3, together with the SEM image of free-standing bunches of evenly grown nanowires after the removal of the template. It is to be noted that the achievement of a uniform pore filling is highly desired, especially in the case when the nanowire growth is not homogeneous but compositionally modulated. The demand for even pore filling is an important criterion of the sample preparation. Once a pore is overfilled, the difference between the filling ratio of the retarded and overgrown pores increases with time because of the enhanced precursor supply of the shorter empty pore section above the overgrown nanowires. When some of the pores are filled and an end cap is formed at the opening of these pores, the growth of the nanowires not reaching the top of the template practically stops. Therefore, when the nanowires are separated from the template after the synthesis by dissolving the template, the common pursuit is to fill the pores up to at most 80% of the pore length.

The electrode potential applied is a crucial parameter in the optimization of the nanochannel filling. If the potential applied is to negative, the decomposition of the solvent takes place which is accompanied with gas evolution, and the pores become blocked. The partial pore blocking adversely impacts the uniform growth rate of the nanowires. It is often found that not all the pores can be filled but the filling ratio is limited to at most 85–90% of the channels if a side reaction may take place. If composition depth profiling is carried out for samples in which several nanochannels are overfilled and the surface is already covered with a continuous deposit, the filling ratio as a function of depth has a minimum just beneath the film formed at the

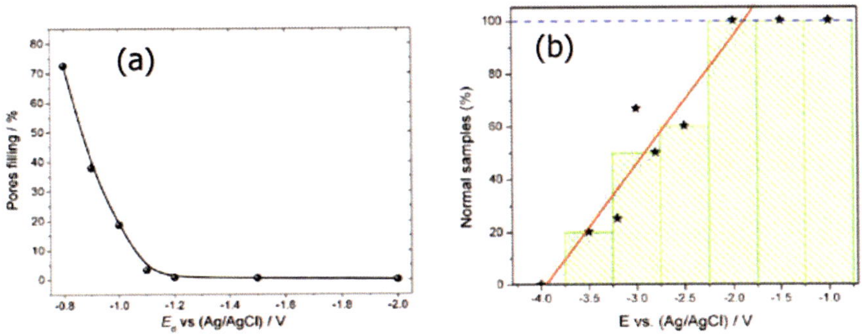

Fig. 11.4 Comparison of the pore filling ratios for Ni electrodeposition in porous anodic alumina membranes of approximately 50-nm pore diameter. **a** $c(Ni^{2+}) = 0.7$ mol dm^{-3} [28]; **b** $c(Ni^{2+}) = 1.33$ mol dm^{-3} [32] (factors other than the reactant concentration may also be partly responsible for the difference observed). Reprinted from [28, 32], respectively. Copyright (2011) and (2012), with permission from Elsevier

solution side of the membrane [31]. The impact of the pore blocking is represented in various ways in the relevant works. Two representation modes of the impact of the electrode potential on the portion of the regularly grown Ni nanowires are shown in Fig. 11.4. This comparison also shows that an optimization is necessary even for the same deposit material and nearly the same pore diameter since various baths can yield different current efficiencies at a particular electrode potential. (Additionally, technical parameters can also affect the potential dependence of the pore filling ratio such as the application or the avoidance of the ohmic compensation.)

Pulsed electrodeposition is often used in the deposition of homogeneous nanowires for tuning their structural properties. The chronoamperometric track during the pulse deposition bears different diagnostic fingerprints as that observed for constant-potential deposition. The current profile within a pulse is indicative for the deposition–dissolution process if the more positive pulse leads to dissolution. If the charge passed is displayed for entire cycles, the pore filling characteristics are the same as shown in Fig. 11.2 [33].

Nevertheless, the optimization of the pulse parameters (pulse frequency, current densities in each pulse, duty cycle) is rather heuristic. The trial-and-error method is customary since no theoretically established trends can be related to the optimization process. Low duty cycle is usually favourable for even pore filling [34]. Pulsed electrodeposition is often used to produce single-crystalline deposits. Another field where the pulsed electrodeposition method is important is the deposition of either alloys or compounds for which the ratio of the surface concentration of the reactants is important to stabilize the deposit composition. The same holds for deposits whose composition is pH-sensitive and the damping of the pH deviation from the desired value requires an off-time period during the deposition.

It has to be noted that a.c. electrodeposition has been used extensively as well as successfully for nanowire synthesis in several works. However, the background of the alternating voltage applied has not been sufficiently clarified. Neither the elemental processes taking place in each half-cycle could be elucidated, nor the voltage amplitudes applied fall in the classical regime of the electrochemical measurements ($10\ V < U < 50\ V$). Therefore, the discussion of the alternating current method was completely excluded from the present work.

11.2.3 Electrodeposition into Nanocavities of High Aspect Ratio: Models and Calculations

The basic assumption for the calculation of the temporal evolution of the nanowire growth process is as follows. A cavity with its open environment at the pore ending is taken with hemispherical symmetry (see Fig. 11.5). Besides the pore diameter and pore length, an effective diffusion channel is defined in the free solution. The diameter of the diffusion zone in the free solution is determined by the pore density but can be taken into account in different ways when the pore arrangement is either regular or random. The calculation may differ in at which distance from the pore ending the reactant concentration is taken as equal to the bulk one (i.e., at a finite distance or at the infinity). The equation to solve is Fick's second law under hemispherical conditions, taken into account also the special stepwise spatial boundary at the pore ending. The mathematics of the problem is identical to the calculation of the current response of a recessed microelectrode [35]. All models calculate with a laterally homogeneous nanowire growth until the nanowire reaches the template–bulk solution boundary. This is much in accord with the experience since laterally uneven growth, commonly taken into account for microtrench templates [36], does not occur for nanochannel templates.

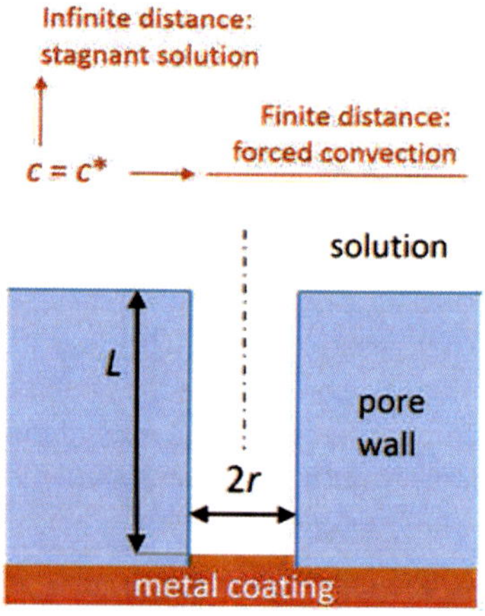

Fig. 11.5 Cross-sectional scheme of the nanopore arrangement and the notation of the geometric parameters of an ideal nanopore. *L*: nanopore length, *r*: nanopore radius.

Solution of the mass transport problem is often given for two limiting cases:

(i) Short-time approximation. Here, the initial reactant concentration is the same as the bulk one. Neither the diffusion field can extend up to the pore ending when the polarization starts, nor the nanochannel can be shortened significantly due to the growth of the deposit within. In this case, a Cottrell-type current transient can be obtained:

$$i\,(t) = zFr^2c^*\left(\frac{D\pi}{t}\right)^{1/2} \tag{11.1}$$

(ii) At long times, the diffusion field is steady state, and the total reactant flux is the same at the tip of the growing nanowire and at the pore mouth. If the diffusion fields of the neighbouring nanopores in the bulk solution do not overlap and no convection is applied, one can obtain for the steady-state current of a single nanopore that

$$i = \frac{zFr^2D\pi\, c^*}{L + \pi\, r/4} \tag{11.2}$$

If the overlap of the diffusion fields outside the pores is to be considered and the pore arrangement is random (i.e., track-etched membranes are used), the Avrami theorem can be used to calculate the mean effective area that provides the material supply of a single pore [37]. This means that the effective "collection surface area" of a single pore can be calculated with the help of an exponential weighting factor instead of a linear one, which is due to the randomness of the pore arrangement.

The models for calculation use a single diffusion coefficient. It is to be noted that in accord with the measurements, the diffusion coefficient may vary with the pore diameter [20], and it is not identical to that measured in the bulk solution either.

The results of a series of complex calculations of the filling process of regularly arranged nanopores have been published [38]. The thickness of the diffusion layer in the free solution was a variable determined by the hydrodynamic conditions, and this was taken into account through a geometric factor depending also on the porosity of the template. The kinetics of the electrode reaction was considered with a single Tafel-type exponential function. Chronoamperometric functions and pore filling times are given for various deposition potentials and reactant concentrations. In the second part of the work [39], the inhomogeneity of the pore filling was addressed. It was shown that the initial inhomogeneity of the pore lengths leads to a self-enhancing differentiation of the nanowire growth. The widening of the nanowire length distribution curve proved to the higher, the larger the deposition rate (i.e., the more negative the electrode potential).

11.2.4 Electrodeposited Homogeneous Metallic Nanowires

Nanowires composed of a single metallic element or alloys have been synthesized in a great variety. A selection of representative examples is presented in Table 11.2. The baths used for nanowire electrosynthesis are in most cases very similar to those that can yield a corresponding planar film with a good quality. An important difference can stem from the increased demand for nanochannel wetting. For a complete filling of the nanochannels with the plating solutions, nanochannel membranes are often soaked in distilled water for a long time, lubricated with a few drops of ethanol prior to the application, or the lubricant is added to the bath. This is the reason why baths used for template filling often contain some organic compound like alcohols or gelatine.

The key factor to produce compact nanowires is to adjust the deposition rate of the nanowire to the transport of the solute(s) to the reaction front though the solution in the nanocavity above the nanowires being grown. For this reason, the growth process of homogeneous nanowires can be just partly diffusion-controlled. Pulse plating strategies are often applied both for limiting the depletion of the electrolyte solution and to replenish the reactant supply.

Cyclic voltammetry is often used to characterize the potential dependence of the deposition rate and to choose the potential regime suitable for electrodeposition. It is to be noted that cyclic voltammograms recorded on a free planar substrate and for a nanochannel template are much different. Although the onset potentials of the deposition process are mostly identical, the potential window in which the mass transport impact is limited can significantly differ for the two electrode geometries.

Table 11.2 Representative collection of works dealing with electrodeposited homogeneous metallic nanowires

Metal	References	Metal	References	Alloy type	References
Ag	[40–42]	Ni	[19, 20, 28, 31, 32, 58, 59, 66–75]	Alloys of the iron group metals	[25, 84–94]
Au	[26, 27, 43–46]	Pb	[76–78]	Iron group metals–platinum group metals	[95–103]
Bi	[47–50]	Pd	[18, 29, 79, 80]	Noble metal alloys	[104]
Co	[19, 51–59]	Pt	[33, 81]	Metastable alloys of immiscible metals	[105–109]
Cu	[21, 23, 60–62]	Sb	[82]	Iron group metal–refractory metal	[34, 110, 111]
Fe	[58, 63–65]	Sn	[30, 78]	Heusler alloys	[112, 113]
In	[13]	Zn	[78, 83]		

This is to be taken into account during the optimization of the templated deposition processes.

11.2.5 Electrodeposited Non-metallic Nanowires

Although the electrodeposition of a large number of non-metallic materials has been established, some deposits are not compact but rather porous. The interesting feature of the template systems is that the strong spatial constraint makes it possible to obtain these materials as compact nanowires. Among the motivation of the electrodeposition of non-metallic nanowires, we have to highlight the application of semiconductors in optoelectronic devices. The band gap of the semiconductors can be tuned by the size of the deposit element, which opens a way to prepare custom-designed devices. Another important property is the thermoelectricity which can be potentially exploited in both refrigerating and waste heat harvesting devices. Below, a short summary of the materials suitable for nanowire preparation will be given with a short description of the conditions of their electrodeposition.

Oxides and hydroxides. Recipes elaborated for reductive electrodeposition of various oxides and hydroxides in the form or either films or porous coatings work with minor modification for nanochannel templates, too. Among others, such processes are suitable to produce ZnO [114], Cu_2O [115, 116] and $Cd(OH)_2$ [117] nanowires. All above-mentioned processes are based on the production of alkaline reagents by the reduction of a suitable co-reactant other than the metal ions in an aqueous solution. For ZnO, an alternative of the deposition from aqueous solution is the application of dimethyl sulphoxide as solvent and the alkalization of the media by reducing the dissolved molecular oxygen [118]. Several oxide nanowire arrays reported were produced in PAA membranes [115–118]. It seems that the pH increase needed for the production of the oxide/hydroxide nanowires is moderate enough so that it does not lead to the destruction of the alumina structure.

Electrodeposition of oxide nanowires involving metal atoms of higher oxidation number is carried out by oxidizing dissolved metal ions of lower oxidation number. Such anodic processes can be used to synthesize PbO_2 [119] and MnO_2 [120] nanowires. For PbO_2, the careful selection of the reactants and that of the pH of the solution lead to the formation of various crystalline forms of the oxide, while MnO_2 is obtained in an amorphous state.

Besides the direct synthesis of oxide nanowires, there is also an indirect synthesis route. An example is the deposition of Cu nanowires and their transformation to CuO nanowires by oxidation in air at high temperature [62]. Although there is a small roughness increment during the oxidation, the shape of the metal nanowires could be retained during the oxidation process.

Compounds of metals with the elements of Groups 5A and 6A. The nature of the precursor materials used for the electrosynthesis of compound semiconductors differs for aqueous and non-aqueous solutions. Therefore, a better insight can be given if these methods are discussed separately, irrespectively of the fact that some materials can be obtained with both process types. The precursor materials of the metallic component for both solvent types are essentially the same.

For aqueous solutions, the component of higher electronegativity can be introduced in an oxoionic form. The reactive species hence will be $HTeO_2^+$ (from TeO_2), $HSeO_3^-$ (from SeO_2), $S_2O_3^{2-}$, etc. Synthesis based on this approach was reported for the electrodeposition for various nanowires like Bi_2Te_3 [22, 121, 122], CdTe [123], Cu_3Se_2 [124], SnS [125], etc. Less commonly, a halogenide compound of both components can be used (e.g., for InSb [126]). The synthesis by using a compound with simple anions (like S^{2-}) is also possible if the metal ion is used in a complex form in order to prevent the precipitation formation. This case was demonstrated for the electrosynthesis of CdS [127] where sodium sulphide and the EDTA complex of the Cd^{2+} ions were used as precursor materials.

The electrosynthesis of similar binary compounds can be carried out with non-aqueous solvents, too. The most common solvent is dimethyl sulphoxide. This route opens new possibilities since the more electronegative component can be used in an elemental form. The list of examples includes the synthesis of CdS [128–130], CdSe [131], Bi_2S_3 [132], ZnS [133], etc. When a compound of the Group 5A and 6A elements is used, their halogenides proved to be suitable like $SbCl_3$ for BiSb [134] and $SeCl_4$ for Ag_2Se [135].

The potential window for the synthesis of binary compound semiconductors is usually very narrow (see also Chap. 4 and the relevant details on the EC-ALD processes). This is particularly true if a ternary compound should be synthesized. The pursuit to produce ternary semiconductors originates from the necessity of the optimization of the deposit properties such as their Seebeck coefficient and photochemical response. It is possible to produce ternary semiconductors with one or two metallic components ($Bi_2Sb_{0.6}Te_3$ [136], $Bi_{2-x}Sb_xTe_3$ [137], $Bi_2Te_{3-y}Se_y$ [138] and Bi_2Te_2Se [139] represent the former and ZnCuTe [140] the latter type). For ternary compounds, not only the narrow potential window for the deposition is a difficulty but the preparation of the suitable electrolyte solution requires special care due to the incompatibility of some of the components without a suitable complex formation.

11.2.6 Electrodeposition of Compositionally Modulated Nanowires

Similarly to the compositionally modulated films, compositionally modulated nanowires can be prepared with either the multiple-bath or the single-bath methods. The forthcoming discussion will follow this classification.

Multiple-bath method. The bath change process is somewhat more delicate for nanowire preparation than in the case of the deposition of a planar film because the electrolyte change within the nanopores has to be completed. Therefore, the multiple-bath method is used if at least one of the following criteria prevails: (i) the segment lengths are relatively large, and the number of the segments is limited; (ii) no single-bath solution of the task is available due to the chemical incompatibility of the bath components to be used. The examples to be shown below will highlight a few cases where the dimensional constraint applied by the nanowire geometry has some significance in the process.

Various nanowires synthesized with the multiple-bath methods were used to study the properties of the junction within the nanowire. Concerning the metal–metal interfaces, it was demonstrated that the crystalline form of Co strongly determines the coherence of the heterojunction connection to Au, and the fcc–fcc connection proved to be crack-free as opposed to the hcp–fcc junctions [141]. For metallic heterojunctions between Au and Sn, an asymmetry was demonstrated for the Au–Sn and Sn–Au deposition order, showing the formation of different intermetallic compounds by the spontaneous interdiffusion of the components [142]. In these studies, nanowires proved to be advantageous because the junction of different metals could be studied without an additional sample preparation step (like cross-sectional cutting and polishing), hence eliminating the possibility of sample preparation artefacts.

Nanowires also make an ideal geometry to study the junction properties between organic materials and metals. For instance, photoelectrochemical response of the metal–polymer junction is difficult to detect for planar films, but it is easily accessible in nanowires [143]. The study of the molecular conductivity is based on the same principle. If nanowires partly filling a template are covered with a self-assembled molecular layer and this layer is plated with another metal, a good "metal 1–organic layer–metal 2" type sequence is obtained in a well-contained nanocavity whose conductivity is easy to study [144]. The synthesis of such structures is possible with the multiple-bath method only.

The double-bath synthesis is a suitable tool in bulk nanowire production, too. If the nanowires needed are much shorter than the available template, the alternating deposition of the wire segments with desired composition separated by a less noble sacrificial metal leads to the preparation of a precursor material containing the nanoobjects needed. After the template synthesis, the sacrificial metal has to be dissolved, and batches of monodispersed nanorods can be obtained. The demonstration study of this method applied Au as nanorod material and Ag as thin sacrificial wire segments [145].

Another production-related field of nanowire synthesis is the construction of multisegmented nanowires with a magnetic segment and one or several non-magnetic segments that can be used for functionalization. Such nanowires are a promising candidate for biomedical applications by using the guiding of the nanowires with external magnetic field for targeted drug delivery.

Double-bath-based nanowire electrodeposition is an interesting synthesis approach when the co-deposition from a single bath with a sufficient composition control cannot be solved. This was demonstrated for Bi–Tl [146] and Pb–Bi nanowires [147]. When segmented Pb–Bi nanowires plated from alternating baths were melted within the PAA membrane, homogeneous Pb–Bi nanowires were obtained after solidification, in spite of the equilibrium miscibility gap of the components in bulk form. Similar processes have a great perspective to produce materials at the nanoscale that cannot exist as bulk alloys.

Table 11.3 Representative list of works dealing with compositionally modulated electrodeposited nanowires obtained with the single-bath method (The less noble segment/layer always contains the more noble metal as a minor component.)

Major type	Composition	References
Iron group metal(s)/Cu	Ni/Cu	[148, 149]
	Co/Cu	[150–154]
	NiFe/Cu	[150, 155]
	CoNi/Cu	[156–159]
Iron group metal/platinum group metal	Ni/Pt	[160]
	Fe/Pt	[161]
	Co/Pt	[162]
	Fe/Pd	[163]
Other	Pb/Cu	[164]

Single-bath method. The application of the double potential pulse deposition method is widespread for the deposition of multilayered and segmented nanowires form a single bath. The majority of the examples refer to magnetic/non-magnetic multilayers (see Table 11.3) since the magnetic and magnetotransport properties of these materials can be easily adjusted by the change of the layer/segment lengths. The principles of the deposition of compositionally modulated nanowires are fully in accord with that described in Chap. 5.

Annealing of Co/Pt nanowires [162] offers a method that is the inverse what was demonstrated for homogeneous Pb–Bi nanowires [147]. Since the direct synthesis of Co–Pt alloys with the desired highly ordered phase with low oxygen impurity is difficult, the multilayer approach combined with annealing offers a suitable way for the synthesis of an otherwise hardly available nanowire composition.

11.2.7 Miscellaneous Properties of Nanowires

Nanowires have peculiar physical and chemical properties for various reasons. They usually overperform their bulk counterparts in catalytic and electrocatalytic properties. For optical properties, their reflectivity or plasmonic behaviour stems from the regularity of the structure and the composite nature of materials of a different refractive index. Since these properties are not peculiarly related to the high aspect ratio of the electrodeposited nanoobject, the discussion below is restricted to the properties that do not occur for systems of other type.

Component distribution. The solution in the nanochannels cannot be agitated. Therefore, the composition gradient of the precursor materials may sufficiently change as the nanowire grows, and this can lead to a composition variation along the nanowire. The composition gradient along the nanowires is seldom measured. Where such data were reported, the results showed that the compositional homogeneity of the nanowires can be ensured with a little optimization effort (see, e.g., Refs. [155] for Ni–Fe and [121] for Bi_2Te_3).

Morphology, crystallinity, structure and texture. It is generally believed that the nearly cylindrical side surface of the compact nanowires is a good replica of the template that it is embedded in. Although the above statement can be taken as a sufficient first approximation, a great variety of information has been published in the relevant literature that makes this picture rather sophisticated. For instance, nanowires deposited into track-etched polymer membranes were found not to by cylindrical but "cigar-shaped", having a central section thicker than the ends [20, 24, 109]. When the nanopores were just partly filled, a thinner diameter was found at the substrate side [114]. It is not fully clear whether this shape is a consequence of either the membrane preparation (overetching the central part during the pore widening) or the shape change of the soft template during the deposition process. Since the diameter measured for electrodeposited nanowire is sometimes significantly larger than the nominal diameter of the nanopores in which they were synthesized [114], it is suspected that polymer membranes can be either mechanically deformed or further etched during some electrodeposition processes.

It is also common that the side surface roughness of the nanowires composed of the same material differs as a function of their deposition potential, and nanowires of different materials in the same template exhibit various side surface morphologies [165]. The side surface roughness of the electrodeposited nanowires nearly always changes if potentiostatic mode is replaced with pulse plating, so much that even diameter-modulated sectioned nanowires can be synthesized from the same material (Pt) [33]. It is also possible that the nanowire (here, Bi) diameter is constant but significantly smaller than that of the nanochannels as pulse-plating parameters are varied [49].

The morphological change along the nanowire length is even more obvious when the nanowire is multilayered or segmented. In the case of segmented nanowires, the formation of smooth side surfaces [141, 149, 163], a small change in diameter at the boundary of layers [146, 148, 157, 160, 166] and periodical diameter oscillations in accord with the composition change [161, 162, 164, 167] are all possible. Since the periodic oscillation occurs equally for both soft polymer and PAA templates, this phenomenon cannot be attributed to the local deformation of the template; rather, a different adhesion of the components to the template side wall may be behind the diameter change. For multilayered nanowires, the layers are seldom planar but the nanowire is undulated, lending a "potato chip pack in a cylindrical box" morphological character to the nanolayer stacking [154, 155, 157, 159].

Nanowires composed of one single metallic element are crystalline, but both the crystallite size and the preferred orientation show a great variety. The success of the production of single-crystalline nanowire can be a function of, e.g., the bath used [43]. Even though the application of the potentiostatic deposition can also lead to single-crystalline nanowires [61], the application of both pulse [54, 55, 72] and reverse pulse plating [43, 63] is very common to achieve the growth of single-crystalline nanowires. The change in the deposition mode is nearly always accompanied with the modification of the preferred growth direction [60]. The variation of the texture and grain size are also related to the pore diameter [42], which means that nearly each system needs a unique optimization means to achieve the desired growth mode.

The constraint imposed on the nanowire dimension has such a fundamental impact on the growth characteristics of the nanowires that even trends established for electrodeposited films become invalid. For instance, the deposition of single-crystalline and polycrystalline Pb nanowires was successful at large and small overpotential, respectively [76], even though the trend for open geometry is just the opposite. Similar experience was also published for Co–Ni alloys [88], while the trend for plain films was found to be valid for Ni nanowires [28].

Besides the growth modes, the defect structure in the nanowires also shows a great variety. In spite of the geometrical constraints, twin boundaries perpendicular, angled or parallel to the nanowire axis can occur [168].

Concerning the grain size of alloys, the general rule is that the larger the unit cell size (and hence, the larger the degree of long-range ordering), the smaller the grain size. Hence, alloy nanowires are seldom single crystalline.

Non-metallic nanowires also show a good crystallinity and a strongly preferred orientation that can be adjusted by tuning the deposition conditions. The exceptions are ZnO for which the high crystallinity is coupled with the lack of preferred orientation [114], and MnO_2 that can be obtained in an amorphous state [120].

Magnetization and coercivity. The magnetic studies of electrodeposited nanowires make one of the largest portions of the field. This is why magnetic nanowires in highly ordered templates were considered to be a prime candidate as perpendicular recording media due to the inherent large shape anisotropy of the nanowires with their axis perpendicular to the sample surface. For a magnetic recording media, a high remanence and a moderately large coercivity are required so that the bit element is sufficiently resistant to magnetization reversal and provides large enough stray field for the read-out of the bit information.

As for the magnetization properties, the material of the nanoporous template has no substantial impact; rather, the magnetic properties depend on the nanowire only [59]. The decrease in diameter of homogeneous nanowires tends to increase the coercive field, the remanence and the squareness of the magnetization curve [53, 57, 58, 66], and the increase of the aspect ratio has a similar effect [64]. This is because the shape anisotropy of the homogeneous nanowires is the dominant factor that determines the magnetization reversal properties. When the separation distance between the neighbouring nanowires is decreased by applying a pore widening process in PAA membranes, the coercivity is diminished, indicating the onset of the magnetostatic interaction between the neighbouring nanowires [65]. A comparison of the magnetization behaviour of single-crystalline and polycrystalline nanowires is shown in Fig. 11.6.

The optimization process of the deposition of Co and Ni nanowires often refers to the achievement of the production of single-crystalline nanowires. The explanation is that the lack of crystal boundaries that serve as domain wall pinning centres ensures the synchronized magnetization reversal of the entire nanowire with no possibility of counterdomain formation within the same wire. The single-crystalline nature of magnetic nanowires, together with the coincidence of the magnetocrystalline easy axis with the nanowire axis, is the key for both the high squareness of the magnetization curve and the high coercivity. These two parameters can often be achieved

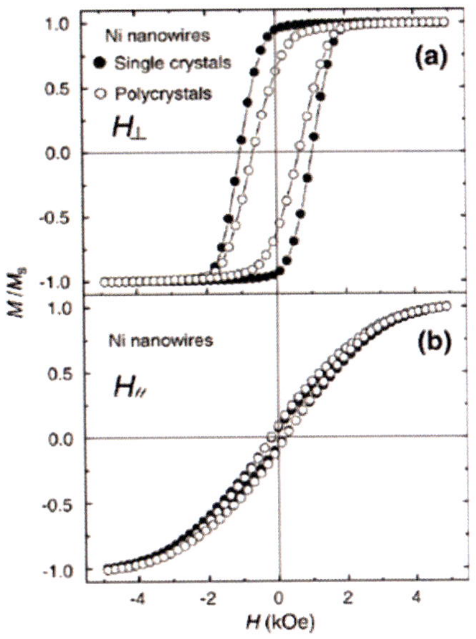

Fig. 11.6 Comparison of the magnetic behaviour of single-crystalline and polycrystalline Ni nanowires (see filled and empy symbols, respectively). The external magnetic field is: **a** parallel to the nanowire axis (i.e., perpendicular to the plane of the template); **b** perpendicular to the nanowire axis (i.e., in-plane with respect to the substrate). Reprinted from [71]. Copyright (2007), with permission from Elsevier

with either a through bath pH regulation [51, 52, 74] or with optimized electrodeposition parameters in direct current [28, 58, 69, 73–75] and pulse-plating [54, 55, 72] modes. As a result of the optimization of the nanowire structure, the coercivity of the nanowire arrays along the nanowire axis direction can be as high as 4 kOe [101, 169], more than an order of magnitude larger than the coercivity of their bulk counterpart. For Co–Pt alloys obtained with annealing of Co/Pt multilayered nanowires, coercivity in the wire axis direction can be larger than 10 kOe [162].

The elucidation of the magnetization behaviour of compositionally modulated nanowires is more difficult than that of their homogeneous counterparts. The shape anisotropy of segments with lower aspect ratio affects the coercivity, but the direction of the coercivity change is ambiguous. The squareness of the magnetization curves decreases with the nanowire segmentation [19, 150]. If the nanowire is multilayered with thickness much less than the wire diameter for both the magnetic and non-magnetic layers, the competition of various anisotropy contributions usually leads to a more even magnetization for in-plane and perpendicular-to-plane directions than for homogeneous wires. If the thin magnetic layers are separated by non-magnetic segments much longer than the wire diameter, the shape anisotropy of the single magnetic layers becomes dominant [150].

Electrical transport properties. The study of the electrical transport properties of nanowires has always been important because thin nanowires are a very good approximation of one-dimensional materials. Hence, their study is crucial for testing theories referring to one-dimensional systems.

The study of the electrical transport properties of nanowires can be carried out in three ways, namely as nanowire array within the template, single nanowire within the template and single nanowire after being removed from the template. For measuring the transport properties of nanowire arrays, the back side electrical contact used as working electrode connection during the nanowire deposition can serve as one of the electrical contacts, while the solution side of the nanowires can be contacted either through the percolating overgrown deposit or with electrically conducting glue. This method makes it possible to measure the average of the transport properties of a large number of nanowires. If the template is removed, single nanowires can be contacted with a micromanipulating system in a SEM device, mostly on lithographically patterned electrical contact pads. Below, we will pay attention to methods dealing with single nanowire still within the template because the preparation of the systems to be measured involves an electrochemical step in excess to the nanowire deposition. The schemes of these methods are shown in Fig. 11.7.

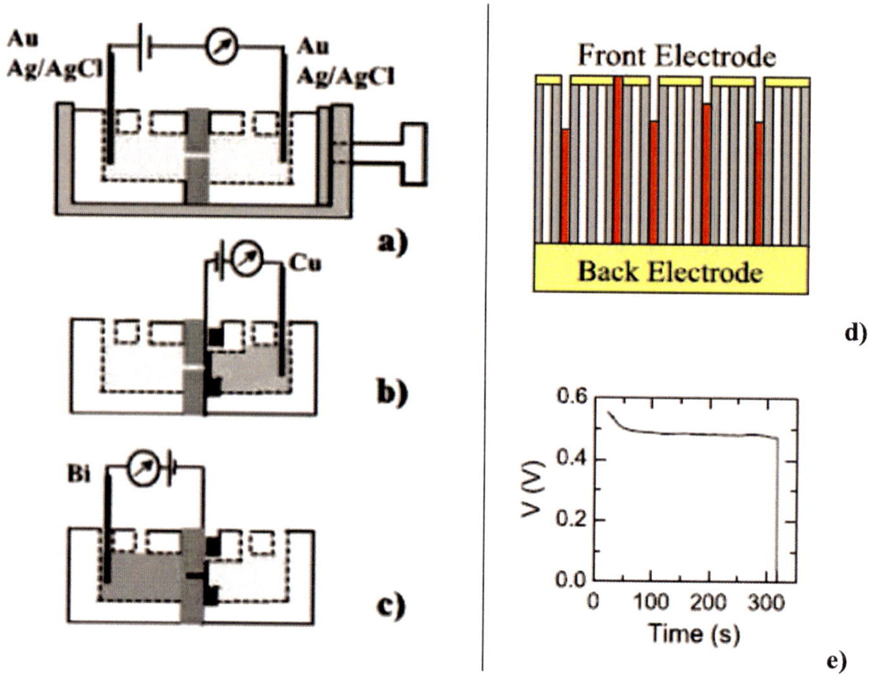

Fig. 11.7 Schemes of the methods used for the preparation of templates to carry out single nanowire transport measurement without a template removal step. **Left**: **a** Permeation type cell to detect the conductivity change upon the chemical etching of a foil treated with single-ion irradiation. The pore opening/widening is followed by the change of the current through the nanochannel. **b** Thickening of the back side contact of the single-track template by electroplating after *ex-situ* coating with a thin conductive layer. **c** Electrodeposition of one single nanowire into the monochannel template. Adapted with permission from [47]. Copyright (2004), American Chemical Society. **Right**: **d** Template configuration for the detection of the penetration of the first nanowire to the solution side by using an additional front contact metal layer. **e** Chronopotentiometric response with a sharp drop of the voltage at the moment of the front side contact formation. Adapted from [78]. Copyright (2008), with permission from Elsevier

The preparation of the membrane with one single-ion track is based on the limitation of the cross section of the irradiating ion beam and the detection of the first ion passing through the polymer membrane [47, 109]. Then, the pore opening/widening process is carried out in a permeation cell. The electrodes in the two chambers of the permeation cell are used as conductivity sensors (see the left panel in Fig. 11.7). Then, the back side contact formation and the electrodeposition of the nanowire can be performed with the usual process. The major difference is the current level observed during the nanowire formation ($0.05 \text{ nA} < I < 2 \text{ nA}$). An alternative method includes the preparation of a nanowire array but with stopping the deposition process once the first wire penetrates through the whole membrane and contacts the front side electrode (see the right panel in Fig. 11.7 [78]).

Superconductivity in Pb nanowires was detected nearly at the same transition temperature as for bulk Pb when the nanowires were single crystalline, while a decrease in the critical temperature was shown for polycrystalline samples [76, 77]. If nanowires are made from materials of lower critical temperature than the end contacts (like Sn with Pb contacts, Zn with either Sn or Pb contacts), the contacting metal can cause an induced superconductivity nearly with the same critical temperature as the contact metal [78]. It was demonstrated that the interdiffusion in the superconductor–contact metal region impacts the resistivity change of the nanowires [142]. The study of superconductivity in nanowires is important due to the possible quantum confinement. For instance, it was found that 70 nm is the critical diameter for Zn nanowires below which the bulk-like behaviour changes to one-dimensional one [83].

Magnetoresistance can also be fundamentally different for multilayered nanowires than for thin films of similar composition modulation. This is because in multilayered nanowires, the measurements can be naturally carried out with the current-perpendicular-to-plane geometry, while the current-in-plane configuration is customary for thin films. The key parameters determining the magnetoresistance ratio are also different for the two geometries. These studies are important for the verification of the theories of magnetoresistance as well as for the possible construction of magnetoresistive memory elements. Measurements for both homogeneous [48, 59, 75, 107] and compositionally modulated nanowires [150, 151, 153–155] have been published for a variety of nanowire compositions.

Thermoelectricity. The thermoelectric properties of semiconductor nanowires have been optimized in a large number of works. The reason for the interest in the thermoelectricity of nanowires stems from the fact that nanowires are well organized within the template, and hence, they can be potentially integrated into thermoelectric devices. The conversion efficiency of such devices is characterized with the figure of merit (Z) of the thermoelectric material applied. Z is defined by the following equation:

$$Z = \frac{S^2\sigma}{\kappa} \tag{11.3}$$

where S is Seebeck coefficient, σ is the electrical conductivity and κ is the thermal conductivity. (The relationship is often given in a form by multiplying both sides with the absolute temperature.)

Although quantum confinement effects lead to a power factor enhancement if the nanowire radius is around 10 nm, the decrease of the thermal conductivity has a measurable contribution to the power factor enhancement with diameters below 100 nm. Also, the template synthesis can easily produce single-crystalline nanowires, which is a great advantage due to the elimination of the grain boundary electron scattering, hence leading to an electrical conductivity improvement as compared to polycrystalline bulk materials. It was shown for various thermoelectric materials that the power factor enhancement predicted on the basis if the spatial confinement can indeed be achieved [22, 136].

11.2.8 Metallic Nanotubes Obtained by Electrodeposition into Nanochannel Templates

It was commonly observed that nanotubes can be grown by electrodeposition into nanochannel templates. As it was described in a comparative study [170], the chronoamperometric transient measured during the nanotube formation is qualitatively very similar to that observed for the nanowire growth. The difference is that the nanotube formation is a much faster process, simply because it requires the transport of a much smaller amount of precursor material. Regarding the deposition conditions, it must be mentioned that the elecrosynthesis of templated nanotubes often takes place with current control instead of potential control. This is because of the necessity of the maintenance of either a large electric field or a fully diffusion-controlled process.

Figure 11.8a shows a mass transport scheme for the cross section of a growing nanotube within a template channel. It can be seen that the filling of the internal cavity already formed in a nanotube is not possible since once the diffusing reactant can reach the top zone of the nanotube, it reacts in the close vicinity of the pore ending. Therefore, the internal cavity within the nanotube is fully depleted with respect to the reactant. Two examples for electrodeposited nanotubes are shown in Fig. 11.9. It can be seen in both images that the nanotubes are composed of several grains not only as tube segments but also along the tube perimeter.

Concerning the circumstances of the nanotube formation, a large number of factors are mentioned in the literature. The role of each factor has been plausibly evidenced; therefore, it is possible that in some cases, the synergy of more than one single factor prevails, resulting in the formation of nanotubes instead of nanowires. The relevant experimental parameters are given below.

Interaction of the growing deposit with the pore walls of the template. This method was introduced in 1991 [173] for the preparation of gold nanotubes in membranes by using the functionalization of the internal pore walls with 2-cyanoethyl terminal groups that serve as anchoring centres for the growing nanotube [174–176]. For

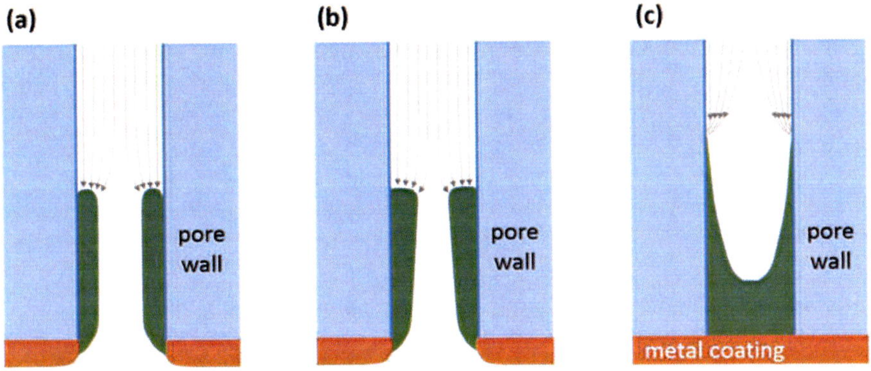

Fig. 11.8 Modes of tubular growth in nanocavities. **a** Steady-state growth of a nanotube; **b** growth of a nanotube with gradually closing internal cavity, indicating the partially sidewise growth at the opening of the nanotube; **c** field-driven change of the nanowire-like to the tubular growth as a result of the preferential interaction with the pore wall. Arrows show the current lines

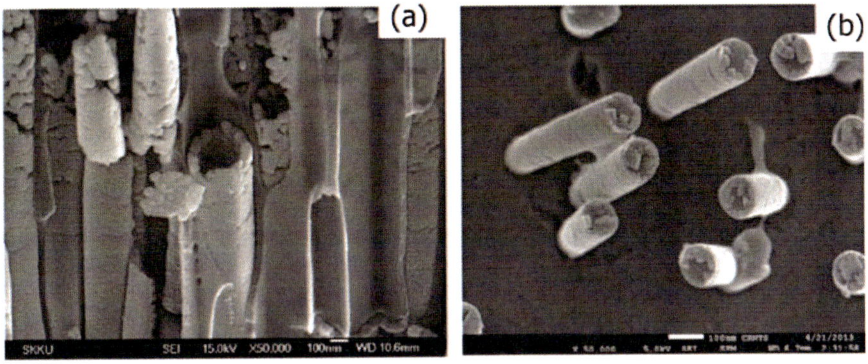

Fig. 11.9 SEM images of electrodeposited nanotubes. **a** Pt nanotubes within a PAA membrane [171]; **b** Cu nanotubes reaching out from a partially removed polymer template [172]. The multigrain structure of the nanotubes is to be observed in both images. Reprinted from [171, 172], respectively. Copyright (2008, 2014), with permission from Elsevier

the deposition of nickel and cobalt nanotubes based on template wall adherence enhancement, methyl-γ-diethylenetriaminopropyl-dimethoxysilane proved to be the proper surface modifier due to the strong interaction of these metals with the amino groups [177, 178]. 3-aminopropyltriethoxysilane was equally good for Au and Ni [176]. It is common for electropolymerization in nanochannels of plastic membranes that polymer nanotubes are produced instead of polymer nanocolumns [174, 175], which is also explained with the preferential interaction of the growing polymer with the highly hydrophobic pore walls. Spontaneous cylindrical phase separation during the electrodeposition of alloy nanowires can also be explained with the preferential interaction of one of the components with the side wall (Ni–Cu [179], Ni–Ag [106];

Ni is the major shell component in both cases). The core–shell type nanowires thus produced make it possible to obtain nanotubes by selectively dissolving the core component [179].

Coverage of the template by a thin enough conductive strike layer. The big majority of works in which the thoughtful electrosynthesis of nanotubes in template was carried out employed a thin electrically conducting coating on one side of the membrane. The thickness of the deposit is at least half of an order of magnitude smaller than the pore diameter; hence, the pores remained open after the coating process and a conducting rim occurred around the perimeter of the nanopores only. If the deposition rate was high enough to achieve a fast depletion of the centre of the pores with respect to the reactant, a steady-state nanotube growth could be achieved (see Fig. 11.8a). This strategy was successful for the deposition of a great variety of nanotubes like Au [173], Pt [171, 180], Pd [181] Co [171], Fe [182, 183], Ni [80, 184–186], Bi [187], Co–Pt [171, 188], ZnO [189], etc.

Hydrogen evolution. The strong deviation of the deposition efficiency from 1 was also found to be related to the formation of nanotubes [190–193]. In some other cases, the current efficiency was not measured, but the potential applied fell beyond the stability regime of water, and the hydrogen evolution was hypothesized as the factor responsible for the nanotube formation [172]. It cannot be established systematically in which cases the hydrogen evolved blocks the pores and when they modify the deposition mode from dense to tubular. It seems to be a criterion that the nanopore diameter should be at least 200 nm. It was found for various materials that the decrease in pH and the application of a more negative deposition potential may be enough for either inducing the growth of nanotubes instead of nanowires; if such changes are applied in the nanotube growth regime, the pore wall of the nanotubes becomes thinner. The latter trends are in accord with the theorem that the gas evolved blocks the centre of the nanopores.

Nanotube-to-nanowire transition during the pore filling. It has been reported for various metallic [194–196] and non-metallic [197, 198] materials that the initial nanotube growth turns into nanowire growth merely by the larger time elapsed after the start of the electrodeposition. The resulting objects in the membrane pore are either hybrid particles having an axial cavity at the substrate side and a compact structure at the solution side or nanowires with a protrusion at the perimeter at the bottom side.

The transition mechanism is visualized in Fig. 11.8b. If the nucleation at the substrate size is restricted to the perimeter but the growth is not diffusion-limited, the sufficient reactant supply can lead to a partly sidewise deposition ranging to the internal cavity of the nanowire. After a sufficiently long time, the cavity can close, which is in accord with the energetic minimum due to the disappearance of a surface element with very high curvature. The closure of the initially formed nanotube cavity can be associated with a drop in the electrical current in the chronoamperometric curves [196]. The initially formed cavity in compact nanowires can be visualized by partially dissolving the template and imaging the bottom side of the nanowires [199]. As an extreme example, the full closure of nanotubes formed all along the nanocavity by the electrodeposition process without an end-bump formation has also been reported [200].

Switch of the nanowire growth to nanotube formation. The morphology of the nanoobject being deposited may change abruptly *in situ* during the deposition into a specific porous membrane. This morphological change, an opposite phenomenon of the nanotube closure with time, is visualized in Fig. 11.8c. The sudden turnover to nanotube formation is a result of the combined impact of the diffusion-limited nature of the growth and the high electrical field. In such cases, it might seem that the deposit appears at the top of the porous layer without an obvious pore filling due to a very fast penetration of the tiny nanotube along the pore wall [201].

It is easy to understand that the change in the deposit morphology can be observed as a function of the current density. Even if hydrogen evolution is not considered, the rod-like growth mode at a small current density can switch to a tube-like growth mode if the current density is high enough and the deposition is largely impacted by both the diffusion limitation and the electrical field [202]. However, the current density effect is not uniform since one can find examples also for the case when the low and high current density leads to nanotube and nanowire formation, respectively [203]. Since the diffusion and electrical field effects must not work in this case, one can speculate that such a current density dependence of the morphological transition is related to the anomalous change of the current efficiency (which is possible for the Ni–P alloy produced in the above-mentioned study).

It was also shown, however, that the large electric field is by far not the single important factor in the nanowire-to-nanotube transient. Instead of this transient, the large negative electrode potential may lead to anomalously fast nanowire growth instead of nanotube formation [32].

Effect of the pore diameter. It seems that there is a threshold nanopore diameter above which nanochannels can accommodate not only nanowires but also nanotubes [204]. One can hardly find data for electrodeposited nanotubes within templates of pore diameter less than 150 nm.

Chemical aspects of the deposition process. In the case when the quality of the substrate covering the pores continuously has an impact on the growth mode, it can also be assumed that the high nucleation overvoltage on the conducting substrate plays a role in the inclination to nanotube growth [205]. The same effect may stand behind the variation of the growth preference when the modification of the growth mode can be achieved with an additive commonly used for the deposition of metal films [206].

11.2.9 Templates Prepared from Diblock Copolymers

Diblock copolymers are composite materials made from two polymers that are prepared separately and are not miscible. The morphological features of diblock copolymers include a range of segregated structures, such as spherical micelles, cylindrical micelles, vesicles, fcc- and bcc-packed spheres, hexagonally packed cylinders, gyroids as well as simple, modulated and perforated lamellae [207]. For electrochemical applications, diblock copolymers with cylindrical guest material embedded in the host polymer are of highest importance. Templates prepared from such soft composites are analogous to nanoporous track-etched and PAA membranes, but their characteristic size parameters fall in a different range. Besides, micellar [208] and gyroid [209] templates have also been used.

The notation of the diblock copolymers follows the B–b–X scheme in which B is the percolating and matrix-forming majority component, whereas X is the dispersed (granular or fibrous) constituent, and "b" stands for identifying the first list element as block. For template formation with nanochannels, electrically conductive materials are coated with a thin layer of the components of the diblock copolymers. The formation of the thin surface layer usually takes place with spin coating followed by the solvent evaporation. The structure of the diblock copolymer does not form spontaneously as the mixture is drained but requires annealing above the glass formation temperature of both components. For the orientation of the nanocolumn-forming minority component, an electric field perpendicular to the surface is often applied. Then, the chemical bonds in the material of the nanofibres are partly destroyed either chemically or with irradiation. This makes it possible to dissolve the fibres with a suitable solvent. An advantage of the process is that the further steps do not involve any etching process by using highly reactive media that can destroy the substrate; hence, metals other than noble ones can be used as substrate. The template thus produced is mesoporous since the diameter of the polymer fibres to be removed usually falls in the diameter range of at most a few tens of nanometres.

Concerning the typical materials used for the preparation of diblock copolymer templates (DCTs) [210–214], polystyrene (PS, 55–70%) and poly(methylmetacrylate) (PMMA) as block-forming and fibre-forming components, respectively, are commonly used. Key parameters for the formation of a uniform diblock structure are the molecular weight of the polymers and their polydispersity (desired values: ~1–5 $\times$ 10^4 Daltons and <1.1, respectively). For the spin-coating step, a dilute solution of the polymeric ingredients in toluol is used (concentration: 0.5–3 wt.%). The subsequent annealing process at 150–190 °C for 1–2 days is carried out with the application of an electric field perpendicular to the substrate. The field strength applied is around 30–400 V μm^{-1}. Without an electric field, lamellar segregation may take place instead of cylindrical domains. The exposure the diblock copolymer to UV light has a twofold impact since it degrades the PMMA domains and simultaneously cross-links the PS matrix. Thereafter, the partly destroyed PMMA can be dissolved, e.g., in acetic acid, hence developing the nanoporous structure with a 15–35-nm nanopore diameter and an interwire spacing about 150% of the pore diameter. Various other copolymer components are also applied for preparing nanoporous systems for electrochemical applications [215–217]. Some surface images of the thus produced templates can be seen in Fig. 11.10. Due to both the soft nature of the templates and their sub-micrometre thickness, cross-sectional SEM images are too difficult to make for either the empty of metal-filled templates obtained from diblock copolymers.

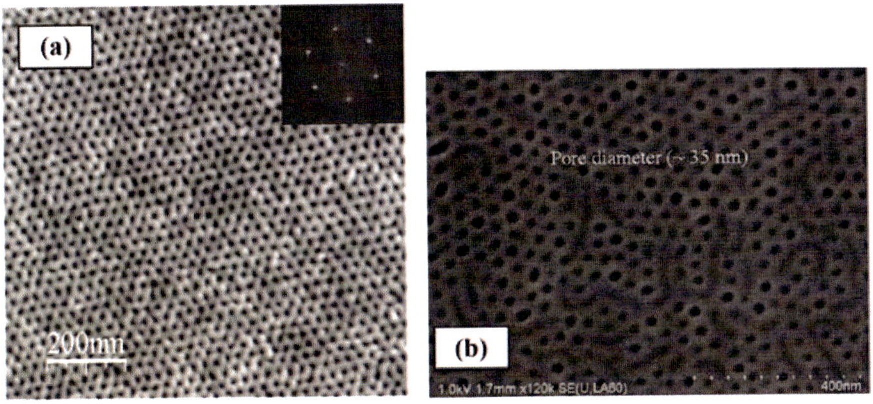

Fig. 11.10 Top-view images of two nanoporous templates prepared from diblock copolymers by removing their component which formed cylinder-shaped nanorods. **a** AFM image; original copolymer: polystyrene-b-poly(4-vinylpyridine) and 2-(4′-hydroxybenzeneazo) benzoic acid [216]. **b** SEM image; starting copolymer: PS-b-PMMA [213]. Reprinted from [216] and [213], respectively. Copyright (2009, 2008, resp.), with permission from Elsevier

Since the *in situ* detection of the nanopore formation is difficult for the pore diameter range of diblock copolymers, electrochemical methods can be conveniently applied for detecting the pore opening process and pore volume ratio [209, 212, 218, 219]. A wide literature background of the electrochemical manipulation of templates obtained from DCTs can be found in a recent review [220]. The electroplating baths used for DCTs are based on those that can be applied for electrodeposition of continuous films, with the difference that they are often mixed with an organic solvent (like methanol) for improving the wetting of the narrow channels in the template. Electrodeposition with either potentiostatic or galvanostatic conditions is customary, although both pulse and reverse pulse deposition proved to be suitable for the optimization of the magnetic properties of the deposits [221].

The following materials electrodeposited into block copolymer templates obtained special attention:

Magnetic metals. The interest in the electrodeposition of magnetic materials (such as Co [210, 211, 221], Ni–Fe [221] and Co–Pt [216]) into DCTs stems from the dependence of the coercivity of magnetic nanowires on their diameter. Although the specific surface area of the nanowires of 10–40 nm in diameter is very large, they can resist oxidation for a sufficiently long time due to the protection provided by the template [215]. The magnetization behaviour of the DCTs filled with magnetic nanowires strongly depends on the filling ratio and, hence, on the shape of the objects and their interaction. Due to the relative mechanical weakness of the narrow nanowires, the removal of the template strongly impacts the magnetization behaviour because of the deformation of the nanowires [216]. Lithographically patterned DCTs are particularly suitable for magnetoresistance measurement of the nanowire arrays in various magnetic field orientations [211].

Conducting polymers. Electrodeposition of extremely narrow conducting polymer fibres into DCTs provides a unique opportunity to measure the conductivity of the polymer fibres along their axis [214]. These experiments revealed that the high degree of polymer chain orientation leads to a much higher conductivity in the case of the ultranarrow polymer fibres than that of the polymer films with a higher chain disorder.

11.2.10 Templates Obtained from Directionally Solidified Fibrous Eutectic Metallic Systems

Eutectic systems exhibit a composition corresponding to the lowest temperature where the eutectic system may exist as a melt. When a eutectic system is solidified, the composition of the solid phases being crystallized is far in composition from the eutectic melt. Solidification of eutectic melts hence results in a phase mixture of defined composition. A classical example for eutectic systems is the Cu–Ag mixture (~28 wt.% Cu) in which the solubility of the components in each other is less than 5 at.% at the eutectic temperature (779 °C, a temperature much lower than the melting point of either of the components). Pseudobinary eutectics are systems in which one of the components is an alloy of fixed composition and the other component is a metallic element. A typical example for pseudobinary eutectics is NiAl–X where X is Mo, Re or W. The mole fraction of X in the latter eutectic systems is less than 2 at.%.

When a eutectic melt is solidified at a low rate and without a significant temperature gradient, interpenetrating random fibrous texture can form. Instead, if the solidification is *directionally oriented* (DO) with a temperature gradient of a few tens of K cm^{-1} and with the motion of the solidification front at the rate in the range of 10–3000 mm h^{-1} to the same direction as the temperature gradient (Bridgman-type growth), a fibre-like morphology is formed in which the fibre axis direction coincides with the normal of the solidification front. The diameter of the fibres formed varies between 100 and 900 nm from one system to another, but the wire diameter distribution is narrow for each particular sample. The relative position of the nanowires is random. In contrast to other template formation methods, the fibres are sometimes not cylindrical because they are single crystalline and have a well-defined polygonal cross section [222, 223].

If the chemical/electrochemical reactivity of the matrix-forming and fibre-forming materials is different enough, selective dissolution can be applied for structuring the solidified eutectic system. If the reactivity of the matrix is higher than that of the embedded wire in a suitable etching medium, self-supporting nanowires can be obtained, while in the opposite case, a porous template with nanoholes can be prepared. The illustration of all these processes together with the images of the resulting nanostructures can be seen in Fig. 11.11.

Regarding the NiAl–X systems, the passivation of the NiAl phase makes it possible to dissolve the embedded metal wires. Regardless of the additional metal that makes

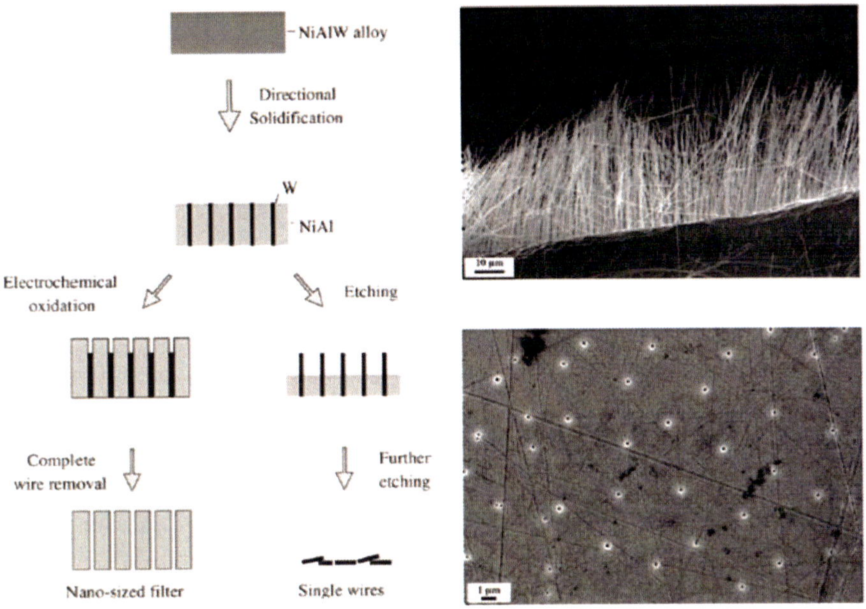

Fig. 11.11 Nanostructures prepared from DO eutectic alloys. Left: scheme of the processes leading to either porous templates (left route) or free nanowires (right route). Right, top figure: W nanowires after the dissolution of the NiAl matrix. Right, bottom figure: porous NiAl matrix after the electrochemical dissolution of the W nanowires. Reprinted from [224]. Copyright (2006), with permission from Elsevier

the single-element nanowire (Re [225–227], Mo [222] or W [223, 224, 228–230]), the selective electrochemical dissolution of the nanowires can be achieved in a nearly neutral acetate buffer solution at a potential depending on the material of the nanowire. During this electrochemical treatment, the surface of the NiAl matrix is passivated if the potential is higher than 0.7 V vs SHE [228], which makes it electrochemically inactive when the template is polarized to negative potentials and the nanochannels formed are filled up by electrodeposition with another metal. The filling material tested was mostly gold [225, 227, 228], and the resulting gold-decorated passive NiAl matrix could be used and as nanoelectrode array. One study [230] dealt with the dependence of magnetic behaviour of Co nanodots electrodeposited into the cavities as a function of the deposition parameters.

Another eutectic system suitable for the preparation of DO solidified nanostructure is Ag–Cu. With this system, both lamellar and fibrous structures are available. By using selective electrochemical dissolution, both Cu nanowires and Ag nanoporous matrix can be prepared [231]. Hypereutectic alloys prepared with DO solidification may also form fibrous and lamellar structures that can be etched selectively to obtain template-like materials. And example is the Ni–Si system [232] in which the metallic Ni phase can be selectively removed with electrochemical etching, leaving behind the Ni_3Si matrix with channels of a diameter of several micrometres.

11.3 Templates Obtained Through the Self-assembly of Particles

11.3.1 Template Preparation from Uniform Solid Particles

The self-assembly of uniform colloidal particles on a flat surface, often termed as "nanosphere lithography", has long been considered as an economic alternative of conventional lithographic surface patterning. When this technique was first mentioned, it was used for non-lithographic mask for substrate etching [233]. Although the self-assembly of uniform spherical particles is indeed a low-cost method that can easily be implemented in any laboratory, it is less versatile than lithography since the pattern obtained is always hexagonal (or, in other words, 111-oriented fcc) if periodicity can be achieved, as opposed to a custom-defined lithographic pattern. Nevertheless, it is sufficient or even desired for many applications, especially for photonic crystals [234]. Even though some publications report dense particulate layers with no obvious regularity, it was commonly observed that a regular periodic system can be achieved at the length scales of several hundreds of micrometres along the surface. A wide range of preparation methods for three-dimensionally ordered macroporous (3DOM) solids has been developed [235] shortly after the suitable monodisperse particles became commercially available, although the monodispersed particles are often home-prepared even nowadays.

The simplest method to prepare colloidal templates is the sedimentation of the colloidal particles onto to surface of the metal to be used as electrode in the forthcoming electrodeposition step [236, 237]. Sedimentation is a suitable template formation tool for neutral particles. Sedimentation is a time-consuming process that requires a template formation period longer than a day. However, unlike some other methods detailed below, the thickness of the particulate template is limited by the amount of the particles only. The latter feature of the sedimentation process makes it possible to electrodeposit porous structures of a thickness up to a few tens of micrometres [236]. Nevertheless, the organization level of the sedimented particulate multilayers is usually smaller than that available with other methods.

Dip coating followed by solvent evaporation is an alternative route with a smaller time necessity [238]. Another simple mechanical process, spin coating can also be applied after a careful adjustment of the substrate rotation speed and the suspension concentration in order to obtain ordered surface structures, by which monolayer coverage with particles is also available [239]. A compact assembly of template-forming particles can also be obtained by filtering them through a filter of sufficiently small pore size. An advanced method of filtering is the collection of the dispersed particles directly on the cathode of a thin-layer flow-through electrochemical cell [240] in which a side wall is porous in order to release the excess solvent. The latter method is suitable for the synthesis of particulate templates up to 100 μm.

Electrophoretic pattern formation on electrode surfaces from colloidal solutions [241] can provide a nearly defect-free surface coverage on a large lateral scale-up to about 20 particle layers. Since electrophoresis is a field-driven process, the particles to be collected at the surface must be charged, and the electrode surface charge has to be of the opposite sign. Theoretical works showed [242, 243] that the loss of the repulsion of the similarly charged particles has a combined background of electrohydrodynamic and capillary forces. Hence, several densely packed particle layers can be built up onto each other by electrophoresis. Due to the field-driven nature of electrophoretic particulate template formation, it is particularly suitable for non-planar substrate like microwires [244]. It is to be noted that electrophoretic deposition of metallic particles is suitable for preparing self-assembled metal dot surface structures, too [245].

Various colloidal template formation procedures are based on the self-assembly of particles on the surface of a liquid [246, 247]. The so-called floating self-assembly route [248] involves the formation of the arranged particulate multilayer on a liquid surface followed by the solvent evaporation. Another similar method of particulate template preparation is the transfer of a self-organized particle assembly from the surface of a liquid onto a metal substrate (which can serve as electrode during the electrochemical sample preparation). This technique, which is partly analogous to the Langmuir–Blodgett technique of monolayer transfer from a liquid surface [249], is particularly popular due to its relative simplicity. As the first step, a relatively dense particle suspension is prepared on the surface of which the self-organization of the particles can take place. The collector metal sheet is immersed into the solution either vertically or in a tilted way so that it can serve as a platform for holding the self-assembled surface layer (or multilayer) of particles as the solvent evaporates. For facilitating the evaporation, the suspension usually contains some volatile component besides water (e.g., ethanol). With the appropriate adjustment of the particle concentration and evaporation rate of the solvent, an intermittent surface structure, exhibiting wire-like strips of assembled particles parallel to the liquid surface, can also be synthesized [250].

For essentially all methods listed above, the particles have to be fixed before using the holder as electrode since the self-assembled layer can be easily destroyed by simply pouring a solution onto it. Either a mild heat treatment for partially melting the contacting areas of the particles or a chemical cross-linking is required so that a durable solid porous template can be obtained. The particulate template thus formed is often termed as "opal template", while the cavity system between the particles (of the material filling this space) is of an inverse opal structure. An example for the opal structure is shown in Fig. 11.12. It often occurs that the top particle layer exhibits a nearly perfect ordering but the layers underneath have many growth defects (vacancies, dislocations, etc.).

The composition of the particles used for opal template construction ranges from polymers (latex, polystyrene /PS/, poly(methylmetacrylate) /PMMA/) through metal salts to silica. Since the majority of the application of the filled opal structures is related to its photonic properties, the lattice constant (and, thus, the diameter of the spherical particles) is in the order of the wavelength of visible light. Although the latter distance is about several hundreds of nanometres, the thickness of the deposit in neck part of the inverse opal structure is often just a few tens of nanometres. Nevertheless, the method of opal template preparation is uniform for any particle

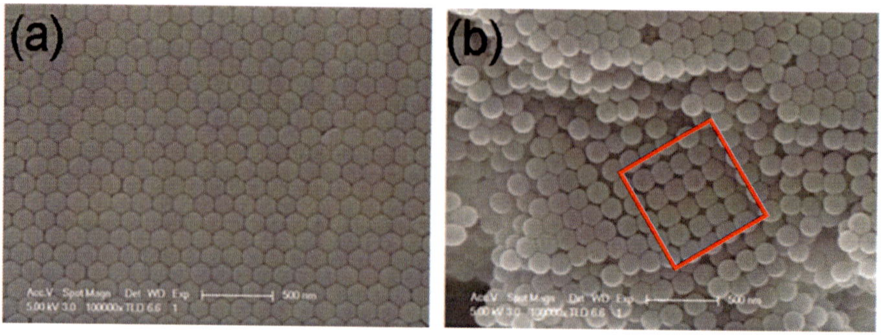

Fig. 11.12 SEM images of a self-assembled colloidal crystal template formed from polystyrene microspheres of 140-nm diameter. **a** Top view; **b** top and partial cross-sectional view (upper edge and the rest of the image, respectively). The middle zone of the image labelled with a rectangle shows a diminished packing density with a cubic local structure. Reprinted from [247]. Copyright (2011), with permission from Elsevier

diameter, and spherical nanoparticles are commercially available nowadays down to a few tens of nanometres.

The formation of the colloidal layers on a substrate is a sensitive process that can be adjusted by changing the wetting properties of the surface with, e.g., the application of a surfactant. This can be exploited in particular if a patterned surface with heterogeneous surface composition is used for substrate of the self-assembly of particles [251].

11.3.2 Electrodeposition of Metals into Particulate Templates

Electrodeposition of metals from conventional galvanic baths into the cavities of particulate templates is a relatively easy process. The resulting metal structure can be revealed after dissolving the template-forming particles in an appropriate solvent (HF for silica and a suitable organic solvent like toluene, chloroform or tetrahydrofuran for PS and PMMA). The application of organic solvents requires a good care because the swelling of the colloidal particles before their dissolution may damage the resulting structure [252], which rationalizes a combustion-based particle removal. The visual appearance of the thus produced inverse opal metal structure strongly depends on both the filling level and the interaction of the growing metal with the particulate template. If the number of the particle layer is large, the cross-sectional image of the metal foam is characteristic of a porous material with spherical cavities where the broken surface does not indicate the same ordering level as the top view (see Fig. 11.13).

The growth modes depicted in Fig. 11.14 are neither related to a particular metal nor can they be associated with a particular metal–particle type pair. Instead, the

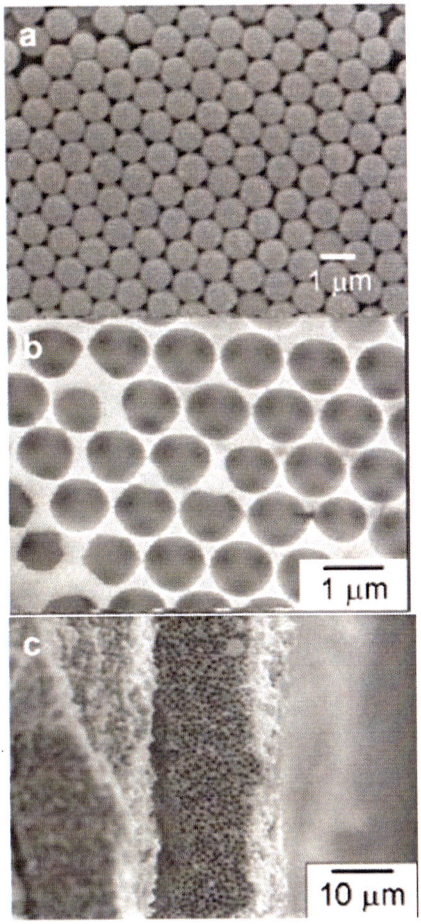

Fig. 11.13 SEM images of a dense inverse opal metal structure (Ni–Sn alloy). **a** Self-assembled colloidal template; **b** Electrodeposited inverse opal structure after the dissolution of the polymer particles; **c** Low-magnification cross-sectional image of the broken porous metal layer. Reprinted from [253]. Copyright (2009), with permission from Elsevier

growth mode is determined by all circumstances of the experiments, including the current density. The dense (even) growth mode can be seen in Fig. 11.14a. It was reported for Ni [254] that by applying the same bath, the even growth mode is dominant at relatively large current densities, while the preferential growth along particle surface could be seen for low current density. Similar growth mode transition for Ni was reported in various works [255, 256]. When the growth prefers a surface-conformal mode (Fig. 11.14c), the covering part becomes eggshell-roofed if the metal overgrows the last particle layer. However, if the growth is terminated in the right phase, triangular cavities can be seen in the pore wall [255]. Another example for the preferential growth along the particle surface and the resulting closed hollow spherical metal structures can be seen in Fig. 11.15 for Cu. Palladium deposition was also shown to lead to hollow spherical structure with the difference that the Pd shell of ~200 nm in diameter was composed of grains with a diameter less than 30 nm [257].

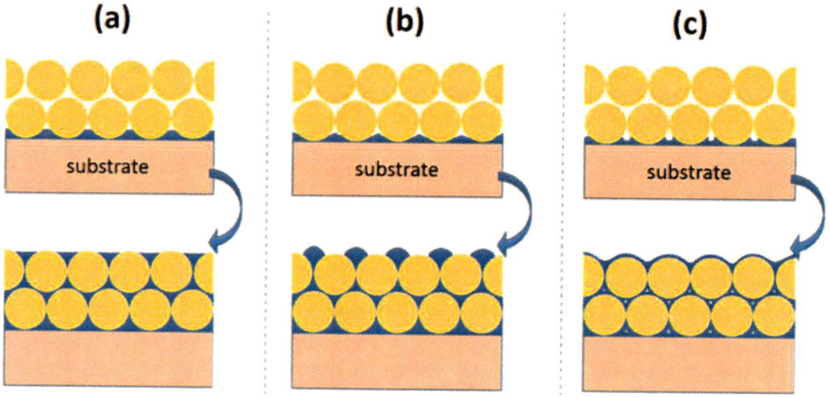

Fig. 11.14 Growth modes of metals in colloidal crystal templates. **a** Even growth; **b** Preferential growth in the cavities; **c** Preferential growth along the surface of the particles, possibly leaving voids behind the growth front in the tetrahedral cavities between the template-forming particles. See also Chap. 7.2.4

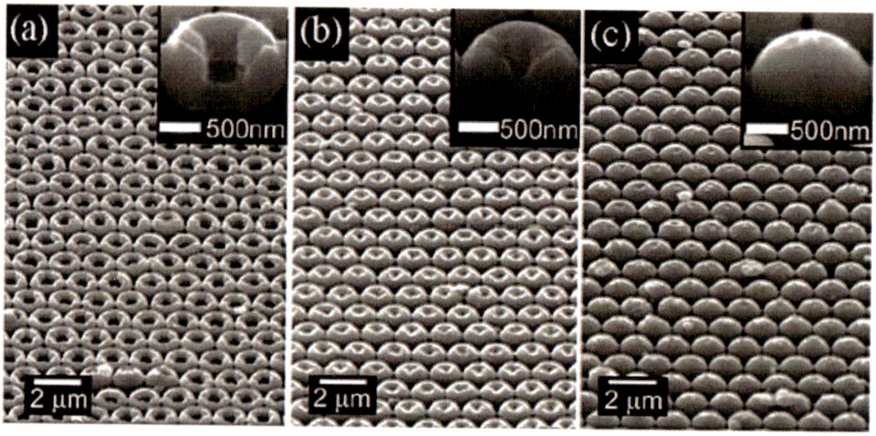

Fig. 11.15 Tilted SEM image of a porous Cu layer growing along a PS colloidal template. The growth was terminated **a** close to the 3/4 of the particle radius; **b** in the phase when the Cu layer almost closed on the top of the colloid particles; and **c** after the closure of the hollow Cu shell along the particles. Reprinted from [258], with the permission of AIP Publishing

Concerning the explanation for the growth preferred along the surface of the particles, the surface charge of the particles, the nature of the terminal groups and the adsorption of the metal ions of the particle surface were considered. It was shown, however, that surfactant-mediated wetting of the interparticle space can also change the hydrophobic character of the template surface to hydrophilic, which improves the ionic transport between the particles [259]. Another sort of growth preference is the inclination of the metal being deposited to fill up the so-called drying cracks, i.e., the areas not completely covered by the colloidal templates. These cracks are formed due to the shrinkage of the colloidal template upon drying, and their width is several times larger than the diameter of the template-forming particles. It was demonstrated that various metals either tend or do not tend to fill up these gaps, while the interparticle space could be filled up continuously in both cases [260]. If the metal overgrows the opal structure, its starts developing crystal shapes that are characteristics of its deposition onto a flat substrate.

The controlled parameter during the electroplating of metals into a colloidal template can be either the current or the electrode potential, and both are suitable for tuning the deposition preference. As far as it can be established from the occasionally published data, the deposition rate has to be chosen so that the deposition remains in the kinetically controlled range and not diffusion-limited. Since the diffusion length between the colloidal particles can be quite long for multilayered templates, the kinetic control practically means low current densities. Hence, the fill-up time of a multilayered template with a handful colloidal particle layers may range from hours to tens of hours [240, 261, 262], a period about two orders of magnitude longer than that normally required for the formation of equally thick continuous coatings. Although a few works are available that present $I(t)$ curves for potentiostatic metal deposition into colloidal templates [240, 244, 255, 256, 262–264], their results are quite controversial, so much that even the classification of the $I(t)$ curves published is challenging. For multilayered colloidal templates, some works claim that the $I(t)$ curves exhibit an oscillatory character, corresponding to the actual active surface area of the growing metal in the interparticle space [244, 262, 263]. Even though a very regular periodicity of the current maxima and minima was presented [263], the variation of the mean current makes it very dubious that the oscillation can be quantitatively related to the growth phase. Another typical observation is that the current tends to increase slightly as the growth front passes a colloid particle layer [255, 256]. However, even the slowly growing mean current cannot be taken as a general observation since decreasing currents [263], followed by a constant-current period after a long time [240], have also been reported. The character of the current transient is rather clear for monolayer templates [264, 265] where the smallest cathodic current can be associated with the growth phase when the deposit reaches the radius of the colloidal particles.

Below, a short summary is presented with metals (elements and alloy families) whose inverse opal structure was extensively synthesized. The specific features that make the deposit useful are discussed together with the metal types.

Gold. Due to its high stability, gold is a very popular material for colloidal template-based synthesis. Its deposition can be carried out with the help of various baths: slightly acidic $HAuCl_4$ solution [239, 252], complex bath with K_2HPO_4 either as a single additive [266] or combined with Na_2SO_3 and other chelating agents [239], and commercial bath formulations [264, 265]. The surface area of the inverse opal Au structure can be conveniently measured by determining the charge for the reduction of the oxide layer on Au formed during the anodic-going part of the CV curves [267].

If the diameter of the colloidal template particles is in the order of the wavelength of the light, the reflectance spectrum of the inverse opal metal structure mostly depends on the filling ratio of the monolayered template [268]. Inverse opal Au layer is often applied in surface-enhanced Raman spectroscopy (SERS) combined with electrochemistry. When the substrate is ITO and the strike layer (usually also from gold, evaporated onto the ITO substrate) is thin enough, the electrode of high surface area is transparent and can be conveniently studied with a light beam passed through it. The adsorption of various organic materials can be effectively studied with this technique. Besides, the resonance mode of SERS can be determined by the filling ratio of the colloidal monolayer, which makes its deposition a versatile tool to produce optically active materials [264, 266].

The transferability of the inverse opal metal layers from the electrode onto which it was deposited to another substrate was demonstrated with gold. If the deposition is carried out with an ITO substrate not coated with an adhesive layer prior to the colloidal template formation, the inverse opal layer can detach spontaneously [239] (the same effect was found for Ni deposits [254]). If a less noble metal is used first for partly filling the colloidal template and then the deposition is continued with gold, the first metal layer can be used as a sacrificial one after removing the colloidal particles [267]. This procedure leads to self-supporting electrodeposited inverse opal layers. Since the electrical resistance of such macroporous Au layers is highly determined by the finite size effect, mostly by the "neck" parts of the porous layer, the resistance of the layer is sensitive to adsorbed chemical species. For this reason, such free-standing structures can be used as chemisorption-based resistive sensors [267].

Platinum group metals. Both Pt [265, 269] and Pd [257] macroporous layers were deposited from relatively simple baths containing H_2PtCl_6 and $PdCl_2$, respectively (occasionally with some supporting electrolyte). While Pd tends to cover the nanospheres conformally and to form hollow spheres, Pt seems to form an open structure. The primary application of the macroporous platinum metal structures is the electrochemical oxidation of organic compounds, taking advantage of the enlarged activity due to the enhanced electrode surface area.

Nickel, cobalt and various ferromagnetic alloys. Electrodeposition of nickel into colloidal templates has been carried out with solutions of highly varying metal salt concentrations. Application of conventional (i.e., Watts type) solution with $c(Ni^{2+})$ > 0.5 mol dm^{-3} is widespread [240, 255, 256, 270, 271], but custom-made low-concentration baths proved to be equally sufficient [254, 262]. Commercial nickel baths also appear in a few works [261, 272]. Some properties of the macroporous Ni films are similar to that of Au films of similar structure (i.e., filling ratio-dependent reflectivity [272]). An advantage of Ni as compared to Au is that the macroporous structure can be etched after the removal of the template-forming particles, which allows one to apply thinner metal ligaments that would naturally belong to the original template [272]. A mechanical study of macroporous Ni film showed that the hardness is nearly proportional to the filling ratio [271].

Magnetic properties of inverse opal structures have been studied for various compositions such as Ni [240, 254, 261, 262, 273], Ni–Fe [273–275] and Co [274–276]. Concerning the magnetic properties of the macroporous films, it was established that their coercive field is much larger, sometimes nearly by an order of magnitude, than that of plain films [240, 254], and it exhibits a maximum as a function of the diameter of the template particles. The coercivity enhancement is common for size-constrained magnetic systems and is partly explained with domain wall pinning. The in-plane magnetization curves of the macroporous films show an improved squareness as compared to their planar counterparts of the same composition. The anisotropy of macroporous films is usually smaller than that of the corresponding planar films.

An interesting phenomenon is the oscillation of the coercive field with the filling height of the metal [275, 277]. The coercivity maxima coincide with the growth phase when the metal ligaments are the narrowest in between the touching points of the nanospheres, and the gap between the top features of the inverse opal structure stray field is the largest. This corresponds to the maximum of the magnetostatic energy due to the stray field. In consequence, the coercivity minima are observed when the thickness of the macroporous layer is an integer multiple of the nanosphere diameter, i.e., when the growing metal surface becomes nearly closed. The oscillatory behaviour of the coercive field as a function of the filling ratio could be successfully modelled by taking into account the domain wall pinning.

Sn alloys. Sn-based intermetallic compounds are prominent candidates for lithium intercalation electrodes; however, their poor cyclability prevents one from their application in a bulk form. In order to overcome the high expansion/compression ratio of these alloys upon the insertion/de-insertion of lithium, Sn–Ni [253, 278–280], Sn–Co [281] and Sn–Cu [282] alloys were synthesized in macroporous form by using colloidal templates. Although the performance of these macroporous electrode materials was much better than that of their bulk form, their cycling performance is yet to be improved significantly for considering them as practically applicable lithium-ion battery components.

Metal–phosphorous alloys. Electrodeposition of metal–phosphorous alloys is a relatively new pursuit that can provide materials with a special combination of properties not available for analogous bulk samples. Macroporous Fe–P alloys [283] allow a simultaneous tuning of wettability, magnetism and electrocatalytic properties towards oxygen evolution reaction (concerning the latter feature, some analogy can be seen with macroporous Ni films [270]). Macroporous Ni–P alloys [284] proved to exhibit a higher specific capacitance in supercapacitor applications.

11.3.3 Electrodeposition of Non-metallic Materials into Particulate Templates

The principles of template formation and electrodeposition of non-metallic materials are essentially the same as those of metals. Different features of the overall sample preparation procedure may originate from the redox reaction taking place, the media used and the post-electroplating treatment. However, for all system in which the deposition of a metal oxide or hydroxide is the initial step, the maximum achievable thickness may be limited by the relatively small conductivity of the metal hydroxide plated. Hence, the cell voltage increases with time if the deposition is carried out galvanostatically (and, vice versa, the deposition rate decreases with time at constant electrode potential).

Cathodic deposition of metal oxides and hydroxides. The common feature of the direct electrodeposition of several sorts of metal hydroxides and oxides is that the reduction process does not modify the oxidation number of the metal ions; instead, a reactant (mostly hydroxide ion) is produced for a precipitation reaction.

ZnO can be produced in colloidal template with the usual method of generating $Zn(OH)_2$ that is spontaneously dehydrated and yields ZnO. The procedures used differ in what reactant provides the alkalization of the solution (reduction of nitrate to nitrite ions [285, 286], H_2O_2 [287] or O_2 [288]). Nanostructured ZnO exhibits unique optical emission and reflectance properties [285, 288].

The cathodic deposition of macroporous WO_3 can be carried out with acidic Na_2WO_4 solution containing H_2O_2 [289, 290]. Hydrogen peroxide and hydroxonium ions are reduced, and the WO_3 is finally deposited onto the cathode in a process of complicated kinetics, involving also oxygen evolution. Macroporous WO_3 exhibits improved electrochromic properties as compared to flat films.

The reduction of sodium permanganate provides an example when a metal oxide is deposited cathodically with the oxidation number change of the metal ion. If MnO_2 is deposited in macroporous form from a $KMnO_4$ solution [291], the electrochemical capacitance of the resulting film is much larger than that of the corresponding continuous film, and the capacity improvement is accompanied with an outstanding cycling stability.

Another typical material that is deposited with the partial reduction of the metal ion is Cu_2O [292–295]. All works cited above adopted the well-known bath used for the deposition of free-standing Cu_2O (0.4 mol dm^{-3} $CuSO_4$, 3 mol dm^{-3} lactic acid, pH = 9–12). While the photoelectrochemical properties of thus obtained layers are noteworthy [293], a phenomenon not seen for other materials was observed for 3DOM Cu_2O materials. Namely, two different growth modes could be distinguished [292, 294]. On the one hand, especially for well-ordered templates, the deposit exhibits a typical inverse opal structure. On the other hand, it may occur that the envelope shape of the growing particle is identical to that observed for free growth, with its internal structure being characteristic of the usual colloid-related porosity. An example for the latter phenomenon is shown in Fig. 11.16. The size of the internally porous parti-

Fig. 11.16 SEM images of Cu_2O deposits. **a** Free growth; **b** template-assisted growth at the same electrode potential as used for the preparation of the sample shown in (**a**); **c** magnification of the image (**b**). Reprinted by permission from Springer [294], copyright (2013)

cles was found to be related to the area on which a single Cu_2O crystal could be nucleated.

Electrodeposition of metals and metal salts followed by annealing. If the colloidal template is filled up with a metal deposit, it can be transformed to an oxide by annealing in air. This process is often coupled with the combustion-based removal of the organic colloidal template. This is a possible route for obtaining inverse opal haematite structures [296]. The advantage of macroporous Fe_2O_3 as compared to bulk one lies in its improved photoelectrochemical properties.

In other systems, the electrodeposition leads to a metal hydroxide, not a metal. Examples include the deposition of nickel hydroxide [297], cobalt hydroxide [298] and mixed nickel aluminium hydroxide [299]. In all above-mentioned systems, the hydroxide precipitation was achieved with a solution in which the nitrate reduction provides the alkalization of the medium. In templated systems, the application of a hydroxide-generating reactant is beneficial as compared to the simple water decomposition because a hydrogen evolution in the latter case may lead to the entrapment of the bubbles and even destroy the template. The annealing in air can efficiently transform the hydroxides to oxides without damaging the structure achieved in the template. No data available on the volume change of the macroporous structure during the dehydration process. The motivation of the preparation of all these hydroxide-mediated macroporous oxide systems was the achievement of improved electrochromic properties with fast switching and stable cycle performance.

A very similar approach was demonstrated for the preparation of colloid-templated macroporous titania films [300]. Here, the solution was made with an ethanol–water mixture in order to improve the stability of the $Ti(SO_4)_2$ solution and provide a sufficient wetting of the colloidal template. In addition, both KNO_3 and H_2O_2 were applied for hydroxide-generating reactant. A similar reaction pathway can be applied for the deposition of hydrated zirconia that can be transformed to ZrO_2 by annealing [301].

Anodic deposition of metal oxides. The advantage of the one-step anodic deposition of metal oxides is the simplification of the process and the possibility to omit the annealing step. This method works for metals with varying oxidation state and having easily soluble and insoluble compounds of small and high oxidation numbers, respectively. Examples include the deposition of MnO_2 from $MnSO_4$ precursor material [302], TiO_2 from $TiCl_3$ [303] and V_2O_5 from $VOSO_4$ [304], all from aqueous solutions. Concerning the improvement of properties due to the macroporous nature, MnO_2 was shown to exhibit better supercapacitance than flat films. Although very little is known about the impact of the deposition conditions on the resulting macroporous anodic oxides, the study of the deposition of V_2O_5 revealed that both the substrate and the deposition potential can be optimized for the desired target property [304].

Electrodeposition of semiconductors. 3DOM Si [305] and Si–Ge with varying composition [306] can be electroplated into colloidal templates from ionic liquids by using $SiCl_4$ and $GeCl_4$ as precursor materials. The treatment of all components of the bath requires glove box environment to keep oxygen and water vapour away, and as-prepared samples of high surface have to be cleaned thoroughly in order to avoid oxidation. A slight Ge excess was found in Si–Ge samples as compared to the component ratio in the bath.

II-VI type semiconductors, CdSe [237, 307–309] and CdS [237], were also synthesized as 3DOM materials. Both CdSe and CdS can be synthesized from aqueous solution with some excess of Cd^{2+} as compared to the calcogenic precursor compounds (SeO_2 and $Na_2S_2O_3$), while CdS can also be obtained from organic solutions in which elemental sulphur is the source of the chalcogenide component. Semiconductor 3DOM structures are all peculiar for photonic properties, especially due to their light absorption in the infrared range.

Polymer deposition into colloidal templates. Colloidal templates make it possible to deposit polymer films with a great variety. Apart from the various polymer compositions (such as polypyrrole [310], poly(para-phenylene) [311] and poly(3,4-ethylenedioxythiophene) [312], etc.), the different filling ratios applied allow a morphological variety. Interestingly, polymers electrodeposited into multilayered colloidal templates can be obtained in self-supporting form after the removal of the templating particles, in spite of the softness of the polymer materials as compared to metals.

11.3.4 The Double-Templating Method Based on Nanosphere Lithography

In the methods discussed in the present chapter, the primary template is made from colloidal particles. This template is filled up with a suitable material that makes an inverse opal structure. Then, the primary template is removed, and the resulting empty inverse opal structure is used as a secondary template. These procedures vary in which step applies electrochemical deposition (the first, the second or both). The scheme of the double-templating process and a few images of the resulting nanostructures are presented in Fig. 11.17.

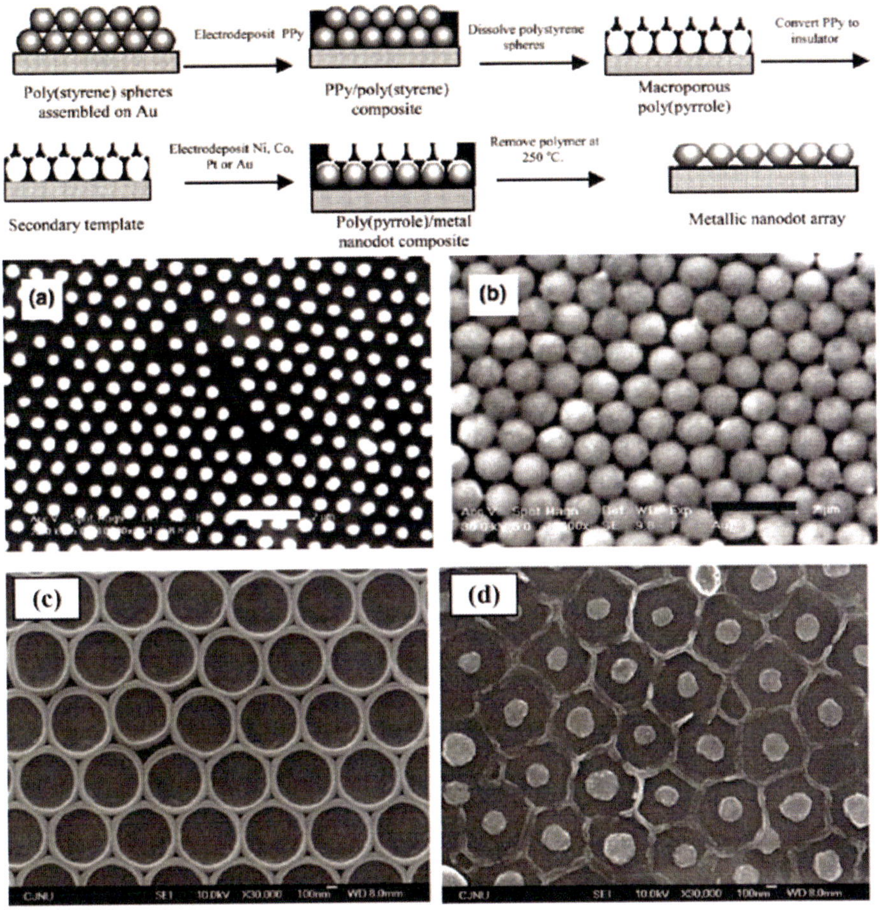

Fig. 11.17 Top: Scheme of the double-templating method involving both intermediate colloid and inverse opal polymer templates. [313] **a** SEM image of a Pt nanodot array with dot height less than half of that of the primary colloid template [313]. **b** SEM image of an Au nanodot array with dot height about the half of that of the primary colloid template [313]. **c** SEM image of a polypyrrole secondary template with much smaller height than the primary colloidal template [314]. **d** ZnO nanodot deposits into the template shown in (**c**) [314]. Note the modification of the secondary polymer template during the deposition of the nanodots. Reprinted from [313, 314], as marked. Copyright (2004) and (2007), respectively, with permission from Elsevier

The double-templating method was first applied [313] by depositing polypyrrole into the nanosphere template cavities. After the dissolution of the nanospheres, the macroporous polypyrrole scaffold was made insulating by cyclic voltammetry with a negative limit where the polymer is reduced irreversibly. Then, the resulting spherical voids can be filled up by electroplating a metal (e.g., Ni, Au and Pt [315]) with a thickness that does not exceed the diameter of the original template particles. Therefore, the metal dot system can be produced with nearly analogous geometry to the original template. With the same templating principle, ZnO nanodots can also be synthesized [314].

In a variation of the double-templating process, the secondary template is prepared with a much smaller thickness than the particle diameter of the primary template. If the thickness of the first deposit is relatively small, it does not determine a spherical cavity for the secondary deposit but rather provides a surface pattern at which the second deposit can nucleate and then grow. With this method, polymer-separated nanodots can be deposited along a hexagonal pattern of the same unit cell size as that of the primary template [315, 316].

In contrast to the above-mentioned method in which the goal is the surface patterning with a thin layer, thick inverse opal structures with a thickness of several times the colloidal particle diameter can be used as substrate. If the plating of the second deposit into the inverse opal structure is carried out under carefully adjusted circumstances, the second deposit can nucleate and grow along the entire surface of the secondary template. The feasibility of this method was evidenced for the deposition of Si coatings on Ni inverse opal secondary template [317]. If conformal deposit cannot form, the inverse opal structure hosts seeds of the secondary deposit, such as it was demonstrated with Fe_2O_3 deposition onto macroporous Ni scaffold [318].

In nanosphere lithography, chemical and electrochemical steps can be combined. For instance, the primary colloidal template can be filled up with a salt, namely $Al(NO_3)_3$ by solvent evaporation, whose annealing results in an Al_2O_3 secondary template that can be used in the forthcoming electroplating process [319]. If the inverse opal structure is electroplated, it can be used as secondary template for chemical synthesis of spherical particles that is hard to produce as free nanoparticles without a size-limiting template [320].

11.4 Soft Templates Formed by Molecular Self-assembly Processes

11.4.1 Lyotropic Systems Suitable for Templates in Electrodeposition Processes

Lyotropic systems consist of amphiphilic molecules as solute and a solvent. For the framework of this monograph, it is an acceptable simplification that water will be considered as solvent in each case. The liquid crystalline properties of the water–amphiphilic solute systems stem from the ordering of the molecules in the lyotropic mixture [321]. This is governed by the preferred interaction of the hydrophilic part of the amphiphilic molecules with water and the hydrophobic (or oleophilic) part, usually a long apolar chain, with the similar parts of other species present in the system. As the amphiphilic solute-to-water ratio increases, the following phases can form (see also Fig. 11.18 and note that Arabic and Roman numbers both occur in various sources as type identifiers):

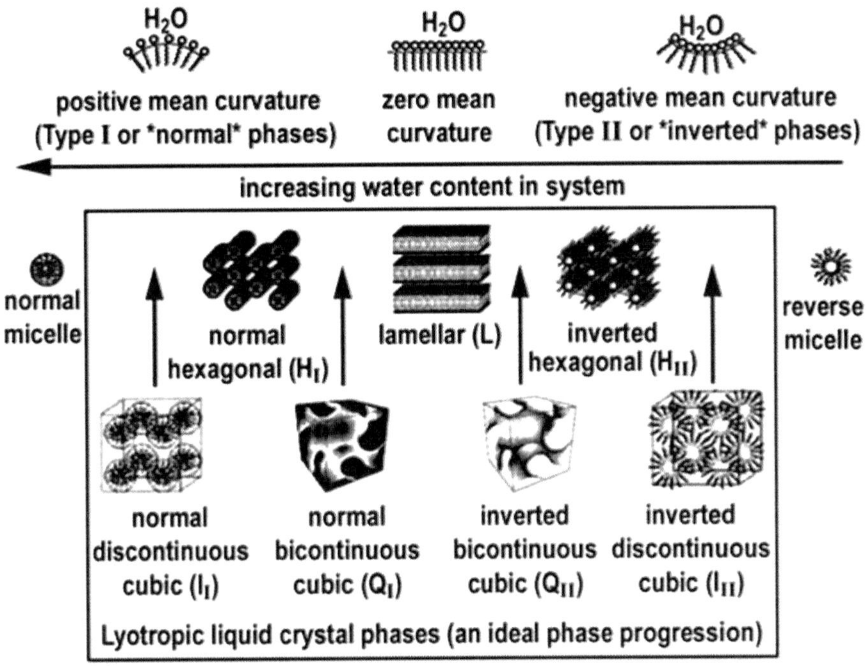

Fig. 11.18 Qualitative overview of the structural change of a lyotropic system as a function of the component ratio. Reprinted from [322]. Copyright (2018), with permission from Elsevier

- Real solution of the material with amphiphilic molecules;
- Micellar solution of the amphiphilic solute in water (L_1 phase);
- Cubic arrangement of the discontinuous amphiphilic micelles in water (I_1 phase);
- Hexagonally arranged micellar column of the amphiphilic molecules in water (H_1 phase);
- Bicontinuous cubic phase with water as the majority component (Q_1 phase);
- Lamellar phase with bilayers of the amphiphilic molecules (L_α phase);
- Inverse bicontinuous cubic phase with the amphiphilic material as the majority constituent (Q_2 phase);
- Reverse hexagonal phase in which the polar molecules are inside of the inverse columnar micelles (H_2 phase);
- Cubic arrangement of the discontinuous polar (water-containing) micelles in a continuous phase containing the amphiphilic molecules (I_1 phase);
- Reverse micelle structure (L_2 phase).

Apart from the above phases, lyotropic systems can also form *vesicles*, i.e., objects similar to either micelles or lamellae in the external shape and the double layer of the amphiphilic molecules at their boundary, respectively (like living cells). However, vesicles have no importance in templated electrodeposition.

The phase diagrams of the lyotropic systems show a great variety but exhibit a few uniform features [321, 323]. The critical concentration of the amphiphilic material at which the L_1 phase is formed increases with temperature. In most of the lyotropic systems used for electrodeposition templates, the Q_1 and Q_2 phases are rather a narrow boundary region between the neighbouring phases. It may occur that some of the phases are even missing. The general rule is that with the increase in temperature, the concentration range in which either the I, H or Q phases are stable get narrowing and finally become vanishing, and two liquid phases can coexist. For the most commonly used non-ionic amphiphilic material in electrodeposition experiments, octaethyleneglycol monohexadecyl ether (often abbreviated as $C_{16}EO_8$), the H_1 phase is stable over a ~20 wt.% range around room temperature, which makes it relatively easy to prepare.

Even though the phase diagrams are known for the mixtures of major polar and amphiphilic molecules, the dissolution of the ionic components necessary for electrodeposition modifies the phase diagrams to significant extent. Hence, the phase diagrams of the water–amphiphilic systems can be taken as a rough approximation only, and the determination of a new phase diagram for the electrolyte solutions is needed. In these cases, the quasi-binary approach is often applied. This means that the aqueous component is not pure water but a solution of the electrolytes with fixed concentrations, even though the distribution of the electrolytes between the possible phases is not studied but assumed that the composition of the aqueous phase does not change as various phases are formed. The identification method of the phases includes polarized light microscopy and small-angle X-ray diffraction. For the latter, the periodicity of either the lyotropic system itself of the nanostructured deposit obtained can be identified with a Bragg peak at $2\Theta < 2°$ diffraction angle, corresponding to the 6–15 nm periodicity.

Since lyotropic systems are in a dynamic equilibrium and are non-solid by nature, they are easy to deform during the deposition process. Hence, the overall shape and the internal structure of the lyotropic templates constantly change during the deposition process. This is the reason why deposits much thicker than the characteristic length scale of the internal composition modulation in the lyotropic media often exhibit either some surface undulation or a large-scale growth morphology unrelated to the lyotropic material. However, the small-scale morphology within the large-scale morphological units bears the fingerprint of the lyotropic template. This is the reason why the SEM method is suitable for the observation of the large-scale morphology, while the template-induced small-scale morphology with <15 nm periodicity can be imaged with TEM only in most cases. An example for both the top and side views of the pore structure in a deposit obtained from a H_1 type template is shown in Fig. 11.19. For a similar comparison of Pt deposits, see Fig. 4 in Ref. [324].

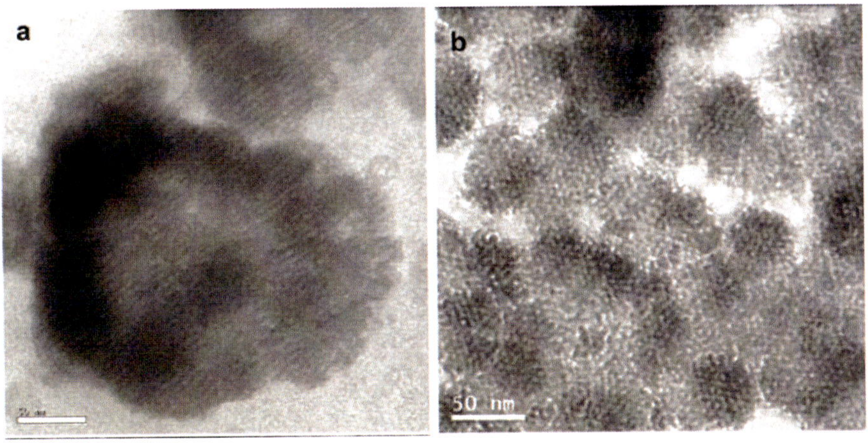

Fig. 11.19 TEM images of a PbTe deposit obtained with the H_1 phase of a mixture containing 50 wt.% $C_{16}EO_8$ and 50 wt.% aqueous solution of $Pb(CH_3COO)_2$, TeO_2 and HNO_3 [325]. **a**: Cross-sectional view and **b**: top view of the nanochannels. Reprinted from [325]. Copyright (2009), with permission from Elsevier

11.4.2 Nanoporous Metals and Oxides Obtained from the H_1 Phase of Lyotropic Liquid Crystals

Platinum group metals and their alloys. Platinum [324, 326–332] and its alloys [333, 334] are the most common mesoporous deposit type in H_1 phase of lyotropic systems (also termed as H_1-e Pt). The explanation is the high catalytic activity of the deposit. The surface roughness factor achieved ranges to several hundred as measured by the area of peaks corresponding to the hydrogen adsorption. The hydrogen adsorption peaks can be measured with CV at several hundred mV s^{-1} sweep rate from acidic solutions, and no hindrance of the accessibility of the hexagonal pores can be detected [324, 326]. However, if a dilute solution of a relatively slowly diffusing solute (like Cu^{2+}) is applied for the surface area estimation by using the UPD peak(s), a sufficiently long penetration time is necessary to provide a uniform coverage of the pore walls [328].

Concerning the deposition process, hexachloroplatinic acid is the common precursor material for mesoporous Pt deposits. A good adherence to the substrate can be achieved in spite of the contact with the lyotropic template. Both the deposition potential and temperature have a great impact on the deposition efficiency and the undulation of the external surface, even though the real surface area is proportional to the volume of the deposit [327].

The electroanalytical features of the lyotropic H_1 phase-mediated Pt deposits are as follows: The modification of Pt microelectrode with a mesoporous Pt layer was shown to be the most feasible strategy to apply such Pt layers in electroanalytical measurements [328, 330–332]. Although the CV recorded on a porous Pt microelectrode was nearly identical to a flat microelectrode (except for the drastic difference in the double layer capacitance) when potassium ferricyanide was reduced [328], the porous microelectrodes proved to be excellent amperometric sensors for a number of compounds with much larger linearity range than either flat or even platinized platinum electrodes of identical geometric surface area. Examples range to hydrogen peroxide [330], ascorbic acid [331] and formic acid [332]. The unique combination of the high temporal stability of the mesoporous structure, the high accuracy and the good reproducibility of the measurements makes such electrode an outstanding candidate for commercial amperometric sensors.

While pure Pt and Pt–Ru [334] mesoporous structures can be deposited and retain their morphology, stable mesoporous Pt–Au alloys can be synthesized if Pt is the majority component [333]. Hence, the composition of mesoporous deposits is not arbitrary. The experience with the Pt–Au alloys is much in accord with that reported for Cu, Ag, Au and Zn [335], namely, that the electrodeposition of the mesoporous form of these metals was unsuccessful. The reason behind the instability of the mesoporous structure of these metals can lie in their high self-diffusion coefficient (see also Sect. 6.2 about the self-annealing process related to nanocrystalline Ag and Cu).

The deposition of H_1-e Pd is very similar to that of Pt and so is its stability. However, in contrast to Pt, the peaks in the voltammogram of H_1-e Pd is H_2SO_4 solutions is much better resolved in the H-adsorption region than for bulk Pd samples [336, 337], which was explained by the nanostructured morphology of H_1-e Pd. The formation of both the α and β hydride phases was observed for H_1-e Pd at the corresponding cathodic potentials, and the hydrogen absorption took place much faster than planar films due to the shorter solid phase diffusion pathways in the nanostructured electrodes. With a suitable *in situ* pre-filling of the porous Pd structure with hydrogen so that the two hydride phases are in equilibrium, the Pd microelectrodes can be used as low-response-time pH sensors [337]. H_1-e Pd was shown to be suitable for combustion-based gas sensor in micromachined planar pellistors with high temporal stability at 500 °C [338].

An interesting approach was applied to deposit multilayered metal coating with the porosity of the H_1 lyotropic phase and by applying the double-bath method [339]. A Ru layer was deposited first, and it was followed by a Pd layer after a solution change. Interestingly, the nanoporosity of the deposit could be retained and a morphologically homogeneous structure was achieved even if the template-forming solution was very carefully removed from the nanopores, indicating that the Ru pores served as seeding points for the nucleation of molecular columns once the second solution was applied.

Oxides and hydroxides of various metals. The pursuit of obtaining mesoporous metal hydroxides and oxides stems from the supercapacitor properties of these materials. The high porosity increases the accessibility of the entire active layer by the electrolyte solution and the cyclability of the layer at a high rate. All supercapacitor materials to be mentioned below proved to be stable under repeated cycling.

The synthesis of H_1-e metal hydroxides was performed with the usual nitrate ion-based method with cathodic polarization in which the alkalization of the media leads to the hydroxide precipitation. This route was feasible for the preparation of mesoporous $Ni(OH)_2$ [339, 340] and $Co(OH)_2$ [341–343]. Due to the precipitation-based deposit formation, the process proved to be quite sensitive to the deposition conditions, most prominently, the electrode potential. The mesoporous $Ni(OH)_2$ deposits were morphologically stable for moderate annealing that was a part of the optimization of supercapacitor performance. The annealing of $Co(OH)_2$ led to the formation of mesoporous Co_3O_4 with good electrochromic performance [343]. Supercapacitor properties similar to H_1-e $Ni(OH)_2$ can be achieved with H_1-e Ni after oxidation in alkaline solutions [344].

Mesoporous RuO_x supercapacitor can be obtained with a combination of two electrochemical steps, the deposition of mesoporous Ru and its subsequent oxidation to RuO_x [345, 346]. The specific surface area of the H_1-e RuO_x was more than an order of magnitude larger than Ru-black obtained with conventional methods. Another supercapacitor material, MnO_2 could be deposited anodically in a single step from Mn^{2+} salts [347].

11.4.3 Semiconductors Obtained with Hexagonal Lyotropic Phases

The electrodeposition of several mesoporous semiconductor materials from the H_1 lyotropic phase has been reported (CdS [348], CdTe [348–350], PbTe [325], Cu_2O [351] and ZnO [351]). For CdTe, the band gap of the mesoporous form was found to be exactly the same as the bulk one, while birefringence and photoelectrochemical conversion efficiency improved as a result of the mesoporous structure. The surface of the mesoporous CdTe obtained with the template-assisted deposition was smooth and defect-free. However, this behaviour cannot be generalized since the band gap of mesoporous Cu_2O was higher than that of its bulk form, although similar to flat electrodeposited Cu_2O layers [351]. A much more pronounced band gap enhancement ("blue shift") due to the mesoporous nature of the deposit was found for PbTe [325]. All mesoporous semiconductor materials had a direct band gap. For ZnO, the optical properties were strongly influenced by the cracked nature of the mesoporous films formed [351], in contrast to the optimal smoothness observed for CdTe. The explanation behind this difference may lay in the deposition potential of the specific materials. CdTe is deposited from $CdSO_4$ and TeO_2 at potentials where no hydrogen evolution takes place, and the gas evolution may be a significant factor where the ZnO deposit formed (−1 V vs. Ag/AgCl) by precipitation from cathodically alkalinized nitrate-containing media.

The H_2 phase of lyotropic liquid crystals was seldom used in electrodeposition experiments. A study for the deposition of Cu_2O applied sodium bis(2-ethylhexyl) sulphosuccinate as the continuous phase with p-xylene and a $CuCl_2$ solution for the inverse hexagonal phase [352]. Since no other component was applied (e.g., lactic acid that is customary in Cu_2O deposition processes to provide OH^- ions during its reduction), the deposition mechanism is not fully clear. The difficulty of the process is indicated by the experimental constraint that the anode and the cathode had to be aligned in a closely facing position with a gap of 0.5–1.0 mm in order to ensure a high enough electrical field for the good phase ordering. With this method, cuprite nanowires of high aspect ratio could be obtained with a good crystallinity.

11.4.4 Electrodeposited Polymers by Using Various Lyotropic Templates

The electrodeposition of polymers into H_I-type mesoporous templates is in some cases analogous to metal and semiconductor deposition in the sense that the polymer is formed in the aqueous phase [217, 353]. Although the porosity of the resulting polymer films can be well evidenced with electrochemical measurements, the inverse H_I structure of the resulting polymer coating is difficult to image due to the soft nature of the polymer. Nevertheless, the low-angle X-ray diffraction indicates the same periodicity of the resulting film as that observed for the template.

However, a deposition mechanism dissimilar to the above detailed pattern has been described, too [354]. In this case, the polymerization of aniline was reported to take place in the apolar phase whose components may also catalyse the polymerization process. Since both the monomers to be polymerized and the amphiphilic component(s) of the lyotropic systems were different in the works cited above, the difference in the mechanisms can be explained with the different properties of the systems studied. At the time being, it cannot be unambiguously categorized which deposition mode is relevant to a specific system.

Another synthesis strategy is the polymerization of a monomer in a lyotropic liquid crystal template and the use of the resulting composite without removing the template. An example is the polymerization of 3,4-ethylenedioxythiophene in the hydroxypropyl cellulose/water system [355]. The additional benefit of this template is the chiral nematic nature of the liquid crystal that is transferred to the polymer embedded. The result shows that the electrodeposited polymer remains electrochemically switchable even in the template, and the electrochemical and optical properties are strongly interrelated.

11.4.5 Dual Template Methods Involving Lyotropic Liquid Crystals

Dual template methods apply two fundamentally different template methods in the same deposition system in parallel. For the successful merge of two templating methods, the characteristic pore size of the two template methods must differ by at least about an order of magnitude. This condition is easily fulfilled if a lyotropic system with the H_1 phase is used as the fluid medium in a solid template of larger pore size, i.e., either an AAO template [356–361] or a colloidal solid particle template [362, 363]. The preparation of self-supporting porous materials requires two steps, corresponding to the dissolution conditions of both templates. The resulting deposit exhibits a hierarchical porous structure in which the macropores are determined by the solid template and the diameter of the mesopores is in accord with the column diameter of the H_1 lyotropic phase. While the macropores directly reflect the pore structure orientation of the larger template, the orientation of the mesopores seems to be rather diverse. While the data in several works indicate a random orientation of the mesopores with respect to the growth direction of the porous deposits, a few examples can be found for nanowires grown in AAO template that the cylindrical pores related to the lyotropic secondary template are ordered in parallel to the nanowire axis [358, 360]. Figure 11.20 presents a few characteristic images obtained for materials obtained with dual template synthesis.

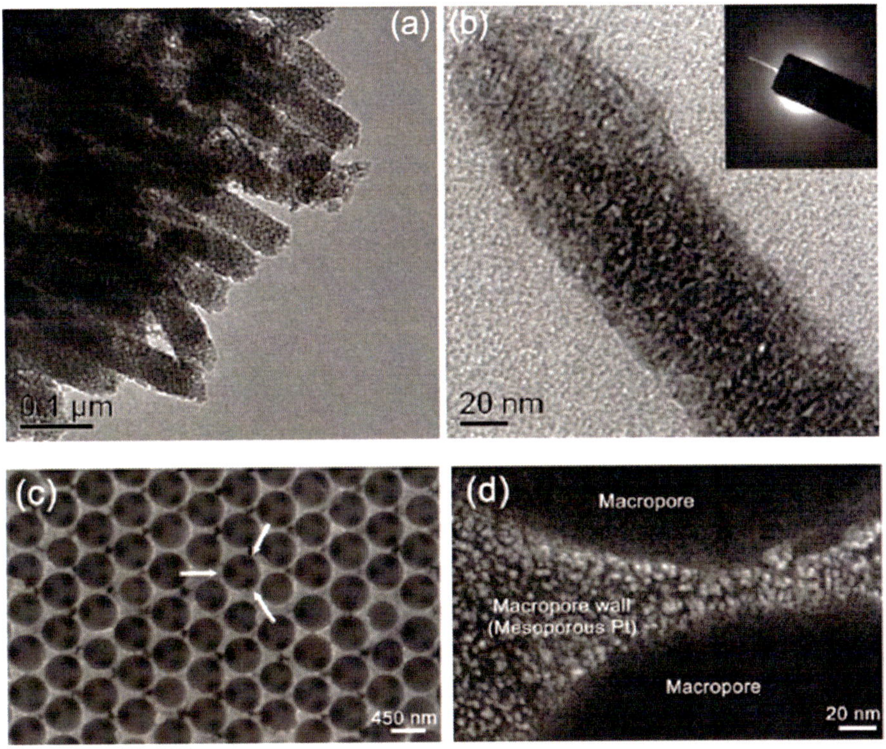

Fig. 11.20 **a** TEM image of MnO_2 nanowire bundle with hierarchical porous internal structure and **b** a magnified TEM image of a single MnO_2 nanowire with an electron diffraction pattern in the inset. Reprinted by permission from Springer [359], copyright (2010). **c** SEM image of the inverse opal structure of a porous Pt deposit prepared with colloidal particle template and **d** a magnified image showing the porosity, corresponding to the lyotropic media used for the deposition. Reprinted from [362]. Copyright (2006), with permission from Elsevier

The materials used for dual template synthesis are essentially the same as used in other lyotropic deposition methods. Pt with high surface area due to the hierarchical pore system obtained an outstanding interest because of its catalytic properties [357, 363]. The high surface area and the hierarchical pore structure of transition metal oxides (like MnO_2 [359]) and hydroxides (like $Co(OH)_2$ [360, 361]) are beneficial for the supercapacitor properties. The analysis of dual template-synthesized Ni alloys showed that besides the composition, the particle size of the alloys plays a fundamental role in the determination of the electrochemical and corrosion properties of these alloys [356, 358].

11.4.6 Micelle Templates

The category of micelle templates involves a large variety of materials. The group of micelle templates ranges from those assembled from either relatively small surfactant molecules or large multiblock copolymers to colloidal particles precipitated *in situ* in the plating solutions. The latter can be distinguished from the previous sorts of micelles with the feature that no dynamic equilibrium with the solution prevails; hence, their shape and size are determined by the precipitation conditions and can be taken as constant. In this respect, they show some resemblance to the colloidal particle templates but differ from them in size and aggregation mechanism. Finally, reverse micelles must be listed here as finite domains that can regulate the special constraints of the electrodeposition process. A few examples will be shown below for each category. The micelle-forming material is a sacrificial component in each case that can be removed by washing the deposit with and appropriate solvent and that no longer has any influence on the deposit properties after being removed. The electrodeposition process is usually very simple, involving potentiostatic polarization for a fixed time. The deposition potential was often reported to have a significant impact on whether the deposit showed a reverse micelle mesoporous structure or not; therefore, the optimization of the deposition process is indispensable.

Micelle templates formed by the association of small surfactant molecules. Anionic, neutral and cationic surfactant all can form micelles in aqueous solutions above the critical micelle-forming concentration but below the limit of the occurrence of the H_1 phase. The most commonly applied materials in each category are sodium dodecylsulphate [364, 365], octaethylene glycol monohexadecyl ether [366] and cetyltrimethylammonium bromide [367], respectively. The impact of the micellar solutions on the deposition process has been demonstrated in various ways. The change in the micrometre-scale morphology of the deposits may be an indicator of the micelle templating process, even if the micelle imprints at the sub-10 nm scale cannot be detected with SEM observations [364]. When TEM observations are available, the mesoporous nature of the deposits can be identified, showing that a disordered pore arrangement was formed [365–367]. A transition from random to hexagonally ordered deposits with the increase of the surfactant concentration is demonstrated in Fig. 11.21.

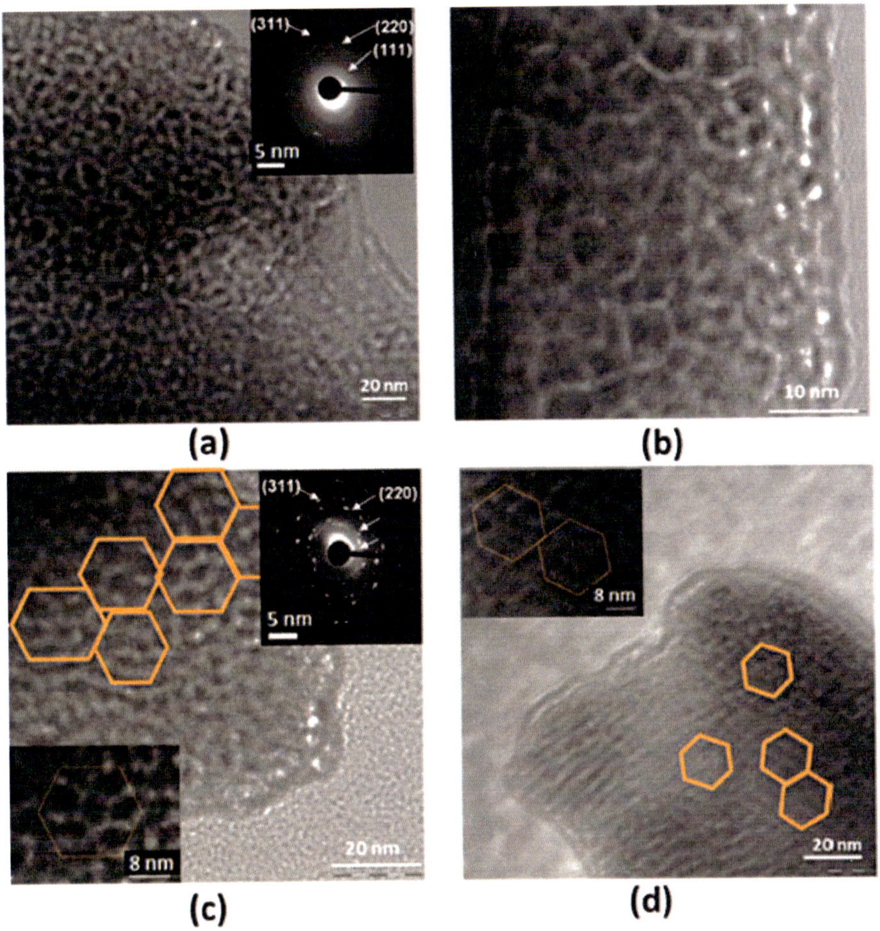

Fig. 11.21 TEM images of mesoporous nickel samples deposited from solutions of cetyltrimethylammonium bromide solutions with different concentrations: **a** 10, **b** 20, **c** 30 and **d** 50 wt.%. The transition from the random granular pore structure (corresponding to a micellar solution) to the hexagonally ordered pore system (corresponding to the H_1 phase) is to be observed. Reprinted from [367]. Copyright (2019), with permission from Elsevier

Micelle templates from triblock copolymers. A typical symmetrical triblock copolymer used in electrodeposition experiments is made up from chain sections of poly(ethylene oxide) at the line ends and poly(propylene oxide) at the centre, $H[OC_2H_4]_n[OCH(CH_3)CH_2]_m[OC_2H_4]_nOH$ with $n \approx 20$ and $m \approx 70$. This compound is often referred to as Pluronic-123 with its commercial name. While the *diblock copolymers* discussed in Sect. 11.2.9 are a mixture of two polymers whose templating capability stems for the segregation of the dissimilar phases, *triblock copolymers* mentioned in the present chapter are pure compounds with a well-defined chemical composition. This notation is analogous to the occurrence of these names in the relevant literature.

The concentration range of micellar solution of Pluronic-123 is 0.001 wt.% $< c <$ 30 wt.%, and the average micelle diameter about 14 nm, although the pore diameter of the resulting deposits was found to be smaller, 11 and 7 nm for Pd [368] and FePt [369] deposits, respectively. The treatment of the micellar solution of Pluronic-123 does not require any special step due to the lack of any orientation preference in the random micellar system. The addition of metal salts necessary for electrodeposition does not change significantly either the limiting compositions of the phases or the micelle diameter.

Irreversibly coagulated micelles as template. The periodicity of the self-assembled templates prepared from particles *ex situ* and prior to the deposition process is limited due to the interaction strength of the particles in the self-organization processes. While the particle templates have a lower limit for the particle size (>100 nm), micelles formed with the association of either tenside or polymer molecules in equilibrium with a true solution exhibit an upper limit for the particle size (<20 nm). Polymer particles coagulated *in situ* by a sol–gel process fill up a size gap between the above-mentioned template types.

The initial step of the particle formation is the dissolution of a diblock copolymer in a suitable solvent. For the commonly used PS-b-PEO diblock copolymer (where PEO stands for polyethylene oxide), tetrahydrofuran (THF) is a suitable solvent that dissolves the polymer without any micelle formation. Then, ethanol and the plating solutions to be used are added successively to the polymer solution. The gradual shift of the solvent polarity leads to the precipitation of core–shell type polymer particles in which the more hydrophobic component (here, PS) forms the core that is covered with the more hydrophilic PEO as shell. The particle size can be controlled with the concentration of the THF solution and the molar weight of the polymers in the 20–60 nm range. In accord with the description of the method [370], the electrodeposited cage-like metal framework and the polymer particle network serving as template are produced in parallel in electrodeposition and electrophoretic assembly, respectively. This can be the reason why the deposition potential (and hence, the deposition rate) has a crucial role in the formation of the mesoporous structure [371]. A deposition rate below the desired value leads to the formation of a bulky deposit since it pushes ahead the polymer particles instead of growing in between them, whereas a too high deposition rate leads to random nucleation and a layer without mesopores.

The deposit types reported for the precipitated diblock copolymer templates show some great variety. The surface image of the deposits is in each case characteristic of a random porous structure in which both the cavities and the walls percolate, regardless of the material deposited (Au [370, 372, 373], Au–Cu alloy [374] or Ni [371]). However, the cross-sectional images can differ in whether the pores are randomly situated or form a vertical channel due to the linear arrangement of the polymer particles during the deposition process [374]. With a purposefully chosen substrate, hemispherical mesoporous Au particles can be grown in which the polymer particles are the pore-directing agent [373].

Electrodeposition from reverse micellar (L_2) systems. Various successful trials have demonstrated that electrodeposition is possible from L_2-type systems containing reverse micelles, although the phase containing the reactant is not continuous. However, the choice of the relatively non-polar continuous phase is usually such that it has a limited ion transport capability, which makes it possible both to pass electrical current through the cells and to measure potential difference across the system. Non-aqueous phases suitable for the reverse micelle electrodeposition include Triton X-100 (i.e., p-octyl polyethylene glycol phenyl ether) as single component [375, 376], and mixture of hexanol, hexane and Triton X-100 [377, 378], and cetyltrimethylammonium bromide with 1-butanol as co-surfactant [379, 380]. The water content of the Triton X-100 phase, for instance, was reported to be 13 wt.%, and its proton mobility was identified to be responsible for the electrical conduction [375].

The deposits formed from various L_2 systems show a large morphological variety. Platinum deposited from L_2-type lyotropic systems was mesoporous, i.e., morphologically analogous to those obtained with H_1 systems of polymer micelle solutions. Nickel formed either cracked deposit layers or nanodots, depending on the water-to-surfactant ratio of the system [379], and cobalt showed a foamy or lamellar structure [380]. The morphology of the metallic Ni and Co deposits does not bear a feature size characteristic of the micelle size in the L_2 phase. The latter dissimilarity also holds for non-metallic deposits such as ZnO [377] and $Ni(OH)_2$ [378], although these deposits developed a significant porosity due to the exclusion effect of the less polar phase in the L_2 system used. We can draw the conclusion that neither the diameter of the reverse micelles nor their mean distance in the solution can be taken as a directly applicable design factor for tuning the deposit morphology and porosity, although the porosity itself is easily achievable with some optimization. Hence, the apolar phase in the L_2 systems is not a comparable "template" to other systems listed in the previous chapters in the sense that its mean thickness (the micelle-to-micelle distance) usually cannot be recognized in the deposit.

11.4.7 Self-assembled Monolayers (SAMs) as Templates

Electrodes modified with self-assembled monolayers have long been studied in relation to electrodeposition processes. For monolayers formed from molecules anchored to the electrode through a sulphur atom (such as n-alkane thiols [381–384] and mercapto-n-alkanoic acids [385]), two deposition mechanisms were identified, namely deposition in the UPD state *under* the monolayer and deposition *onto* the monolayer. In addition, it was revealed that the bulk deposition starts at the defect sites of the substrate (e.g., terrace edges and crystal boundaries) and that the deposition can be accelerated by electrochemically induced desorption of the monolayer. In each case, the experience was that the nuclei formed were fine-grained as compared to the uncoated substrate. The experience for surfaces templated with a mixed layer of thiolated β-cyclodextrin and alkanethiols was partly similar. Relatively large scattered grains of CdSe could be deposited into the monolayer defects [386], but the internal cavity of the β-cyclodextrin was the preferred deposition site for Te [387]. For monolayered formed on dithiols, the nucleation took place exclusively on the top of the SAM [388]. The thoughtful design of the nucleus density and the nucleus size could be implemented with the above-mentioned methods in exceptional cases only.

Binary SAMs can be used as deposition template if one of the components can be selectively stripped off and the remaining layer is stable under the circumstances of the subsequent deposition step(s). This behaviour was well identified for the mixed monolayer composed of 3-mercaptopropionic acid and 1-dodecanethiol [389, 390]. By using Ag(111) as substrate, it was shown that 3-mercaptopropionic acid can be stripped off cathodically at less negative potentials than 1-dodecanethiol, and the patterned dodecanethiol sub-monolayer obtained with the removal of the co-adsorbent can be used as deposition template. Since the thiolated alkane monolayer cannot be exposed to very negative potentials by retaining its intactness, the deposition probe was CdS by using the EC-ALD technique (see Chap. 4). The deposit thickness was proportional to the number of the EC-ALD cycles. As an obvious consequence of the template formation technique, the nucleus density is proportional to the surface occupation ratio of the monolayer component stripped off, and both the shape and the relative position of the nanodots deposited are randomly situated at the surface.

References

1. Piraux L, Dubois S, Demoustier-Champagne S (1997) Nucl Instrum Meth B 131:357–363
2. Cao G, Liu D (2008) Adv Coll Interf Sci 136:45–64
3. Lai M, Riley DJ (2008) J Colloid Interf Sci 323:203–212
4. Li Y, Koshizaki N, Cai W (2011) Coordin Chem Rev 255:357–373
5. Tiwari JN, Tiwari RN, Kim KS (2012) Prog Mater Sci 57:724–803
6. Pérez-Page M, Yu E, Li J, Rahman M, Dryden DM, Vidu R, Stroeve P (2016) Adv Colloid Interfac 234:51–79
7. Davydov AD, Volgin VM (2016) Russ J Electrochem 52:806–831
8. Li C, Iqbal M, Lin J, Luo X, Jiang B, Malgras V, Wu KCW, Kim J, Yamauchi Y (2018) Acc Chem Res 51:1764–1773
9. Serrà A, Vallés E (2018) Appl Maters Today 12:207–234
10. Toimil-Molares ME (2012) Beilstein J Nanotechnol 3:860–883
11. Lupu N (ed) (2010) Electrodeposited nanowires and their applications. Intech, Vukovar
12. Piraux L (2020) Appl Sci 10:1832(1–25)
13. Possin GE (1970) Rev Sci Instrum 41:772–774
14. Williams WD, Giordano N (1984) Rev Sci Instrum 55:410–412
15. Sun L, Chien CL, Searson PC (2000) J Mater Sci 35:1097–1103
16. Muench F, Kunz U, Wardenga HF, Kleebe HJ, Ensinger W (2014) Langmuir 30:10878–10885
17. Rauber M, Alber I, Müller S, Neumann R, Picht O, Roth C, Schökel A, Toimil-Molares ME, Ensinger W (2011) Nano Lett 11:2304–2310

18. Frydrychewicz A, Biegański AT, Jackowska K, Tsirlina GA (2008) J Solid State Electrochem 12:1085–1091
19. Whitney TM, Jiang JS, Searson PC, Chien CL (1993) Science 261:1316–1319
20. Schönenberger C, van der Zande BMI, Fokkink LGJ, Henny M, Schmid C, Krulger M, Bachtold A, Huber R, Birk H, Staufer U (1997) J Phys Chem B 101:5497–5505
21. Konishi Y, Motoyama M, Matsushima H, Fukunaka Y, Ishii R, Ito Y (2003) J Electroanal Chem 559:149–153
22. Wang W, Huang Q, Jia F, Zhu J (2004) J Appl Phys 96:615–618
23. Riveros G, Gómez H, Cortes A, Marotti RE, Dalchiele EA (2005) Appl Phys A 81:17–24
24. Motoyama M, Fukunaka Y, Sakka T, Ogata YH, Kikuchi S (2005) J Electroanal Chem 584:84–91
25. Wu CG, Lin HL, Shau NL (2006) J Solid State Electrochem 10:198–202
26. Soleimany L, Dolati A, Ghorbani M (2010) J Electroanal Chem 645:28–34
27. Bahari Mollamahale Y, Ghorbani M, Dolati A, Hosseini D (2018) Mater Chem Phys 213:67–75
28. Napolskii KS, Roslyakov IV, Eliseev AA, Petukhova DI, Lukashin AV, Chen SF, Liu CP, Tsirlina GA (2011) Electrochim Acta 56:2378–2384
29. Kartopu G, Habouti S, Es-Souni M (2008) Mater Chem Phys 107:226–230
30. Linpé W, Harlow GS, Evertsson J, Hejral U, Abbondanza G, Lenrick F, Seifert S, Felici R, Vinogradov NA, Lundgren E (2019) ACS Appl Nano Mater 2:3031–3038
31. Vázquez M, Hernández-Vélez M, Pirota K, Asenjo A, Navas D, Velázquez J, Vargas P, Ramos C (2004) Eur Phys J B 40:489–497
32. Proenca MP, Sousa CT, Ventura J, Vazquez M, Araujo JP (2012) Electrochim Acta 72:215–221
33. Rauber M, Brötz J, Duan J, Liu J, Müller S, Neumann R, Picht O, Toimil-Molares ME, Ensinger W (2010) J Phys Chem C 114:22502–22507
34. Tsyntsaru N, Silkin S, Cesiulis H, Guerrero M, Pellicer E, Sort J (2016) Electrochim Acta 188:589–601
35. Bond AM, Luscombe D, Oldham KB, Zoski CG (1988) J Electroanal Chem 249:1–14
36. Haghdoost A, Pitchumani R (2011) Electrochim Acta 56:8260–8271
37. Valizadeh S, George JM, Leisner P, Hultman L (2001) Electrochim Acta 47:865–874
38. Bograchev DA, Volgin VM, Davydov AD (2013) Electrochim Acta 96:1–7
39. Bograchev DA, Volgin VM, Davydov AD (2013) Electrochim Acta 112:279–286
40. Choi J, Sauer G, Nielsch K, Wehrspohn RB, Gösele U (2003) Chem Mater 15:776–779
41. Riveros G, Green S, Cortes A, Gómez H, Marotti RE, Dalchiele EA (2006) Nanotechnology 17:561–570
42. Dalchiele EA, Marotti RE, Cortes A, Riveros G, Gómez H, Martínez L, Romero R, Leinend D, Martin F, Ramos-Barrado JR (2007) Physica E 37:184–188
43. Dobrev D, Vetter J, Angert N, Neumann R (2000) Electrochim Acta 45:3117–3125
44. Foss CA, Hornyak GL, Stockert JA, Martin CR (1992) J Phys Chem 96:7497–7499
45. van der Zande BMI, Böhmer MR, Fokkink LGJ, Schönenberger C (2000) Langmuir 16:451–458
46. Tan LK, Chong ASM, Tang XSE, Gao H (2007) J Phys Chem C 111:4964–4968
47. Chtanko N, Toimil Molares ME, Cornelius T, Dobrev D, Neumann R (2004) J Phys Chem B 108:9950–9954
48. Liu K, Chien CL, Searson PC (1998) Phys Rev B 58:R14681–R14684
49. Li L, Zhang Y, Li G, Zhang L (2003) Chem Phys Lett 378:244–249
50. Li L, Zhang Y, Li G, Wang X, Zhang L (2005) Mater Lett 59:1223–1226
51. Darques M, Piraux L, Encinas A, Bayle-Guillemaud P, Popa A, Ebels U (2005) Appl Phys Lett 86:072508(1–3)
52. Xua J, Huang X, Xie G, Fang Y, Liu D (2005) Mater Lett 59:981–984
53. Cho JU, Wu JH, Min JH, Ko SP, Soh JY, Liu QX, Kim YK (2006) J Magn Magn Mater 303:e281–e285
54. Zhang J, Jones GA, Shen TH, Donnelly SE, Li G (2007) J Appl Phys 101:054310(1–5)
55. Huang XH, Li GH, Dou XC, Li L (2009) J Appl Phys 105:084306(1–4)

56. Ren Y, Liu QF, Li SL, Wang JB, Han XH (2009) J Magn Magn Mater 321:226–230
57. Cattaneo L, Franz S, Albertini F, Ranzieri P, Vicenzo A, Bestetti M, Cavallotti PL (2012) Electrochim Acta 85:57–65
58. Thongmee S, Pang HL, Ding J, Lin JY (2009) J Magn Magn Mater 321:2712–2716
59. Ohgai T, Gravier L, Hoffer X, Lindeberg M, Hjort K, Spohr R, Ansermet JP (2003) J Phys D Appl Phys 36:3109–3114
60. Maurer F, Brötz J, Karim S, Toimil Molares ME, Trautmann C, Fuess H (2007) Nanotechnology 18:135709(1–8)
61. Gao T, Meng GW, Zhang J, Wang YW, Liang CH, Fan JC, Zhang LD (2001) Appl Phys A 73:251–254
62. Li X, Wang Y, Lei Y, Gu Z (2012) RSC Adv 2:2302–2307
63. Dobrev D, Vetter J, Angert N, Neumann R (2001) Appl Phys A 72:729–733
64. AIMawlawi D, Coombs N, Moskovits M (1991) J Appl Phys 70:4421–4425
65. Menon L, Zheng M, Zeng H, Bandyopadhyay S, Sellmyer DJ (2000) J Electron Mater 29:510–515
66. Chu SZ, Wada K, Inoue S, Todoroki S (2002) Chem Mater 14:4595–4602
67. Hernández-Vélez M, Pirota KR, Pászti F, Navas D, Climent A, Vázquez M (2005) Appl Phys A 80:1701–1706
68. Meng G, Cao A, Cheng JY, Vijayaraghavan A, Jung YJ, Shima M, Ajayan PM (2005) J Appl Phys 97:064303(1–5)
69. Pan H, Liu B, Yi J, Poh C, Lim S, Ding J, Feng Y, Huan CHA, Lin J (2005) J Phys Chem B 109:3094–3098
70. Xu CL, Li H, Zhao GY, Li HL (2006) Appl Surf Sci 253:1399–1403
71. Yu CY, Yu YL, Sun HY, Xu T, Li XH, Li W, Gao ZS, Zhang XY (2007) Mater Lett 61:1859–1862
72. Xu J, Wang K (2008) Appl Surf Sci 254:6623–6627
73. Yue G, Xu Q, Meng G, He X, Han F, Zhang L (2009) J Alloy Compd 477:L30–L34
74. Cortés A, Riveros G, Palma JL, Denardin JC, Marotti RE, Dalchiele EA, Gómez H (2009) J Nanosci Nanotechnol 9:1992–2000
75. Chen SF, Wei HH, Liu SP, Hsu CY, Huang JCA (2010) Nanotechnology 21:425602(1–5).
76. Yi G, Schwarzacher W (1999) Appl Phys Lett 74:1746–1748
77. Zhang XY, Dai JY (2004) Nanotechnology 15:1166–1168
78. Liu H, Ye Z, Luo Z, Rathnayaka KDD, Wu W (2008) Physica C 468:304–309
79. Tasaltın N, Öztürk S, Kılınc N¸ Yüzer H, Öztürk ZZ (2010) Nanoscale Res Lett 5:1137–1143
80. Wang XW, Yuan ZH, Fang BC (2011) Mater Chem Phys 125:1–4
81. Zhao GY, Xu CL, Guo DJ, Li H, Li HL (2007) Appl Surf Sci 253:3242–3246
82. Zhang X, Ding Y, Zhang Y, Hao Y, Meng G, Zhang L (2007) J Therm Anal Calorim 89:493–497
83. Wang JG, Tian ML, Kumar N, Mallouk TE (2005) Nano Lett 5:1247–1253
84. Zhu H, Yang SG, Ni G, Yu DL, Du YW (2001) J Magn Magn Mater 234:454–458
85. Zhu H, Yang SG, Ni G, Tang SL, Du YW (2001) J Phys Condens Matter 13:1727–1731
86. Zhu H, Yang S, Ni G, Yu DL, Du YW (2001) Scripta Mater 44:2291–2295
87. Dragos O, Chiriac H, Lupu N, Grigoras M, Tabakovic I (2016) J Electrochem Soc 163:D83–D94
88. Thongmee S, Pang HL, Yi JB, Ding J, Lin JY, Van LH (2009) Acta Mater 57:2482–2487
89. Atalay FE, Kayaa H, Atalay S, Tari S (2009) J Alloy Compd 469:458–463
90. Talapatra S, Tang X, Padi M, Kim T, Vajtai R, Sastry GVS, Shima M, Deevi SC, Ajayan PM (2009) J Mater Sci 44:2271–2275
91. Aravamudhan S, Singleton J, Goddard PA, Bhansali S (2009) J Phys D Appl Phys 42:115008(1–9)
92. Salem MS, Sergelius P, Zierold R, Montero Moreno JM, Görlitz D, Nielsch K (2012) J Mater Chem 22:8549–8557
93. Rosa WO, Vivas LG, Pirota KR, Asenjo A, Vázquez M (2012) J Magn Magn Mater 324:3679–3682

94. Ghemes A, Dragos-Pinzaru O, Chiriac H, Lupu N, Grigoras M, Shore D, Stadler B, Tabakovic I (2017) J Electrochem Soc 164:D13–D22
95. Chu SZ, Inoue S, Wada K, Kurashima K (2005) Electrochim Acta 51:820–826
96. Chu SZ, Inoue S, Wada K, Kanke Y, Kurashima K (2005) J Electrochem Soc 152:C42–C47
97. Gao TR, Yin LF, Tian CS, Lu M, Sang H, Zhou SM (2006) J Magn Magn Mater 300:471–478
98. Dahmane Y, Cagnon L, Voiron J, Pairis S, Bacia M, Ortega L, Benbrahim N, Kadri A (2006) J Phys D Appl Phys 39:4523–4528
99. Xu CL, Li H, Xue T, Li HL (2006) Scripta Mater 54:1605–1609
100. Fei XL, Tang SL, Wang RL, Su HL, Du YW (2007) Solid State Commun 141:25–28
101. Gao JH, Sun DL, Zhang XQ, Zhan QF, He W, Sun Y, Cheng ZH (2008) Appl Phys Lett 92:102501(1–3)
102. Shamaila S, Sharif R, Riaz S, Ma M, Khaleeq-ur-Rahman M, Han XF (2008) J Magn Magn Mater 320:1803–1809
103. Khurshid H, Huang YH, Bonder MJ, Hadjipanayis GC (2009) J Magn Magn Mater 321:277–280
104. Liu L, Huang Z, Wang D, Scholz R, Pippel E (2011) Nanotechnology 22:105604(1–6)
105. Li GR, Tong YX, Kay LG, Liu GK (2006) J Phys Chem B 110:8965–8970
106. Srivastava C, Rai RK (2013) Chem Phys Lett 5758:91–96
107. Garcia-Torres J, Gómez E, Vallés E (2012) J Phys Chem C 116:12250–12257
108. García J, Prida VM, Vivas LG, Hernando B, Barriga-Castro ED, Mendoza-Reséndez R, Luna C, Escrige J, Vázquez M (2015) J Mater Chem 3:4688–4697
109. Enculescu I, Toimil-Molares ME, Daub M, Westerberg L, Neumann R, Spohr R (2007) Appl Phys A 86:43–48
110. Ohgai T, Fujimaru T, Tanaka Y (2014) J Appl Electrochem 44:301–307
111. Yoo E, Moon JH, Jeon JS, Kim Y, Ahn JP, Kim YK (2020) Mater Charact 166:110451(1–9).
112. Safeer A, Ahmad N, Khan S, Azam LA, Bashir D (2019) J Appl Phys 125:034302(1–10).
113. Galdun L, Szabo P, Vega V, Barriga-Castro ED, Mendoza-Reséndez R, Luna C, Kovac J, Milkovic O, Varga R, Prida VM (2020) ACS Appl Nano Mater 3:7438–7445
114. Leprince-Wang Y, Wang GY, Zhang XZ, Yu DP (2006) J Crystal Growth 287:89–93
115. Oh J, Tak Y, Lee J (2004) Electrochem Solid State Lett 7:C27–C30
116. Inguanta R, Piazza S, Sunseri C (2008) Electrochim Acta 53:6504–6512
117. Santamaria M, Bocchetta P, Di Quarto F (2009) Electrochem Commun 11:580–584
118. Gomez H, Riveros G, Ramirez D, Henriquez R, Schrebler R, Marotti R, Dalchiele E (2012) J Solid State Electrochem 16:197–204
119. Inguanta R, Piazza S, Sunseri C (2008) J Electrochem Soc 155:K205–K210
120. Xu CL, Bao SJ, Kong LB, Li H, Li HL (2006) J Solid State Chem 179:1351–1355
121. Frantz C, Stein N, Zhang Y, Bouzy E, Picht O, Toimil-Molares ME, Boulanger C (2012) Electrochim Acta 69:30–37
122. Li L, Yang Y, Huang X, Li G, Zhang L (2006) Nanotechnology 17:1706–1712
123. Zhao AW, Meng GW, Zhang LD, Gao T, Sun SH, Pang YT (2003) Appl Phys A 76:537–539
124. Wang Q (2009) Mater Lett 63:1493–1495
125. Yue GH, Wang LS, Wang X, Chen YZ, Peng DL (2009) Nanoscale Res Lett 4:359–363
126. Hnida K, Mech J, Sulka GD (2013) Appl Surf Sci 287:252–256
127. Bicer M, Aydın AO, Şişman I (2010) Electrochim Acta 55:3749–3755
128. Routkevitch D, Bigioni T, Moskovits M, Xu JM (1996) J Phys Chem 100:14037–14047
129. Xu D, Xu Y, Chen D, Guo G, Gui L, Tang Y (2000) Chem Phys Lett 325:340–344
130. Sun H, Li X, Chen Y, Li W, Li F, Liu B, Zhang X (2008) Nanotechnology 19:225601(1–8)
131. Xu D, Shi X, Guo G, Gui L, Tang Y (2000) J Phys Chem B 104:5061–5063
132. Peng XS, Meng GW, Zhang J, Zhao LX, Wang XF, Wang YW, Zhang LD (2001) J Phys D Appl Phys 34:3224–3228
133. Sun HY, Li HX, Li W, Li F, Liu BT, Zhang XY (2007) Nanotechnology 18:115604(1–4)
134. Martín-González M, Prieto AL, Knox MS, Gronsky R, Sands T, Stacy AM (2003) Chem Mater 15:1676–1681
135. Chen R, Xu D, Guo G, Gui L (2003) Electrochem Commun 5:579–583

136. Mannam RS, Davis D (2010) Electrochem Solid State Lett 13:P15–P17
137. Martín-González M, Prieto AL, Gronsky L, Sands T, Stacy AM (2003) Adv Mater 15:1003–1006
138. Martín-González M, Snyder GJ, Prieto AL, Gronsky R, Sands T, Stacy AM (2003) Nano Lett 3:973–977
139. Li XL, Cai KF, Yu DH, Wang YY (2011) Superlattice Microst 50:557–562
140. Kumar S, Vohra A, Chakarvarti SK (2012) J Mater Sci Mater Electron 23:1793–1797
141. Jang B, Pellicer E, Guerrero M, Chen X, Choi H, Nelson BJ, Sort J, Pané S (2014) ACS Appl Mater Interfaces 6:14583–14589
142. Wang JG, Tian ML, Mallouk TE, Chan MHW (2004) Nano Lett 4:1313–1318
143. Bangar MA, Hangarter CM, Yoo B, Rheem Y, Chen W, Mulchandani A, Myung NV (2009) Electroanal 21:61–67
144. Cai LT, Skulason H, Kushmerick JG, Pollack SK, Naciri J, Shashidhar R, Allara DL, Mallouk TE, Mayer TS (2004) J Phys Chem B 108:2827–2832
145. Burdick J, Alonas E, Huang HC, Rege K, Wang J (2009) Nanotechnology 20:065306(1–6)
146. Jaleh B, Nasri A, Kakuee O (2015) Superlattice Microstr 85:768–775
147. Wang XW, He ZC, Li JS, Yuan ZH (2014) J Alloy Compd 595:221–225
148. Chen M, Searson PC, Chien CL (2003) J Appl Phys 93:8253–8255
149. Guo YG, Wan LJ, Zhu CF, Yang DL, Chen DM, Bai CL (2003) Chem Mater 15:664–667
150. Piraux L, Dubois S, Duvail JL, Ounadjela K, Fert A (1997) J Magn Magn Mater 175:127–136
151. Elhoussine F, Vila L, Piraux L, Faini G (2005) J Magn Magn Mater 290–291:116–119
152. Cho JU, Liu QX, Min JH, Ko SP, Kim YK (2006) J Magn Magn Mater 304:e213–e215
153. Huang X, Tan L, Cho H, Stadler BJH (2009) J Appl Phys 105:07D128(1–3).
154. Fedorov FS, Mönch I, Mickel C, Tschulik K, Zhao B, Uhlemann M, Gebert A, Eckert J (2013) J Electrochem Soc 160:D13–D19
155. Kok KY, Hangarter CM, Goldsmith B, Ng IK, Saidin NB, Myung NV (2010) J Magn Magn Mater 322:3876–3881
156. Evans PR, Yi G, Schwarzacher W (2000) Appl Phys Lett 76:481–483
157. Tang XT, Wang GC, Shima M (2007) J Magn Magn Mater 309:188–196
158. Böhnert T, Niemann AC, Michel AK, Bäßler S, Gooth J, Tóth BG, Neuróhr K, Péter L, Bakonyi I, Vega V, Prida VM, Nielsch K (2014) Phys Rev B 90:165416(1–11)
159. Zsurzsa S, Pellicer E, Sort J, Péter L, Bakonyi I (2018) J Electrochem Soc 165:D536–D542
160. Liang HP, Guo YG, Hu JS, Zhu CF, Wan LJ, Bai CL (2005) Inorg Chem 44:3013–3015
161. Wang H, Wu YC, Zhang L, Hu X (2006) Appl Phys Lett 89:232508(1–3)
162. Liu LF, Xie SS, Zhou WY (2009) J Phys D Appl Phys 42:205002(1–5)
163. Jeon B, Yoon S, Yoo B (2010) Electrochim Acta 56:401–405
164. de Horne FM, Piraux L, Michotte S (2005) Appl Phys Lett 86:152510(1–3)
165. Tian M, Wang J, Kurtz J, Mallouk TE, Chan MHW (2003) Nano Lett 3:919–923
166. Chen M, Chien CL, Searson PC (2006) Chem Mater 18:1595–1601
167. Wang W, Zhang G, Li X (2008) J Phys Chem C 112:15190–15194
168. Wang J, Tian M, Mallouk TE, Chan MHW (2004) J Phys Chem B 108:841–845
169. Ren Y, Wang J, Liu Q, Dai Y, Zhang B, Yan L (2011) J Mater Sci 46:7545–7550
170. Proenca MP, Sousa CT, Ventura J, Vazquez M, Araujo JP (2012) Nanoscale Res Lett 7:280(1–9)
171. Fu J, Cherevko S, Chung CH (2008) Electrochem Commun 10:514–518
172. Kumar N, Kumar R, Kumar S, Chakarvarti SK (2014) Curr Appl Phys 14:1547–1552
173. Brumlik CJ, Martin CR (1991) J Am Chem Soc 113:3174–3175
174. Brumlik CJ, MenonVP, Martin CR (1994) J Mat Res 9:1174–1183
175. Hulteen JC, Martin CR (1997) J Mater Chem 7:1075–1087
176. Mollamahalle YB, Ghorbani M, Dolati A (2012) Electrochim Acta 75:157–163
177. Bao J, Tie C, Xu Z, Zhou Q, Shen D, Ma Q (2001) Adv Mater 13:1631–1633
178. Bao J, Xu Z, Hong J, Ma X, Lu Z (2004) Scripta Mater 50:19–23
179. Wang Q, Wang G, Han X, Wang X, Hou JG (2005) J Phys Chem B 109:23326–23329

180. Zhang X, Dong D, Li D, Williams T, Wang H, Webley PA (2009) Electrochem Commun 11:190–193
181. Song YJ, Lee YW, Han SB, Park KW (2012) Mater Chem Phys 134:567–570
182. Xu XJ, Yu SF, Lau SP, Li L, Zhao BC (2008) J Phys Chem C 112:4168–4171
183. Xu X, Huang J, Shao M, Wang P (2012) Mater Chem Phys 135:6–9
184. Wang XW, Yuan ZH, Sun SQ, Duan YQ, Bie LJ (2008) Mater Chem Phys 112:329–332
185. Kozlovskiy AL, Shlimas DI, Shumskaya AE, Kaniukov EY, Zdorovets MV, Kadyrzhanov KK (2017) Phys Met Metallogr 118:164–169
186. Shumskaya A, Bundyukova V, Kozlovskiy A, Zdorovets M, Kadyrzhanov K, Kalkabay Gd, Kaniukov E (2020) J Magn Magn Mater 497:165913(1–5)
187. Yang D, Meng G, Xu Q, Han F, Kong M, Zhang L (2008) J Phys Chem C 112:8614–8616
188. Žužek-Rožman K, Šturm S, McGuiness PJ, Kobe S (2009) IEEE Trans Magn 45:4413–4416
189. Li L, Pan S, Dou X, Zhu Y, Huang X, Yang Y, Li G, Zhang L (2007) J Phys Chem C 111:7288–7291
190. Davis DM, Podlaha EJ (2005) Electrochem Solid State Lett 8:D1–D4
191. Fukunaka Y, Motoyama M, Konishi Y, Ishii R (2006) Electrochem Solid State Lett 9:C62–C64
192. Mishra R, Podlaha EJ (2006) Electrochem Solid State Lett 9:C199–C202
193. Motoyama M, Fukunaka Y, Ogata YH, Prinz FB (2010) J Electrochem Soc 157:D357–D369
194. Cherevko S, Fu J, Kulyk N, Cho SM, Haam S, Chung CW (2009) J Nanosci Nanotechnol 9:3154–3159
195. Li X, Wang Y, Song G, Peng Z, Yu Y, She X, Li J (2009) Nanoscale Res Lett 4:1015–1020
196. Motoyama M, Fukunaka Y, Sakka T, Ogata YH (2007) Electrochim Acta 53:205–212
197. Dou X, Li G, Huang X, Li L (2008) J Phys Chem C 112:8167–8171
198. Xia H, Feng J, Wang H, Lai MO, Lu L (2010) J Power Sources 195:4410–4413
199. Kim K, Kim M, Cho SM (2006) Mater Chem Phys 96:278–282
200. Wang HJ, Zou CW, Yang B, Lu HB, Tian CX, Yang HJ, Li M, Liu CS, Fu DJ, Liu JR (2009) Electrochem Commun 11:2019–2022
201. Yoo WC, Lee JK (2004) Adv Mater 16:1097–1101
202. Liu H, Wang F, Zhao Y, Liu J, Park KC, Endo M (2009) J Electroanal Chem 633:15–18
203. Liu J, Wang F, Zhai J, Ji J (2010) J Electroanal Chem 642:103–108
204. Ertan A, Tewari SN, Talu O (2008) J Exp Nanosc 3:287–295
205. Atalay FE, Kaya H, Yagmur V, Tari S, Atalay S, Avsar D (2010) Appl Surf Sci 256:2414–2418
206. Chowdhury T, Casey DP, Rohan JF (2009) Electrochem Commun 11:1203–1206
207. Förster S (2003) Top Curr Chem 226:1–28
208. Li X, Tian S, Ping Y, Kim DH, Knoll W (2005) Langmuir 21:9393–9397
209. Crossland EJW, Cunha P, Ludwigs S, Hillmyer MA, Steiner U (2011) ACS Appl Mater Interfaces 3:1375–1379
210. Thurn-Albrecht T, Schotter J, Kästle GA, Emley N, Shibauchi T, Krusin-Elbaum L, Guarini K, Black CT, Tuominen MT, Russell TP (2000) Science 260:2126–2129
211. Bal M, Ursache A, Tuominen MT, Goldbach JT, Russell TP (2002) ApplPhys Lett 81:3479–3481
212. Li Y, Maire HC, Ito T (2007) Langmuir 23:12771–12776
213. Sudha Rani V, Yoon SS, Parvatheeswara Rao B, Kim CG (2008) Mater Chem Phys 112:1133–1136
214. Lee JI, Cho SH, Park SM, Kim JK, Kim JK, Yu JW, Kim JC, Russell TP (2008) Nano Lett 8:2315–2320
215. Seifarth O, Krenek R, Tokarev I, Burkov Y, Sidorenko A, Minko S, Stamm M, Schmeißer D (2007) Thin Solid Films 515:6552–6556
216. Khatria MS, Schlörb H, Fähler S, Schultz L, Nandan B, Böhme M, Krenek R, Stamm M (2009) Electrochim Acta 54:2536–2539
217. Crossland EJW, Cunha P, Scroggins S, Moratti S, Yurchenko O, Steiner U, Hillmyer MA, Ludwigs S (2010) ACS Nano 4:962–966
218. Maire HC, Ibrahim S, Li Y, Ito T (2009) Polymer 50:2273–2280
219. Perera DMNT, Pandey B, Ito T (2011) Langmuir 27:11111–11117

220. Liu SSY, Ludwigs S (2020) Macromol Rapid Commun 41:1900485(1–15)
221. Ursache A, Goldbach JT, Russell TP, Tuominen MT (2005) J Appl Phys 97:10J322 (1–3)
222. Frankel D, Milenkovic S, Smith AJ, Hassel AW (2009) Electrochim Acta 54:6015–6021
223. Gao J, Zhao Z, Wei L, Cui K, Wang S, Li N, Guo J, Chen S, Hu Z, Liu Y (2018) Metall Mater Trans A 49:3575–3583
224. Hassel AW, Smith AJ, Milenkovic S (2006) Electrochim Acta 52:1799–1804
225. Bello Rodriguez B, Schneider A, Hassel AW (2006) J Electrochem Soc 153:C33–C36
226. Hassel AW, Milenkovic S, Schürmann U, Greve H, Zaporojtchenko V, Adelung R, Faupe F (2007) Langmuir 23:2091–2094
227. Bello Rodriguez B, Hassel AW (2008) J Electrochem Soc 155:K31–K37
228. Bello Rodriguez B, Smith AJ, Hassel AW (2008) J Electroanal Chem 618:11–16
229. Gao J, Zhao Z, Wei L, Cui K (2017) J Electrochem Soc 164:C474–C480
230. Gao J, Luo J, Geng H, Zhong J, Han G, Cui K, Zhao Z, Liu L (2020) Mater Chem Phys 244:122594(1–12).
231. Brittman S, Smith AJ, Milenkovic S, Hassel AW (2007) Electrochim Acta 53:324–329
232. Wei L, Zhao Z, Gao J, Cui K Chen S, Guo J, Liu L (2018) J Electrochem Soc 165:E45–E49
233. Deckman HW, Dunsmuir JH (1982) Appl Phys Lett 41:377–379
234. Gaulding EA, Liu G, Chen CT, Löbbert L, Li A, Segev G, Eichhorn J, Aloni S, Schwartzberg AM, Sharp ID, Toma FM (2017) J Mater Chem A 5:11601–11614
235. Stein A, Schroden RC (2001) Curr Opin Solid St M 5:553–564
236. Luo Q, Liu Z, Li L, Xie S, Kong J, Zhao D (2001) Adv Mater 13:286–289
237. Braun PV, Wiltzius P (2001) Adv Mater 13:482–485
238. Jiang P, Bertone JF, Hwang KS, Colvin VL (1999) Chem Mater 11:2132–2140
239. Sun F, Cai W, Li Y, Cao B, Lu F, Duan G, Zhang L (2004) Adv Mater 16:1116–1121
240. Eagleton TS, Searson PC (2004) Chem Mater 16:5027–5032
241. Trau M, Saville DA, Aksay A (1996) Science 272:706–709
242. Trau M, Saville DA, Aksay A (1997) Langmuir 13:6375–6381
243. Yeh SR, Seul M, Shraiman BI (1997) Nature 386:57–59
244. Lai CH, Huang YJ, Wu PW, Chen LY (2010) J Electrochem Soc 157:P23–P27
245. Teranishi T, Hosoe M, Tanaka T, Miyake M (1999) J Phys Chem B 103:3818–3827
246. Li W, Fu T, He S (2006) Mater Sci Eng A 441:239–244
247. Zhang J, Wang M, Ge X, Wu M, Wu Q, Yang J, Wang M, Jin Z, Liu N (2011) J Coll Interf Sci 353:16–21
248. Liu Y, Wang S, Lee JW, Kotov NA (2005) Chem Mater 17:4918–4924
249. Deepak Kallepalli LN, Constantinescu C, Delaporte P, Utéza O, Grojo D (2015) J Coll Interf Sci 446:237–243
250. Masuda Y, Itoh T, Itoh M, Koumoto K (2004) Langmuir 20:5588–5592
251. Guo Q, Arnoux C, Palmer RE (2001) Langmuir 17:7150–7155
252. Wijnhoven JEGJ, Zevenhuizen SJM, Hendriks MA, Vanmaekelbergh D, Kelly JJ, Vos WL (2000) Adv Mater 12:888–890
253. Nishikawa K, Dokko K, Kinoshita K, Woo SW, Kanamura K (2009) J Power Sources 189:726–729
254. Duan G, Cai W, Li Y, Li Z, Cao B, Luo Y (2006) J Phys Chem B 110:7184–7188
255. Chung YW, Leu IC, Lee JH, Yen JH, Hon MH (2007) Electrochim Acta 53:1703–1707
256. Chung YW, Leu IC, Lee JH, Yen JH, Hon MH (2007) J Electrochem Soc 154:E77–E83
257. Dai L, Jiang LP, Abdel-Halim ES, Zhu JJ (2011) Electrochem Commun 13:1525–1528
258. Chen Z, Dong H, Pan J, Zhan P, Tang C, Wang ZL (2010) Appl Phys Lett 96:051904(1–3)
259. Lan D, Wang YR, Yu Y, Ma WJ, Li C (2007) Chinese Phys 16:468–471
260. Zhu R, McLachlan M, Reyntjens S, Tariq F, Ryan MP, McComb DW (2009) Nanoscale 1:355–359
261. Xu L, Zhou WL, Frommen C, Baughman RH, Zakhidov AA, Malkinski L, Wang JQ, Wiley JB (2000) Chem Commun 997–998
262. Hao Y, Zhu FQ, Chien CL, Searson PC (2007) J Electrochem Soc 154:D65–D69
263. Sumida T, Wada Y, Kitamura T, Yanagida S (2002) Langmuir 18:3886–3894

264. Abdelsalam ME (2009) Cent Eur J Chem 7:446–453
265. Ismail IMI (2010) J Saudi Chem Soc 14:351–356
266. Lu H, Liu Y, Han GG (2011) Appl Surf Sci 257:7030–7035
267. Hung D, Liu Z, Shah N, Hao Y, Searson PC (2007) J Phys Chem C 111:3308–3313
268. Bartlett PN, Baumberg JJ, Coyle S, Abdelsalam ME (2004) Faraday Discuss 125:117–132
269. Chen C, Li Y, Liu S (2009) J Electroanal Chem 632:14–19
270. Huang YJ, Lai CH, Wu PW, Chen LY (2010) J Electrochem Soc 157:P18–P22
271. Liao CH, Hung PS, Cheng Y, Chang SY, Wu PW (2018) Mater Lett 215:152–156
272. Yu X, Lee YJ, Furstenberg R, White JO, Braun PV (2007) Adv Mater 19:1689–1692
273. Han R, Pan W, Shi S, Wu H, Liu S (2007) Mater Lett 61:5014–5017
274. Zhukov AA, Ghanem MA, Goncharov AV, de Groot PAJ, El-Hallag IS, Bartlett PN, Boardman R, Fangohr H (2004) J Magn Magn Mater 272–276:1621–1622
275. Zhukov AA, Goncharov AV, de Groot PAJ, Ghanem MA, El-Hallag IS, Bartlett PN, Boardman R, Fangohr H, Novosad V, Karapetrov G (2005) J Appl Phys 97:10J701(1–3)
276. Li Z, Liu Y, Liu P, Chen W, Feng S, Zhong W, Yu C (2012) RSC Adv 2:3447–3450
277. Zhukov AA, Goncharov AV, de Groot PAJ, Ghanem MA, El-Hallag IS, Bartlett PN, Boardman R, Fangohr H, Novosad V, Karapetrov G (2006) Appl Phys Lett 88:062511(1–3)
278. Ke F, Huang L, Jiang H, Wei H, Yang F, Sun S (2007) Electrochem Commun 9:228–232
279. Woo SW, Okada N, Kotobuki M, Sasajima K, Munakata H, Kajihara K, Kanamura K (2010) Electrochim Acta 55:8030–8035
280. Jin Y, Munakata H, Okada N, Kanamura K (2013) J Nanomat 937019(1–6)
281. Ke FS, Huang L, Wei HB, Cai JS, Fan XY, Yang FZ, Sun SG (2007) J Power Sources 170:450–455
282. Ke FS, Huang L, Cai JS, Sun SG (2007) Electrochim Acta 52:6741–6747
283. Golvano-Escobal I, de Paz-Castany R, Alcantara CCJ, Pané S, García-Lecina E, Sort J, Pellicer E (2019) Electrochim Acta 313:211–222
284. Fan T, Sun P, Zhao J, Cui Z, Cui G (2019) J Electrochem Soc 166:D37–D43
285. Cao B, Sun F, Cai W (2005) Electrochem Solid State Lett 8:G237–G240
286. Fu M, Zhou J, Li B, Huang X, Wang Y, Li L (2008) J Mater Chem 18:5986–5992
287. Zhou Q, Zhao J, Xu W, Zhao H, Wu Y, Zheng J (2008) J Phys Chem C 112:2378–2381
288. Elias J, Lévy-Clément C, Bechelany M, Michler J, Wang GY, Wang Z, Philippe L (2010) Adv Mater 22:1607–1612
289. Zhang J, Tu JP, Cai GF, Du GH, Wang XL, Liu PC (2013) Electrochim Acta 99:1–8
290. Qi CX, Tan Z, Feng ZH, Yu LPJ (2014) Mater Sci: Mater Electron 25:1553–1558
291. Benedetti TM, Gonçales VR, Petri DFS, Córdoba de Torresi SI, Torresi RM (2010) J Braz Chem Soc 21:1704–1709
292. Li X, Jiang Y, Shi Z, Xu Z (2007) Chem Mater 19:5424–5430
293. Li X, Tao F, Jiang Y, Xu Z (2007) J Coll Interf Sci 308:460–465
294. Ding Y, Yang L, Li Y, Wu X, Lu Y, Zhao J (2013) J Porous Mater 20:601–605
295. Kim J, Kim HS, Choi JH, Jeon H, Yoon Y, Liu J, Park JG, Braun PV (2014) Chem Mater 26:7051–7058
296. Shi X, Zhang K, Shin K, Moon JH, Lee TW, Park JH (2013) Phys Chem Chem Phys 15:11717–11722
297. Yuan YF, Xia XH, Wu JB, Chen YB, Yang JL, Guo SY (2011) Electrochim Acta 56:1208–1212
298. Xia XH, Tu JP, Zhang J, Huang XH, Wang XL, Zhao XB (2010) Electrochim Acta 55:989–994
299. Martin J, Jack M, Hakimian A, Vaillancourt N, Villemure G (2016) J Electroanal Chem 780:217–224
300. Yan H, Yang Y, Fu Z, Yang B, Xia L, Xu Y, Fu S, Li F (2006) Chem Lett 36:864–865
301. Pan L, Xu H, Sun Y, Zhao J, Li Y (2016) Crystals 6:76(1–9)
302. Nakayama M, Kanaya T, Inoue R (2007) Electrochem Commun 9:1154–1158
303. Kuo CY, Lu SY (2007) J Am Ceram Soc 90:1956–1958
304. O'Hanlon S, McNulty D, O'Dwyer C (2017) J Electrochem Soc 164:D111–D119
305. Liu X, Zhang Y, Ge D, Zhao J, Li Y, Endres F (2012) Phys Chem Chem Phys 14:5100–5105
306. Xin W, Zhao J, Ge D, Ding Y, Li Y, Endres F (2013) Phys Chem Chem Phys 15:2421–2426

307. Braun PV, Wiltzius P (1999) Nature 402:603–604
308. Yeo SH, Teh LK, Wong CC (2006) J Porous Mater 13:281–285
309. Neale NR, Lee BG, Kang SH, Frank AJ (2011) J Phys Chem C 115:14341–14346
310. Park JY, Dutta P, Advincula R (2011) Soft Matter 7:3775–3779
311. Al Zoubi M, Al-Salman R, Li Y, Endres F (2011) Z Phys Chem 225:393–403
312. Zhong Q, Xu H, Ding H, Bai L, Mu Z, Xie Z, Zhao Y, Gu Z (2013) Colloid Surface A 433:59–63
313. Ghanem MA, Bartlett PN, de Groot P, Zhukov A (2004) Electrochem Commun 6:447–453
314. Choi B, Myung N, Rajeshwar K (2007) Electrochem Commun 9:1592–1595
315. Tiu BDB, Pernites RB, Foster EL, Advincula RC (2015) J Coll Interf Sci 459:86–96
316. Wang J, Duan G, Liu G, Li Y, Xu L, Cai W (2015) J Mater Chem C 3:5709–5714
317. Kim DY, Suk J, Kim DW, Kang Y, Im SH, Yang Y, Park OO (2014) J Mater Chem A 2:6396–6401
318. Wang J, Zhou H, Nanda J, Braun PV (2015) Chem Mater 27:2803–2811
319. Duan G, Cai W, Luo Y, Li Z, Lei Y (2006) J Phys Chem B 110:15729–15733
320. Fu M, Deng L, Zhao A, Wang Y, He D (2010) Opt Mater 32:1210–1215
321. Alfutimi A, Curtis R, Tiddy JT, Lyotropic Surfactant Liquid Crystals: Micellar Systems. In: Goodby JW, Collings PJ, Kato T, Tschierske C, Gleeson HF, Raynes P (eds) (2014) Handbook of liquid crystals, Chapter 12.3, Wiley-VCH Verlag & Co., Weinheim
322. Libster D, Aserin A, Garti N (2018) Oleogels based on nonlamellar lyotropic liquid crystalline structures for food applications. In: Marangoni AG, Nissim G (eds) Edible oleogels. Structure and health implications, 2nd edn. Elsevier
323. Jákli A, Saupe A (2006) One- and two-dimensional fluids. Properties of smectic, lamellar and columnar liquid crystals. In: Series in condensed matter physics. Taylor and Francis Group, New York and London
324. Guo R, Zhang B, Liu X (2007) Appl Surf Sci 254:538–543
325. Li X, Nandhakumar IS, Attard GS, Markham ML, Smith DC, Baumberg JJ (2009) Micropor Mesopor Mater 118:403–407
326. Attard GS, Bartlett PN, Coleman NRB, Elliott JM, Owen JR, Wang JH (1997) Science 278:838–840
327. Elliott JM, Attard GS, Bartlett PN, Coleman NRB, Merckel DAS, Owen JR (1999) Chem Mater 11:3602–3609
328. Elliott JM, Birkin PR, Bartlett PN, Attard GS (1999) Langmuir 15:7411–7415
329. Gollas B, Elliott JM, Bartlett PN (2000) Electrochim Acta 45:3711–3724
330. Evans SAG, Elliott JM, Andrews LM, Bartlett PN, Doyle PJ, Denuault G (2002) Anal Chem 74:1322–1326
331. Nie M, Elliott JM (2005) J Mater Sci Technol 21:863–865
332. Daniele S, Bragato C, Battistel D (2012) Electroanal 24:759–766
333. Yamauchi Y, Tonegawa A, Komatsu M, Wang H, Wang L, Nemoto Y, Suzuki N, Kuroda K (2012) J Am Chem Soc 134:5100–5109
334. Chen SA, Liang YC, Lu KT, Pao CW, Lee JF, Lin TL, Chen JM (2014) Phys Chem Chem Phys 16:3939–3945
335. Bender F, Mankelow RK, Hibbert DB, Gooding JJ (2006) Electroanal 18:1558–1563
336. Bartlett PN, Gollas B, Guerin S, Marwan J (2002) Phys Chem Chem Phys 4:3835–3842
337. Imokawa T, Williams KJ, Denuault G (2006) Anal Chem 78:265–271
338. Bartlett PN, Guerin S (2003) Anal Chem 75:126–132
339. Zhao D, Zhou W, Li H (2007) Chem Mater 19:3882–3891
340. Zhao DD, Xu MW, Zhou WJ, Zhang J, Li HL (2008) Electrochim Acta 53:2699–2705
341. Zhou W, Zhang J, Xue T, Zhao D, Li H (2008) J Mater Chem 18:905–910
342. Zhou WJ, Xu MW, Zhao DD, Xu CL, Li HL (2009) Micropor Mesopor Mater 117:55–60
343. Yuan YF, Xia XH, Wu JB, Gui JS, Chen YB, Guo SY (2010) J Membrane Sci 364:298–303
344. Nelson PA, Elliott JM, Attard GS, Owen JR (2002) Chem Mater 14:524–529
345. Sugimoto W, Makino S, Mukai R, Tatsumi Z, Fukuda K, Takasu Y (2012) YamauchiY. J Power Sources 204:244–248

346. Makino S, Yamauchi Y, Sugimoto W (2013) J Power Sources 227:153–160
347. Dong B, Xue T, Xu CL, Li HL (2008) Micropor Mesopor Mater 112:627–631
348. Braun PV, Osenar P, Twardowski M, Tew GN, Stupp SI (2005) Macroscopic nanotemplating of semiconductor films with hydrogen-bonded lyotropic liquid crystals. Adv Funct Mater 15 (11):1745–1750
349. Markham ML, Baumberg JJ, Smith DC, Li X, Gabriel T, Attard GS, Nandhakumar I (2005) Appl Phys Lett 86:011912(1–3)
350. Li X, Nandhakumar IS, Gabriel T, Attard GS, Markham ML, Smith DC, Baumberg JJ, Govender K, O'Brien P, Smyth-Boyle D (2006) J Mater Chem 16:3207–3214
351. Luo H, Zhang J, Yan Y (2003) Chem Mater 15:3769–3773
352. Huang L, Wang H, Wang Z, Mitra A, Zhao D, Yan Y (2002) Chem Mater 14:876–880
353. Bender F, Chilcott TC, Coster HGL, Hibbert DB, Gooding JJ (2007) Electrochim Acta 52:2640–2648
354. Guo R, Li ZC, Liu TQ (2006) J Polym Sci Pol Chem 44:2388–2394
355. Goto H, Akagi K (2006) Chem Mater 18:255–262
356. Foyet A, Hauser A, Schäfer W (2007) J Electroanal Chem 604:137–143
357. Zhong Y, Xu CL, Kong LB, Li HL (2008) Appl Surf Sci 255:3388–3393
358. Foyet A, Hauser A, Schäfer W (2008) J Solid State Electrochem 12:47–55
359. Wei Z, Feng ZX, Ze LL, Er JN (2010) J Porous Mater 17:253–257
360. Xue T, Wang X, Lee JM (2012) J Power Sources 201:382–386
361. Xue T, Lee JM (2014) J Power Sources 245:194–202
362. Yamauchi Y, Kuroda K (2006) Electrochem Commun 8:1677–1682
363. Zhang C, Yang H, Sun T, Shan N, Chen J, Xu L, Yan Y (2014) J Power Sources 245:579–582
364. Michaelis E, Wöhrle D, Rathousky J, Wark M (2006) Thin Solid Films 497:163–169
365. Nam HS, Jang KS, Ko JM, Kong JM, Kim JD (2011) Electrochim Acta 56:6459–6463
366. Al Abass NA, Denuault G, Pletcher D (2014) Phys Chem Chem Phys 16:4892–4899
367. Nasirpouri F, Barzegar S, Samardak AY, Ognev AV, Zubkov AA, Stancu A, Samardak AS (2019) Appl Surf Sci 471:776–785
368. Iqbal M, Li C, Wood K, Jiang B, Takei T, Dag Ö, Baba D, Nugraha AS, Asahi T, Whitten AE, Hossain MSA, Malgras V, Yamauchi Y (2017) Chem Mater 29:6405–6413
369. Isarain-Chávez E, Baró MD, Alcantara C, Pané S, Sort J, Pellicer E (2018) ChemSusChem 11:367–375
370. Li C, Dag Ö, Dao TD, Nagao T, Sakamoto Y, Kimura T, Terasaki O, Yamauchi Y (2015) Nat Commun 6:6608(1–8)
371. Baba D, Kim J, Henzie J, Li C, Jiang B, Dag Ő, Yamauchi Y, Asahi T (2018) Chem Commun 54:10347–10350
372. Li C, Jiang B, Chen H, Imura M, Sang L, Malgras V, Bando Y, Ahamad T, Alshehri SM, Tominaka S, Yamauchi Y (2016) Nano Res 9:1752–1762
373. Lim H, Nagaura T, Kim M, Kani K, Kim J, Bando Y, Alshehri SM, Ahamad T, You J, Na J, Yamauchi Y (2020) RSC Adv 10:8309–8313
374. Nugraha AS, Malgras V, Iqbal M, Jiang B, Li C, Bando Y, Alshehri A, Kim J, Yamauchi Y, Asahi T (2018) ACS Appl Mater Interfaces 10:23783–23791
375. Park S, Lee SY, Boo H, Kim HM, Kim KB, Kim HC, Song YJ, Chung TD (2007) Chem Mater 19:3373–3375
376. Park S, Song YJ, Boo H, Han JH, Chung TD (2009) Electrochem Commun 11:2225–2228
377. Hu L, Yu Z, Song Y, Liu C, Han Y, Cao X, Huang R, Jiao S (2015) Electrochem 83:715–720
378. Hu L, Yu Z, Hu Z, Song Y, Zhang F, Zhu H Jiao S (2015) Electrochim Acta 174:273–281
379. Sultana S, Saha S, Islam MM, Rahman MM, Mollah MYA, Susan MABH (2013) J Electrochem Soc 160:D524–D529
380. Saha S, Sultana S, Islam MM, Rahman MM, Mollah MYA, Susan MABH (2014) Ionics 20:1175–1181
381. Hagenström H, Schneeweiss MA, Kolb DM (1999) Langmuir 15:7802–7809
382. Langerock S, Ménard H, Rowntree P, Heerman L (2005) Langmuir 21:5124–5133
383. Pattanaik G, Shao W, Swami N, Zangari G (2009) Langmuir 25:5031–5038

384. Bayrakçeken Nişancı F, Demir Ü (2012) Langmuir 28:8571–8578
385. Petri M, Kolb DM, Memmert U, Meyer H (2003) Electrochim Acta 49:183–189
386. Choi SJ, Woo DH, Myung N, Kang H, Park SM (2001) J Electrochem Soc 148:C569–C573
387. Woo DH, Choi SJ, Han DH, Kang H, Park SM (2001) Phys Chem Chem Phys 3:3382–3386
388. Escalera-López D, Gómez E, Vallés E (2015) Phys Chem Chem Phys 17:16575–16586
389. Innocenti M, Bellassai S, Bianchini C, Carlà F, Loglio F, Polonia L, Vizza F, Foresti ML (2010) Electrochim Acta 55:2550–2554
390. Foresti ML, Loglio F, Innocenti M, Bellassai S, Carlá F, Lastraioli E, Pezzatini G, Bianchini C, Vizza F (2010) Langmuir 26:1802–1806

Chapter 12
Localized or Spatially Selective Electrodeposition Methods

12.1 Electrodeposition in Ultrathin Solution Layers Along the Edge of the Cell

12.1.1 Transport and Fluid Motion in Thin-Layer Cells of Various Orientations

A correct view of the transport phenomena in thin-layer electrochemical cells is essential to understand the processes taking place and the growth dynamics of deposits forming therein. Below, a short summary is devoted to this topic for underpinning the forthcoming sections. This overview is restricted to the case when the current in the cell (and, hence, the direction of the growth) is parallel to the solution layer plane. The two basic possible arrangements are the horizontal and the vertical solution layers, which terms also refer to the current direction. The electrodes fill the entire cell cross section horizontally along the edge of the cell at the initial phase of the deposition in the case of both arrangements. It is also common in the two cases that, in addition to the concentration gradient, a density gradient (which is, in turn, proportional to the concentration gradient in the first approximation) may lead to a gravity-induced convection.

In horizontal thin-layer cells, it is very tempting to treat the transport as a one-dimensional diffusion and migration problem. It has long been recognized that this one-dimensional treatment is not necessarily acceptable [1]. The conclusion of the above-mentioned work is summarized here shortly. It has been shown that two distinct regimes may be present, depending on the relative rates of the horizontal expansion of the diffusion field and gravity-induced convection. The latter effect modifies the horizontal composition gradient to a tilted or vertical one by a circular flow of the solution within the cell. It was shown that, after applying the current, convection should first appear. This period was called the "immiscible-fluid regime", indicating that the alleviation of the concentrations due to diffusion was negligible. The immiscible-fluid regime is present to layer thicknesses down to 70 μm. The circular gravity-induced

L. Péter, *Electrochemical Methods of Nanostructure Preparation*,
Monographs in Electrochemistry, https://doi.org/10.1007/978-3-030-69117-2_12

flow could be visualized with the help of a solution containing tracer particles [2, 3]. The second period called the "diffusion-hindered regime" is effective at later times. The theoretical approach to the transport means was quantitatively verified by copper electrodeposition studies in thin-layer cells. The spatial and temporal concentration variation was nearly symmetrical at the anode and the cathode. However, the depletion near the cathode along with the occurrence of a small-conductivity layer around the cathode surface may give rise to electroconvection, unlike at the (sacrificial) anode where the concentration increases. The complete formulation of the phenomena can be found in [1] and in the literature cited therein.

For vertical thin-layer cells with horizontal electrode at the top and the bottom of the cell, it was shown that a stable transport and constant concentration gradient can be achieved if the cathode is the upper electrode (the anode in these studies was a sacrificial electrode of the same material as the cathode) [4]. If the metal salt is consumed at the upper part of the cell, the density of the solution decreases there, and the solution layering eliminates any buoyancy-driven convection. If the current density is below the diffusion-limited value, the concentration gradient observed along the vertical cross section of the cell increases approximately proportionally to the current density [5]. As the diffusion-limited current density is approached, instability occurs, and oscillation starts (cell voltage oscillation under galvanostatic conditions and vice versa) [5]. Even if the deposition at the top-situated cathode becomes dendritic near the limiting current, the growth front remains fairly straight across the cell, even though the solution between the dendrites is fully depleted for the reactant. This situation is completely the opposite if the cathode is the bottom electrode. Then, the buoyancy force gives rise to the occurrence of turbulence-looking fluid motion. The vortex tubes detaching from the electrodes expand towards each other and later merge. This leads to a mixing much exceeding the concentration equalization effect of diffusion, and the concentration gradients tend to be attenuated. Experimental observations based on two-dimensional light absorption maps were strongly supported with numerical simulations leading to essentially the same concentration gradients and fluid motion.

Ramification of the growing metal is particularly common in thin-layer cells, especially when two conditions are fulfilled: (i) the cell is operated at deposition rates close to the mass transport-limited one; and (ii) no supporting electrolyte is used. As dendrites start growing, the solution between the dendrites becomes practically depleted with respect to the electrolyte. Concerning the solution concentration along the direction of the mean electrical field (i.e., perpendicular to the edges which comprise the electrodes), various zones follow each other from the cathode to the anode [6]. Due to the nearly complete depletion and the difference between ion migration rates, there is a space charge layer with a thickness of at most 10 μm, which is followed by a quasi-neutral (~100 μm thick) and a bulk neutral solution layer. At the close vicinity of the tip of the growing dendrites, electroconvection is the major force acting. Between the growing end zones of the dendrites, the electrical forces give rise to counter-rotating fluid vortices, whose relative contribution to the overall transport much exceeds that of the diffusion if the growth rate is high as compared to the classical diffusion-limited current density.

The difference in the concentration along or between the growing dendrites can lead to highly unexpected phenomena. Due to the concentration difference, local concentration cells can form, and the local current density can be inverted as compared to the mean one. This means that dissolution may happen locally much behind the growth front (or, in other words, relatively far from the outer envelope curve of the advancing deposit). This phenomenon was observed for Cu deposition in a thin-layer cell by using the analysis of consecutive images of the growing deposit [7]. This cannot be observed for thick deposits due to the lack of an appropriate *in situ* cross-sectional imaging technique for bulk materials. The observation of the local dissolution draws the attention to that the evolution of the entire deposit structure and local density at a particular depth can change also after the outer deposition front passes the depth of interest. If a disintegrated stand-alone nanowire (dendrite) is studied in the solution in which it was deposited previously, a parallel deposition and dissolution can be seen under appropriate conditions, due to which the wire appears to migrate into the field direction [8]. Such electric field-driven phenomena may occur in thin-layer cells because the potential difference across a nanoobject can be large due to the large resistance of the solution.

Finally, it has to be noted that the information available for thin-layer electrodeposition cells are far from being complete. Since many of the studies were performed with easy-to-handle and fairly reversible systems like the Cu^{2+}/Cu, less is known about irreversible deposition processes. Therefore, one has to be cautious by the generalization of the result for all possible systems.

12.1.2 Deposition of Metal-Based Structures Under Cryogenic Conditions

The big majority of the nanostructure preparation methods in thin-layer cells refer to the deposition of Cu/Cu_2O mixtures with varying oxide content [9–16]. In one kind of the experimental setup, a horizontal thin-layer cell is built with a solution layer thickness of about 40–500 μm. The schematic image of the rectangular cell can be seen in Fig. 12.1. The anode is a sacrificial Cu wire at one edge of the cell, which faces the cathode whose material is of less importance (e.g., carbon). The top cover layer of the cell is the *substrate.* This is very unusual since the cathode and the substrate are usually the same part of a cell, unlike here. The surprising experience is that the deposit penetrating through the cell from the cathode exhibits a strong enough adherence to the substrate, although it is not the electron-conducting part but a boundary surface only. The bottom part of the cell, a glass plate, is attached to a Peltier element that allows one to cool the system. The filling solution at the stage of the cell assembly is a relatively dilute copper sulphate solution. Upon cooling the cell, ice nucleation takes place at the chilled bottom face, which leads to the thinning of the liquid layer. In the ideal case, the residual liquid layer is only 200–300 nm thick as estimated from the thickness of the deposit that later formed during the

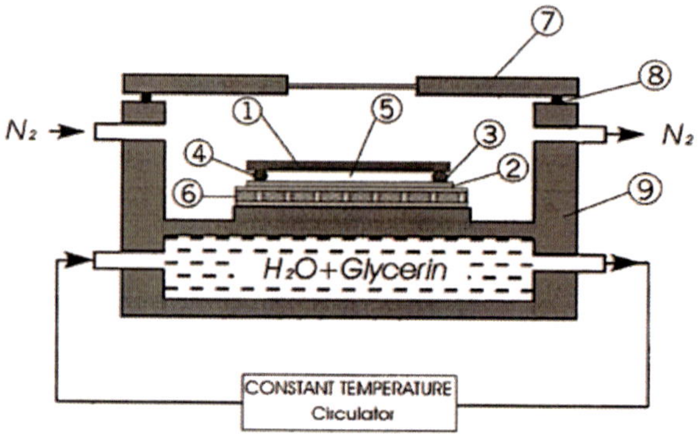

Fig. 12.1 Cross-sectional view of the structure of the cryostatic cell for deposition experiments with ultrathin solution layers. 1: silicon wafer, 2: bottom glass plate, 3: cathode, 4: anode 5: trapped solution between the parallel plates (before ice formation), 6: Peltier element, 7: top cover, 8: sealing, 9: thermostated chamber. Reproduced from [9]. Copyright (2001) The Physical Society of Japan

electrodeposition process. Although the *in situ* control of the cell with the thickness, concentration, and lateral homogeneity detection in the trapped liquid layer was not published, it was estimated that at the temperature applied (–2...–5 °C) the solution is just near to the saturation but the precipitation of the solute does not take place. It was stressed in the studies dealing with this technique that the ice layer has to be very smooth, which can be achieved with several preliminary temperature cycles of freezing and partial melting. In the ideal case, the frozen layer is like a planar ice single crystal. Formation of dendritic ice crystals is detrimental for the electrodeposition of any regular structure. The electrochemical process was driven typically by applying a constant voltage or constant current across the two-electrode cell without measuring the electrode potential of either the cathode or the anode.

In essentially all experiments, it was found that needle-like copper was grown with some Cu_2O content [9, 12, 15, 16]. The width of the filament varied between 300 nm and 2 μm, and bundles of the filaments also formed. The distribution of the cuprous oxide was random in some cases [12, 16], while other studies reported the occurrence of the Cu_2O crystallites at the surface of the growing Cu wire at the side contacting with the icy solution surface only [9, 15]. In some cases, the growing filaments were composed of alternating segments of Cu and Cu_2O [17]. The non-metallic impurity present in the wires was sensitively indicated by the specific resistivity of the deposits, and the Cu_2O phase was directly identified with diffraction techniques. The structural analysis also showed that the wires were assembled from randomly orientated crystallites.

The deposition process was reported to be accompanied with the oscillation of the non-regulated electrical parameter of the cell (potential oscillation at current control and vice versa) if the Cu_2O distribution was either random or highly periodical. In these cases, a pearl chain-like structure was observed. The oscillation frequency, the thickness of the copper filaments and the cuprous oxide content could be regulated

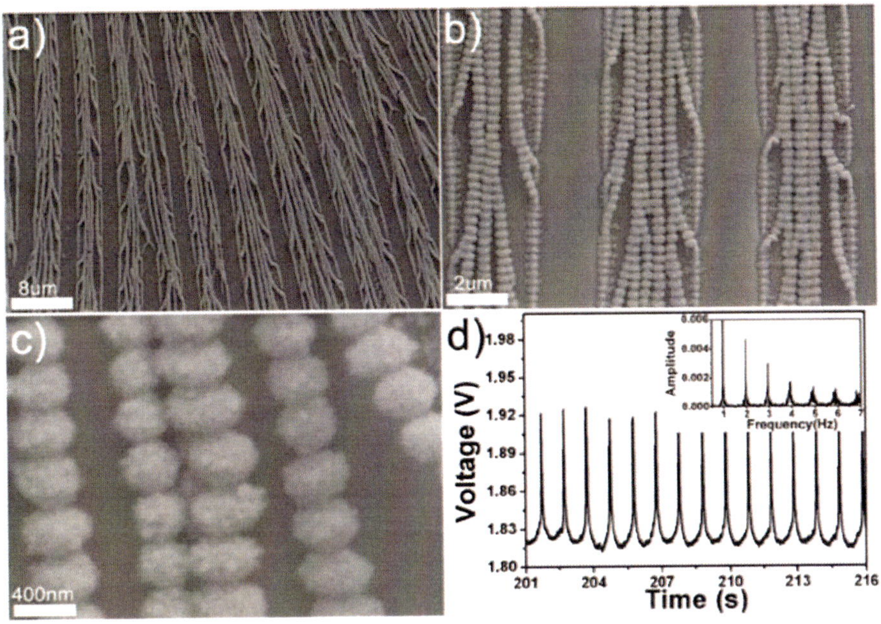

Fig. 12.2 Electrodeposited copper structure obtained from 0.05 M $CuSO_4$ solution iced at –4 °C with –30 μA in a cell with 1 cm anode–cathode separation. **a** Scanning electron micrograph of the bundles of copper filaments; **b** The bundles observed with higher magnification, showing the periodic corrugations along the filaments. **c** High magnification SEM images showing the beads of copper crystallites. **d** Oscillation pattern of the cell voltage, with the inset showing the Fourier transform of the signal measured. Reprinted with permission from [16]. Copyright (2010) by the American Physical Society

with the initial $CuSO_4$ concentration, the temperature and the current (or cell voltage) applied. An example for the structures formed and the oscillation characteristics can be seen in Fig. 12.2.

The thin-layer configuration offers a straightforward explanation why the filaments are deposited in a self-organized compact manner without a fractal formation. However, the explanation for the occurrence of the oscillation and the related Cu_2O formation seems yet incomplete. The attempts for the explanation deal with the cathode process only, attributing neither of the factors giving rise to the oscillation to the anode behaviour, even though the electrode potentials were not recorded. A major point of the explanation is that the Cu^{2+} concentration at the tip of the growing filament is very small, and hence, the Nernst potential for Cu deposition increases above the deposition potential of Cu_2O. As the oxide deposition process is slow, it allows the repeated Cu^{2+} accumulation at the growth front with a parallel OH^- ion concentration drop due to the Cu_2O formation, leading to the nucleation of metallic copper again. The explanation offered did not take into account the hydrolysis of the cupric ions that may lead to a pH change as copper is being deposited. As the cathode potential with respect to the hydrogen electrode is not known and the gas evolution rate could not be measured in the thin-layer cell, the pH variation related to a side process could not be evidenced.

Another cell configuration with a cylindrical arrangement and with a tip cathode connected to the centre of the upper closing plate and with an anode ring was also reported [10, 11]. In this setup, the cover plate was prepared from glass which made possible the visual observation of the processes, while other parts of the cell were essentially unchanged. By using a glass top cover, the adherence of the deposit to the solid surface was not of interest. A non-negligible difference as compared to the rectangular version is that the envelope around the end of the growing spikes is expanding during the deposit formation, which naturally leads to the branching of the filaments and a fractal-like overall appearance. It is the branching that provides the nearly equal lateral density of the filaments along the radius. In many other aspects of the growth, however, the observation was similar to the rectangular cells at the microscopic scale (oscillation during the growth, formation of periodic corrugated structures, Cu_2O formation during the copper deposition).

After the early works describing the basic characteristics of the chilled thin-layer cells, the technique was significantly improved, allowing a thoughtful design of nanostructures. All further works discussed were prepared with rectangular cells. Parallel nanowire arrays with homogeneous component distribution and adjustable areal density but without either ramification or corrugation were reported [13]. The technical background of the latter work is highly common with that of the studies mentioned above. The only apparent difference is that the cathode surface was a silicon wafer with a native oxide at the surface covered with a hydrophobic octadecyltrichlorosilane monolayer. Although a surprisingly good uniformity ranging to millimetre length scale was reported, the reason leading to this improvement was not clearly revealed. Another interesting improvement of the thin-layer technique came from the discovery that the metal deposition is preferred at an artificial three-phase boundary line [14]. The silicon substrate commonly used in the thin-layer experiments was covered with poly(methyl-metacrylate) (PMMA) strips that were perpendicular to the electrodes. The deposition in this case led to a copper wire formation at the concave corner of the solution–silicon–PMMA boundary line. The growth preference at the three-phase boundary (triple junction) was so strong that it could easily overwrite the impact of the electric field also in the cases when the PMMA pattern was not perpendicular to the electrodes due the occurrence of micrometre-scale hemispherical defects. The width of the longitudinally growing Cu wire was as small as 50 nm.

The freezing technique used for the deposition of various Cu structures was adapted for the deposition of zinc and zinc oxide [18]. The solution used contained solely $ZnSO_4$ without any supporting electrolyte. Galvanostatic deposition led to potential oscillation whose frequency increased with both the current density and the pH. The deposit consisted of well-adherent needle structures on the cover glass plate of the cell. The diameter of the needles was less than 2 μm. The growing Zn tips exhibited a ramification-rich tree-like structure with regular 60° corners between the branches; therefore, the deposits were much less regular than that of the Cu-containing nanowires. The structural study indicated the presence of Zn and ZnO.

It was speculated that the formation of ZnO is related to the oxygen content of the solution. However, the commonality of the oscillation pattern with the Cu deposition system indicates that a common reason may stand behind the oscillatory behaviour that is related to the lack of the supporting electrolyte, to the hydrolysis of the cations, to the deposition rate beyond the diffusion-limited regime and to the high negative potential of the cathode.

Finally, it is to be mentioned that the chilled thin-layer cell technique with the combination of a post-deposition etching step was adapted for copper nanowire deposition on non-patterned silicon surface [19]. Here, the cell voltage was modulated during the deposition. The authors chose a semi-sine waveform for the cell voltage. The pronounced opinion of the author of this book is, however, that the waveform is of little importance if the extremes of the cell voltage comprise the desired deposition regimes at the cathode. Namely, at small cell voltage copper deposition takes place with the formation of dense copper, and it is believed that the copper deposition rate is below the diffusion-limited one. The visual observation during the growth process confirmed that the low cell voltage is what results in the deposition of dense but relatively thick copper. When the cell voltage is high, the advancement rate of the growth front increases, which corresponds to the formation of a network of tiny copper filaments (called membrane in the original work [19]). In this growth regime, the impact of the electric field becomes important, and the resulting spongy layer is very thin as compared to the slowly grown dense copper. The scheme of this deposition technique is shown in Fig. 12.3a, and the consistency of the deposit obtained during each section of the deposition cycle is visualized with a microscopic observation of the as-received deposit.

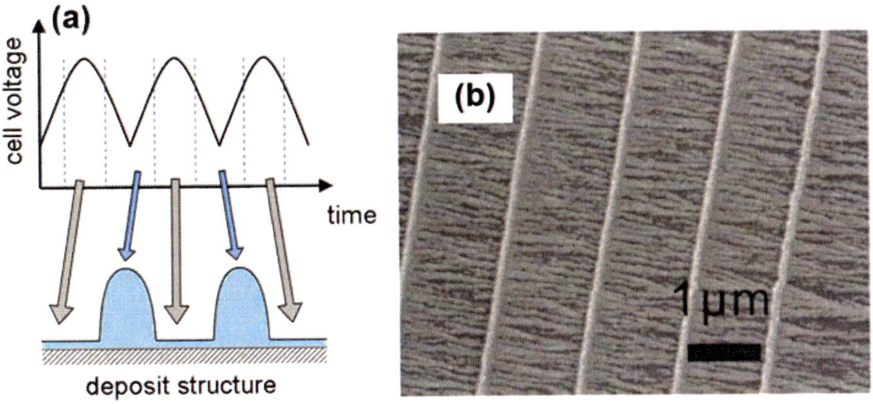

Fig. 12.3 **a** Scheme of the deposition procedure by using a modulated potential waveform for obtaining modulated deposit structure along the growth direction. **b**, A microscopic image of the thus obtained deposit. It is to be observed that the current is not linearly proportional to the cell voltage, and hence, the lateral growth rate of the alternating structure element can be very different. Growth direction: from left to right in both images. Adapted **(a)** and reprinted **(b)** with permission from [19]. Copyright (2009) American Chemical Society

The final step of the nanowire preparation is the removal of the thin Cu deposit of high specific surface area grown during the high-voltage intervals. The etching process has to be mild enough and can last for a short time only to leave the wires intact while removing the thin layers in between.

The combined technique of alternating cell voltage with post-deposition etching is tuneable is various ways. The thickness of the copper wires can be adjusted with the charge passed while the deposition process in within the diffusion-limited low-voltage regime, while the interwire distance is determined by the charge passed during the rest of the deposition cycle. The potential limits are obviously of key importance for setting the appropriate deposition regimes, and the surface density of the wires finally obtained is roughly proportional with the frequency of the cell voltage variation.

The modulated-voltage deposition technique was adapted to the deposition of parallel cobalt nanowires, too [20]. It was shown that the set of parallel nanowires obtained after the deposition and etching process covered the substrate uniformly at large surface areas even in the case when the substrate was intentionally corrugated in various ways, either with randomly aligned flat areas of with highly periodic undulations. An excellent agreement was found between SEM observations and magnetic force microscopic images.

12.1.3 Miscellaneous Deposits Obtained in Cryostatic Thin-Layer Cells

The precursor deposit of a micrometre-periodicity CuO/In_2O_3 semiconductor heterostructure array was prepared with the freezing-based thin-layer technique [21]. The starting solution contained $Cu(NO_3)_2$ and $In(NO_3)_3$ in a concentration of a few tens of mM, and HNO_3 was added to the solution for adjusting the pH to 3. The cell voltage was modulated between two values, the lower and the higher one leading to a copper-containing and an indium-containing deposit, respectively. In accord with the authors, the key cathode reaction was the reduction of the nitrate ions, hence leading to the alkalization of the vicinity of the cathode (similar to the processes already discussed in Chap. 8.2.3 for ZnO formation):

$$NO_3^- + 2e + H_2O \rightleftarrows NO_2^- + 2OH^- \tag{12.1}$$

The lower value of the cell voltage was 0.6 V, which was claimed to lead to Cu_2O formation. This is strange in the light of that deposition of metallic Cu from solutions of higher pH was possible at cell voltages in the range of 3 to 10 V [9–11, 13], and the deposition of Cu should precede the nitrate reduction on the basis of the standard potentials. The higher cell voltage (1.2–1.5 V) resulted in the formation of $In(OH)_3$ due to the nitrate reduction and the forthcoming precipitation. The penetration of the deposition front was higher during the indium(III) hydroxide formation, hence

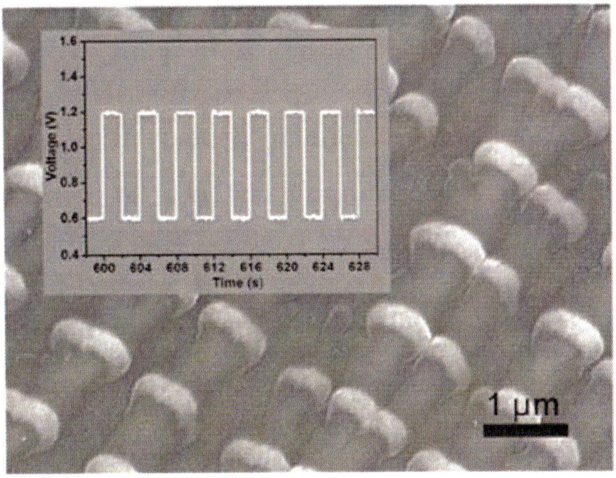

Fig. 12.4 Scanning electron micrograph of a CuO/In_2O_3 semiconductor heterostructure array (after the annealing of the primary electrodeposited precursor exhibiting the desired component modulation). The inset shows the cell voltage modulation program. Reprinted from [21]. Copyright (2013), with permission from Elsevier

leading to thinner deposit. The final composition was achieved by a gentle annealing in air, which caused the transformation of both parts of the deposit (Cu_2O oxidation to CuO and $In(OH)_3$ dehydration to In_2O_3). The final structure was an assembly of occasionally interconnected wires with well-separated segment, CuO looking like a bump at the end of the elongated In_2O_3 segments, as shown in Fig. 12.4. The resulting structure was shown to exhibit a reversible resistivity change up to 25% for gas exposure, especially for hydrogen, although the selectivity was not high.

Later, Cu_2O/SnO_2 heterostructure was electrosythesized on the basis of the same principle [22]. The hydroxide ions were produced by the electrochemical reduction of the nitrate ions. Cuprous oxide was produced as wires perpendicular to the growth direction at low cell voltages, whereas a thin layer of stannic oxide was deposited at higher cell voltage. The production of this heterostructure did not require any heat treatment. The heterostructure thus produced exhibited a significant photoresponsivity.

Another example for the electrolytic deposition on quasi-one-dimensional heterostructure is the formation of Pb/PbTe [23]. Again, the ultrathin solution layer was provided with the above-discussed partial solution layer freezing method. The concentration ratio of $Pb(NO_3)_2$ and TeO_2 was 10. The heterostucture formation was achieved with a cell voltage modulation. The resulting deposit was a strip with periodic bumps, exhibiting some branching. Although the presence of both desired components were detected by both XRD and local area diffraction in TEM, the morphological regions of the strips were not assigned to the components.

12.2 Two-Plate Thin-Layer Cells for Electron Microscopic Studies of Electrodeposition Processes

12.2.1 Electrochemical Cells in High-Vacuum Environment

Although the TEM observations in thin-layer cells do not make a part of the preparative works related to nanostructure formation, it was found that this emerging field cannot be omitted when thin-layer techniques are discussed. Many details of this technique are already available in various reviews and textbooks [24–27], and the electrical engineering aspects of accurate electrochemical measurement under TEM conditions have also been addressed [28]. Therefore, a few important aspects related to electrodeposition and electrochemical nanofabrication will only be summarized below.

One phase in electrochemical experiments is usually a liquid, which is not compatible to the conventional TEM environment where the high-vacuum condition is a prerequisite for imaging with the electron beam. *In situ* electrochemical TEM studies can be, however, performed with droplets of ionic liquids whose vapour pressure is small enough not to disturb the imaging with the electron beam. The application of ionic liquids is the only case when electrochemical TEM studies can be performed without a well-sealed cell. Even with the application of ionic liquids, the sufficient transparency for the electron beam has to be provided. This means from the practical point of view that the total thickness of the layer that the electron beam has to pass through should not exceed 2 micrometres, while a few hundred nanometre thickness is rather ideal.

If an electrochemical cell is necessary, the key part of the cell is the window that is thin enough for being transparent for the electron beam as well as strong enough to endure the pressure difference between the internal and external parts of the cell. This pressure difference can be the atmospheric pressure or higher when the cell is not completely sealed, and this is the case when flow-cell conditions are applied (the pumping system is always outside the TEM instrument). The window is made exclusively from silicon nitride (Si_3N_4). The lateral dimension of the Si_3N_4 window is below the millimetre scale. The *in situ* electrochemical cell is usually made from a bottom and a top part, each having a Si_3N_4 window facing with each other. The nitride layer is manufactured on a silicon wafer with usual microelectronic techniques, and then the rest of the silicon is etched away from the window area. The two parts of the cell are pinned at the desired distance with the help of strips also prepared directly at the silicon surface. Since the thickness of the cell is very small, manual adjustment of a thin sealing layer is highly undesired.

From the electrochemical viewpoint, a problem of high importance is the arrangement of the electrodes. Due to the miniaturized nature of the cell, electrodes (metals, mostly Au, Pt and Ti; or carbon) are pre-deposited onto the desired surface of the cell wall (mostly bottom plate) during the manufacturing process. For creating the external contacts, the metal deposition can reach to the outside of the solution-filled volume of the cell. An elegant contact mode is to apply a polycrystalline silicon layer with very high doping level sandwiched between the electrode metal and a SiO_2 insulating layer [29]. This electrical contact is driven outside the cell for creating an external electrical contact point. This connection made has a great advantage for a multiple electrode arrangement.

The arrangement of the electrodes may vary depending on the goal of the experiment. Figure 12.5 offers a basic overview of the systems commonly found in the relevant literature. It is noteworthy that *in situ* synchrotron-based soft X-ray microscopy

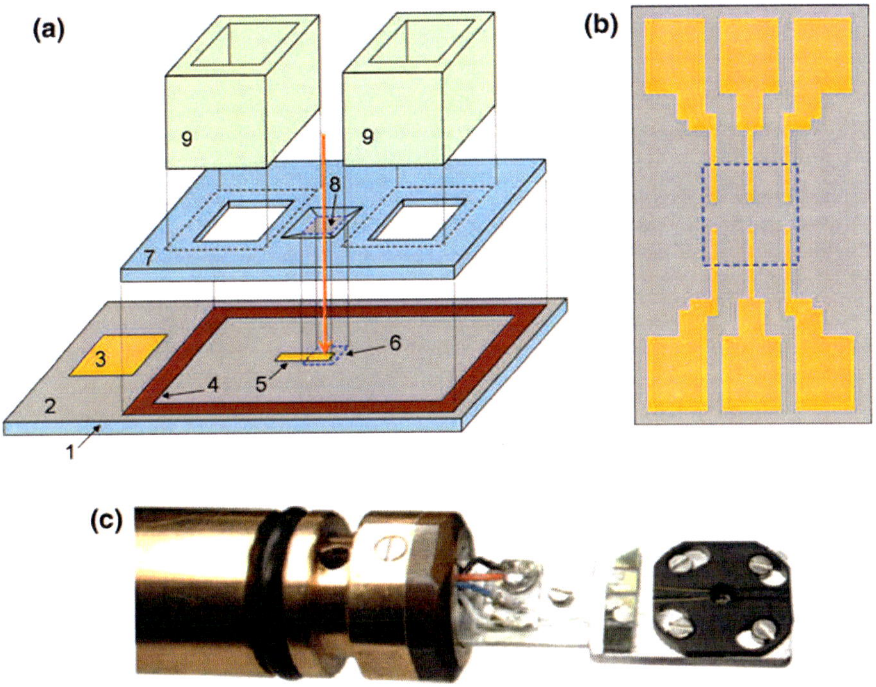

Fig. 12.5 **a** Thin-layer TEM cell structure for nucleation studies where the electron beam is passed through the surface of the working electrode. 1: bottom Si wafer (with high doping level for good conductivity), 2: Si_3N_4 layer on the bottom wafer (electrically insulating), 3: external contact for the working electrode, 4: spacer layer with the thickness equivalent to the solution layer thickness above the working electrode, 5: working electrode, Au, with ~20 nm thickness (there is no Si_3N_4 layer between the working electrode/external contact point and the Si wafer for providing the electrical contact), 6: observation window with 30–80 nm thick Si_3N_4 layer only at the bottom wafer, 7: top Si wafer, 8: Si_3N_4 observation window on the top part, 9: top chambers for counter/reference electrode and/or for connectors to the solution flow system. Figure was sketched after [31–34]. **b** Example for the electrode arrangement when the current direction is in the plane of the thin solution layer (top view). Here, the electrodes can be a few tens of micrometre wide, several hundred nanometre thick and are not transparent for the electron beam. The dashed line shows the Si_3N_4 window. Figure was sketched after [35]. **c** Photograph of a fully assembled *in situ* electrochemical TEM cell. The width of the black end part can be between 6 and 15 mm for various devices. Figure 12.5c is reprinted from Ref. [36]. Copyright (2012), with permission from Elsevier

studies of electrodeposition apply very similar setups [30], though X-ray microscopy has a lower resolution.

If the phenomenon of interest is the nucleation, a lateral view of the working electrode is to be observed. In this case, one of the Si_3N_4 windows (usually the bottom one) is covered with a continuous but thin and electron-transparent conducting layer. The counter and the reference electrodes are outside the observation windows. This makes the cell asymmetrical, which is generally unfavourable for the current uniformity; nevertheless, this is a necessary compromise for the undisturbed observation by using an electron beam. Although the direction of the electric current corresponding to the electrochemical process is determined by the relative position of the anode and the cathode, it can be regarded as if the current were perpendicular to the substrate, at least at the length scale of the nuclei being deposited. It is to be noted that this geometry is not used in thin-layer cells of macroscopic dimensions. The transmission electron images thus obtained make it possible to describe the population of nucleating grains during electrodeposition processes. Studies utilizing this arrangement will be dealt with in Sect. 12.2.2.

Another typical goal during *in situ* TEM studies is the observations of either the phase boundary or the propagation of the front during layer/grain growth. For such purposes, well-separated electrode strips are pre-manufactured on the bottom plate of the electron-transparent cell. These electrodes do not have to be electron transparent since the newly emerging phase at their edge will be the target of the observation, and the transparency change as compared to the neighbouring solution will provide the detection signal. Although the pre-manufactured electrodes are typically pure gold strips, their modification (e.g., by electroplating) can make them suitable for reference electrode. Plating Cu in a Cu^{2+}-containing solution and Ag/AgCl in chloride-containing solution provides a reference electrode of the first and second kind, respectively. For short-term observation in the absence of any potential-determining reactions, the bare Au strips themselves can serve as a quasi-reference electrode. The direction of the current related to the electrochemical process is in the plane of the thin liquid layer entrapped between the TEM observation windows. Sect. 12.2.3 will deal with studies where this geometry was used.

Regardless of the electrode arrangement, an unbeatable advantage of the *in situ* TEM observation is the high imaging rate combined with a large lateral resolution. As compared to optical imaging, an ultrafast camera is capable of an image recording rate several orders of magnitude higher than the normal video rate (which is 30 frame/second). However, the detection limit of the optical devices is about half a micrometre, and there is no opportunity to image nanoscale objects even as a whole, not to mention their subtle details. Scanning probe techniques are capable of imaging surfaces down to the atomic resolution, but this is accompanied with a slow recording rate like 2–3 image/minute. *In situ* TEM observation covers an intermediate regime, making it possible to obtain a few tens or hundreds of images per second with a resolution at the nanometre scale. The lateral resolution is still less than that for stand-alone solid crystalline samples, which is due to the irregular scattering background of the surrounding liquid and the cell windows. At the same time, construction of reconstructed diffraction images of nanoscale segments of the electrode surface is possible, hence opening the way for the *in situ* identification of the phase and its orientation.

A factor that is very unusual in macroscopic electrochemical experiments may arise in all *in situ* TEM studies. Namely, the electron beam itself is capable of generating or modifying chemical reactions. Here, we encounter something similar to quantum mechanical observations where the probe method itself modifies the system to be measured. Therefore, whenever it is possible, this has to be taken into account and results separated from beam-induced effects should be given, unless the chemical changes originating from the electron beam are exploited purposefully. The latter method as a special field of electrochemistry will be dealt with in Sect. 12.2.4.

12.2.2 In Situ *Electrochemical Nucleation Studies with TEM*

The first fundamental study with *in situ* electrochemical nucleation used Cu deposition as a model system [31]. The solution concentration was 0.3 M, and the deposition experiments were carried out with constant current density of either 5 or 50 mA cm^{-2}. The voltammogram in the thin-layer cell was equivalent to that obtained with a conventional cell arrangement, the only difference being the impact of the uncompensated ohmic drop. It was carefully controlled that the surface inhomogeneity of the polycrystalline Au substrate does not influence the nucleation in subsequent experiments; i.e., nucleation sites in repetitive experiments with the same substrate were uncorrelated. The result of the experiments was that the areal density of the Cu grain grew steadily in the first period of about 2.5 s, but later practically no nucleation event was observed. This is good agreement with what is generally thought about nucleation. Namely, even if nucleation is progressive, the depletion of the media for the reactant will increase the nucleation barrier so much that after a certain period only grain growth can be observed.

After the proof-of-concept trial, the electrochemical manipulation was improved, and *in situ* TEM observations were accompanied also by detailed electrochemical data [32, 37, 38]. The quantitative analysis of the cluster radius r as a function of time yielded the result that the nucleation and growth process can be divided into periods, as indicated by the two subsequent linear sections in the log r vs log t function. The exponents found experimentally for the start of the nucleus formation could be correlated to the $r \propto t^{1/2}$ law originating from the diffusion-limited growth of the individual nuclei without the overlap of the hemispherical diffusion field around each particle. The complete expression for the radius is [37, 39, 40]:

$$r = \left(1 - \frac{c_S}{c_B}\right)^{1/2} \left(\frac{2D c_B t}{V_M}\right)^{1/2} \qquad (12.2)$$

(where V_M is the molar volume of the deposit, D is the diffusion coefficient, and indices S and B refer to the surface and the bulk, respectively). If potentiostatic control is applied and hence c_S can be taken independent of time, the time dependence of the radius directly follows. At higher times, the following considerations are valid. From the volume of the hemispherical nucleus, one can write the $r(Q)$ function by using Faraday's law:

$$r = \left(\frac{3V_M}{2\pi zF}Q\right)^{1/3} \tag{12.3}$$

The above relationship can be combined with the integrated Cottrell equation, expressing the charge passed as a function of time:

$$Q = 2zFAc\left(\frac{Dt}{\pi}\right)^{1/2} \tag{12.4}$$

then one can get that

$$r = \left(\frac{3V_M Ac}{\pi^{3/2}}\right)^{1/3}(Dt)^{1/6}. \tag{12.5}$$

This was in good agreement with the direct TEM observation for long times [37, 38]. The importance of the observation originates from the fact that it was not possible to detect these data directly, only the agreement of the shape of the current transient with the theoretical approach yielded a rather indirect confirmation.

Although the $r(t)$ function was qualitatively fully in line with the theories, it also turned out from the *in situ* TEM study that the experimental areal density of nuclei and the particle radius itself showed a correlated deviation from the result obtained from the fit of the current transients observed in parallel to the TEM observation. Namely, the nucleus density was about 3 orders of magnitude larger than the value obtained from the current transient. Due to the mass conservation, the radius ratio had to be one order of magnitude smaller than expected, which was indeed found in the analysis of the TEM images. The dependence of the nucleus density on the electrode potential showed the same trend, should it have been obtained from either current transients or TEM images. These observations showed that the nucleation theories have yet to be improved.

An *in situ* TEM nucleation study of Pt on glassy carbon [41] indicated that the growth is fractal-like in the diffusion-limited potential regime. The fractal dimension of the deposit fell between 1.6 and 1.8, increasing slightly with the solution concentration of the reactant. The fractal dimensions are in good agreement with the theories referring to diffusion-limited aggregation. Saturation cluster density was found to increase with solution concentration, while the nucleation rate constant did not show concentration dependence. Concerning the preferred site for nucleation, thicker glassy carbon areas were more prone to serve as nucleation sites than thinner ones. The saturation island density decreased with substrate layer thickness, while the trend concerning the nucleation rate constants was the opposite.

12.2.3 In Situ *Electroplating Studies with TEM*

The first study dealing with the *in situ* TEM observation of the continuous deposit formation with the "current in solution layer plane" configuration referred to Ni deposition from $NiCl_2$ solution [36]. The voltage was controlled in a two-electrode system during deposition and subsequent dissolution of Ni. The experimental finding was difficult to elucidate because of the artefact that the silicon nitride window did not prove to be a good enough insulator. This led to the unexpected finding that Ni nucleation could take place in advance of the growth front of the continuous deposit being present on the cathode itself. Hence, the "electrode" and "substrate" was not essentially the same. The propagation of the growth front during the non-inhibited growth of the deposit showed a large lateral inhomogeneity, consistently with the dendritic morphology obtained. Upon dissolution, the elctropolishing conditions could be observed in the sense that the dissolution front propagation was much more uniform than the growth front.

The *in situ* TEM observation of Pb growth and dissolution was partly contradictory to the Ni deposition/dissolution study [42]. Here, the goal was the study of dendrite formation. What was found is that the dendrite growth and dissolution were asymmetrical in the sense of the thickness and length changes. Upon growth, the dendrites grew preferentially in the current direction with nearly proportional thickening. However, the dissolution took place through a remarkable thinning before being shrunk along the perpendicular-to-substrate direction. This observation is highly in contrast to electropolishing theories and can probably be explained with the solution concentration changes during the process. Since the solution used was saturated with respect to lead nitrate, the partial depletion between the dendrites might have been provided more favourable conditions for dissolution than the end of the dendrite tip where the saturation concentration could be spontaneously restored much more easily. Similar results were obtained in another study [33] where the reactant concentration was far from the saturation level and surfactant was also applied to regulate the deposition. The detailed analysis of the branching dendrite showed that the deposition and dissolution process of the side branches are nearly symmetrical concerning the length/width ratio, but the main dendrite anchoring directly on the substrate tends to be thinned before being shortened. It was also shown that the decrease in the reactant concentration promotes both the dendrite formation and the branching.

In studies dealing with Cu deposition, the general trends established for surface roughness evolution were confirmed also at the nanoscale by imaging about 2-μm-wide sections of the growing deposits [43, 44]. Deposits obtained with a relatively small d.c. current were less rough that obtained with a rate close to the diffusion-limited growth. Pulse plating was shown to suppress the development of the roughness even at a high rate, and $PbSO_4$ applied as an inhibitor was shown to almost completely eliminate the increase of surface roughness.

12.2.4 In Situ *TEM Studies of Nanostructures Related to Batteries*

In situ TEM studies gained recently an important role in the investigation of various nanostructures that are produced electrochemically in various battery constituents, especially in lithium ion batteries [45]. It has long been known that Li is prone to develop dendrites upon battery charging that are not consumed upon the battery discharge. This leads to the formation of a direct electrical contact between the electrodes upon battery cycling, causing an obvious battery failure. The details of the lithium growth and dissolution process were revealed with *in situ* TEM studies [46, 47]. In accord with general nucleation and growth theories, it was directly imaged that higher Li deposition rate tends to lead to a dendrite-rich growth. The spatially selective dissolution of the Li dendrites near their root results in separated Li needles that can contribute to the battery failure.

Lithium growth and dissolution is always coupled with the formation of a layer around the cathode called the solid electrolyte interface (SEI). SEI is produced during the partial solvent reduction and has a protective nature against the spontaneous reaction between Li and the solution components, making possible Li^{+} ion transfer at the same time. The competition between Li growth/dissolution and SEI-related electrochemical processed has been shown to regulate the overall electrochemical behaviour of Li electrodes [48, 49]. The direct *in situ* imaging of SEI in sealed thin-layer cells revealed the thickness distribution of SEI, the nature of the processes due to the breaks in a fragile SEI layer and the impact of bubble formation under the SEI. It is practically impossible to study such processes *ex situ* due to the very high reactivity of lithium. The study of intercalation-type electrodes in lithium ion batteries can also be performed with *in situ* TEM. Here, the volume and structural change of the lithiation/de-lithiation gained the major interest [50, 51].

Unlike in Li batteries, in Mg batteries the dissolution of the metal is the problematic part of the battery operation. Magnesium forms a smooth deposit with no dendrite formation, but the dominant anodic process is the formation of an organometallic compound that passivates the remaining metallic magnesium underneath. A combination of direct *in situ* TEM and scanning transmission X-ray microscopy (STXM) was successfully applied for revealing the details of the electrochemical behaviour of magnesium under typical battery conditions [52].

12.2.5 *Chemical Changes in Thin-Layer Cells Induced by the Electron Beam*

Electron beam in an *in situ* TEM cell can be used, besides imaging, for the modification of electrochemical processes or for the initiation of electron-transfer reactions (i.e., reduction of ions). The term used for such processes is focused electron-beam-induced deposition using bulk liquid precursors (LP-EBID). Although such processes are seldom considered to be the core part of electrochemistry but are rather classified as radiolysis, it seems to be important to deal also with such an emerging field among other thin-layer cell techniques.

Concerning the chemical impact of the electron beam, one has to consider the absorption characteristics of the high-energy electrons. As it is well known from the EDS measurements, absorption characteristics of high-energy particles cannot be simply described with formulas similar to the Lambert–Beer law; instead, characteristic collision cascades have to be considered and the mean depth of energy absorbance has to be taken into account. As it has been recently shown [53], for a liquid cell filled with water, the presence of interfaces between water and the surrounding surfaces causes a spatial variation in the energy absorbed by the water near the walls. The mechanism for this effect is that the walls act as a source of secondary and backscattered electrons which diffuse and deposit energy in the water nearby. This increased dose rate then changes the local concentrations of reactive species. The dose rate can increase by several times within tens of nanometres of a water/metal interface, locally increasing the concentrations of species such as the hydrated electron. In accord with the electrode arrangement, such interface-influenced species density can cause lateral and in-depth variations in the composition of the liquid medium, which may modify the conditions of the reactions studied.

The discrimination of irradiation-induced deposit from electrodeposited crystals proved to be rather easy for gold [54]. A clear bimodality was observed for the particle size distribution. Larger particles were obtained on the carbon electrode used for electrodeposition, while smaller particles appeared on all illuminated surfaces. Not surprisingly, the morphology of the particles also differed. While electrodeposited particles were nearly hemispherical (indicating a mixed diffusion-activation control during the agglomeration process), the illumination-induced deposit showed an acicular morphology, which is characteristic of diffusion-limited aggregation. Since the growth rate of the illumination-induced particles was smaller than that of the electrodeposit, it is clear that the diffusion limitation must come from the transport of a dilute reactant that is produced by the electron beam.

Electron beams can also be used to assist the deposition which is purely electrochemical in nature. Due to the focusing of the electron beam, this offers a tool of performing laterally selective electrodeposition; in other words, writing with the electron beam. This was shown for Cu deposition on Au [55]. When the electrochemical behaviour of the nucleation zone was studied, it turned out that there is a pronounced difference in the electrochemical nucleation rate in the activation-controlled potential regime in the absence and presence of chloride ions. In the potential interval where chloride ions present in a few tens of ppm concentration, nucleation of Cu on Au was hindered, which could be overcome by applying the illumination with an electron beam. Hence, the electron illumination area could be covered selectively without patterning the substrate.

When the only source of the reducing agent is the electron beam, the preferential deposition site is again the wall of the cell, whose quality has an impact on the deposition process. This was exemplified for Pt deposition in an *in situ* liquid SEM system [56]. Here, the focus of the electron beam is adjusted to the solution side of the observation window. Therefore, the area to which the electron beam is focused determines the reaction zone, allowing a locally selective deposition without patterning the substrate itself. The *in situ* obtained image refers to the wall side of the species forming, while the solution side can be imaged *ex situ* after opening the liquid microcell. The major conclusion of the above-mentioned studies was that a threshold charge is necessary to initiate the growth of Pt by using the electron beam for the reduction of the Pt(IV) precursor. The preliminary step before the Pt deposition itself was a beam-induced modification of the substrate surface. Since the reaction zone was confined to the irradiated area, the feature size of the nanostructures produced depended only on the focusing condition of the electron beam, which could lead to 25 nm dot sizes in the best case [57, 58].

The electron beam can induce the reduction of metal ions where nanoparticles are present in the solution. In this case, the typical growth mode is the formation of a shell on the already existing particles, as it was shown for the case of Pd deposition in the presence of Au nanoparticles [59]. The great advantage of the *in situ* TEM observation of nanoparticle-seeded growth is that the impact of the electron beam can be well separated from the impact of the particles itself. It was found that the growth rate of particles decreases with their distance from the illuminated area, indicating that the transport of the beam-induced reactive particles (most likely the hydrated electrons) is a rate-determining step in the shell formation around the nanoparticles. Depending on the nanoparticle size, the morphology of the external deposit changes from a compact shell to dendrite-decorated external layers as the size of the Au nanoparticle increased.

12.3 Deposition at the Liquid–Liquid Interface

12.3.1 Basic Overview of the Appropriate Systems

This book follows the classification of electrochemical nanostructure formation processes on the basis of the configuration of the phases involved and the structure of various elements of the electrochemical cell, in particular the substrate, and secondarily on the energetics of the deposition process. Another means of the classification of electrodeposition processes is to take into account the consistency of the phases involved in the process. From this point of view, the possible pairs are solid–solid, solid–liquid, liquid–liquid, solid–gas and liquid–gas. The treatment of the deposition systems in accord with the phases involved can be read in another review [60]. The list above enumerates the phases involved on the basis of their density, mentioning the phase with higher density first; however, this is different from the classification where the emphasis is on the charge carrier. For instance, in plasma processes (that will be discussed later in Chap. 14), the gaseous phase is a mixed conductor with a high electron conductivity. It is just shortly mentioned

Table 12.1 Properties of the solvents used in electrodeposition studies at the liquid–liquid interface listed in the order of the density ρ. ε_r: relative dielectric constant

Solvent	ρ/g cm^{-3}	ε_r	Reference(s)
Chloroform	1.49	4.81	[63]
Dichloromethane	1.33	9.1	[64, 65]
1,2-dichlorobenzene	1.30	9.9	[66–68]
1,2-dichloroethane (DCE)	1.25	10.6	[66, 69–91]
Nitrobenzene	1.20	34.8	[92–95]
n-buthylacetate	0.88	5.0	[96, 97]
Toluene	0.87	2.4	[98]
4-methyl-2-pentanon	0.80	13.1	[99]
n-heptane	0.68	1.9	[100]

that nanostructure deposition is possible at a solid–solid interface, too, a prominent example being the silver whisker growth at silver/silver halogenide interface. Therefore, deposition at the liquid–liquid interface will be discussed below only. Electrodeposited nanomaterials produced at the liquid–liquid interface just make a minor part of the larger material family, whose properties and synthesis opportunities in a wider scope have been summarized elsewhere [61].

The principle of the formation of a new phase at the boundary of two liquids with the assistance of the electric current is as follows. The two-phase system has to be created by using two liquids with negligible miscibility. For the reaction site in such systems, the common terminology that can be found in the literature is the "Interface between two immiscible electrolyte solutions" (ITIES). One of the phases is aqueous in all studies published so far, and the other solvent can be either a polar or apolar organic molecular liquid. A summary of the key physical properties of the molecular liquids applied as the less polar phase is given in Table 12.1 with the list of the works in which they were used. Water-insoluble ionic liquids also appeared recently as a constituent of such systems [62], having also the advantage that the required miscibility gap can be coupled with a large ionic strength.

Depending on the density ratio of the two contacting phases, the aqueous solution can be either the bottom or the top phase (the former being typical for hydrocarbons as organic solvent and the latter holds for nitrobenzene and dichloroethane). Due to the difference in density, the interface between the two liquids is horizontal and, very importantly, planar. In exceptional cases, the organic phase is applied as a thin wetting layer on a substrate [92] when the interfacial tension as interaction overwrites the impact of the gravitation.

In order to provide a large partition coefficient of the components in the two-phase system, and hence, to keep the reactant in the desired phase, there must be a large difference in the properties of the ionic solutes. While the typical supporting electrolyte in the aqueous phase is a usual ionic compound with compact ions (such as alkaline sulphates or halogenides), the organic phase contains solutes with bulky ions having larger affinity to the apolar phase than to the aqueous one. For the latter, salts containing tertaalkylammonium cations with relatively long carbon chains are typical, and the anion is usually also large such as tetrakis(pentafluorophenyl)borate.

It has to be kept in mind that not only is the liquid–liquid interface polarized but charge transfer has to take place therein at the same time. This means that one of the contacting phases contains the reagent to be deposited and the other phase also must contain an electron donor/scavenger compound. The variety of the redox couples largely depends on which of the contacting phases contains the reactant. If the reactant is present in the aqueous phase (e.g., $PdCl_2$ when Pd nanoparticles are to be deposited at the interface), the reduced form of a redox couple has to be dissolved in the organic phase. Butylferrocene and decamethylferrocene are common that can be oxidized easily, and they can provide the electron for the metal ion reduction. In the opposite case when the reactant is in the low-polarity phase, another compound of the same role is to be dissolved in the aqueous phase. For the reduction of $[AuCl_4]^-$ that is initially present in the organic phase, $K_4[Fe(CN)_6]$ is a suitable electron exchange partner that is present in the aqueous phase. The electron transfer partner compound in the phase containing no precursor compound for the deposition may vary as a function of the nature of the entire system. However, similarly to the supporting electrolytes, they also have to exhibit a large partition coefficient between the contacting phases.

The features of the deposition processes taking place at the *bare* liquid—liquid interface without the assistance of any electrode (electron conductor) will be discussed in Sects. 12.3.2–12.3.3. Beside these conventional deposition systems, there are additional methods making use of the liquid–liquid interface in some other means, also with a special electrode configuration. Sect. 12.3.4 will enumerate the studies where a macroscopic planar electron conductor crosses the liquid–liquid interface vertically, but the reaction zone is restricted to the triple-phase boundary line. Finally, in Sect. 12.3.5, the cases will be discussed when either a tip or a filament is applied that touches or crosses the liquid–liquid interface, respectively, but the apolar phase has no role in the electrochemical reactions apart from forming the growth-regulating interface with the aqueous electrolyte solution.

12.3.2 Particle Deposition at the Liquid–Liquid Interface

The main reason of the particle deposition studies at ITIES is the opportunity of the controlled size regulation and the availability of small particles without the necessity of their release from a surface. The typical cell arrangement can be seen in Fig. 12.6. The cell contains two electrodes for passing the current which are the counter electrodes in the two contacting liquid phases. These are usually made of platinum gauze. The exact position of the counter electrodes is not crucial; nevertheless, the symmetry of the cell and the even current distribution has to be taken into account. In order to measure the potential difference across the interface, one needs two reference electrodes which remain unpolarized during the entire experiment. For reducing the ohmic drop contribution that arises between the junction point of the end of the reference capillary and the interface, the endings of the capillary of the

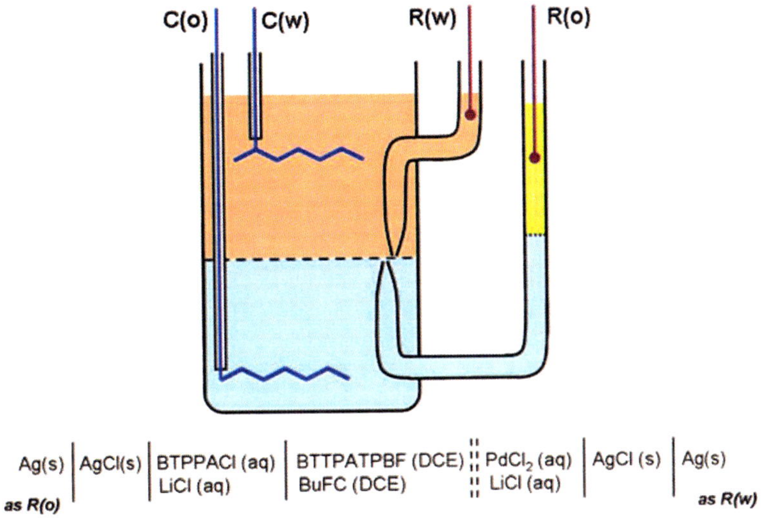

Fig. 12.6 **Top**: Schematic picture of a cell used for electrochemical studies of the liquid–liquid interface. Observe the difference between the two aqueous phases, i.e., the top part of the main cell and the right arm connecting the reference electrode for the organic phase, R(o). **Bottom**: Cell diagram with the indication of a few typical phase components. BuFC: buthyl-ferrocene; BTTPA: Bis-(triphenylphosphor-anylidene)ammonium cation; TPBF: tetrakis-(pentafluoro)phenyl borate anion

reference electrode containers has to be as close to the interface as possible. The independent (regulated) variable during a voltammetric experiment is the difference between the potentials of the reference electrodes, while the current passing through the counter electrode is the dependent variable. The cell arrangement is functionally identical to those applied for the study of membranes.

Attention has to be paid to the choice of the appropriate reference systems. Concerning the aqueous phase, the usual commercial reference electrodes can be applied (silver/silver chloride or calomel electrodes). Here, we have to consider that a phase boundary is always created if such a closed ready-to-use commercial electrode is immersed into the test solution. When the reference electrode is created by the immersion of, e.g., a silver chloride-coated silver wire, no new solution–solution interface occurs.

The reference electrode in the organic phase appears to be rather non-conventional. Theoretically, there is an opportunity to apply a redox electrode with dissolved ferrocene/ferrocenium redox couple; nevertheless, this mode has never appeared in the literature of electrodeposition studies. Instead, the reference arm of the cell contacting the organic phase was completed with another aqueous solution containing a common ion with the organic phase studied, and a usual reference system was applied for this phase, too. A possible cell diagram for such a cell is depicted in the bottom part of Fig. 12.6. (and similar ones can also be found in a great number of relevant studies [66, 70, 73, 74, 77, 79, 81, 82, 85, 86, 91]). The equation of the interfacial reaction for the interface defined in the above-described cell arrangement when a Pd particle is growing is:

$$Pd^{2+}(w) + (Pd)_m(s) + 2BuFC(DEC) \rightleftarrows 2\,BuFC^{+}(DEC) + (Pd)_{m+1}(s) \quad (12.6)$$

where the subscript m refers to the number of atoms forming the Pd particle growing. In cases when the metal salt in the aqueous phase is complexed, the release of the anion(s) during the deposition process can also be indicated. In the case of equilibrium, a Nernst-type equation can be written for the interface in the following manner:

$$\Delta\Phi_o^w = E_o - E_w = \left(E^0_{BuFC^+/BuFC} + \frac{RT}{F}\ln\frac{a_{BuFC^+}}{a_{BuFC}}\right) - \left(E^0_{Pd^{2+}/Pd} + \frac{RT}{2F}\ln a_{Pd^{2+}}\right) \quad (12.7)$$

In the above formula, E^0 refers to the standard potential of the relevant redox system as measured against a standard hydrogen electrode in the same solvent. The activity of the metal, being composed of a pure element, is taken unity as usual. The impact of the size of the particle is often neglected, although the equilibrium at surfaces of high curvature is always different from the truly planar phases. The non-electrochemical analogy is the vapour pressure of a droplet or the solubility of small particles as opposed to bulk ones. The general treatment of the problem shows [101] that the size-dependent factor in the equilibrium potential is at most some 20 mV for the particle size range of interest, and a measurable effect can be anticipated for particles with diameter less than 1 nm (see Fig. 12.2 in Ref. [81]).

The concentration of the precursor compound for the deposition usually falls in the 0.25 … 10 mM range. Traditionally, the positive current direction is when positive ions in the aqueous solution move towards the interface and are discharged therein. In the voltammetric study of the liquid–liquid interface, the polarization limits are usually defined by different factors than on solid electrodes. Instead of the decomposition of the solution, the polarization limit is defined by the onset of the field-induced transfer of the ions from one phase to the other. The positive polarization limit can be defined by the transfer of the most hydrophobic cation in the aqueous phase to the organic one or by the transfer of the most hydrophilic anion in the other direction (for the negative limit, the same holds with the reversal of the ion charges). The polarization window of the system can be determined with double-blank solution pairs (neither of which contains any redox-active species but the supporting electrolytes are present only). The polarization limits obtained with single-blank systems (in which one redox-active species is present in one of the phases only) are usually in very good agreement with the double-blank case.

As the reactants are added to both phases, the voltammetric curves change. In the cyclic voltammograms, classical diffusion-limited peaks arise due to the electrochemical reaction taking place at the interface. The peak current varies in a linear manner with the square root of the sweep rate, at least for the positive-going (onward) sweep [69, 72], indicating that the reactions are purely diffusion limited.

The range of polarization in any aqueous/non-aqueous liquid pair studied so far is around 1 V, which imposes a strong limitation onto the particles that can be deposited. Examples studied so far include Pd [70, 73, 74, 86], Pt [81], Ag [92] and Cu [88, 91] when the precursor was in the aqueous phase and Au when the precursor was in the organic phase [68, 69, 85, 87]. A detailed overview of the systems studied, including the reducing compounds applied in the organic phase, has been published elsewhere [102].

The reaction pathway of the formation of Au (where the precursor ion is $[AuCl_4]^-$ in the organic phase) differed significantly from the case when the precursor ion was in the aqueous phase. For Au deposition, the first step was the transfer of the Au-containing complex across the interface and its reduction to $[AuCl_2]^-$ that can be reduced later to obtain nanoparticles. This was strongly evidenced with combined spectroelectrochemical studies [66, 68].

The theory of the particle nucleation at the free, large surface area liquid–liquid interface has been elaborated in detail [70, 73, 74]. Special attention was given to the role of the interfacial tension on the nucleation behaviour [74]. This is particularly important for the nucleation at the liquid–liquid interface since the relative surface area of the particle floating at the interface strongly depends on the wetting ability of both liquid phases. Having no energetically favourable surface sites for the nucleus formation at the liquid–liquid boundary, the results for the nucleation have an unusually large scatter [81], and the importance of the purity during the experimental work is to be emphasized [85].

Shortly after the appearance of the studies of electrodeposition at ITIES, a few studies showed up where the deposition at the liquid phase boundary took place in porous nanochannel membranes [75, 77, 79, 82] instead of an open surface of large surface area. In the case of the membrane-supported ITIES, the mass transport conditions are clearly different, and the hemispherical diffusion problem prevailing around the particle formed at the open surface is reduced to a one-dimensional linear one. The application of the membrane was also hoped to prevent the aggregation of the particle that highly influences the overlap of the diffusion fields at open surfaces and hence impacts also the chronoamperometric transients. Beside the ease of the theoretical approach, the advantage of the membrane-supported systems was that a regular set of particles could be obtained within a well-defined support but without the necessity of another metal as electrical lead. It was shown that the relative position of the liquid phases with respect to the template was controlled by the surface wetting properties of the liquids, rather than gravity [75]; hence, the particles could be formed at the orifice of the channels of the membrane. The membrane as a support made it easy to collect particles for TEM studies, which showed that the agglomerates produced were composed of particles of 3 to 5 nm diameter [77]. This was much smaller than the pore diameter applied (~100 nm). This was in an approximate agreement with the results obtained for Au particle deposition at ITIES of large surface area [87].

Another interesting field within the ITIES-related deposition studies is the parallel electrochemical reaction of polymer-forming monomers and metal salts resulting in nanoparticle formation during their reduction. This approach may have various advantageous features, including that the size range of the polymer-coated particles may be limited since the metal nucleation on the polymer surface is much hindered. An attempt to exploit this opportunity was given for silver particles and polyphenylpyrrole [72]. In this case, *N*-phenylpyrrole was not only the polymer-forming monomer but the electron donor that facilitated both the Ag^+ ion transfer across the interface and its reduction to silver. Unfortunately, the direct observation of the particles was not performed in this case. Further cases of the codeposition of metals and polymers will be discussed in the next chapter.

Finally, it is worthwhile mentioning that the particles deposited at ITIES are not necessarily supportless. If electrically conductive nanoobjects like carbon nanotubes or carbon platelets containing a few layers of graphene sheet are accumulated (adsorbed) at the ITIES, they can serve as a nucleation and accumulation platform for the electrodeposited nanoparticles. While such a deposition method results in an assembly of supported particles that are easier to collect than free-standing ones, the electrosynthesis at the ITIES can still lead to ultrasmall particles with a narrow size distribution [89, 91], hence retaining the general advantage of the particle electrodeposition at the ITIES. Possible catalytic application of such deposits has also been anticipated [88].

12.3.3 Electrodeposition of Various Films at the ITIES

In the systems described in the previous chapter, an electrochemical activation (i.e., polarization of the interface) was necessary to initiate the nucleation of metallic clusters. Since the experimental conditions were chosen in a way so that the partition coefficient of the salts applied in both phases was extreme, the metal deposition could not take place spontaneously. However, if the anion in the aqueous phase could enter easily into the organic phase and the standard potential of the redox reactions are close enough, the spontaneous metal deposition at the ITIES can also happen. This was found for silver deposition at the water/n-octanol interface by using ferrocene as reducing agent in the organic phase [103]. Depending on the experimental conditions (e.g., by tuning the concentration ratio of the reactants), the deposition of silver whiskers and smooth films covering practically the entire ITIES was possible. Like any other electroless deposition reaction, the mechanism was explained with electrochemical principles, although no external current source was applied for the deposition itself. It is worth noting that no similar reaction was achieved by electrochemical means for the liquid–liquid interface. Particle formation was dominant, even if the coalescence of the particles and the formation of a spongy loose network could be kept track of in some studies [88].

Although metals could not be electroplated at the ITIES as continuous films, polymers may adopt this growth mode upon electropolymerization. Free-standing tertiophene films were obtained with this method [80]. Since the deposition mode at the ITIES was found to differ from the usual heterogeneous nucleation at metal surfaces, the speculation was made that the intermixed solution layer may play a role in the process [71]. It was also shown that by tuning the Galvani potential difference between the two liquids in a thoughtful manner, the dimerization of the monomer (resulting in sexithiophene) is the dominant reaction, hence opening a synthesis route for this compound.

Electropolymerization is a feasible tool to prepare polymer/carbon nanotube composites at the ITIES. The same way as it was used for metal deposition [89], the accumulation of the nanotubes at the ITIES can serve as both an electron-transfer medium for the charge transfer between the phases and for the support of the polymer being formed. This method was evidenced by using pyrrole as monomer and the $[IrCl_6]^{2-}/[IrCl_6]^{3-}$ redox pair in the aqueous phase [67]. Spectroscopic evidence was obtained for that the polymer grew onto the carbon nanotube. As compared to the free-standing polypyrrole film, the nanotube-supported film was of higher density. The film obtained was easily transferable from the ITIES to another support.

The number of metal/monomer pairs tested so far for electrochemical composite film formation at the ITIES is very small. The parallel reaction of the precursor metal ions and the monomers can lead to a composite along the ITIES, hence resulting in nanoparticle-decorated polymer networks, which may gain a role in catalysis. A short series of papers is available in this field [78, 83, 84] for tetraoctylammonium tetrachloroaurate and tyramine dissolved in organic and aqueous phases, respectively. The direct observation of the particle size distribution confirmed [84] that the mean particle size can be easily regulated by the Galvani potential difference between the two liquid phases.

12.3.4 *Electrodeposition Near the Solid–Liquid–Liquid Three-Phase Boundary*

Sections 12.2.3–12.4 discussed the nanostructure preparation methods that used no electrically conducting material under potential control in the vicinity of the liquid–liquid interface to obtain a nanostructure. In this chapter, various methods are listed where an electrically conducting plate crosses the ITIES and it is connected as the working electrode. The reference and the counter electrodes are immersed into the aqueous phase, and the cell is regulated electrically as a conventional three-electrode cell. From this point, we shall see a bifurcation depending on which phase contains the electroactive material and whether the deposition process is an electrochemical one or it is only electrochemically assisted. The forthcoming discussion will follow this logical order.

In the studies where a truly electrochemical deposition process takes place in the vicinity of the three-phase boundary line, the reactant to be reduced is initially present in the organic phase. In the concept-proofing work [62], the authors applied $HAuCl_4$ dissolved in an ionic liquid, 1-decyl-3-methylimidazolium bis(trifluoromethylsulfonyl)imide. Later, the same technique was applied for toluene as non-aqueous phase [98], hence inverting the density ratio of the contributing phases as compared to the earlier ionic liquid study. The supporting electrolyte was KPF_6 in both cases, although it was mentioned that the experiment would work with either $KClO_4$ or KCl, too; the important factor is the non-reacting nature of the supporting electrolyte. Upon passing current through the cell, the following net reaction takes place:

$$[AuCl_4]^-(o) + 3e \rightleftarrows Au(s) + Cl^-(o) + 3Cl^-(aq). \qquad (12.8)$$

Note that the transfer of the chloride ions to the aqueous phase is indispensable for maintaining the charge neutrality of the organic phase. It was speculated that the electrochemical process may be followed by another ion exchange process between $Cl^-(o)$ and $[PF_6]^-(aq)$ because of the different solvation free energy of these ions in the contacting phases, but this reaction does not play a role in the deposition process itself.

The major reason why the deposition takes place exclusively in the triple-phase boundary zone is that the current pathway with the least possible integral resistance is only feasible, which condition can only be fulfilled if the charge transfer takes place in the close vicinity of the triple-phase boundary. Although not directly evidenced, it was also assumed that the deposition takes place in microemulsion droplets near the phase boundary. This concept was brought because the particle size did not seem to vary with deposition time, rather the surface coverage with particles could be increased [98]. In both relevant studies, it was found that the deposition zone is restricted to a strip of about 20 micrometre near the triple-phase junction and the particle shape was angular, as opposed to typical globular structures obtained with conventional electrodeposition. At the same time, the surface coverage in the deposition zone was higher than for similar electrodes used in conventional systems.

In another type of setup [93–95], the aqueous solution contains a reactant which undergoes the electron transfer at the electron conductor–aqueous solution phase boundary, but the product of this reaction does not make a deposit. Up to this point, the entire setup is identical to a conventional three-electrode cell. However, the product of the electrochemical process is an *in situ* generated reaction partner (catalyst) for a purely chemical reaction that takes place at the liquid–liquid interface with the participation of a reaction partner in the organic phase. The electrochemical process taking place at the working electrode was an oxidation process leading to acidification of the solution:

$$SO_3^{2-} + H_2O \rightleftarrows SO_4^{2-} + 2H^+ + 2e. \qquad (12.9)$$

The organic phase contains a compound that undergoes a condensation reaction. The compounds used can be an alkyl-trialkoxysilane or a tetraalkoxysilane (the ones whose condensation leads to three-dimensional network). An elementary reaction step, essentially the same as used in many other sol–gel processes, is as follows:

$$2R'{-}Si(OR'')_3 + H_2O \rightleftarrows R'{-}Si(OR'')_2{-}O{-}Si(OR'')_2{-}R' + 2\,HOR'' \qquad (12.10)$$

where R′ and R″ mean various alkyl chains. It is to be noted that the reaction product of the electrochemical process in the condensation of the alkoxysilane compound is not a net reaction partner but a catalyst. This electrochemically assisted reaction pathway was shown to lead to various nanostrucures: (i) hydrophobic films at the ITIES if n-octyltriethoxysilane is used as a reactant in the organic phase [93]; (ii) a hydrophobic but electrically conducting spongy nanocomposite if the ITIES contains carbon nanotubes [94]; and a few nanometre-thick silicate film on conducting electrodes if the electrode is slowly withdrawn from the organic phase during the gel formation process [95]. The latter method is particularly promising to achieve a large coverage of an electrode with a layer containing bulky functional groups.

12.3.5 Tip- and Filament-Assisted Deposition at the Liquid–Liquid Interface

Making use of the properties of ITIES is also possible if an electrode is applied for defining the nucleation spot of the newly formed phase. In this case, the nucleation is followed by the growth of the new phase. However, in contrast to the classical deposition systems with metal–solution interfaces, the ITIES here has a growth-regulating impact, and the growth often follows the interface. The hydrophilicity of the interface close to the nucleation site of a growing film was also demonstrated for polymer-coated metal electrodes [104]. The experience was that the polypyrrole film exhibited a preferred growth direction along the three-phase boundary where the non-conducting surface close to the growth front was hydrophobic, and the growth rate of the polymer near the hydrophobic surface much exceeded the rate far from the growth-regulating surface. A similar impact can be seen for ITIES, which works surprisingly not only for polymer films but sometimes for metals, too.

In the studies where the ITIES served merely as an interface-forming component to act as a growth regulator, the organic phase does not have to contain any electrolyte. Hence, the properties of the organic phase are not of importance, except for that it should exhibit a negligible specific conductivity. This is a strongly distinguishing feature of the tip-assisted metal deposition studies at the ITIES as compared to systems discussed in the previous chapters. However, if electropolymerization is carried out at the interface, the organic liquid may contain the monomer. Therefore, methods dealing with the deposition of either metals or polymer layers will be discussed separately. In spite of the substantial difference between the metal and polymer growth, their common feature is the orientation effect of the interface concerning the growth direction of the phase being deposited.

Concerning the *metal deposition methods* at the ITIES, the aqueous phase can contain the reactant metal salt at a high concentration without any supporting electrolyte. The organic solution is always the top one if tip-assisted metal deposition is in the focus of interest. The metal tip serving as cathode must be positioned accurately at the surface approaching from the organic (non-conducting or upper) phase so that no bulk deposition is initiated. The setup is hence technically somewhat problematic since the proper distance from the interface (or the immersion depth) cannot be defined either in general or as a function of some physico-chemical parameter of the interface. The tip diameter was reported to vary between 0.1 and 0.5 mm. In most cases, the tip cathode was placed nearly at the centre of an anode ring for the sake of an even field distribution [96, 99, 100]. If an anode plate was used instead of the cylindrical arrangement, the metal growth was rather unidirectional after a short nucleation period [97], pointing towards the anode. This clearly shows that, beside the presence of the ITIES, the electric field has an impact on the metal layer growth.

Lateral metal growth along the ITIES was studied for Zn [99] and Cu [100]. In both cases, the deposition was carried out by applying an electrode potential at which the layer growth is either much influenced by mass transport or pronouncedly diffusion limited.

Zn deposition was reported to be plate-like at the beginning and then adopting a fractal-like branched pattern, similarly to diffusion-limited aggregates [99]. Upon the application of a magnetic field perpendicular to the ITIES, the magnetohydrodynamic effect manifested itself by leading to a spiral fractal growth. The thickness of the zinc layer forming was not reported. In two other papers [96, 97], the growth of the Zn whisker at the tip was reported to be accompanied by a correlated oscillation of the current and the meniscus, indicating a wetting/de-wetting phenomenon as the interfacial conditions changed.

Cu deposition at ITIES was well documented [100]. The smaller was the concentration of the Cu^{2+} ions and the lower was the overpotential of the deposition, the smoother edge could be obtained and, at the same time, the upper view of the growing plate appeared to be less and less wavy. This is presented in Fig. 12.7. Under ideal circumstances, a nearly mirror-like upper surface was obtained as the growth front followed the ITIES (c.f. the result obtained for electroless metal growth at the ITIES as shown in Fig. 12.6 of Ref. [103]). The morphology of the film made it clear that the mass transport limitation of the layer growth was approximately hemicylindrical and planar near the penetrating edge and at the plate facing the aqueous solution, respectively. This led to a current density ratio between 10^5 and 10^7 depending on the deposition conditions, and the thickness of the 3-cm disc obtained was a few tens of micrometres only. As it can be seen in Fig. 12.7d, the morphology of the two sides of the film was clearly different. Since the plate growth took place at the bottom side, the depleted and, hence, lower-density electrolyte solution layer was not mixed (no gravitational instability occurred). This probably provided an ever-expanding diffusion field throughout the deposition period, which led to a typical branching of the deposit obtained with mass transport control. This filament-like morphology is obvious from the cut-away SEM image of the plate produced.

In contrast to the setup applied for metal layer growth along the ITIES, the electrode configuration suitable for the dynamic three-phase interline electropolymerization (D3PIE) is quite not crucial [63–65, 90]. Since the conditions for the electropolymerization are appropriate only at the three-phase boundary line, the metal wire can cross the interface with no strict restriction concerning its end position. Interestingly, but apparently not necessarily, the organic phase chosen was always of

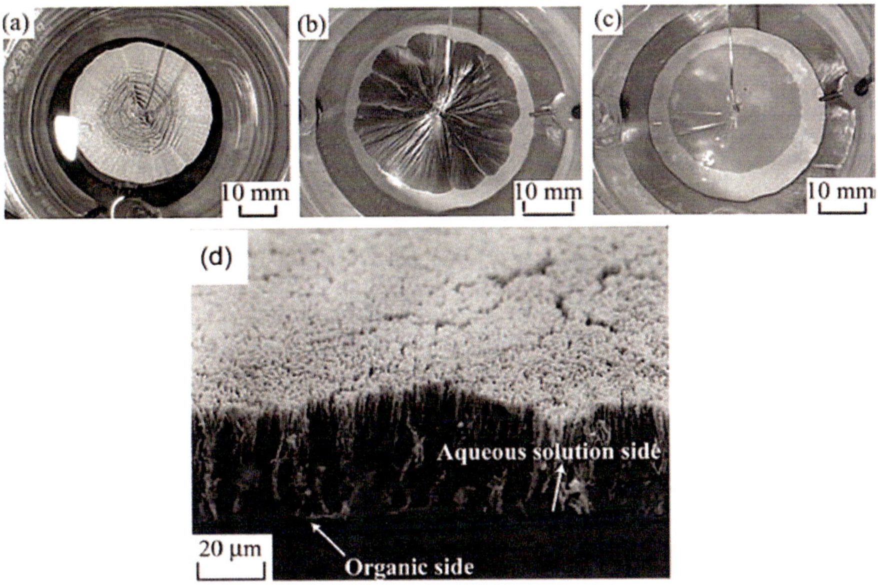

Fig. 12.7 Top view of the Cu platelets deposited at the heptane/aqueous solution interface with 1.0 M **a**, 0.1 M **b** and 0.01 M **c** $CuSO_4$ concentration in the aqueous solution. SEM figure **d** indicates the filament growth at the aqueous side of the film. Republished with permission of The Electrochemnical Society, from [100]; permission conveyed through Copyright Clearance Center, Inc.

larger density than water; hence, the organic phase was the bottom one. The counter and the reference electrodes were immersed into the aqueous phase that contained the supporting electrolyte. The concentration of the monomer in the organic phase was larger than the usually applied concentrations in the single-solution electropolymerization works, partly because of the difference in its solubility in water and in the organic solvent. As large as 0.1 M (EDOT or pyrrole in dichloromethane [64]) and 0.5 M (pyrrole in dichloroethane [90]) monomer concentrations were reported for D3PIE studies.

For understanding the D3PIE method and the role of each phase in the spatial restriction of the polymer growth, the following can be considered. Electropolymerization takes place with the release of protons, the incorporation of one monomer leading to the release of two protons and two electrons. Using EDOT as an example and showing terminal hydrogen atoms for both the monomer and the polymer, we can write that:

$$\mathrm{H{-}EDOT{-}H(o) + H{-}(EDOT)_{\mathit{n}}{-}H(ITIES) \rightleftarrows H{-}(EDOT)_{\mathit{n}+1}{-}H(ITIES) + 2e + 2H^{+}(aq)} \tag{12.11}$$

Hence, the source of the monomer is the organic phase while the absorbent of the protons produced is the aqueous phase. In addition to the above reaction, since the polymer is formed in an oxidative route, the polymer chain itself is also oxidized, leading to a charged polymer chain for which the aqueous electrolyte solution provides the counterions:

$$\mathrm{H{-}(EDOT)}_n\mathrm{{-}H(ITIES)} + k\mathrm{An}^-\mathrm{(aq)} \rightleftarrows \left[\mathrm{H{-}(EDOT)}_n\mathrm{{-}H}\right]^{k+} \cdot \mathrm{An}_k^-\mathrm{(ITIES)} + ke \quad (12.12)$$

where the maximum of the *n/k* ratio is around 2…4 for various types of polymers. The anion doping during the oxidation also makes the interfacial zone preferred for the polymer formation site. The decrease in the hydration free energy of the anion An^- in the aqueous phase promotes the ingress of this anion into either the interfacial "mixed" solution layer or the polymer film being formed. Hence, the decrease in the hydration free energy of the counterion as well as the decrease its concentration favours the lateral growth of the self-standing polymer film as opposed to its thickening.

The two basic approaches of the D3PIE reaction are similar to that of conventional systems; i.e., either constant-potential polymerization (coupled with chronoamperometric record) or cyclic voltammetry is used. For the constant-potential polymerization, current increases with time in a manner defined by the experimental conditions. For PEDOT formation by D3PIE, a nearly time-proportional *surface area* increment was reported [64], while the *diameter* of the film showed a linear increase rate for polypyrrole growth [63], which obviously means a quadratic surface area increment with time. The determining factors of the growth mode are yet unclear. However, the indication of the identical quality of the film along its diameter is shown by the fact that the mass vs. polymerization charge functions are linear for the constant-potential growth [63].

For the D3PIE films prepared with cyclic voltammetry, the current change during the cycles may be anomalous in the sense that it is larger during the negative-going sweep that in the preceding positive-going sweep (anomalous hysteresis [63–65]). This is because the growth rate along the perimeter (with positive current contribution) is large enough for compensating the current originating from the redox and capacitive properties of the film (that have negative current contribution). The anomalous hysteresis becomes more pronounced if the electrolyte(aq) vs. monomer(o) concentration ratio increases [63].

The morphology of D3PIE-prepared films varies with the preparation method. If CV is used, the film shows a tree-ring pattern due to the variation of the polymerization conditions with time [65], but constant-potential polymerization leads to a laterally homogeneous structure [65, 90]. Regardless of the preparation method, the morphology difference between the sides of the films is the same; moreover, it shows the same trend as that was established for metal films. Namely, the film side right at the ITIES is rather smooth, but the growing plane contacting the solution that contains the precursor material is porous [65, 90]. The growing face of the polymer film shows rather an open-pore structure as opposed to the dense filament array shown for the metal films. The proper choice of the organic solvent and the decrease in temperature improves the film smoothness at the ITIES side [90].

The D3PIE method was also adopted for interfaces where carbon nanotubes were accumulated. The current increase rate for constant-potential polymerization is much faster for the interface already covered with electrically conducting nanowires since the interconnections formed between the nanowires by the *in situ* growing polymer yield a faster expansion of the eclectically connected area than in the absence of the nanowires [65]. The conclusion of the latter study was that at first, the surface of the nanowires is covered with the polymer, and the growth of the polymer towards the monomer-containing organic phase starts only thereafter. Nevertheless, the evolution of the morphological features after the full coverage of the nanowire network is identical to that of the simple D3PIE-prepared samples, which is rationalized by the same processes taking place after the growth of the reaction front beyond the nanotube network [65].

12.4 Deposition with the Help of Tips

12.4.1 Deposition of Microcolumns by Using Microtips

Manufacturing of micropillars by using microtips [105–114] is a method related to the family of localized electrochemical deposition (LECD) techniques. Although the technique of obtaining micropillars of about 10 micrometre minimum width will not directly lead us to the preparation techniques of nanoscale object, it is didactically important to shortly discuss its experimental conditions and limitations for getting a complete view on all related techniques, in particular for the sake of comparison with nanopipette-based methods to be shown in Sect. 12.4.3.

The basis of this technique is that a conducting substrate is used as working electrode that will later host the micropillar. The solution used is a conventional plating bath with relatively large metal salt concentration and also with various deposition-regulating components. The anode is a microtip made of a noble metal (e.g., Pt) that *does not* serve as a source of the reactant for the deposition. Instead, the reaction at the anode (counter electrode) is the decomposition of the solvent (e.g., oxygen evolution in aqueous solutions). The diameter of the counter electrode is a few tens of micrometres. The active surface is the bottom-facing microdisc with the sides of the electrode being sealed. Reference electrode is seldom applied in microcolumn-preparation setups [106, 114]; and if so, it is an indicator electrode only but is not a part of the regulation circuit of either the working electrode potential or the cell voltage. The cells are operated with a cell voltage of a few volts.

One of the key parameters in the microcolumn formation is the distance between the substrate and the anode. The interelectrode distance is comparable to the diameter of the anode disc. This distance is much larger than the Debye length of the electrical double layer in the fairly concentrated solutions. The close proximity of the cathode and the anode concentrates the electric field between the electrodes, and this determines the preferential reaction sites. As the micropillar starts growing, its thickening due to the metal deposition at the column side becomes negligible.

For the growth of a constant-diameter microwire, the optimal anode–cathode distance must be maintained during the wire growth. This is carried out with an automated anode guide that can be operated by using an electrical control unit monitoring either the current passing through the cell or its variation with the inter-electrode distance as detected during an oscillatory motion of the anode (current gradient method [108]). Occasionally, this electronic control is completed with a video system. The maintenance of the optimal anode–cathode distance is essential to keep the morphology under control.

Studies published so far dealt with the deposition of Cu [105, 107, 108, 110, 112, 114], Ni [106, 111] and Cu–Ni [113] microcolumns. Irrespectively of the composition, experience related to microcolumn growth can be summarized as follows. The columns are polycrystalline and the external surface shows a granular structure which can be suppressed with the application of suitable additives. The current and the growth rate both increase as the anode–cathode distance decreases or as the cell voltage increases. However, the decrease in the interelectrode separation has a shadowing effect for the component transport, which causes a dense-to-hollow structural transformation at too short distances [107]. Hollow (microtubular) structures can also be achieved with the increase in cell voltage [111]. Columns with bending can be synthesized if the anode is not only raised but also displaced laterally during the column growth; i.e., the growing column follows the anode due to the strong local electric field [106]. By applying pulse plating, the decrease in duty cycle results in more compact and dense structures (i.e., it has the same impact as the decrease in growth rate achieved by any other means) [114].

12.4.2 STM and AFM Tips in Electrochemical Nanomanufacturing

Tips with ultrasmall end curvature were originally developed for surface imaging. Shortly after their introduction to the scientific practice, a number of studies appeared that used these nanoscale tips for surface manufacturing. The review of Penner [115] summarized the early non-electrochemical manipulation methods at the nanoscale, while a fairly recent collection of tip-assisted nanopatterning methods is also available [116]. Below, a classification of the electrochemical methods will be shown in which the pure mechanical applications of the tips will be rather neglected (for a few examples of such tip applications, see Chap. 10). The schematic organization chart of the tip-assisted *electrochemical* fabrication methods is presented in Fig. 12.8. Reviews with different classification systems are also available in the relevant literature [117, 118]; here, the primary point of view was the nature of the electrochemical

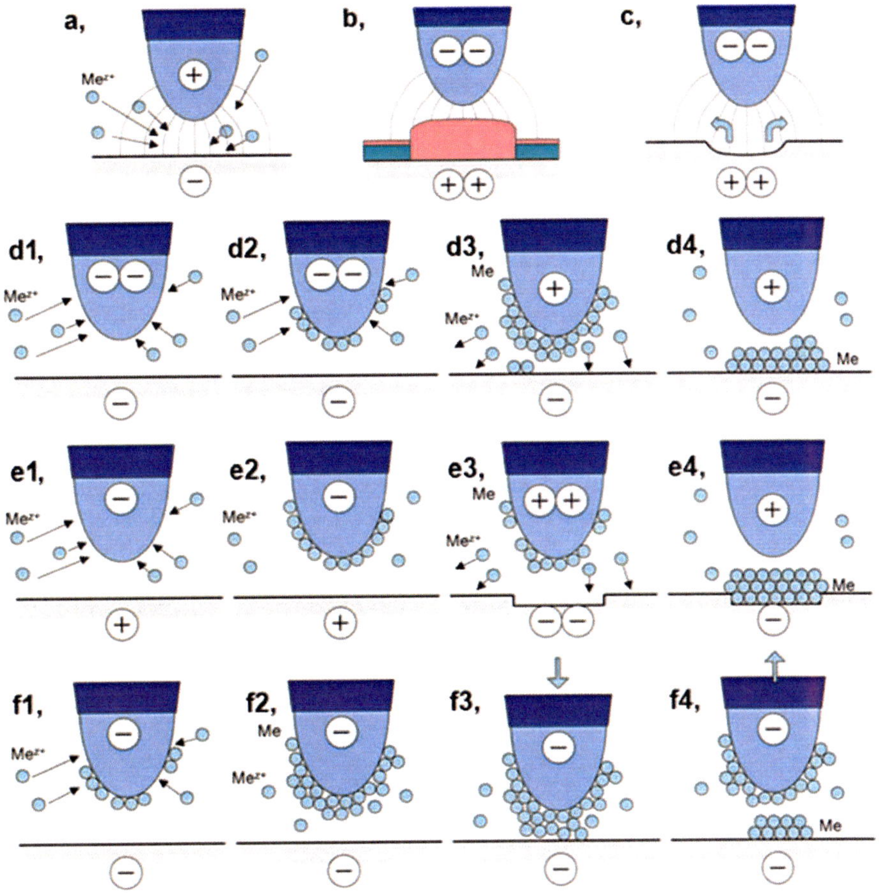

Fig. 12.8 Schematic overview of the application of STM tips in various electrochemical nanomachining processes. **a**. Field-assisted local electrodeposition on the substrate merely with a counter electrode reaction taking place on the tip; **b**. Application of the tip for high-voltage local oxidation of a coating for producing an oxide nanodot; **c**. Cavity formation on a substrate with high-voltage pulse; **d**. 1 and 2: collection of the metal to be deposited onto the tip, 3 and 4: reversing the tip bias for dissolving the metal and depositing it at the same time onto the substrate; **e,** Same as d, with applying a short high-voltage pulse in step 3 for etching an active spot on the substrate; **f**. 1 and 2: electrodeposition of a metal onto the tip, 3: approaching the tip to the surface until contacting the surface, 4: depositing a metal dots without changing the tip bias by tearing apart the deposit bridge formed between the tip and the substrate with upward tip movement

process involved in the process of interest. A common advantage of all methods to be described is that the electrochemical modification of the substrate can be performed with very high spatial selectivity along a predefined pattern. When the tip itself is used for electrodeposition in a step of a sequential process (see Fig. 12.8d–f), the working distance of the tip is at least an order of magnitude larger than the tunnelling or force-sensing distance for STM and AFM, respectively. The *in situ* imaging with the same tip after the nanostructure fabrication is also customary.

The simplest application of ultrasmall conducting tips for electrochemical nanopatterning is essentially analogous to the system described in Sect. 12.4.1 for nanopillar fabrication, with the non-negligible difference that the electrode size is 3 to 4 orders of magnitude smaller (see Fig. 12.8a). In both cases, the metal ions to be deposited are the solutes, and neither deposition nor dissolution takes place on the tip itself which serves as anode. This method is often termed as field-induced deposition, referring to that the current can flow in the close proximity of the point-like tip electrode only. In an early studies of this technique [119] 60 μs pulses of 4 V were applied to plate Ag clusters from AgF solution onto HOPG surface, resulting in nanodots of about 7 nm height and 25 nm width. The latter aspect ratio is typical for nanotip-assisted deposition processes. Later, a nanoscopic electrochemical cell was implemented by subsequent field-induced electrochemical deposition of Ag and Cu nanodots serving as cathode and anode, respectively [120]. With the opposite polarization of the tip and the substrate, the oxidation of the substrate surface can be achieved locally under the tip (Fig. 12.8b). A demonstration of this technique by using an AFM tip for the oxidation of thin Nb films showed [121] that not only Nb_2O_5 nanodots but surface patterns of essentially arbitrary shapes can be produced, also with a complete in-depth oxidation of the Nb film.

There is also an opportunity for metal cluster formation on a substrate when the field is yet small below the tip and the tip itself is not used for the generation of a flush of the reactant. This intermediate case between the field-induced and collector-generator role of the tip is exemplified by the deposition of Cu clusters on Au when the electrode potential of both the tip and the substrate is just slightly negative as compared to the Cu^{2+}/Cu equilibrium potential [122].

Nanotips can be used as a source of the material to be deposited onto a substrate [123–126]. In this method, the tip, which is made of an inert material, is under potential control. First, the negative polarization of the tip makes it possible to collect a deposit from the solution. Then, the tip polarization is changed to a sufficiently positive potential at which the dissolution of the previously collected metal takes place. This dissolution step is often called "burst" in the literature. During the entire collection–emission sequence on the tip, the substrate potential is invariant and just slightly positive relative to the deposition potential of the same metal from the *bulk* solution. However, at the large ion concentration generated locally by the burst, the metal deposition becomes thermodynamically possible, which makes the deposition process locally selective to the close vicinity of the tip. The process can be repeated sequentially for patterning the surface with the dots of the deposit, and the constant potential of the substrate stabilizes the previously deposited islands. One cycle of the electrochemical process takes a few tens of milliseconds, the majority of which is the collection time for the metal deposition onto the STM tip, and the typical length of a burst is about 5 ms. Calculations show [124] that the metal ion concentration as a result of the dissolution achieves a steady-state value within less than 200 μs, which means that the metal ion release takes place under diffusion-controlled conditions. Due to the shape of the diffusion field around the tip, the metal ions released on the side remain ineffective for the deposition onto the substrate, and the ions produced in the round-shaped end of the tip can be re-deposited onto the substrate. Hence, the radius of the metal dot plated onto the substrate is about twice of the tip curvature radius near the end. Due to the small efficiency of the metal deposition at the substrate as referred to the charge corresponding to the collection step at the tip, the latter parameter can be used for the process regulation only by measuring an empirical recollection efficiency for a particular tip at a fixed distance. Literature data are available for Co deposition on Au substrate [123–125] and Pb deposition onto Ag(111) [126]. Tip current in the subsequent intervals of a cycle and nanodots imaged in the same STM system that is used for the sample preparation with the burst method are shown in Fig. 12.9.

Production of silver nanodots on HOPG was demonstrated in a process where burst was combined with substrate etching (see also Fig. 12.8e) [127]. No potentiostatic control was used, but a single power source was applied in voltage-control mode in a two-electrode cell. Since metal nucleation is much hindered on the defect-free surface of HOPG, the metal collection on a Pt tip (formation of an Ag UPD layer) was followed by a high-voltage pulse of opposite polarity, which induced the formation of a defect on the HOPG surface just under the tip. The shallow (nearly monoatomic) pit was formed in the first 5 μs of the polarization with at least 4 V cell voltage. The defect lied exactly at the area where the large metal ion emission due to the positive tip potential is prone to induce the nucleation of the metal at the substrate; nevertheless, the key factor in the deposit formation was clearly the breakdown of the graphene structure on the HOPG surface. The time period required for the silver

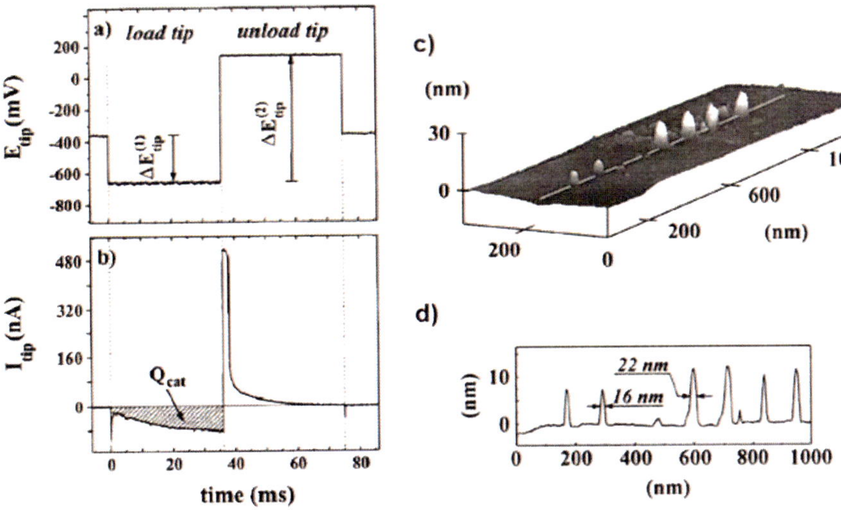

Fig. 12.9 Co nanodot preparation with STM tip by using the burst method. **a** Potential of the tip during the load (deposition) and unload (dissolution) period. **b** Variation of the tip current during the polarization program shown in (**a**). **c** Series of nanodots produced with sequential bursts as shown in an STM image. **d** Cross-sectional height curve along the line shown in (**c**). Reprinted from [124] with permission. Copyright (2001) The Electrochemical Society

nanodot deposition was a few tens of μs only. A reproducible defect initiation was well detected in subsequent cycles; nevertheless, the amount of Ag deposit produced was always the largest in the first cycle and then it gradually decreased with cycle number because of an unknown reason.

Since the fully electrochemical processes did not prove to be reproducible at the desired level, a half-electrochemical process was proposed by Kolb et al. [128] in which the STM tip serves as cathode for the collection of metal but its relocation to the substrate does not involve any charge transfer (see Fig. 12.8f). The Au substrate was held at a potential where bulk Cu deposition is not possible, but the surface is covered with a Cu UPD layer. This corresponds to about +10... + 30 mV potential as compared to the reversible Cu^{2+}/Cu potential in the same solution. After the collection of the sufficient amount of Cu on the tip held in the bulk Cu deposition potential regime, the electronics drives the tip towards the surface until reaching a "jump-to-contact" level in the tunnelling current, when the tip is withdrawn. However, the contact is not broken where it is formed. Since the mobility of Cu is sufficient for a rearrangement, the system tends to decrease the shape and curvature asymmetry, and a neck is formed from the Cu atoms that originally belonged to the tip. Therefore, a well-defined amount of Cu is "dropped" onto the substrate surface when the neck is torn apart upon the withdrawal of the tip. Meanwhile, the tip potential did not need any adjustment as a function of the distance from the surface.

Although the cluster left behind on the surface by the light contact with the Cu-covered tip is by far thicker than the UPD layer already being present therein, the clusters proved to be stable at the potential of the deposition, although the opportunity for a long-term *ex situ* investigation was quite limited [129]. It required a few hundred mV more positive potential to strip off the mechanically deposited clusters together with the underlying UPD layer. This made it possible to manufacture very large arrays of copper nanodots on Au (111) surface. The largest reported Cu nanodot array covered a $1.1 \times 1.1\ \mu m^2$ surface area [130]. Later, it was argued [131–133] that the stability of the clusters may originate from alloying effects, especially when various Au–Cu alloys were used as substrate. However, the alloying theory is in contradiction to the observation that dissolution of the Cu clusters leave behind an apparently perfect Au surface without any trace of reconstruction [122, 134]. The controversy of the stability issue is yet to be resolved.

A detailed scrutiny of the tip-assisted transfer process revealed that the Cu cluster height strongly depends on the distance of maximum approach to the substrate surface [128, 135]. Surprisingly, a closer approach within a 1-nm distance led to a decrease of the cluster height, whose maximum was about 0.8–0.9 nm at a 1-nm closest tip approach. Concerning the substrate dependence of the process, it was revealed that Ag(111) does not serve as a good receptive substrate for cluster deposition but a monoatomic Cu layer is formed, although no UPD of Cu is possible on this surface. Construction of Cu clusters on Ag(111) surface could be carried out with sequential deposition cycles by holding the tip in identical lateral position [135]. The comparison of a few structures produced with the jump-to-contact method is presented in Fig. 12.10.

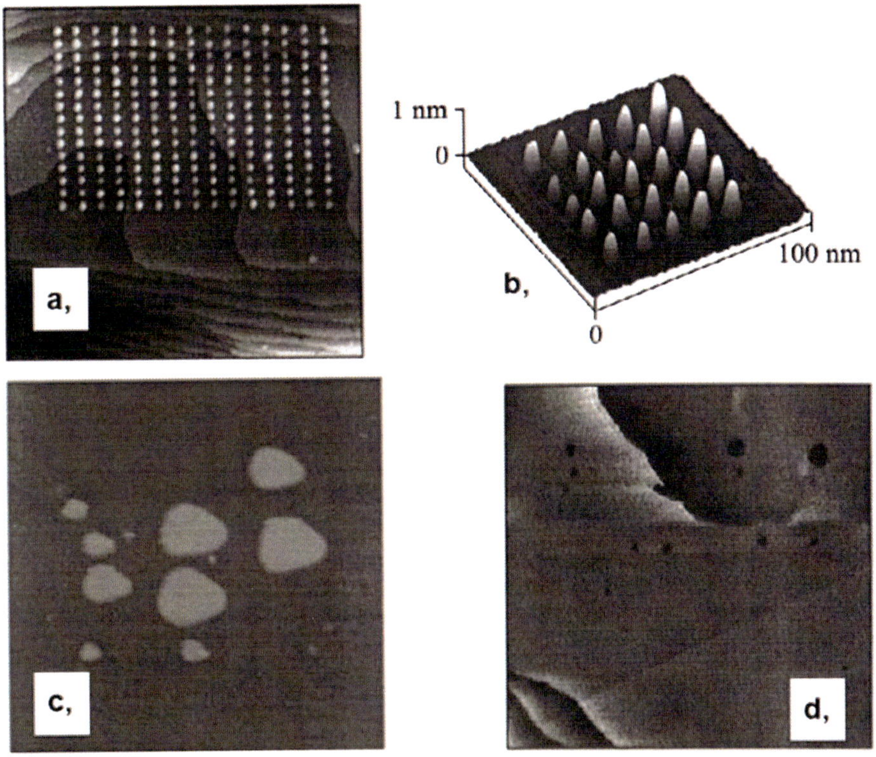

Fig. 12.10 Comparison of various surface structures produced with the jump-to-contact method. **a** Cu clusters deposited on Au(111) surface; image area: 197 × 197 nm^2. **b** Ag nanoclusters on Au(111) surface. **c** Cu layers formed on Ag(111) surface; image area: 150 × 150 nm^2. **d** Defects induced on Au(111) surface by the contact with a Ni-plated tip; image area: 200 × 200 nm^2. Reproduced from [134] (a, b and d) and [135] (c). Copyright (2000) and (1998), respectively, with permission from Elsevier

While Ag as a metal transferred proved to be suitable for building nanocluster arrays on Au(111), Ni-coated tips caused a surface defect. Upon the trial to transfer Ni onto the surface, the metal transfer was of the opposite direction [134]. Although the adherence between the deposited and substrate metals was effective, the metal of smaller stiffness and higher atomic mobility could be transferred only. By applying Ni-coated tips for Au substrate, the jump-to-contact method resulted in nanocavities at the point of contact (see also Fig. 12.10d).

It is to be noted that the jump-to-contact method was adapted for Zn cluster deposition onto Au(111) surface in ionic liquid [121]. A commonality with the previously discussed systems is that Zn forms a UPD layer on Au(111). The mean diameter and height of the clusters were 4.3 and 0.51 nm, respectively. By using Fe for the metal transferred to Au surface in the jump-to-contact method under an ionic liquid , the transfer is possible, although iron does not show UPD on the gold surface [136].

Interestingly, the Fe cluster height was about 2-3 monolayers, similarly to other metals, but the cluster diameter was rather high, around 120 nm.

When the STM tip near the substrate surface is used to etch the surface, a large variety of nanoscale surface damage or nanocavity can be initiated, as shown in the scheme of Fig. 12.8c. Relevant examples are the cavity formation on gold [137–140], copper [141] or H-terminated Si [142]. While a relatively large voltage is necessary to induce the nanocavity formation on gold, in the rest of the examples the substrate voltage is in the range where no dissolution is expected on a flat surface without another electrode in the close vicinity. Therefore, it was assumed that the overlap of either the electrical double layers or the other Helmholtz planes plays an important role in the dissolution/oxidation of the substrate [141]. The direct electron tunnelling between the substrate and the tip was also assumed, hence leading to localized substrate dissolution. It was also a common experience that the tip-induced nanocavity was a preferred site for the nucleation of nanodots formed from a foreign metal as compared to the rest of the substrate surface [138, 140], which was explained with both structural and electronic effects.

Another interesting application of the scanning probe tips is the so-called electrochemical nanolithography. In this method, the tip is scanning the surface that is covered with a layer, made from either an organic monomer [143] or a metal [144]. The tip potential is negative for both the oxidative polymerization of the monomer coating and during the anodic etching of the substrate. In both cases, the substrate potential is in the range where the oxidation process would also take place on a macroscopic electrode, too, and the localization of the process at a particular part of the substrate is highly due to the vicinity of the point-like anode tip.

Finally, it has to be noted that tips are suitable to initiate electrochemical polymerization of various monomer coatings. Such a reaction may work with an STM tip [145] or with an AFM tip with conductive coatings [146]. In neither of these cases was an electrolyte solution used, although the monomer film contacting the tip might form a liquid meniscus between the tip and the substrate surface. The cell voltage of a few volts with polarization direction corresponding to the monomer properties was applied to polymerize the film just under the tip, hence forming a conducting covalent network attached to the surface. The feature size in this method depends primarily on the tip size. Manipulation of a self-assembled monolayer (SAM) is also possible with an STP tip under dry conditions [147]. In the latter study, an Ag-coated STM tip was used. The first step of the process was the removal of the SAM by the anodic and cathodic polarization of the substrate and the tip, respectively. Then, by reversing the polarization conditions, Ag was directly transferred from the tip to the bare part of the substrate where SAM was previously removed. These examples clearly show that electrochemical reactions by using tips can be carried out under circumstances that are not convenient in electrochemistry, although the processes taking place are electrochemical in nature.

12.4.3 Nanopipette-Based Methods

In this chapter, the following setup types are discussed. The tip end is a micro- or nanochannel that contains an electrolyte solution. The wall of the tip itself is non-conducting; hence, the tip material itself is not under electrochemical control. The counter electrode (and, in most of the cases, a reference electrode) is placed in the liquid held by the tip channel. The tip assembly with the electrolyte solution and the electrodes hence provides a two-fold means of control of the electrochemical process. On the one hand, the geometrical constraint due to the narrow conducting channel localizes both the diffusion zone of the reactant and the electric field near the orifice of the tip. And, on the other hand, the potential control of the electrodes (e.g., by pulses) makes it possible to limit the amount of deposits. Figure 12.11 offers an overview on the variety of electrode and tip arrangements that can be used for tip-assisted electrochemical nanomanufacturing. It is interesting to note that works published so far in this field all deal with deposition processes.

When a solid surface is patterned by means of electrodeposition with the help of a liquid-containing tip, the process is often called electrochemical printing (ECP). Since the end of the tip is too narrow for accurate visual observation and it is also too fragile for manual positioning, an automated electronic control is essential for monitoring the approach of the tip to the substrate surface. The controlling unit often combines a unit for microscopic observation as well as a piezostage, a controller used also in scanning probe methods. In this case, the tip approach to the surface can be detected by damping the oscillation by the force exerted by the substrate surface.

Concerning the choice of solutions in the ECP techniques, various approaches can be found in the literature. If the tip is submerged into a solution with the same concentration as within the tip [148], the deposition process is clearly controlled by the field and current flow concentrated to the nanopipette orifice (see Fig. 12.11a). Here, the internal diameter of the capillary was about 200 nm, and the diameter of Cu dots plated at their half height was about 270 nm, indicating that the spread-out of the deposition zone was not significant. The situation is similar to the direct deposition by using an STM tip as anode. The major difference is that as the deposit grows, the distance between the cathode and anode does not change since the counter electrode within the tip is far from the orifice. By using a dilute $CuSO_4$ solution (1 mM), the deposition under the 200 nm nanopipette took place with a current less than 100 pA, but the cell voltage was above 3 V. This high value of the cell voltage is due to the large resistance of the narrow capillary. Similar cell arrangements are also known from electrochemical micromachining techniques with pipette diameter in the micrometre range [149]. The latter works reported that a solution flow could also be applied at micropipette internal diameter of 40 μm, but the tip tends to be clogged at smaller diameters.

In order to avoid submerging the entire substrate into the plating solution, a meniscus-controlled deposition is also feasible by using nanopipette (see Fig. 12.11b). For submicrometer-scale capillary diameters, the capillary force is sufficient to inhibit the solution loss from the capillary. In this configuration, the working

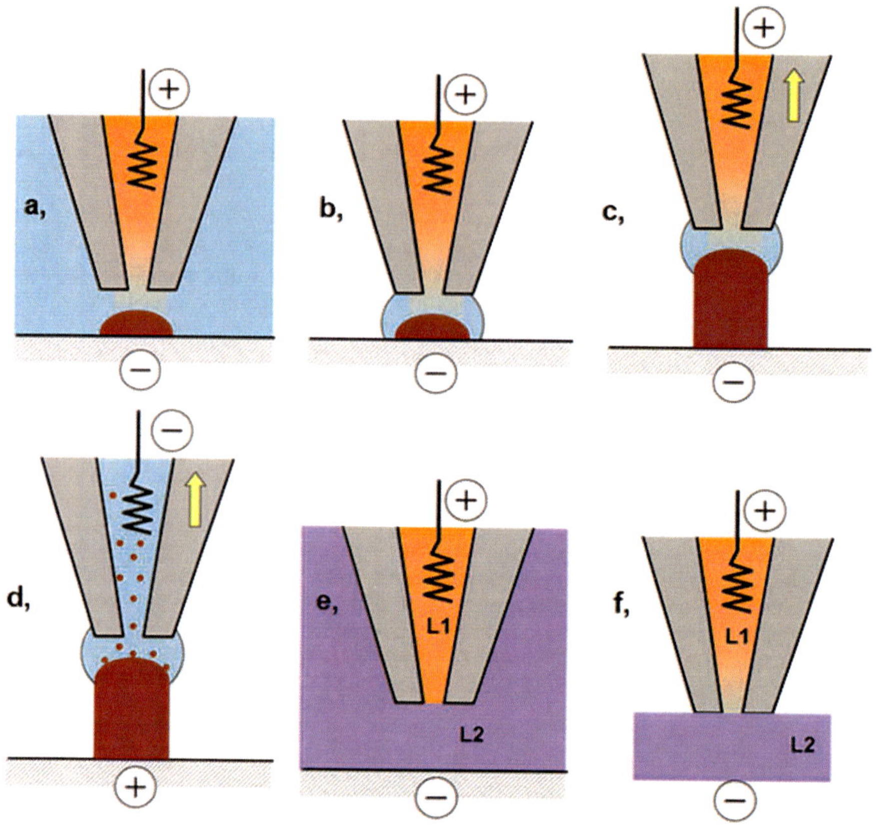

Fig. 12.11 Overview of the nanopipette-based electrochemical methods. **a,** Nanopipette with anode facing with the substrate, both the pipette and the substrate fully submerged into the solution; **b,** Nanopipette contacting with the substrate through a liquid droplet; **c,** Meniscus-confined method with pulling up the pipette as the deposit grows; **d,** Nanopipette filled up with nanoparticle suspension for electrophoretic deposition; **e,** Nanopipette filled with a liquid L1 contacted to another solution L2, by using immiscible liquids and a conductive solid surface in the close vicinity of the tip; **f,** Same as e, with the difference that the substrate is either far from the tip orifice or L2 is a molten metal that itself serves as liquid substrate

and reference electrodes are within the capillary [150]. Since the solution resistance in the capillary is large and ohmic drop correction is quite cumbersome, the current control yields better reproducibility than potential control, although the chronopotentiometric detection can be used for process monitoring in a semi-quantitative manner. It can be assumed that for the meniscus-confined method, the diameter of the wetted area on the substrate is comparable to that of the capillary, but it is difficult to control. Hence, the distance between the capillary ending and the substrate remains an essential parameter. By comparing the fully submerged [149] and meniscus-confined [150] methods, the sizes of the dots deposited are close to each other, and no superiority of either of these methods can be established at the technical level presently available.

It is essential to mention that microtechnology also applies similar electrode arrangements, although the size of the features deposited is above the micrometre. By using a laterally moving micronozzle with the permanent maintenance of the meniscus-confined configuration, microwires can be drawn at the substrate surface [151]. The writing process has to start at a conductive spot of the substrate (which is connected to the working electrode input of the process controller), but the conductivity of the wire already plated to the surface can be taken advantage of later; therefore, the writing process can be extended to non-conductive parts of the substrate surface. Deposition of microdots can be made more efficient by using microcanteliver arrays [152, 153].

Similarly to the methods described in Sect. 12.4.1, micropipette-based methods are also suitable for pillar fabrication. However, as compared to the processes described earlier when the major driving force of the pillar formation was the electric field, pipette-based methods in the meniscus-confined configuration are also suitable for nanowire deposition (see Fig. 12.11c). This method was called as electrochemical fountain pen nanofabrication (EC-FPN) [154]. On the one hand, the deposition setup is very simple, being composed of a glass capillary with about 100 nm diameter, the solution filling it and a counter electrode immersed into its upper part. On the other hand, the actuator of the nanopipette operated in a feedback-controlled mode makes the setup thoroughly controlled. This deposition mode is a complex process in which electrodeposition, fluid dynamics, and mass and heat transfer are simultaneously involved. The environmental humidity, the evaporation rate of the liquid as well as the nozzle diameter and the pulling rate were found to have a crucial importance, as shown in multi-physics simulation of the process [155, 156].

Free-standing Pt nanowires were continuously grown off the substrate by moving the nanopipette away from the substrate while maintaining a stable meniscus between the nanopipette and the nanowire growth front. In the introductory study of this technique [154], the cell voltage was controlled (1 V), and the current was about 0.5 nA. The withdrawal rate of the pipette was 50 nm s^{-1}, which must be strictly equal to the wire growth rate for maintaining the stable meniscus configuration and to avoid a thickness fluctuation due to the varying conditions. Free-standing nanowires up to 30 μm length could be produced with this technique. It was found to be typical that the wire diameter is larger than the internal diameter of the pipette, which is due to the extension of the droplet diameter at the end surface of the growing nanowire. Pt nanowires thus produced were later tested as working electrodes in voltammetric electroanalytical techniques [157].

By improving the meniscus-confined nanopipette pull-up deposition method [158], the deposit choice was extended to Cu. Here, the solution concentration was an order of magnitude larger than earlier applied for Pt, which led to a significant increase in the growth rate. Also, the nanowire diameter could be reduced to 80 nm, and empirical trends could be established for the dependence of the wire diameter on the pulling rate. The wire diameter was inversely proportional to the pulling rate

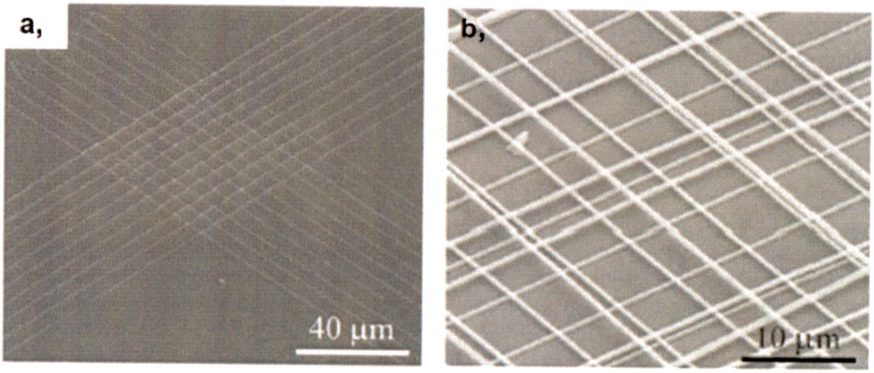

Fig. 12.12 Cu cross-bar deposits obtained with the meniscus-confined nanopipette pull-up method from 50 mM $CuSO_4$ solution. Wire diameters are **a** 200 nm and **b** 400 nm. Reproduced from [158] with permission of John Wiley & Sons, Inc.

for a relatively large aperture diameter of 5 µm, while for a 2-µm aperture diameter, the wire diameter decreased linearly with the pulling rate. Another important factor of improvement was that the process was adapted to wire deposition onto non-conducting surfaces (see Fig. 12.12), also in a cylindrical roll-up configuration.

Another approach to use nanopipettes is based on the electrophoretic deposition of nanoparticles [159]. Here, the solution does not contain an ionic solute. For the deposition, a 2 wt% solution of gold particles of 3 nm diameter protected by mercaptosuccinic acid as a protective agent against particle aggregation was used. Although it was not mentioned in the original work, it is assumed that a weak dissociation of the surface-bound acidic group lends the negative charge to the nanoparticles. This is why the substrate has a positive bias, in contrast to all other deposition methods discussed in this chapter. The nanopillars thus formed seemed to be compact in the SEM images and exhibit a 1 µm lower diameter limit. No chemical analysis was performed to assess the amount of residual anti-aggregation agent in the agglomerated nanowire.

Nanotips have also been used for studying the liquid—liquid interface. A few typical scanning electrochemical microscopy (SECM) studies are mentioned here only for sake of comprehensiveness [160, 161]; however, since these studies are typically not preparative methods, they are not discussed in detail. Let us first deal with a system where the nanopipette is filled up with an aqueous solution containing Ag^+ ions and is brought close to a surface submerged into a nitrobenzene-based solution [162]. Here, the polarization of the system leads to the Ag^+ ions enter from the aqueous phase to the thin nitrobenzene layer under the tip and are deposited onto the substrate surface. The reason why the Ag^+ ions enter the organic phase instead of being deposited, as it was seen in Sect. 12.3.2 dealing with nanoparticle synthesis at ITIES, was that a suitable complexing agent was applied in the organic phase, favouring ion ingress instead of ion discharge. With a synchronized control of the tip motion and the polarization of the system, surface patterning is possible.

In the above nano-ITIES, the metal deposition took place on a substrate. If the metal is deposited onto the ITIES, the nanopipette can be transformed into a nanoelectrode if the nanoparticles being formed can agglomerate and form a solid block at the end of the nanopipette. Two examples are presented for such a process. In the first case, the water/nitrobenzene interface was used without applying a complexing agent that facilitates Ag^+ transfer between the two phases [163]. Another important prerequisite of the electrode preparation was that the outer wall of the glass nanopipette had to be covered with a hydrophobic layer in order to suppress the formation of a water droplet extending outwards from the pipette's internal nanochannel. After the electrode preparation, a contact from the interior of the pipette was formed. Since there was no way to contact the solid ending of the nanopipette mechanically from the interior, a silver wire was immersed into the internal solution. When this unit was used as an *electrode*, it worked as it is expected for nanoelectrodes – however, in fact, *two electrochemical cells* were connected here in serial. The internal cell in the nanopipette was a symmetrical one in terms of the electrochemical reaction (Ag oxidation or Ag^+ reduction), and this was a low-impedance cell (with electrodes of high reversibility and small polarization resistance). The electrode response was determined by the reaction taking place at the external surface of the Ag-filled nanotip, which made it ideal for using as nanoelectrode.

A variation of the above-described technique is when the bottom liquid phase (outside the nanopipette) is a liquid metal. In the pioneering work of this technique [164], eutectic Ga–In (E-GaIn) was used, which dissolves metals much less than mercury. When the nanopipette filled up with a solution of the metal ion to be reduced was brought into contact with E-GaIn, the reduction of the noble metal ions first takes place spontaneously due to the large difference in the standard potentials of the liquid forming metals (i.e., that of the Ga^{3+}/Ga and In^{3+}/In redox pairs) and those to be deposited (i.e., Ag^+/Ag or Au^{3+}/Au). Further electrodeposition of the nanoelectrode-forming metal took place on the membrane formed spontaneously by cementation. When the metal filling was finished, pyrrole was polymerized between the metal tip and an internal wire for providing backside contact. Ion milling of the outside surface of the tip proved to improve the functionality of the nanoelectrodes.

12.4.4 Miscellaneous Electrochemical Preparation Methods of Nanoelectrodes

One of the oldest electrochemical nanoelectrode preparation technique is the etching of STM tips made mostly from tungsten [165–173] and occasionally from other metals (Re [174], Ag [175, 176], Au [177], PtIr [178, 179], Ni [180] and other magnetic metals [181]). The fine details of the tip manufacturing process and the post-treatment modes show a great variety. The typical procedure applies anodic etching in fairly concentrated (2–3 M) MeOH solution (Me: Na, K), but chloride- and cyanide-based solutions are also applied in some cases. The process starts with a tungsten wire that is immersed into the alkaline solution partly by holding the top of the wire. The meniscus getting shaped near the contact of the wire with the solution plays a crucial role in the tip formation. When a bulk solution is used in the manufacturing process, the tip position is often monitored and adjusted during the etching process. In the most popular setup, the wire is surrounded with a thin liquid membrane the holder ring of which serves as a counter electrode and provides the most suitable electric field direction at the same time (Fig. 12.13). As the neck in the meniscus position of the tungsten wire is getting narrow in the etching process and finally the bottom parts drop off, the electronic system used for the process monitoring observes the sudden decrease of the current. Then, the circuit is broken immediately after the break in the tungsten wire for avoiding overetching. The sharpened tip remains at the top part of the wire. The procedure of the STM tip preparation was adapted for the manufacturing of AFM and SECM tips, too [182, 183].

Although electrochemical manipulation with STM tips became popular, as described in Sect. 12.4.2, the STM tip itself cannot be used as an ultramicroelectrode for classical electroanalysis. The reason for the inappropriateness of STM tips as a classical electroanalytical tool is that the side of the tip cannot be covered with the desired precision, and the near-end shape of the tip is not defined well enough. Therefore, it has long remained an unsatisfied need to construct ultramicroelectrodes with a well-defined segment of a single metallic nanoparticle. Various studies showed that suspended nanoparticles can be used as mediators together with microelectrodes with small activity (like carbon). In such systems, a dilute solution of nanoparticles is used with such a small nanoparticle concentration that the collision of the nanoparticles with the electrode is infrequent enough so that the current decay after the sudden "activation" of the electrode as a result of the nanoparticle contact can be measured.

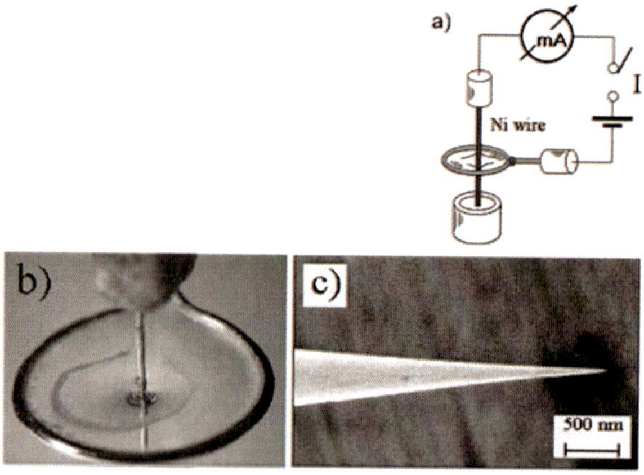

Fig. 12.13 Setup for STM tip preparation by electrochemical etching. **a** Scheme of the cell with the liquid membrane held by PtIr cathode ring and wire to be etched crossing the liquid membrane centrally. **b** Photo of the cell during the operation. **c** SEM picture of the tip prepared. Reprinted from [180], with the permission of AIP Publishing

However, this method leads to unpredictable events, and potential sweeps are clearly not possible to run up to the user's need.

Application of tungsten nanotip for nanoelectrode fabrication is possible if the sharp tip is covered well enough sidewise. This can be carried out by electrophoretic painting [184]. This coating method eliminated any manual step but leaves the very end of the W tip still electroactive for a further deposition of Au particle. The electrode size thus achieved was above 60 nm.

Another important step in the electrochemical nanoelectrode fabrication was when electrodeposition of single metallic nanoparticles onto insulator-coated microelectrodes became possible [185]. The scheme of the preparation method and the resulting nanoelectrode is shown in Fig. 12.14. The Pt microelectrode is coated with a thin enough (~1 nm thick) semiconductor layer through which tunnelling current can still pass. This unit is called the tunnelling ultramicroelectrode (TUME). This electrode is immersed into dilute (0.1-0.01 M) H_2PtCl_6 solution and polarized to a sufficiently negative potential so that platinum nucleation can take place. Due to the small surface of the ultramicroelectrode, one single nucleation event can take place only. The nucleation can be followed by the onset of a negative current of the pA level. The size of the nanoparticle sitting at the TUME tip can be regulated with the cut-off time (or charge) of the polarization after the current onset in the hexachloroplatinate solution. Pt particle diameter between 1 and 40 nm was achieved with this technique. This electrode size made it possible to measure electrochemical rate constants of much larger orders of magnitude than with formerly available techniques.

Nucleation of single nanoparticles on low-activity ultramicroelectrode was demonstrated from very dilute solutions of the reagents by observing the current

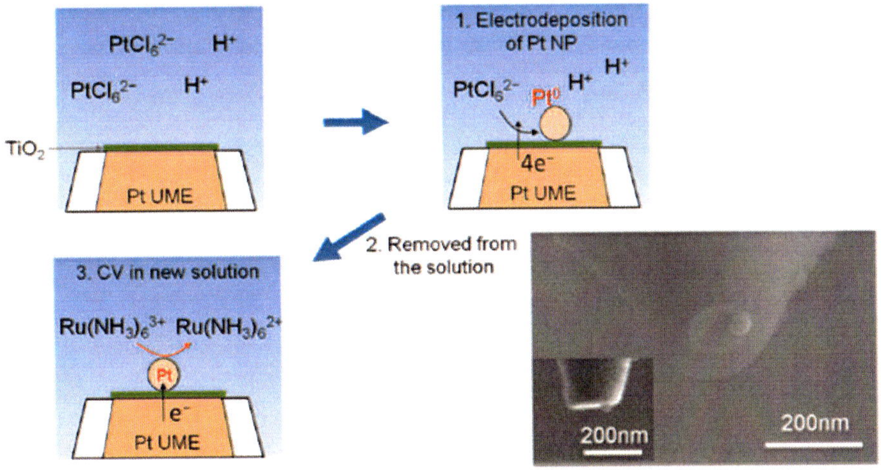

Fig. 12.14 Electrodeposition of a single nanoparticle at a tunnelling microelectrode. The inset at right shows the SEM image of the resulting nanoparticle sitting on the microelectrode disc embedded into the insulator shielding. Reprinted with permission from [185]. Copyright (2016) American Chemical Society

transient following the particle nucleation [182]. When Pt was deposited onto glassy carbon from acidic *femtomolar* Pt(IV) solution, the particle nucleation at <0.1 Hz frequency was accompanied by the onset of the hydrogen evolution, leading to significantly larger currents than the electrochemical nucleation of the nanoparticle itself. After the activation of the electrode due to the appearance of a new particle with high electrocatalytic activity, the current peak decayed due to the shielding effect of the hydrogen bubble covering the active spot of the electrode. A very similar process was observed upon anodic polarization and nucleation of Co-oxide nanoparticles accompanied by oxygen evolution. In both cases, the nanoparticle growth became limited by the side reaction originating from the locally modified electrochemical activity.

References

1. Chazalviel JN, Rosso M, Chassaing E, Fleury V (1996) J Electroanal Chem 407:61–73
2. Huth J, Swinney H, McCormick W, Kuhn A, Argoul F (1995) Phys Rev E 51:3444–3458
3. Mocskos E, Marschall G (2009) Electron Trans Numer Anal 34:90–101
4. Marshall G, Mocskos E, González G, Dengra S, Molina FV, Iemmi C (2006) Electrochim Acta 51:3058–3065
5. González G, Rosso M, Chassaing E, Chazalviel JN (2007) Electrochim Acta 53:141–144
6. Fleury V, Chazalviel JN, Rosso M (1993) Phys Rev E 48:1279–1299
7. Deng Y, Pritzker M (1992) J Electroanal Chem 336:25–34
8. Sun B, Zou XW, Jin ZZ (2004) Electrochim Acta 50:1031–1037
9. Zhong S, Wang M, Yin XB, Zhu JM, Peng RW, Wang Y, Ming NB (2001) J Phys Soc Japan 70:1452–1455
10. Wang M, Zhong S, Yin XB, Zhu JM, Peng RW, Wang Y, Zhang KQ, Ming NB (2001) Phys Rev Lett 86:3827–3830
11. Zhong S, Wang Y, Wang M, Zhang MZ, Yin XB, Peng RW, Ming NB (2003) Phys Rev E 67:061601(1–9)
12. Zhang MZ, Wang M, Zhang Z, Zhu JM, Peng RW, Minga NB (2004) Electrochim Acta 49:2379–2383
13. Zhang M, Zuo G, Zong Z, Cheng H, He Z, Yang C, Zou G (2006) Small 2:727–731
14. Zhang B, Weng YY, Huang XP, WangM, Peng RW, Ming NB, Yang B, Lu N, Chi L (2009) Adv Mater 21:3576–3580
15. Yu G, Hu X, Liu D, Sun D, Li J, Zhang H, Liu H, Wang (2010) J Electroanal Chem 638:225–230
16. Weng YY, Zhang B, Fu SJ, Wang M, Peng RW, Ma GB, Shu DJ, Ming NB (2010) Phys Rev E 81:051607(1–6)
17. Wu Z, Bao YJ, Yu GW, Wang M, Peng RW, Fleury V, Hao XP, Ming NB (2006) J Phys Condens Matter 18:5425–5434
18. Liu T, Wang S, Shi ZL, Ma GB, Wang M, Peng RW, Hao XP, Ming NB (2007) Phys Rev E 75:051606(1–6)
19. Liu C, Yao B, Wang S, Tian H, Zhang M (2009) J Phys Chem C 113:21303–21307
20. Chen F, Li J, Yu F, Zhao D, Wang F, Chen Y, Peng RW, Wang M (2016) Adv Mater 28:7193–7199
21. Cui G, Gao L, Yao B, Wang S, Zhang P, Zhang M (2013) Electrochem Commun 30:42–45
22. Cui G, Xiao C, Zhang P, Zhang M (2016) Phys Chem Chem Phys 18:10918–10923

23. Zong Z, Liu J, Hu T, Li D, Tian H, Liu C, Zhang M, Zou G (2010) Nanotechnol 21:185302(1–5)
24. de Jonge N, Pfaff M, Peckys DB (2014) Practical aspects of transmission electron microscopy in liquid. In: Hawkes PW (ed) Advances in imaging and electron physics, vol 186. Elsevier, Amsterda, pp 1–37
25. Ross FM (ed) (2017) Liquid cell electron microscopy. Cambridge University Press, Cambridge
26. Xu T, Sun L (2016) Superlattices Microstruct 99:24–34
27. Zheng H, Zhu Y (2017) Ultramicroscopy 180:188–196
28. Fahrenkrug E, Alsem DH, Salmon N, Maldonado S (2017) J Electrochem Soc 164:H358–H364
29. Leenheer AJ, Sullivan JP, Shaw MJ, Harris CT (2015) J Microelectomech Syst 24:1061–1068
30. Bozzini B, D'Urzo L, Gianoncelli A, Kaulich B, Prasciolu M, Sgura I, Tondo E, Kiskinova M (2009) J Phys Chem C 113:9783–9787
31. Williamson MJ, Tromp RM, Vereecken PM, Hull R, Ross FM (2003) Nat Mat 2:532–536
32. Radisic A, Vereecken PM, Searson PC, Ross FM (2006) Surf Scie 600:1817–1826
33. Sun M, Liao HG, Niu K, Zheng H (2013) Sci Rep 3:3227(1–5)
34. Grogan JM, Bau HH (2010) J Microelectromech Sys 19:885–894
35. Sacci RL, Dudney NJ, More KL, Parent RL, Arslan I, Browning ND, Unocic RR (2014) Chem Commun 50:2104–2107
36. Chen X, Noh KW, Wen JG, Dillon SJ (2012) Acta Mater 60:192–198
37. Radisic A, Vereecken PM, Hannon JB, Searson PC, Ross FM (2006) Nano Lett 6:238–242
38. Radisic A, Ross FM, Searson PC (2006) J Phys Chem B 110:7862–7868
39. Scharifker BR, Hills GJ (1983) Electrochim Acta 28:879–889
40. Hyde ME, Compton RG (2003) J Electroanal Chem 549:1–12
41. Wang L, Wen J, Sheng H, Miller DJ (2016) Nanoscale 8:17250–17255
42. White ER, Singer SB, Augustyn V, Hubbard VA, Mecklenburg M, Dunn B, Regan BC (2012) ACS Nano 6:6308–6317
43. Schneider NM, Park JH, Grogan JM, Kodambaka S, Steingart DA, Ross FM, Bau HH (2014) Microsc Microanal 20(Suppl 3):1530–1531
44. Schneider NM, Park JH, Grogan JM, Steingart DA, Bau HH, Ross FM (2017) Nature Commun 8:2174(1–10)
45. Wun F, Yao N (2015) Nano Energy 11:196–210
46. Leenheer AJ, Jungjohann KL, Zavadil KR, Sullivan JP, Harris CT (2015) ACS Nano 9:4379–4389
47. Kushima A, So KP, Su C, Bai P, Kuriyama N, Maebashi T, Fujiwara Y, Bazant MZ, Li J (2017) Nano Energy 32:271–279
48. Zeng Z, Liang WI, Liao HG, Xin HL, Chu YH, Zheng H (2014) Nano Lett 14:1745–1750
49. Sacci RL, Black JM, Balke N, Dudney NJ, More KL, Unocic RR (2015) Nano Lett 15:2011–2018
50. Wang CM, Xu W, Liu J, Zhang JG, Saraf LV, Arey BW, Choi D, Yang ZG, Xiao J, Thevuthasan S, Baer DR (2011) Nano Lett 11:1874–1880
51. Zeng Z, Zhang X, Bustillo K, Niu K, Gammer C, Xu J, Zheng H (2015) Nano Lett 15:5214–5220
52. Wu YA, Yin Z, Farmand M, Yu YS, Shapiro DA, Liao HG, Liang WI, Chu YH, Zheng H (2017) Sci Rep 7:42527(1–9)
53. Gupta T, Schneider NM, Park JH, Steingart D, Ross FM (2018) Nanoscale 10:7702–7710
54. Yang J, Andrei CM, Botton GA, Soleymani L (2017) J Phys Chem C 121:7435–7441
55. den Heijer M, Shao I, Radisic A, Reuter MC, Ross FM (2014) APL Mater 2:022101(1–9)
56. Donev EU, Schardein G, Wright JC, Hastings JT (2011) Nanoscale 3:2709–2717
57. Donev EU, Hastings JT (2009) Nano Lett 9:2715–2718
58. Donev EU, Hastings JT (2009) Nanotechnology 20:505302(1–7)

59. Jungjohann KL, Bliznakov S, Sutter PW, Stach EA, Sutter EA (2013) Nano Lett 13:2964–2970
60. Peppler K, Pölleth M, Meiss S, Rohnke M, Janek J (2006) Z Phys Chem 220:1507–1527
61. Divya V, Sangaranarayanan MV (2015) J Nanosci Nanotechnol 15:6863–6882
62. Kaminska I, Niedziolka-Jonsson J, Roguska A, Opallo M (2010) Electrochem Commun 12:1742–1745
63. Li M, Zhu H, Mao X, Xiao W, Wang D (2013) Electrochim Acta 92:108–116
64. Zhu H, Gao L, Li M, Yin H, Wang D (2011) Electrochem Commun 13:1479–1483
65. Gao L, Mao X, Zhu H, Xiao W, Gan F, Wang D (2014) Electrochim Acta 136:97–104
66. Gründer Y, Mosselmans JFW, Schroeder SLM, Dryfe RAW (2013) J Phys Chem C 117:5765–5773
67. Toth PS, Rabiu AK, Dryfe RAW (2015) Electrochem Commun 60:153–157
68. Booth SG, Uehara A, Chang SY, Mosselmans JFW, Schroeder SLM, Dryfe RAW (2015) J Phys Chem C 119:16785–16792
69. Cheng Y, Schiffrin D (1996) J Chem Soc, Faraday Trans 92:3865–3871
70. Johans C, Lahtinen R, Kontturi K, Schiffrin DJ (2000) J Electroanal Chem 488:99–109
71. Gorgy K, Fusalba F, Evans U, Kontturi K, Cunanne VJ (2002) Synth Met 125:365–373
72. Johans C, Clohessy J, Fantini S, Kontturi K, Cunnane VJ (2002) Electrochem Commun 4:227–230
73. Johans C, Kontturi K, Schiffrin DJ (2002) J Electroanal Chem 526:29–35
74. Johans C, Liljeroth P, Kontturi K (2002) Phys Chem Chem Phys 4:1067–1071
75. Platt M, Dryfe RAW, Roberts EPL (2003) Electrochim Acta 48:3037–3046
76. Evans-Kennedy U, Clohessy J, Cunnane VJ (2004) Macromol 37:3630–3634
77. Platt M, Dryfe RAW, Roberts EPL (2004) Electrochim Acta 49:3937–3945
78. Knake R, Fahmi AW, Tofail SAM, Clohessy J, Mihov M, Cunnane VJ (2005) Langmuir 21:1001–1008
79. Platt M, Dryfe RAW (2005) Phys Chem Chem Phys 7:1807–1814
80. Vignali M, Edwards R, Cunnane VJ (2006) J Electroanal Chem 592:37–45
81. Trojánek A, Langmaier J, Samec Z (2007) J Electroanal Chem 599:160–166
82. Platt M, Dryfe RAW (2007) J Electroanal Chem 599:323–332
83. Lepková K, Clohessy J, Cunnane VJ (2007) J Phys Condens Matter 19:375106(1–12)
84. Lepková K, Clohessy J, Cunnane VJ (2008) Electrochim Acta 53:6273–6277
85. Gründer Y, Ho HLT, Mosselmans JFW, Schroeder SLM, Dryfe RAW (2011) Phys Chem Chem Phys 13:15681–15689
86. Izquierdo D, Martinez A, Heras A, Lopez-Palacios J, Ruiz V, Dryfe RAW, Colina A (2012) Anal Chem 84:5723–5730
87. Uehara A, Hashimoto T, Dryfe RAW (2014) Electrochim Acta 118:26–32
88. Aslan E, Patir IH, Ersoz M (2015) Chem Eur J 21:4585–4589
89. Toth PS, Rodgers ANJ, Rabiu AK, Dryfe RAW (2015) Electrochem Commun 50:6–10
90. Wójcik K, Grzeszczuk M (2015) J Solid State Electrochem 19:1293–1300
91. Aslan E, Akin I, Patir IH (2016) ChemCatChem 8:719–723
92. Mirčeski V, Gulaboski R (2006) J Phys Chem B 110:2812–2820
93. Niedziolka J, Opallo M (2008) Electrochem Commun 10:1445–1447
94. Szot K, Jönsson-Niedziolka M, Palys B, Niedziolka-Jönsson J (2011) Electrochem Commun 13:566–569
95. Niedziolka-Jonsson J, Jonsson-Niedziolka M, Nogala W, Palys B (2011) Electrochim Acta 56:3311–3316
96. Tada E, Hiratsuka T, Kaneki H (2007) Electrochemistry 75:731–733
97. Nakanishi S, Nagai T, Fukami K, Sonoda K, Oka N, Ihara D, Nakato Y (2008) Langmuir 24:2564–2568

98. Kaminska I, Jonsson-Niedziolka M, Kaminska A, Pisarek M, Hołyst R, Opallo M, Niedziolka-Jonsson J (2012) J Phys Chem C 116:22476–22485
99. Nakabayashi S, Aogaki R, Karantonis A, Iguchi U, Ushida K, Nawa M (1999) J Electroanal Chem 473:54–58
100. Tada E, Oishi Y, Kaneko H (2005) Electrochem Solid-State Lett 8:C26–C29
101. Budevski E, Staikov G, Lorentz WJ (1996) Electrochemical phase formation and growth. An introduction to the initial stages of metal deposition. VCH, Weinheim–New York–Basel–Cambridge–Tokyo
102. Dryfe RAW, Uehara A, Booth SG (2014) Chem Rec 14:1013–1023
103. Scholz F, Hasse U (2005) Electrochem Commun 7:541–546
104. Koyama A, Fukami K, Yamauchi T, Nishi N, Sakka T, Kitada A, Murase K (2014) ECS Electrochem Lett 3:G5–G7
105. El-Giar EM, Said RA, Bridges GE, Thomson DJ (2000) J Electrochem Soc 147:586–591
106. Jansson A, Thornell G, Johansson S (2000) J Electrochem Soc 147:1810–1817
107. Seol SK, Yi JM, Jin X, Kim CC, Je JH, Tsai WL, Hsu PC, Hwu Y, Chen CH, Chang LW, Margaritondo G (2004) Electrochem Solid State Lett 7:C95–C97
108. Said RA (2004) J Microeletromech Sys 13:822–832
109. Choo JH, Yeo SH, Tan FF (2004) Microsys Technol 10:127–136
110. Lin CS, Lee CY, Yang JH, Huang YS (2005) Electrochem Solid State Lett 8:125–129
111. Chang TK, Lin JC, Yang JH, Yeh PC, Lee DL, Jiang SB (2007) J Micromech Microeng 17:2336–2343
112. Yang JH, Lin JC, Chang TK, Lai GY Jiang BS (2008) J Micromech Microeng 18:055023(1–8)
113. Lin JC, Chang TK, Yang JH, Jeng JH, Lee DL, Jiang SB (2009) J Micromech Microeng 19:015030(1–10)
114. Lin JC, Chang TK, Yang JH, Chen YS, Chuang CL (2010) Electrochim Acta 55:1888–1894
115. Penner RM (1993) Scanning Microsc 3:805–813
116. Simeone FC, Albonetti C, Cavallini M (2009) J Phys Chem C 113:18987–18994
117. Tao NJ, Li CZ, He HX (2000) J Electroanal Chem 492:81–93
118. Kolb DM, Simeone FC (2005) Electrochim Acta 50:2989–2996
119. Li W, Virtanen JA, Penner RM (1992) Appl Phys Lett 60:1181–1183
120. Li W, Virtanen JA, Penner RM (1992) J Phys Chem 96:6529–6532
121. Heidelberg A, Rozenkranz C, Schultze JW, Schäpers T, Staikov G (2005) Surf Sci 597:173–180
122. Ullmann R, Will T, Kolb DM (1993) Chem Phys Lett 209:238–242
123. Schindler W, Hofmann D, Kirschner J (2000) J Appl Phys 87:7007–7009
124. Schindler W, Hofmann D, Kirschner J (2001) J Electrochem Soc 148:C124–C130
125. Schindler W, Hugelmann P, Hugelmann M, Kärtner FX (2002) J Electroanaly Chem 522:49–57
126. Widmer R, Siegenthaler H (2005) Electrochem Commun 7:421–426
127. Li W, Hsiao GS, Harris D, Nyffenegger RM, Virtanen JA, Penner RM (1996) J Phys Chem 100:20103–20113
128. Kolb DM, Ullmann R, Will T (1997) Science 275:1097–1099
129. Maupai S, Stratmann M, Dakkouri AS (2002) Electrochem Solid State Lett 5:C35–C37
130. Engelmann GE, Ziegler JC, Kolb DM (1998) Surf Sci 401:L420–L424
131. Maupai S, Dakkouri AS, Stratmann M, Schmuki P (2003) J Electrochem Soc 150:C111–C114
132. Zhang Y, Maupai S, Schmuki P (2004) Surf Sci 551:L33–L39
133. Maupai S, Dakkouri AS, Schmuki P (2005) Surf Sci 597:20–25
134. Kolb DM, Engelmann GE, Ziegler JC (2000) Solid State Ionics 131:69–78
135. Kolb DM, Ullmann R, Ziegler JC (1998) Electrochim Acta 43:2751–2760
136. Wei YM, Zhou XS, Wang JG, Tang J, Mao BW, Kolb DM (2008) Small 4:1355–1358
137. Schuster R, Kirchner V, Xia XH, Bittner AM,† Ertl G (1998) Phys Rev Lett 80:5599–5602

138. Xia XH, Schuster R, Kirchner V, Ertl G (1999) J Electroanal Chem 461:102–109
139. Chi Q, Zhang J, Friis EP, Andersen JET, Ulstrup J (2000) Surf Sci 463:L641–L648
140. Solomun T, Kautek W (2001) Electrochim Acta 47:679–687
141. Xie ZX, Kolb DM (2000) J Electroanal Chem 481:177–182
142. Xie ZX, Cai XW, Tang J, Chen YA, Mao YA (2000) Chem Phys Lett 322:219–223
143. Jang SY, Marquez M, Sotzing GA (2004) J Am Chem Soc 126:9476–9477
144. deAbril O, Gündel A, Maroun F, Allongue P, Schuster R (2008) Nanotechnol 19:325301(1–7)
145. Ma LP, Yang WJ, Xie SS, Pang SJ (1998) 73:3303–3305
146. Jegadesan S, Sindhu S, Advincula RC, Valiyaveettil S (2006) Langmuir 22:780–786
147. Zamborini FP, Crooks RM (1998) J Am Chem Soc 120:9700–9701
148. Ito S, Iwata F (2011) Jpn J Appl Phys 50:08LB15(1–6)
149. Whitaker JD, Nelson JB, Schwartz DT (2005) J Micromech Microeng 15:1498–1503
150. Staemmler L, Suter T, Böhni H (2002) Electrochem Solid State Lett 5:C61–C63
151. Lei Y, Zhang X, Xu D, Yu M, Yi Z, Li Z, Sun A, Xu G, Cui P, Guo J (2018) J Phys Chem Lett 9:2380–2387
152. Leïchlé T, Nicu L, Descamps E, Corso B, Mailley P, Livache T, Bergaud C (2006) Appl Phys Lett 88:254108(1–3)
153. Descamps E, Leïchlé T, Corso B, Laurent S, Mailley P, Nicu L, Livache T, Bergaud C (2007) Adv Mater 19:1816–1821
154. Suryavanshi AP, Yu MF (2007) Nanotechnol 18:105305(1–4)
155. Morsali S, Daryadel S, Zhou Z, Behroozfar A, Qian D, Minary-Jolandan M (2017) J Appl Phys 121:024903(1–8)
156. Morsali S, Daryadel S, Zhou Z, Behroozfar A, Baniasadi M, Moreno S, Qian D, Minary-Jolandan M (2017) J Appl Phys 121:214305(1–10)
157. Hu J, Yum K, Tajik A, Minary-Jolandan M, Bang J, Yu MF (2011) Appl Phys Lett 99:053113(1–3)
158. Suryavanshi AP, Hu J, Yu MF (2008) Adv Mater 20:793–796
159. Iwata F, Metoki J (2017) Jpn J Appl Phys 56:126701(1–7)
160. Guo J, Tokimoto T, Othman R, Unwin PR (2003) Electrochem Commun 5:1005–1010
161. Li F, Edwards M, Guo J, Unwin PR (2009) J Phys Chem C 113:3553–3565
162. Yatziv Y, Turyan I, Mandler D (2002) J Am Chem Soc 124:5618–5619
163. Zhu X, Qiao Y, Zhang X, Zhang S, Yin X, Gu J, Chen Y, Zhu Z, Li M, Shao Y (2014) Anal Chem 86:7001–7008
164. Hao R, Zhang B (2016) Anal Chem 88:614–620
165. Bourque H, Leblanc RM (1995) Rev Sci Instrum 66:2695–2697
166. Oliva AI, Romero GA, Peña JL, Anguiano E, Aguilar M (1996) Rev Sci Instrum 67:1917–1921
167. Anwei L, Xiaotang H, Wenhui L, Guijun J (1997) Rev Sci Instrum 68:3811–3813
168. Kazinczi R, Szőcs E, Kálmán E, Nagy P (1998) Appl Phys A 66:S535–S538
169. Kerfriden S, Nahlé AH, Campbell SA, Walsh FC, Smiths JR (1998) Electrochim Acta 43:1939–1944
170. Ekvall I, Wahlström E, Claesson D, Olin H, Olsson E (1999) Meas Sci Technol 10:11–18
171. Guise OL, Ahner JW, Jung MC, Goughnour PC, Yates JT (2002) NanoLett 2:191–193
172. Khan Y, Al-Falih H, Zhang Y, Ng TK, Ooi BS (2012) Rev Sci Instrum 83: 063708(1–8)
173. Silva EL, Silva RF, Zheludkevich M, Oliveira FJ (2014) Rev Sci Instrum 85:095109(1–7)
174. Watanabe MO, Kinno T (1994) Appl Surf Sci 76–77:353–358
175. Dickmann K, Demming F, Jersch J (1996) Rev Sci Instrum 67:845–846
176. Zhang C, Gao B, Chen LG, Meng QS, Yang H, Zhang R, Tao X, Gao HY, Liao Y, Dong ZC (2011) Rev Sci Instrum 82:083101(1–4)
177. Gingery D, Bühlmann P (2007) Rev Sci Instrum 78:113703(1–4)
178. Fainchtein R, Zarriello PR (1992) Ultramicroscopy 42–44:1533–1537
179. Morán Meza JA, Polesel-Maris J, Lubin C, Thoyer F, Makky A, Ouerghi A, Cousty J (2015) Current Appl Phys 15:1015–1021
180. Cavallini M, Biscarini F (2000) Rev Sci Instrum 71:4457–4460

181. Ceballos SF, Mariotto G, Murphy S, Shvets IV (2003) Surf Sci 523:131–140
182. Kim J, Dick JE, Bard AJ (2016) Acc Chem Res 49:2587–2595
183. Slevin CJ, Gray NJ, Macpherson JV, Webb MA, Unwin PR (1999) Electrochem Commun 1:282–288
184. Wusimanjiang Y, Ma Y, Lee M, Pan S (2018) Electrochim Acta 269:291–298
185. Kim J, Bard AJ (2016) J Am Chem Soc 138:975–979

Part IV
High-Voltage Methods of Nanostructure Preparation

Chapter 13
Preparation of Nanoporous Oxides from Metals by Electrochemical Anodization

13.1 Electrochemistry and High-Voltage Processes

Several processes have been discussed in the earlier chapters in which the total cell voltage was unusually high (several tens of volts), which is clearly beyond the potential difference scale that any cell with reversible electrodes in equilibrium may produce. In spite of the relatively high cell voltage in the examples discussed earlier, the electrode processes themselves could be elucidated with the classical terms of electrochemical kinetics, i.e., with the exponential dependence of the reaction rate constants on the electrode potential. In many cases, the large cell voltage was partly due to the high cell resistance, mostly due to either the application of relatively dilute solutions or small cross-sectional surface area of the cell.

The processes to be listed in Chaps. 13 and 14 have the common feature that the elemental steps of the processes can hardly be described in terms of classical electrochemical kinetics; i.e., the equations given in Chap. 2.8 cannot be taken valid. In fact, the description of the impact chain of elemental processes is totally missing in some cases. Hence, practically no quantitative analysis of these processes can be given but one has to rely on experimental findings to correlate the technical parameters of the sample preparation to the properties of the product.

Should these conditions seem to fall quite far from classical electrochemistry, they fulfil the definition that some change in chemical state takes place as a result of current applied. It is to be emphasized that in all processes presented in the present and the next chapters, the nature of the current-induced chemical transformation is highly specific, and no similar transformation can be achieved with any ancillary effect of the high current passed through the system; for instance, the Joule heat of the current alone would not lead to the same effect.

Two major process types are outlined in Part IV. The present chapter offers a summary of metal anodization processes that lead to nanoporous structures. Chap. 14 will deal with electrical arc discharge processes resulting in nanostructure formation, paying particular attention to various carbon nanostructures.

L. Péter, *Electrochemical Methods of Nanostructure Preparation*,
Monographs in Electrochemistry, https://doi.org/10.1007/978-3-030-69117-2_13

13.2 Historical Background: From the Aluminium Industry to the Nanotechnology Laboratories

Anodization of aluminium is a well-know process applied from the mid of the twentieth century, primarily for the improvement of the corrosion resistance of the workpieces and secondarily for colouring them. Already in a paper written as early as in 1953, the description of the suitable solution compositions and structural features of the anodized parts seem to be much in accord with the up-to-date knowledge of the field [1]. Until about 1990, several works were published about various aspects of the field, slowly enriching the available information of aluminium anodization in line with the technical development and the available technical instrumentation. A clear distinction could be given between the conditions of anodic oxidation [1–8] and electropolishing [9]. The following factors of the process could be reliably outlined:

- Under appropriate electrochemical anodization conditions, a porous oxide layer can be grown on aluminium samples.
- The three major bath types (sulphuric acid, oxalic acid and phosphoric acid) were identified.
- The pore axes are primarily perpendicular to the sample surface. The pores being formed behave like vortices repelling each other, which leads to a hexagonal arrangement of the pores.
- The pore diameter is a linear function of the voltage applied for anodic oxidation (for electropolishing conditions, this relationship does not hold). The proportionality factor of about 2 nm V^{-1} was also properly assessed.
- The bottom of the pores is separated from the metal with a barrier layer whose thickness is about the half of the thickness of the walls between two pores.

The above-listed features of porous anodic alumina (PAA, often also mentioned as anodic aluminium oxide, AAO) are presented in Fig. 13.1. Even though the imaging of the nanoporous structures obtained was less perfect than that available nowadays, essentially all basic information were know that later established the development of the field in the mid-1990s. The application of starting specimens of relatively low purity grade well explains the imperfections in the porous structures themselves, while the limitations of the imaging conditions come from the old electron microscopic techniques based on studies of replica structures.

Anodization of technical-grade aluminium leads to a partly disordered pore structure with pore branching, tilting and pore growth ending. The formation of these defects is caused by impurities, inclusions and crystal disorders. Therefore, PAA on technical-grade aluminium was not considered as a material for either technical applications or for further scientific studies.

The spin-up of the scientific interest in nanoporous materials started in the 1990s. At that time, ion-track-etched membranes were already used as templates for both electrodeposition [10] and non-electrochemical filling methods [11]. The opening of the new era of the PAA research is dated to 1995 when Masuda and Fukuda published a method for producing well-ordered, nearly defect-free hexagonal nanoporous alumina [12]. The advantage of their method was that it made possible to detach the nanoporous membrane from the aluminium raw material and hence using it as a

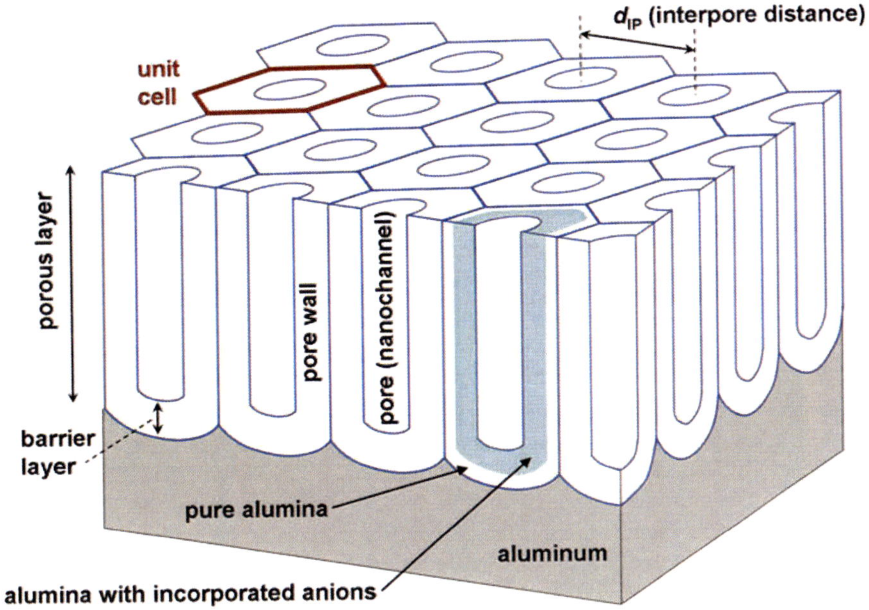

Fig. 13.1 Ideal structure of the porous alumina layer on the aluminium substrate

self-supporting specimen. This opened the way for using it in various ways (filters, luminescent coatings with regular pore spacing, host plate for cell cultures, biosensors and, importantly for electrochemists, as template for deposition as presented in Chap. 11). Preparation and properties of PAA prepared with anodization were reviewed time to time [13–16].

13.3 Traditional Laboratory Technique of PAA Preparation

13.3.1 Aluminium Samples to Be Anodized

A prerequisite of the regularity of the pore system produced on aluminium by anodization is the good quality of the starting material. For the purity of aluminium, this means that 99.97% is the minimum quality requirement that can be hoped to yield a regular pore system upon anodization, but Al with 99.999% purity level also appears in several works. The thumb rule is "the cleaner, the better". This has to be kept in mind in each case, should the starting material purchased as sheet/foil or produced by evaporation onto a suitable substrate.

The sample purity is the first but by far not the only criterion for regular pore formation. Other factors of importance are the residual stress and dislocation density in the specimens to be anodized [17]. Since unrelaxed stress can accumulate in the sample during either the rolling of the aluminium foils or the in-house cutting process prior to the anodization, a relaxation heat treatment is recommended (500 °C was applied in several works). Here, a bifurcation occurs between bulk Al samples and thin layers. Bulk samples contain relatively large crystals with diameter much above the micrometre range, and the relaxation heat treatment affects the long-range pore regularity but not the pore size and the porosity. However, the grain size of as-received thin sputtered layers is comparable to the layer thickness, and one ultrasmall grain may accommodate not more than one single pore. This leads to an apparent layer thickness dependence of the pore diameter, although the real determining factor is the grain size [18]. Annealing thin Al layers leads to grain coarsening, which also leads to a change in pore size and porosity upon anodization.

Concerning the surface quality, the smaller the surface roughness of the aluminium, the more regular pore formation can be expected. Mechanical polishing of the surface may be necessary if the high-purity aluminium sheet purchased is too rough, but electropolishing is practically an indispensable preliminary step prior to the anodization. Besides diminishing the surface roughness, electropolishing removes the surface layer that may be deformed during the preliminary mechanical steps. The most common electropolishing solution contains ethanol and perchloric acid (with 4…8:1 mixing ratio), and the electropolishing time at a fixed current density is about 5 min.

13.3.2 Anodizing Cells: Cell Arrangement, Electrodes, Thermal Management

Laboratory cells used for aluminium anodization are exclusively two-electrode cells. The counter electrode is most often platinum (or a Pt–Ir alloy). Occasionally, inert electrodes made of other materials such as carbon or lead are also applied, even though the overvoltage of the hydrogen evolution on these electrodes can be relatively large. In the big majority of the papers, the anodization is carried out at a preset cell voltage, and this value is reported as the key electrical parameter of the process. Working electrode potential is neither measured nor reported. Although the total cell voltage includes both the ohmic drop across the cell and the cathode overvoltage, the latter terms do not account for more than ± 1 V difference in the overall cell voltage that is in the 20–300 V range, depending on the electrolyte solution. The major voltage drop is across the thin barrier layer of the pore bottom with a local electric field of about 2×10^7 V cm^{-1}; hence, the electric field elsewhere in the cell is small and the voltage drops can be practically neglected.

In the cells for PAA preparation, the aluminium anode is fixed to the bottom of the cell and the cathode is in a facing position. The solution is chilled and stirred in order to lead away the heat produced in the reaction. Since local overheating often leads to the burning of the PAA coating, a backside cooling of the Al anode is customary. The additional heat sink can be a cooled metal block against which the anode is pressed, but the inclusion of a Peltier element between the cooling media and the anode improves the thermostation of the reaction zone significantly. The desired temperature for the preparation of highly ordered PAA samples is between −5 and +15 °C. The interpore distance tends to increase with temperature.

13.3.3 Common Electrolyte Solutions

Three common electrolyte solutions became widespread in the preparation of PAA membranes by mild anodization: aqueous sulphuric acid, oxalic acid and phosphoric acid solutions (the acids are listed in the incremental order of the anodization voltages corresponding to the growth of PAA with each solution) [19]. The typical solution concentrations and their working ranges are listed in Table 13.1.

13.3.4 Voltage and Current Density Ranges Corresponding to the Stages of the Anodization Process

When voltage-controlled anodization of aluminium is carried out, different current density transients can be obtained, depending on the process(es) taking place. The behaviour of the samples in various cases is shown in Fig. 13.2. If the voltage is too small, only a barrier-type oxide can be produced. In this case, the initial current transient is followed by a decay (which is usually exponential). If the voltage chosen is suitable for the production of PAA, periods corresponding to barrier-type oxide formation, pore initiation and pore growth can be seen (Curve 2 in Fig. 13.2). It has been show with *in situ* grazing incidence small-angle X-ray scattering (GISAXS) study that the pore formation starts in Stage III of the chronoamperometric curve (see Fig. 13.2) Finally, if the voltage applied is too high for a stable PAA growth, the

Table 13.1 Major bath types and their working range that yield ordered PAA structures with mild anodization

Acid applied	Concentration	U range V	d_{IP} nm	d_P nm	References
Sulphuric acid	1.7–20 wt%	15–27	30–70	12–30	[20–24]
Oxalic acid	0.3 mol dm^{-3}	20–60	75–160	18–45	[12, 20, 25, 26]
Phosphoric acid	0.1–3.0 mol dm^{-3}	100–195	440–500	140–180	[19, 27–31]

The lower/upper values of the interpore distance (d_{IP}) belong to the corresponding limit of the pore diameter (d_P) range given. The reference list includes works in which the optimization of the PAA preparation was dealt with

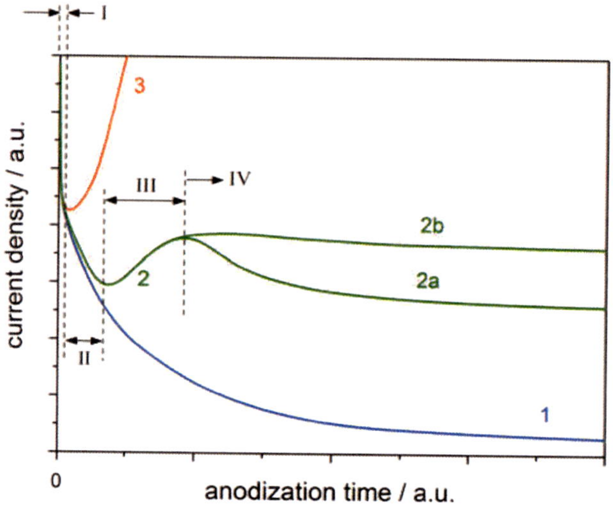

Fig. 13.2 Schematic presentation of the current transient characteristic for the barrier layer formation with no porous oxide growth (Curve 1), for stable PAA growth in mild anodization (Curve 2) and burning at too high voltages (Curve 3). For Curve 2, "a" and "b" indicate the various extent of current decay only that may vary from one system to another. Stages in Curve 2: I initial transient, II barrier oxide formation, III start of the porous oxide growth, IV stable porous oxide growth

samples is "burnt", which means that the stress produced between the oxide layer and the metal leads to a periodic delamination of the oxide particles. This process is accompanied with a vigorous oxygen evolution, and the current density may increase to a few hundred mA cm^{-2} (i.e., to a value about an order of magnitude higher than that obtained during the stable PAA growth that takes place with a current density of at most a few tens of mA cm^{-2}).

During anodization, a volume expansion takes place since the packing density of the aluminium atoms in Al_2O_3 is just half of that in Al metal. The layer thickness ratio of the resulting PAA layer and that of the material consumed is influenced by the packing density ratio and the current efficiency of the anodization. The current efficiency tends to increase with the anodization voltage and ranges to about 45–65% in the PAA-forming regime in both sulphuric and oxalic acids [24], and 65–95% in phosphoric acid [28]. The current efficiency around 66% was confirmed in another study [32] that excluded the chemical dissolution of the alumina formed but suggested the field-driven aluminium oxidation as Al^{3+}(aq).

A self-organized PAA growth takes place at moderate expansion [19, 24]. Very large/small expansion rates at too high/low voltages decrease the tendency for the formation of hexagonal pore arrays. In contrast, a 100% current efficiency was found for porous oxides with dense structure and no regular pore systems, while the volume expansion factor was a linear function of the anodization voltage, indicating the variation of porosity [33]. The current efficiency and the volume expansion factor are strongly interrelated and depend on the acid used for anodization as well as on the anodization current density range [34].

13.3.5 Porosity and Degree of Ordering

With the anodization process parameters shown in Table 13.1, the resulting samples exhibit a nearly uniform 10% overall porosity. This means that the samples prepared under different conditions are similar, as if they were the "magnified" versions of each other. The consequence of this similarity is that the pore diameter over pore axis distance ratio is a uniform parameter of the PAA samples obtained with the conventional anodization processes. The pore radius (r_P), the interpore distance (d_{IP}, the distance between neighbouring pore axes) and the porosity (P) are interrelated and can be described with the following formula [35]:

$$P = \frac{2\pi}{\sqrt{3}}\left(\frac{r_P}{d_{IP}}\right)^2 \tag{13.1}$$

Concerning the interpore distance, the anodization voltage is the key parameter. Taking into account the voltage ranges corresponding to each bath type where regular pore systems can be obtained, a linear relationship is obtained: $d_{IP} = kU$ where the proportionality factor is $k \approx 2.5$–2.8 and 1.9 nm V^{-1} for the so-called mild and hard anodization regimes, respectively.

The pore initiation is a random process in the early phase of the anodization process which results in surface nanocracks of a higher density than the pore density during a stable anodization. This leads to the termination of the pore growth in some cracks, while others keep growing. Hence, the initial pore system in the vicinity of the sample surface is by far not regular. As the pores penetrate into the metal, a regularization process takes place. The ordered domains of pores are surrounded with disordered boundary regions, similar to the crystal boundary regions in polycrystalline materials. Such systems were hence termed as "polycrystalline nanopore arrays" [36].

However, the spontaneous ordering of the pores during prolonged anodization yields the opportunity to overcome the initial randomness of the pore system. If the initially produced pore system is removed by chemical etching, the repeated anodization will start at the seeds left behind after the removal of the first anodization layer. For the oxide layer removal, a mixture of dilute phosphoric and chromic acids is used at slightly elevated temperature (40–50 °C), which dissolves the few micrometre thick porous oxide layer within a few hours but leaves the pore seeds intact. Ellipsometric measurements indicated that the dissolution of the porous layer in aggressive filling solutions takes place evenly along the pores [37]. The longer time elapsed during the first anodization process, the higher degree of regularity the second anodization result in. For the conventional anodization processes, the basic rule was that the first and forthcoming anodization processes are to be carried out under the same conditions (solution composition, temperature and voltage) so that the pore seeds do not need to rearrange and the regularity achieved during the first anodization can be retained.

The pore structure can be observed with various techniques and imaging directions, as presented in Fig. 13.3. In all cases, both the preparation technique and the imaging method may have an impact on the observation (i.e., sample thinning and polishing procedure, AFM tip size and scanning mode, etc.).

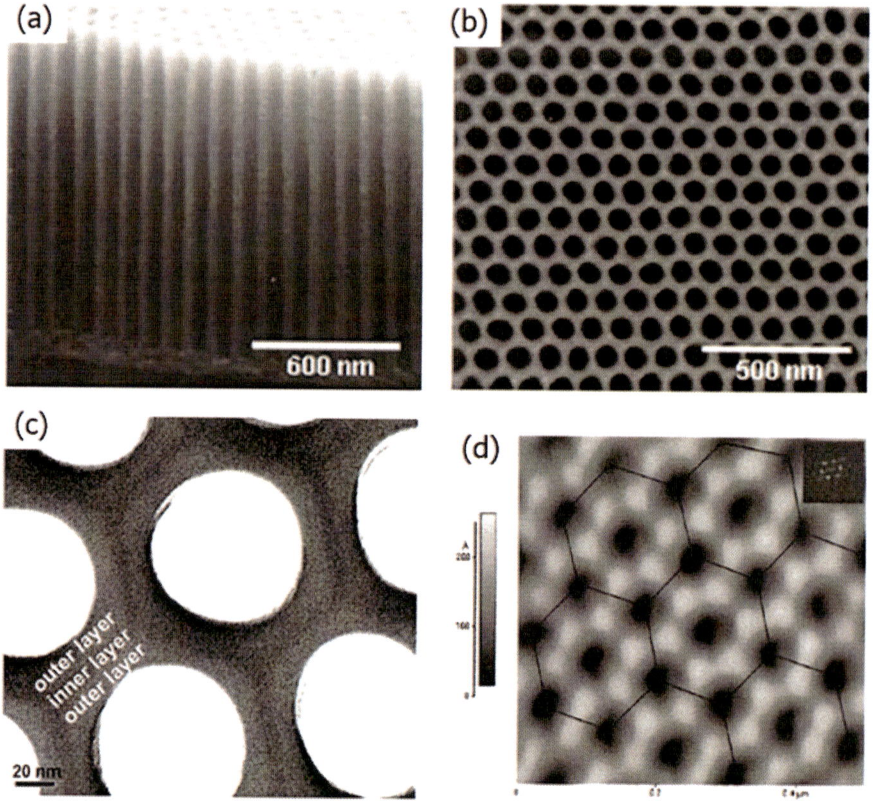

Fig. 13.3 Various images obtained for PAA prepared with oxalic acid anodization and treated with phosphoric acid for pore widening. **a** Cross-sectional SEM image of the pore structure; **b** Top view SEM image of the PAA sample; **c** Scanning transmission electron micrograph of a thinned sample (top view), indicating the zones in the pore walls; **d** AFM image of the bottom part of an anodized sample after the removal of the barrier layer with the indication of the hexagonal arrangement. Reprinted from [38] (**a** through **c**) and [39] (**d**). Copyright (2006) and (2001), respectively, with permission from Elsevier

13.3.6 Composition and Structure of PAA

The standard anodization processes results in nearly pure, slightly hydrated and amorphous aluminium oxide pore walls. Recrystallization of the porous layer can be achieved with a heat treatment. The porous structure is retained at least up to 1200 °C. The annealing process is coupled with both a pore density reduction and pore widening, which is in accord with the shrinkage of the material of the pore walls [40]. Interestingly, the as-anodized PAA membranes transform to γ-alumina first during the heat treatment at about 900 °C. Nanocrystalline regions of α-alumina also appear at this temperature, although it is not the stable form of bulk alumina. PAA samples annealed at 1200 °C consist of the α-alumina phase only [40].

A moderate degree of anion incorporation from the anodization bath takes place; therefore, the signal of the central atom of the anion (i.e., P when phosphoric acid is used) can usually be identified in the EDS spectrum of the samples. The anion incorporation level changes in the order of sulphuric acid (~ 11 wt% $SO_4{}^{2-}$) > phosphoric acid (~7.6 wt% $PO_4{}^{3-}$) > oxalic acid (~2.4 wt% $(COO)_2^{2-}$) > chromic acid (~0.1 wt% $CrO_4{}^{2-}$) [41]. More recent data show that the anion incorporation level strongly changes with the acid concentration and the anodization voltage [42]: for PAA membranes prepared with sulphuric acid, the S/Al atomic ratio varied in the range of 0.062–0.13. Hence, the PAA structures are by far not clean alumina. The anion incorporation is not uniform but regioselective [43, 44], as it is shown in Fig. 13.1 (cross-sectional scheme) and Fig. 13.3c (top-view STEM image). The inner layer is pure alumina, while the outer layer (i.e., that contacting the solution) contains incorporated anions. The inner layer is relatively thin and ranges to some 10–30% of the total pore wall only, and the ion mobility is the lowest here; hence, this is the region where the field strength is the largest. The inner and outer layers yield a pronounced brightness difference in top-view TEM images, which can be seen on SEM images on phosphoric acid anodized samples with large enough wall thickness because of the limited resolution. On samples thinned similar to that shown in Fig. 13.3c, an EDS line scan can yielded a quantitative analysis on local composition and revealed the zone thickness ratio [45].

The presence of anions in the outer layer of the pore walls may suggest that the pore wall itself is also porous. An ellipsometric study revealed a 33% porosity of the walls [46], which did not change during the pore widening process. Although such a high porosity is not in accord with the current efficiency of the anodization process, it indicates that the pore walls are not completely dense. The results on the wall porosity are yet to be confronted with the wall shrinkage observed during annealing [40].

13.3.7 Removal of the Remaining Metal and the Barrier Layer

For using the PAA samples as self-supporting open specimens, the metal layer has to be dissolved from the back side. Nowadays, the customary method is to apply a dilute hydrochloric acid solution with copper(II) chloride (sometimes an acidic $SnCl_4$ solution), while in older works, the more toxic $HgCl_2$ was often used for dissolving Al [36]. The process is self-limiting and does not lead to pore opening. The removal of the metal layer can be visually observed (the sample becomes slightly opaque and the bubble formation stops). Before the pore opening process, AFM images recorded on the back side of the barrier layers show the height modulation [36], in accord with the scheme of Fig. 13.1.

The removal of the barrier layer is said to be the most sensitive step of the sample preparation, and it is seldom carried out in a way so that the process can be detected (or quantified) with an instrumental method. The solution for the removal of the barrier layer contains phosphoric acid (~5 wt%), and the exposure of the back side of the sample to this solution is maintained for a fixed time that is empirically adjusted to the thickness of the barrier layer. The application of an excessively long time for the removal of the barrier layer leads to pore widening. An electrochemical control of the pore opening process is possible if the membrane with the barrier layer is fixed in a permeation cell and the half cell at barrier layer side is filled up with the etching solution [47]. In this case, the pore opening process can be monitored with the conductivity change between the two half cells.

13.3.8 Theoretical Approaches, Modelling of PAA Formation

In the early years of the research on PAA synthesis, the qualitative picture on the process was that the mechanical stress associated with the expansion of the aluminium during oxide formation is the cause of repulsive forces between neighbouring pores being grown. This force leads to a self-organization of the pores, leading to hexagonal pore arrays [24].

Since the first attempt of the qualitative understanding of the pore formation and the occurrence of regularity during anodization, various theories arose that made it possible to give quantitative description of the phenomena and to perform numerical simulations. However, the theories published so far are very different and partly contradictory; hence, we cannot speak about a uniform theory of PAA formation. Since the present work is meant to be a methodology guideline for nanostructure synthesis, the major elements of the works related to the theoretical analysis is listed below, whereas interested readers are referred to the some relevant works and the literature cited therein [42, 48–64]. The modelling attempts take into account the stress occurring in the barrier layer, thermodynamic factors, current distribution, the electric field effects, oxidation kinetics, ion migration and diffusion, varying dissociation equilibrium of the second hydrogen atoms of divalent acids and oxygen evolution. It is usually shown by using linear stability analysis that short-wave instability occurs in certain parameter ranges, which then can lead to the formation of hexagonally ordered pores.

13.4 Patterning Methods in the PAA Preparation Process

The patterning methods involving anodized aluminium have two major goals, either initiating a regular large-scale pattern formation on aluminium or to transfer the pattern formed on anodized aluminium to another material. Several methods have been developed for both goals, and these methods can be combined in some cases. The basic concepts of the pattern transfer methods are summarized in Fig. 13.4.

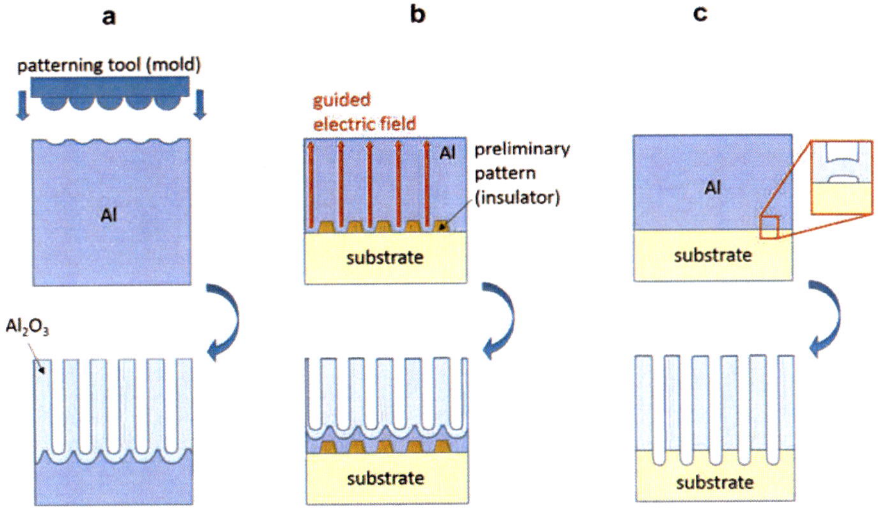

Fig. 13.4 Pattern formation in PAA preparation. Top parts represent the initial state, and the bottom schemes correspond to the result after anodization. **a** Patterning the aluminium work piece to be anodized by using a mould in order to initiate the active spots where the pore formation is to begin. **b** Anodization of aluminium sputtered onto conductive substrate pre-patterned with non-conducting bumps for guiding the electric field. **c** Transfer of the anodization pattern of an aluminium thin layer onto a conductive substrate. The inset shows the cavity formation under the pore if the substrate is less reactive than Al. (Transfer modes can be combined in various ways like A + C or B + C.)

13.4.1 Patterning Methods for Achieving Regular Pore Formation on Aluminium

Since the first successful production of highly ordered PAA preparation in 1995, the PAA template has been considered as suitable host material for individual magnetic entities (nanodots or nanowires), which might establish a new technology for the production of magnetic recording media. However, the polycrystalline nature of the pore formation remained a substantial drawback, even though the multiple-step anodization helped some extent to improve the pore regularity. This problem can be solved by applying a pre-patterning method prior to the anodization. First, it was carried out by applying a SiC mould patterned by electron beam lithography, hence achieving a fully regular area of 3 mm × 3 mm with a hexagonal pattern [65]. The mechanical impact of the mould creates an initiation spot with slightly thinned oxide layer at the desired position where the pore formation starts during the anodization (Fig. 13.4a). As a variation of this indentation-initiated pore formation process, a layer of self-organized array of silica beads on glass also proved to be suitable as pressing mould [66]. Instead of applying a mechanical tool for pre-patterning, direct laser writing combined with chemical etching led to similar high pore ordering levels [67]. These processes all exhibit the advantage that a fixed interpore distance can be achieved which may also differ to a limited extent from that obtained with spontaneous pore formation.

Some attempts have been reported on the trials of overwriting the natural hexagonal arrangement of the pore arrays by applying Nanoimprint lithography using a mould with square-shaped unit cells [68, 69]. Although the pore initiation in accord with the square-shaped pattern was successful, the crossing points of the diagonals always behaved as defect sites even if the natural interpore distance of the anodization matched the period of the imprinted pattern. Since the diagonal distance between the pores in the square pattern is larger than the natural oxide layer thickness, either unoxidized aluminium remained at the defect sites or secondary pore formation took place. Similar structure breaks and non-periodical porous layers were obtained with a strong mismatch of the imprinted pattern and the spontaneous periodicity of the PAA formation [27]. A duplicated nanoindentation patterning with rotated mould positions leads to Moiré pattern in the PAA samples [70].

The above-mentioned methods are common in that the aluminium to be anodized was patterned from "upward", e.g., from the electrolyte side. The sample patterning from "downward" is also possible if a hexagonal array of insulating bumps is deposited onto a conductive substrate which is then covered with a sputtered aluminium layer (see Fig. 13.4b) [71]. The electrically insulating nanobumps serve as field guides, and the spots above the non-covered area will behave as active pore initiation sites during the anodization.

13.4.2 *Anodized Porous Aluminium as a Pattern Transfer Tool*

The more or less regular structure of the PAA membranes makes them suitable for applications as non-lithographic templates [72–75]. They can be used in non-electrochemical processes such as chemical or reactive ion etching and electron beam evaporation. The PAA template can be formed either *in situ* at the surface of the specimen to be patterned or produced *ex situ* and then applied as a self-supporting mask. Below, the attention will be paid, however, to the processes in which the masking functionality of the PAA structure is coupled with the anodization of the underlying material on which the PAA was produced. In such cases, a direct contact between Al and the substrate must prevail, which can be achieved with the coating of the substrate with Al by a physical deposition method. The composition of the interlayer between the substrate and Al may substantially impact the process [76–78]. The advantage of the direct anodization of the substrate through the PAA structure is that the periodicity of the nanostructure being formed underneath is fully determined by the PAA periodicity, totally independently of whether the substrate material is capable of periodic structure formation or not.

When Al is sputtered (or evaporated) onto the surface of a conducting material that is less reactive under the anodization conditions then aluminium, the termination of the Al anodization results in a different barrier layer structure than that obtained for bulk Al. Since the Al layer is fully consumed, the last stage of the anodization proceeds on the residue of Al just under the pore walls when the barrier layer under the pores already reached the substrate material. This situation leads to a different stress field than during steady-state anodization, and a nanocavity is formed just under the pore-covering barrier layers (see the inset in Fig. 13.4c).

A peculiar characteristics of the PAA-based electrochemical pattern transfer is that the resulting pattern caused by the anodization of the substrate strongly depends on the substrate material. When the substrate is *etched* without the formation of a solid reaction product, cavities may form (concave pattern, as it was shown in Fig. 13.4c). For a pattern formation, the modulation of the anodization voltage upon achieving the substrate may be necessary. Such nanocavity structure can be produced on Si when the substrate is etched [79, 80]. However, when the anodization of the substrate leads to the formation of a solid (mostly oxide), the reaction takes place with volume expansion. Hence, nanodots, nanopillars and even nanowires can grow directly into the pores of the alumina (convex pattern). This happens during the oxidation of various substrates: in TiO_2 formation from either TiN [81] or Ti [82–85], Ta_2O_5 formation from Ta [86, 87], Nb_2O_5 formation from Nb [86, 88], ZrO_2 formation from Zr [89] and SiO_2 formation from Si (if etching does not take place).

13.5 Pore Shaping with Modulated Anodization Conditions

All processes leading to non-cylindrical pores are based on non-steady-state anodization. The modulated nature of the anodization process can be spontaneous or purposeful. In most cases, it involves the change in the anodization voltage, but the variation in solution composition has also been reported. Below a logical classification of the pore shape engineering methods is presented.

13.5.1 Pore Branching

For a particular set of experimental conditions, the pore diameter is an unambiguous function of the anodization voltage during steady-state anodization. In all electrolyte solutions, the pore diameter decreases with anodization voltage. This voltage dependence is exploited when the anodization voltage is gradually decreased during the anodization process. When the anodization voltage is lowered, the oxide formation rate decreases as a result of reduced oxidation rate of Al, while the solution conditions at the pore bottom are unchanged. Since the dissolution rate of the barrier layer is determined by the solution environment, the barrier layer thickness is getting lowered until new and narrower pores can nucleate instead of the earlier grown ones. This leads to pore branching. The traditional way of the voltage decrease is exponential (or stepwise with a decay of nearly exponential character) in which at the bottom of each nanopores two smaller pores are nucleated. Since the voltage range from which porous layer with straight nanochannels can be grown in a particular solution is limited, the pore formation is terminated at a threshold voltage. Due to this limitation, tree-like nanopores can be formed with at most four branching levels, which means that one nanopore can be split up into $2^4 = 16$ smaller ones [90–92]. The hierarchy in the pore branching was clear in all works cited, and the planes of pore density doubling could be easily identified.

Although it would sound logical that a suitable voltage increase programme applied for the anodization process might lead to pore merging, no example for such a process has been found. The reverse pore branching structure with merging pores has been implemented by using a four-step anodization process by using different anodizing solutions in each step. The anodization voltage was increased from one step to another, and hence, a four-level structure could be produced [93]. Although the pore merging planes were always well-defined, the number of merging pores shows some scattering laterally.

The above examples all refer to anodization on flat surfaces. However, pore branching or merging always happens due to geometric constraints when the anodization process is carried out with specimen of curved surfaces [94, 95].

13.5.2 Monotonous Pore Diameter Variation

Since the barrier oxide layer cannot be chosen arbitrarily for a particular interpore distance, one cannot grow funnel-like nanopores by simply decreasing the anodization voltage. Funnel-shaped nanopores can be produced with a periodic manufacturing technique involving two consecutive steps: pore growth with anodization and pore widening with chemical etching [96, 97]. The pore growth takes place with identical voltage in each cycle. Hence, the interpore distance and the barrier layer thickness correspond to the ideal self-ordering conditions in every anodization step and no pore branching takes place. The chemical etching steps between the anodization periods is carried out with another suitable solution and impacts the section of the pores already produced; this is why the pore diameter will monotonously increase from the barrier layer to the top of the sample. With a careful design of the lengths of the anodization and chemical etching sections, nearly arbitrary pore shape with monotonous pore diameter variation can be designed.

13.5.3 Periodic Pore Diameter Modulation

The prerequisite of the periodic pore diameter modulation carried out solely by the variation of the anodization voltage is that no pore branching/merging takes place at the "unfavourable" anodization conditions, but the system can relax the stress occurring due to the voltage–interpore distance mismatch. The above criterion can be fulfilled by various means. First, if the cell voltage remains in the anodization regime of pore formation but changes to a limited extent only [98, 99]. When the voltage remains in the same regime, rectangular, sawtooth-like and even more complicated voltage modulations wave forms are possible. Another possibility is that the voltage applied is changed stepwise between the mild and hard anodization regimes (pulse anodization) [100–102]. Obviously, the continuous voltage change is not feasible here since the two voltage domains are separated by a burning region. In such structures, the sections of the normal 10% porosity (mild anodization) alternate with a section of different porosity. The porosity of the hard anodized section may be either higher or lower if the solution contains oxalic or sulphuric acid, respectively.

Hard anodization yields other opportunities for the pore diameter modulation. Since this anodization mode exhibits a regime with spontaneous current density oscillation, this can be used for modulating the pore diameter [103]. Another means is the temperature modulation by periodically chilling the back side of the sample being anodized [104].

Finally, we mention that laterally interconnected, longitudinally periodical pore arrays can be produced in various ways. Sidewise connecting channels can be produced between pores of modulated diameters followed by a chemical etching adjusted to the length so that the narrower pore wall sections can be penetrated while the wider wall sections remain continuous [105]. However, lateral (or transversal) nanochannels can also be induced with high-voltage anodic spikes during an otherwise stable steady-state anodization process [106]. When the voltage during the spikes favours a pore diameter larger than the stable pore diameter plus the pore wall thickness, the walls can be partly destroyed. However, the length of such spikes has to be short enough so that the entire coating is not burnt (i.e., not the entire pore wall is detached from the substrate).

13.6 Pursuits Related to the Extension of the Anodizing Conditions

13.6.1 Anodization of Commercial Aluminium Samples

In the recipes of the preparation of highly ordered PAA samples, the advice concerning the purity of the starting material tells dogmatically that the material of highest possible purity has to be chosen. However, there are various applications such as adsorbents, drug carriers and protective coatings, in which the pore regularity is of no importance, unlike the price of the starting material. Therefore, it is a persistent pursuit to develop suitable anodization procedures for aluminium samples of commercial quality. Grade AA1050 is of particular interest (composition: Al > 99.5%; Si, Fe < 0.25%; other metals < 0.05%) [107–112].

The general trends in the anodization of commercial grade aluminium can be summarized as follows. The pore regularity is much lower than for high-purity starting materials, even though a proper surface pretreatment can improve it significantly [107, 111]. Under the same anodization conditions, the pores are usually somewhat wider [110]. As it was shown for pure aluminium, the pore diameter increases with anodization voltage of commercial grade Al, too. It is common that three-dimensional porous alumina is formed instead of well-directed parallel pore arrays [113, 114]. The growth rate of the PAA structure is lower than pure Al because of its lower corrosion rate due to the impurities that cause chemical dissolution [114]. This is particularly true when the commercial-grade aluminium contains precipitates around which pore branching and pore bending takes place, hence encapsulating the inhomogeneity by fully consuming Al around it [109, 115].

The means of improving the anodization of commercial grade aluminium include various modifications of the conventional processes: (i) tuning the solution composition, anodization voltage and temperature [108, 110]; (ii) optimizing the steps of the sequential anodization and layer removal processes [107]; (iii) application of pulse anodization, partly in order to limit the heat flux in the oxidation zone [116–118]; (iv) application of constant-current anodization [109]; (v) application of various heat treatment steps for stress relaxation and/or sample oxidation [113]. The best results so far were achieved with sulphuric acid solutions more concentrated than in the conventional anodization process [108].

13.6.2 Hard Anodization

Hard anodization was a well-established industrial process much before laboratory-scale mild anodization of regular PAA templates. However, it was notorious of irregular pore formation, and the process was hard to regulate. These problems have been overcome in the last two decades, and hard anodization processes have been developed for some sulphuric acid [119–121] and oxalic acid [100, 122–124] baths. The current density of hard anodization is about an order of magnitude larger than that of mild anodization, which leads to a much faster PAA layer formation. Also, the empirical porosity rule established for mild anodization is not valid for hard anodization; therefore, new pore size and porosity ranges became available. In order to avoid the burning of the samples, the electrical control of the experiments is often different from mild anodization. A gradually increasing voltage and the application of current control instead of the voltage control are often mentioned as a tool to avoid cracking of the samples and to increase the mechanical stability of the resulting PAA layer.

13.6.3 Disparity Between the First and Second Anodization Steps

The ordinary thumb rule for PAA anodization was that the first (and later sacrificed) anodized layer has to be prepared under identical conditions than that applied for the second anodization in order to achieve the maximal regularity in the pore arrangement. Later, various strategies have been elaborated to destroy this rule. It is possible to combine two mild anodization steps with different electrolyte solutions [125], but the process may also involve hard anodization as the first step [126–128]. This leads to a faster sample preparation than the pore ordering process during mild anodization. The interpore distance and regularity of the pore arrangement is determined by the first anodization, while the wall thickness depends on the second anodization step. The natural interpore distance of the two steps must be close enough so that no pore branching takes place in the second step. The advantage of the dissimilar conditions in the first and second anodizations is that the tendency for good ordering in the hard anodization (that is less controllable in the long run) can be coupled with the easily performed mild anodization.

13.6.4 Increase in Anodization Temperature

Extensive efforts have been made in order to extend the anodization temperature range to or above room temperature with retaining the regularity of the pore arrays, especially for oxalic acid solutions [117, 129–137]. The motivation of this pursuit partly originated from the reduction of cost (sparing the installation of a cooling system), but it had the advantage of faster anodization, leading to an about fivefold increase in current between 0 and 20 °C. For providing the sufficient heat absorbance, pulse anodization was recommended [117] in which the pore density and barrier layer thickness correspond to the on-time voltage but the heat absorbance takes place also during the off-time. When the anodization was performed at higher temperature, the increase in acid concentration proved to be beneficial since it decreased the barrier layer thickness and prevented the breakdown of the stable anodization process even without a rigorous temperature control and agitation [133]. While the interpore distance decreased with the increase in the acid concentration (in accord with the reduced barrier layer thickness) [134], the interpore distance was invariant to the temperature change [134, 135].

13.6.5 New Components of the Anodization Baths

Anodizing solutions with various acids—In the standard anodization processes elaborated between 1995 and 2000, three acid solutions were employed (see Table 13.1). Since then, the ever-lasting pursuit to extend the well-established ranges of both the pore diameter and the interpore distance resulted in novel processes that had not been considered to be feasible earlier. A summary of the new bath formulations with a single key component that yielded nanoporous alumina samples with mild anodization is listed in Table 13.2.

For the solutions listed in Table 13.2, the proportionality of the interpore distance and the anodization voltage is nearly the same as for the conventional anodizing solutions (sulphuric, oxalic and phosphoric acids). However, for many of these baths, the pore array is significantly more disordered than in the well-established conventional processes. It is noteworthy that typically multiprotic acids lead to the formation of porous alumina.

Anodizing with mixed solutions of several acids—The conventional mild anodization yields PAA samples discrete intervals in the pore diameter (and correspondingly, discrete intervals for both the interpore distance and the anodization voltage). In order to overcome the limitation originating from the hardly tuneable pore diameter, various successful attempts have been reported for aluminium anodization in acid mixtures. The application of the mixtures of acids that yield nearly overlapping anodization regimes (like sulphuric and oxalic acids) is straightforward [156, 157]. However, such acid mixtures also proved to be applicable whose components alone would yield faraway pore diameters [158–161]. The variation in the acid concentration ratio opens the way for a nearly continuous pore diameter variation.

Anodizing with solutions containing alcohols—Various bath formulations have been developed on the basis of classical anodization bath with various volume ratios of alcohols as co-solvent [128, 162–169]. The list of the alcohols tested as co-solvent includes methanol, ethanol, ethylene glycol and glycerol. Their application allows one to perform anodization at temperatures formerly not available, both much below 0 °C and above 100 °C. Alcohols were found to stabilize the hard anodization process

Table 13.2 List of non-conventional anodizing baths and their operation range yielding nanoporous aluminium oxide

Key component type	Specific bath component(s)	Voltage range V	d_{IP}	d_P	References
Inorganic	Arsenic acid	310–340	690–800	~200	[45]
	Phosphonic acid	110–180	370–440	132 (mean value)	[138]
	Selenic acid	25–48 60–100	85–120 120–160	8–12 60–75	[139, 140] [141]
	Sulphate salts	20–28	50–77	35–50	[142, 143]
Organic	Acetlylenedicarboxilic acid	87.5–95	250	100	[144]
	Cyclic oxocarbonic acids (various)	75–200	200–450	50–100	[145]
	Citric acid	260–450	645–1100	>230	[87]
	Etidronic acid	145–310	400–670	N/A	[146, 147]
	Malic acid	200–450	300–950	100–200	[148, 149]
	Malonic acid	120–140 150–230	240–333 290–490	N/A	[34, 150–152] [153]
	Squaric acid	100–120	200–400	< 100	[154]
	Tartaric acid	195–240	440–650	N/A	[149, 152, 155]

Some of the experiments were performed with constant-current anodization. Where the pore diameter is not available (N/A), the pore structure was typically observed after pore widening only

and impact the dielectric properties of the barrier layer, which makes it possible to widen significantly the voltage range of anodization and also to achieve ultrasmall pore diameters. The effect of alcohols is attributed to various factors, including the efficiency of cooling the reaction zone and the modified viscosity of the bath.

13.7 Anodization of Valve Metals

13.7.1 Titanium

Importance—Anodization of titanium obtained nearly as much attention as that of aluminium. Although titanium is much more expensive than aluminium, its application fields are completely different. Titania is indispensable in photoelectrochemical reactions, including solar cells. Due to the biocompatibility of titanium, nanoengineered titania layers impact the strong adhesion of titanium implants to bones (osseointegration). Although various functionalities of nanoscale titania can be also achieved with particles prepared with other methods, surface-bond nanoporous titania exhibits unique properties that make titanium anodization a very intensively studied field. Several reviews on this field can be found [170–174] that offer a wide overview, while the present chapter can give a short introduction only.

Categorization of the anodization baths used—It is common to divide the anodizing solutions into three generations. The early attempts to obtain porous titanium oxide layer on titanium by anodization applied acidic aqueous solutions with a small concentration of hydrogen fluoride [175–180]. Later, the acidic solutions were replaced with neutral ones, still retaining water as solvent [181, 182]. The application of non-aqueous solvents with a well-controlled HF and H_2O concentration is considered to be the most advanced version [183, 184]. These solutions, containing typically ethylene glycol or glycerol as major solvent, are named as "viscous" anodizing solutions, referring to one of the distinct properties of the baths, although it is debated whether the viscosity itself is that makes them effective or the small solubility of titanium and the small diffusion coefficient of the ions in the bath also play a role.

Whichever solution is used, fluoride ion is an indispensable component of the bath for obtaining nanostructured titanium dioxide. In the absence of the fluoride ion, a non-structured barrier layer can be obtained only. This is very well known from the anodization-based colouring of titanium that relies on variation of the thickness of the barrier oxide layer as a function of anodization voltage.

The role of fluoride ions—Regardless of the specific bath used, the fluoride ions play the same role; namely, they act as a regulator of the dissolution and oxidation. Dissolution is a prerequisite of the formation of a nanostructured oxide. The dissolution process taking place during the anodization is the most important reason why anodized titanium develops nanotubes instead of a nanopore structure similar to PAA.

The cross-sectional scheme of the nanotube formation is shown in Fig. 13.5a. This figure indicates that the preferred site of the dissolution process is the central area of the walls forming. The reason for this site preference is that the freshly anodized nanoporous titania structure is in some sense the inverse of what we saw for PAA. Namely, what was shown in Fig. 13.1 for PAA as a layer with anion incorporation is nearly pure titania in the case of anodized titanium, whereas the pure alumina in the PAA corresponds to a fluoride-doped zone.

The enhanced fluoride ion concentration of the anodized titanium in the central zone of the freshly formed titania wall was detected with either high-resolution scanning Auger electron spectroscopy (AES) [186, 187] and time-of-flight secondary ion mass spectrometry (ToF SIMS) [187]. By using a different approach, the enhanced fluorine content of the barrier layer can be detected with a composition depth profile analysis stated from the pore bottom [188]. Based on the above data, the atomistic mechanism of the nanopore formation is as follows. During the anodization, the cations (Ti^{IV}) migrate from the metal size of the barrier layer towards the solution, while anions migrate to the opposite direction. In contrast to the aluminium oxidation, the fluoride ions reaching the bottom part of the barrier layer move away from the pore axis, and this sidewise motion leads to their accumulation in the far-from-axis parts of the pore walls (this process was also identified as "plastic flow" [186]; see

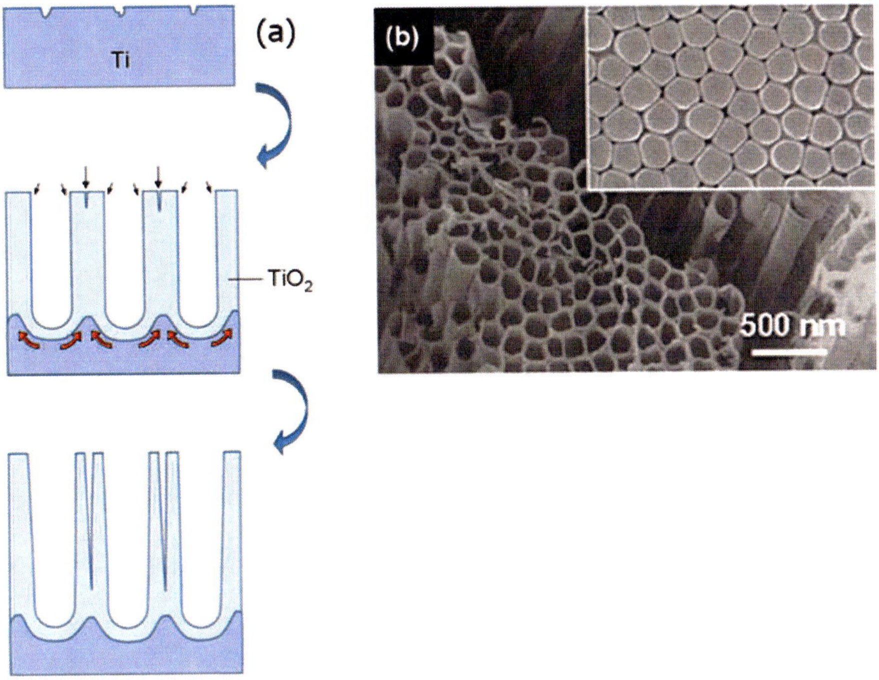

Fig. 13.5 **a** Schematic cross-sectional view of the anodized titanium, showing the sequence of the pore formation followed by oxide dissolution, hence leading to nanotube formation. Red block arrows indicate the flow direction at the bottom of the pore, while thin black arrows show the chemical etching. **b** Top view (main image) and bottom view (inset) of the resulting nanotube array. Reprinted by permission from Springer [185], copyright (2009)

red arrows in Fig. 13.5a). The residual fluoride ions then lead to the chemical (i.e., not electrical field-driven) dissolution of the fluoride-rich part of interpore region, which is highly unrelated to the anodic process taking place in the barrier layer zone. Since the chemical dissolution is delayed as compared to the penetration of the pore during the anodization, the pores have a conical shape with slightly narrowing towards the barrier layer [189]. The parallel dissolution of the interpore region and the chemical etching of the pore walls from the pore axis direction impose a limitation on the pore length since a full dissolution of the top zone may take place during prolonged anodization. This limitation was particularly strong in first-generation anodization baths, allowing the growth of nanotubes up to typically half a micrometre [176]. The second-generation bath leads to an improvement with nearly 3 μm of the maximum tube length [181], and the third-generation highly viscous baths brought the breakthrough with the possibility of the preparation of self-supporting samples.

Electrochemical parameters of the anodization process—The preparation of titania nanububes is technically very similar to the PAA preparation process (two-electrode cells with temperature control). The anodization voltage can be fairly small for acidic aqueous baths, and a 3 V cell voltage can be enough for producing porous titania layers. The cell voltage used for first- and second-generation baths is typically 10–20 V and never exceeds 25 V. Third-generation baths with high organic concentration allow a voltage range of 20–90 V. The current transients are often similar to that shown for PAA formation (Fig. 13.2), and the stable anodization current density for the third-generation bath is in the mA cm^{-2} range. The interpore distance during Ti anodization exhibits a linear dependence on the anodization voltage, in full analogy with PAA formation, though the proportionality coefficient is smaller (1.5–1.7 nm V^{-1}). The volume expansion factor of the porous titania structure strongly depends on the anodization circumstances and varies between 1.3 and 2.8 [190].

Structure and annealing behaviour—The as-received TiO_2 nanotube arrays are mostly amorphous. Upon annealing, a crystallization process takes place, leading to either rutile or anataze (or one after the other as the temperature increases). The formation temperature of the crystalline forms of TiO_2 strongly depends on the formation parameters, similar to the temperature limit at which the tubular structure collapses [191, 192]. In exceptional cases, the as-received samples can be crystalline [193]. The adjustment of the structural properties of the titania nanotubes is important for achieving the desired photoelectrochemical activity.

Various pursuits concerning titanium anodization—It was a challenge to achieve conditions at which the anodized titanium develops pores instead of nanotubes. It was shown that the accurate control of the water content of the electrolyte solution makes it possible to avoid the pore wall splitting shown in Fig. 13.5a since water is an essential component for the dissolution [194]. Galvanostatic anodization was also reported to be a feasible way for nanopore formation [195]. The regularity of the pore system formed upon anodization can be improved with nanoimprint lithography, fully similar to the process shown in Fig. 13.4a [196–199].

Tube shaping and branching with the change of the anodization voltage is possible for titania nanotubes, similarly to what was demonstrated for PAA samples [185, 200]. However, the shape-modulated titania nanotubes show less regular structures that PAA samples.

Another interesting variation of titanium anodization is when well-separated regular nanotube arrays are prepared instead of the closely packed tube arrays [201, 202]. This type of anodization process proved to be successful with diethylene glycol as solvent and with thorough control of the water content of the bath.

13.7.2 Other Valve Metals

Valve metals form a closely related family with regard of the anodization process. The common features of their anodization are as follows: (i) a barrier-type oxide develops on their surface upon anodization in fluoride-free electrolyte solutions. (ii) Nanoporous and nanotubular oxide structures can be produced during their anodization in bath containing fluoride ions. (iii) The specific form of the nanostructure produced with anodization (i.e., pores or tubes) can be tuned with the water and fluoride salt concentration in the solution and the voltage used for anodization.

Table 13.3 Valve metals yielding a porous nanostructure upon anodization with a few representative reference studies

Anodized valve metal	References
Zr	[203–208]
Nb	[209–212]
Hf	[213–216]
Ta	[217–221]
W	[222–227]

(iv) The oxide structure formed is either amorphous or crystalline and the degree of crystallinity can be improved with annealing up to the collapse temperature of the nanostructure. A few examples for works dealing with this field are summarized in Table 13.3, and the transition between nanoporous to nanotubular oxide is exemplified for Zr in Fig. 13.6.

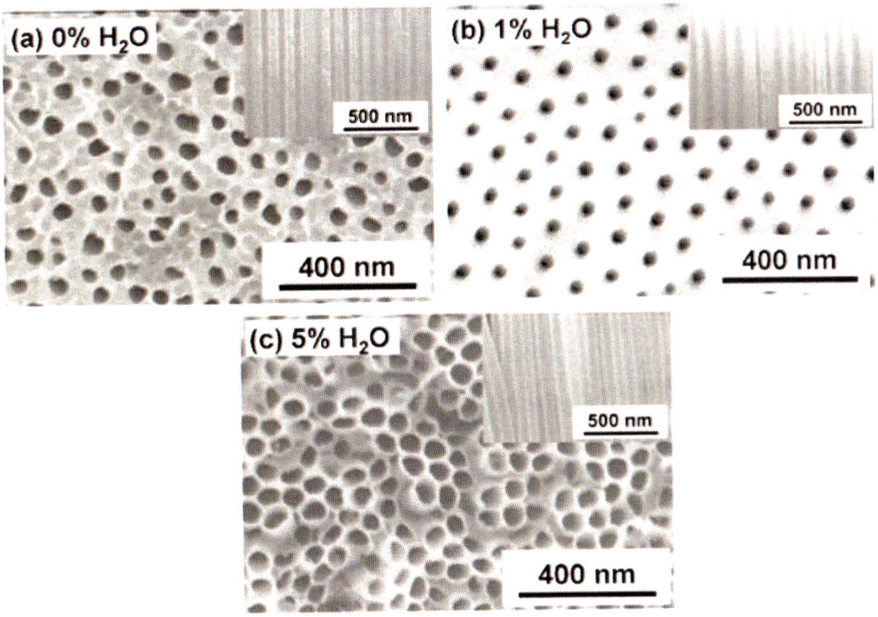

Fig. 13.6 Transition from a porous to nanotubular structure of anodized zirconium with the change of the water content of NH_4F solution in glycerol. Insets show the cross-sectional morphology. Reprinted from [204]. Copyright (2008), with permission from Elsevier

13.8 Nanostructures Prepared with Anodization of Miscellaneous Other Metals

13.8.1 Iron

Nanoporous iron oxide layers became important recently for a number of applications like supercapacitors, electrodes for enzyme-free electroanalytical detection of organic molecules, photoelectrochemical reactions and host materials for electrodes in lithium ion batteries. The varying oxidation state of ion and the diversity of the crystalline forms of the oxides make it a versatile tool in various fields of electrochemistry.

The purity of the anodized species was typically at least 99.9 wt% or better. The anodization solutions that offer to prepare nanostructured oxide layer on iron are versatile and so are the resulting nanostructures. The majority of works report the application of viscous organic solvents with a well-controlled water concentration and ammonium fluoride as solute [228–235]. While the organic baths analogous to those applied for titanium anodization proved to be efficient, aqueous buffer solutions also allowed the preparation of nanostructured surface layer with a thickness not larger than a few dozens of nanometers [230]. This limitation was attributed to the dissolution–reprecipitation nature of the oxide formation in aqueous baths. For organic baths, an NH_4F concentration lower than 0.5 wt% was favourable for vertical pore and nanotube formation, but a higher fluoride salt concentration led to disordered nanoporous structure [228]. Studies dealing with the impact of the water concentration of the organic baths agree in that a 0.1–2.0 mol dm^{-3} water content makes it possible to obtain nanotubes with anodization, and the optimal anodization voltage decreases with the increase of the water concentration [231, 232, 234]. The higher the water content, the thicker barrier layer is produced and the smaller the maximum total layer thickness achieved [234]. The anodization voltage ranges to the 10–100 V interval, and the corresponding anodization current density is between 8 and 50 mA cm^{-2}. The interpore distance increases with the anodization voltage, similar to both Al and Ti. However, the cell size per anodization voltage coefficient strongly depends on the water content of the bath. This parameter was rated as 1.7 and 3.7 nm V^{-1} for ethylene glycol-based solution containing 0.1 mol dm^{-3} NH_4F plus 0.3 and 1.5 mol dm^{-3} H_2O, respectively [234].

When a fluoride salt is applied as solute of the anodizing solution, the as-received nanoporous/nanotubular iron oxide contains not only a detectable but rather high fluorine incorporation [228, 229, 235]. The fluorine content of the porous layer has a maximum in the barrier layer [231, 234]. Since the as-received structure of the nanostructured iron oxide is amorphous, an annealing is commonly applied for achieving the crystalline form. During the annealing process, the fluorine incorporation decreases significantly [228]. The crystal structure of the annealed samples is usually not uniform, but the XRD pattern indicates the presence of various phases [228–230, 233, 235] such as magnetite and maghemite as major crystalline forms and haematite observed occasionally only.

13.8.2 Tin

The importance of the anodization of tin stems from the fact that it forms an optically transparent but electrically conductive oxide that can be used as electrode for various photoelectrochemical processes. Moreover, the anodic polarization limit without significant water decomposition is higher for SnO_2 than for conventional transparent electrode materials like indium tin oxide (ITO) (indium tin oxide) [236].

The anodization of tin is different from that of other metals by various means. Although the anodization of Sn in oxalic acid solution also leads to a porous oxide structure [237–239], the anodization in alkaline solutions became rather widespread [236, 239–242]. The morphological characteristics of the porous tin oxide structures obtained with the two baths are well comparable or even nearly undistinguishable, as it is shown in Fig. 13.7. The pore size distribution curves shown in the original study [239] revealed that the mean pore diameter is 45–50 nm and the distribution curves range to the 15–100 nm interval. The simplicity of the Sn anodization process in NaOH solution is spectacularly indicated by the fact that the mean pore size parameters practically do no depend on the NaOH concentration in the 0.1–1.0 mol dm^{-3} interval, nor was any pore size dependence found on the anodization voltage in the 5–15 V range. Regarding that the purity of the Sn metal specimen to be anodized can be rather low (e.g., 98.8 at% [242]), the Sn anodization cannot be rated as being very demanding.

Since the pore structure produced on Sn specimen by anodization is very irregular, a one-step anodization process is always sufficient. The product of the anodization is amorphous and close in composition to SnO [237] and the Raman spectra verify the presence of this phase in the as-received or moderately annealed samples. For achieving the desired transparency, an annealing process is required at about 500 °C where the porous structure does not collapse.

A difficulty of the synthesis of porous tin oxide structures with in-depth morphological homogeneity and the concomitantly good mechanical stability originates from the vigorous oxygen evolution as side reaction. Even though the growth rate

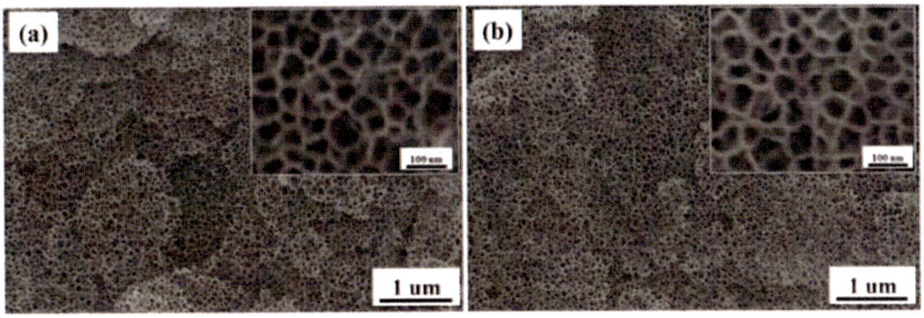

Fig. 13.7 SEM image of the surface of tin samples anodized in **a** NaOH and **b** oxalic acid solution. The insets are the magnification of a relevant area of the same sample. Reprinted from [239]. Copyright (2016), with permission from Elsevier

was obtained as being invariant with the anodization time [240], the stress caused by the oxygen bubbles and their release may lead to a periodically thinned wall structure [237]. To overcome this difficulty, anodization voltage as low as 2 V was suggested.

13.9 Anodization of Alloys

13.9.1 Binary Alloys with Components that Form Regular Anodized Nanostructures

The anodization of binary alloys whose components exhibit a regular anodized nanostructure opens a way to prepare a wider range of nanostructures than what is available with a pure metal. An excellent example is the alloy of Al and Ti [243, 244]. With the careful adjustment of the composition, anodization voltage and fluoride ion concentration of the bath, the anodization process leads to a wide range of nanostructures, including nanoporous and nanotubular ones. The organization chart of such an experimental matrix is presented in Fig. 13.8.

The gradual change of the character and pore size of the anodized nanostructures can be observed for various other alloys such as Ti–Zr [245–247], Ti–Ta [245, 248] and Ti–Nb [245] alloys. In the nanostructured oxide obtained with the anodization of the alloys of some valve metal(s) and/or aluminium, the oxide is composed of both metals in nearly the same ratio as the precursor alloy composition predicts it.

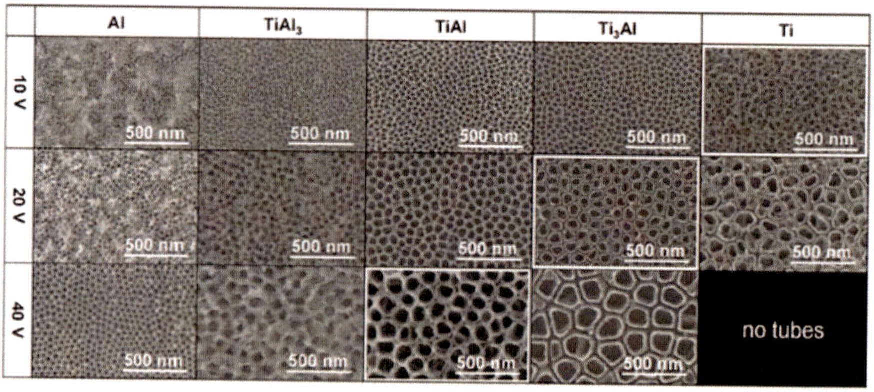

Fig. 13.8 Top view SEM images of the nanostructured oxide layers obtained on Al, Ti and three Al–Ti alloys by anodization is H_2SO_4 solution containing 0.15 wt% HF. Reprinted with permission from [244]. Copyright (2008) American Chemical Society

13.9.2 Miscellaneous Other Alloys

Ternary alloys of nanopore-forming metals—The details of the anodization process of alloys with complex composition are not fully clear from the scientific point of view. The reason for the lack of detailed information is that the motivation of the anodization of such alloys is rather practical, ranging from the decrease of the corrosion resistance, the improvement of the biocompatibility of the metal surface or the acceleration of the biomineralization process by providing a porous surface layer. The examples discussed here refer to Ti–6Al–7Nb [249–251] and Ti–13Nb–13Zr [252] alloys.

The nanopore or nanocolumn structure of the anodized ternary alloys is by far not as regular as that of pure metals (although the regularity is not a pronounced goal here). The disorder partly stems from the phase structure of these alloys. The local crystallographic peculiarities impact the pore formation process, and hence, various pore/tube domains may form or three-dimensional network-like structures are produced. The anodization of the ternary alloys containing aluminium is often performed in alkaline solution that leads to the leaching of aluminium. Therefore, it is not a general rule that the pore/tube walls contain all components of the precursor metal. The release of aluminium is beneficial for the biocompatibility of the base metal.

Ti–Ni alloys—The alloys containing Ti, Ni and occasionally a third component in a small concentration are important for biomedical applications. From electrochemical point of view, the difference as compared to the above discussed other alloys is that anodized Ni alone does not form a porous structure, while a porous surface layer can be achieved with the anodization of Ti–Ni alloys [253–255]. Since there is no recipe for Ni anodization, the baths used differ from those applied for Ti, mostly because higher water content is required and chloride ions must be present for the development of a porous structure. Since the etching of the remaining wall structure composed of mostly TiO_2 does not take place in the absence of the fluoride ions, the upper limit for the porous layer thickness originates from the delamination of the layer from the substrate, not from the layer dissolution.

The chloride-containing solution leads to a materials loss during anodization, and the preferred leaving component is nickel. For this reason, the Ti:Ni ratio in the porous layer is likely to be larger than in the master alloy. Ni can be seen in the EDS spectra of the porous layers, although exact composition is not given.

References

1. Keller F, Hunter MS, Robinson DL (1953) J Electrochem Soc 100:411–419
2. Hunter MS, Fowle P (1954) J Electrochem Soc 101:514–519
3. Diggle JW, Downie TC, Goulding CW (1969) Chem Rev 69:365–405
4. O'Sullivan JP, Wood GC (1970) Proc Roy Soc Lond A 317:511–543
5. Bailey G, Wood GC (1974) Trans IMF 52:187–199

6. Csanády Á, Szontágh E, Romhányi K (1976) Kristall und Technik 11:171–181
7. Csanády Á (1979) J Mater Sci Lett 14:745–747
8. Csanády Á, Imre-Baán I, Lichtenberger-Bajza E, Szontágh E, Dömölki F (1980) J Mater Sci 15:2761–2770
9. Holló MG (1960) Acta Metall 8:265–268
10. Whitney TM, Jiang JS, Searson PC, Chien CL (1993) Science 261:1316–1319
11. Martin CR (1994) Science 266:1961–1966
12. Masuda H, Fukuda K (1995) Science 268:1466–1468
13. Thompson GE (1997) Thin Solid Films 297:192–201
14. Lee W (2010) JOM 62:58–63
15. Asoh H, Ono S (2007) Fabrication of ordered anodic nanoporous alumina layers and their application in nanotechnology. In: Staikov G (ed) Electrocrystallization in nanotechnology. Wiley-VCH, Weinheim
16. Losic D, Santos A (eds) (2015) Nanoporous alumina. Fabrication, structure, properties and applications. Springer, Berlin
17. Yu WH, Fei GT, Chen XM, Xue FH, Xu XJ (2006) Phys Lett A 350:392–395
18. Feil AF, da Costa MV, Amaral L, Teixeira RS, Migowski P, Dupont J, Machado G, Peripolli SB (2010) J Appl Phys 107:026103(1–3)
19. Li AP, Müller F, Birner A, Nielsch K, Gösele U (1998) J Appl Phys 84:6023–6026
20. Masuda H, Hasegwa F, Ono S (1997) J Electrochem Soc 144:L127–L130
21. Sulka GD, Stroobants S, Moshchalkov V, Borghs G, Celis JP (2002) J Electrochem Soc 149:D97–D103
22. Sulka GD, Parkoła KG (2007) Electrochim Acta 52:1880–1888
23. Bwana NN (2008) J Nanopart Res 10:313–319
24. Jessensky O, Müller, Gösele U (1998) Appl Phys Lett 72:1173–1175
25. Sulka GD, Stępniowski WJ (2009) Electrochim Acta 54:3683–3691
26. Choi YC, Hyeon JY, Bu SD (2010) J Kor Phys Soc 56:113–119
27. Li AP, Müller F, Gösele U (2000) Electrochem Solid-State Lett 3:131–134
28. Garcia-Vergara SJ, Habazaki H, Skeldon P, Thompson GE (2007) Nanotechnol 18:415605(1–8)
29. Ramazani A, Almasi Kashi M, Maleki K, Mohammadniaei M (2011) Jap J Appl Phys 50:035203(1–4)
30. Martín J, Manzano CV, Martín-González M (2012) Micropor Mesopor Mater 151:311–316
31. Masuda H, Yada K, Osaka A (1998) Jpn J Appl Phys 37:L1340–L1342
32. Wu Z, Richter C, Menon L (2007) J Electrochem Soc 154:E8–E12
33. Vrublevsky I, Parkoun V, Schreckenbach J, Marx G (2003) Appl Surf Sci 220:51–59
34. Knörnschild G, Poznyak AA, Karoza AG, Mozalev A (2015) Surf Coat Tech 275:17–25
35. Nielsch K, Choi J, Schwirn K, Wehrspohn RB, Gösele U (2002) Nano Lett 2:677–680
36. Li AP, Müller F, Birner A, Nielsch K, Gösele U (1999) J Vac Sci Technol, A 17:1428–1431
37. Stein N, Rommelfangen M, Hody V, Johann L, Lecuire JM (2002) Electrochim Acta 47:1811–1817
38. Ko S, Lee D, Jee S, Park H, Lee K, Hwang W (2006) Thin Solid Films 515:1932–1937
39. Sui YC, Saniger JM (2001) Mater Lett 127–136
40. McQuaig MK, Toro A, Van Geertruyden W, Misiolek WZ (2011) J Mater Sci 46:243–253
41. Thompson GE, Furneaux RC, Wood GC (1978) Corros Sci 18:481–498
42. Mínguez-Bacho I, Rodríguez-López S, Climent A, Fichou D, Vázquez M, Hernández-Vélez M (2015) J Phys Chem C 119:27392–27400
43. Thompson GE, Furneaux RC, Goode JS, Wood GC (1978) Trans IMF 56:159–167
44. Thompson GE, Wood GC (1981) Nature 290:230–232
45. Akiya S, Kikuchi T, Natsui S, Suzuki RO (2017) Appl Surf Sci 403:652–661
46. Kooij ES, Wormeester H, Gâlcă AC, Poelsema B (2003) Electrochem Solid State Lett 6:B52–B54
47. Lillo M, Losic D (2009) J Membrane Sci 327:11–17
48. Singh GK, Golovin AA, Aranson IS, Vinokur VM (2005) Europhys Lett 70:836–842

49. Friedman AL, Menon L (2007) J Appl Phys 101:084310(1–5)
50. Zhu XF, Song Y, Liu L, Wang CY, Zheng J, Jia HB, Wang XL (2009) Nanotechnol 20:475303(1–7)
51. Stanton LG, Golovin AA (2009) Phys Rev B 79:035414(1–7)
52. Van Overmeere Q, Blaffart F, Proost J (2010) Electrochem Commun 12:1174–1176
53. Oh J, Thompson CV (2011) Electrochim Acta 56:4044–4051
54. Cheng C, Ngan AHW (2011) Electrochim Acta 56:9998–10008
55. Van Overmeere Q, Proost J (2011) Electrochim Acta 56:10507–10515
56. Patermarakis G, Moussoutzanis K (2011) J Electroanal Chem 659:176–190
57. Wang M, Liu Y, Yang H (2012) Electrochim Acta 62:424–432
58. Aarão Reis FDA, Badiali JP, di Caprio D (2012) Langmuir 28:13034–13041
59. Orsi A, Riley DJ (2013) J Electrochem Soc 160:D10–D12
60. Pashchanka M, Schneider JJ (2013) Phys Chem Chem Phys 15:7070–7074
61. Bartosika L, Stafiej J, Di Caprio D (2016) Electrochim Acta 188:218–221
62. Pashchanka M, Schneider JJ (2016) J Phys Chem C 120:14590–14596
63. Roslyakov IV, Gordeeva EO, Napolskii KS (2017) Electrochim Acta 241:362–369
64. Mirzoev RA, Davydov AD, Vystupov SI, Kabanova TB (2019) Electrochim Acta 294:276–285
65. Masuda H, Yamada H, Satoh M, Asoh H, Nakao M, Tamamura T (1997) Appl Phys Lett 71:2770–2772
66. Surawathanawises K, Cheng X (2014) Electrochim Acta 117:498–503
67. Stasi V, Cattaneo G, Franz S, Bestetti M, Ubaldi MC, Piccinin D, Pietralunga SM (2007) Photonics and nanostructures—fundamentals and applications 5:136–139
68. Asoh H, Ono S, Hirose T, Nakao M, Masuda H (2003) Electrochim Acta 48:3171–3174
69. Asoh H, Ono S, Hirose T, Takatori I, Masuda H (2004) Jap J Appl Phys 43:6342–6346
70. Choi J, Werhspohn RB, Gösele U (2003) Adv Mater 15:1531–1534
71. Nasir ME, Allsopp DWE, Bowen CR, Hubbard G, Parsons KP (2010) Nanotechnol 21:105303(1–6)
72. Liang J, Chik H, Yin A, Xua J (2002) J Appl Phys 91:2544–2546
73. Chik H, Liang J, Cloutier SG, Kouklin N, Xua JM (2004) Appl Phys Lett 84:3376–3378
74. Wang YD, Chua SJ, Sander MS, Chen P, Tripathy S, Fonstad CG (2004) Appl Phys Lett 85:816–818
75. Crouse D, Lo YH, Miller AE, Crouse M (2000) Appl Phys Lett 76:49–51
76. Wu MT, Leu IC, Hon MH (2004) J Mater Res 19:888–895
77. Foong TRB, Sellinger A, Hu X (2008) ACS Nano 2:2250–2256
78. Liu P, Singh VP, Rajaputra S (2010) Nanotechnol 21:115303(1–8)
79. Asoh H, Matsuo M, Yoshihama M, Ono S (2003) Appl Phys Lett 83:4408–4410
80. Asoh H, Sasaki K, Ono S (2005) Electrochem Commun 7:953–956
81. Chen PL, Kuo CT, Tsai TG, Wu BW, Hsu CC, Pan FM (2003) Appl Phys Lett 82:2796–2798
82. Chu SZ, Inoue S, Wada K, Hishita S, Kurashima K (2005) Adv Func Mater 15:1343–1349
83. Chang YH, Lin HW, Chenz C (2012) J Electrochem Soc 159:D512–D517
84. Chu SZ, Inoue S, Wada K, Hishita S, Kurashima K (2005) J Electrochem Soc 152:B116–B124
85. Sjöström T, Fox N, Su B (2009) Nanotechnol 20:135305(1–6)
86. Mozalev A, Sakairi M, Saeki I, Takahashi H (2003) Electrochim Acta 48:3155–3170
87. Mozalev A, Mozaleva I, Sakairi M, Takahashi H (2005) Electrochim Acta 50:5065–5075
88. Mozalev A, Vázquez RM, Bittencourt C, Cossement D, Gispert-Guirado F, Llobet E, Habazaki H (2014) J Mater Chem C 2:4847–4860
89. Chu SZ, Wada K, Inoue S, Segawa H (2011) J Electrochem Soc 158:C148–C157
90. Cheng W, Steinhart M, Gösele U, Wehrspohn RB (2007) J Mater Chem 17:3493–3495
91. Zaraska L, Kurowska E, Sulka GD, Jaskuła M (2012) J Solid State Electrochem 16:3611–3619
92. Sousa CT, Apolinario A, Leitao DC, Pereira AM, Ventura J, Araujo JP (2012) J Mater Chem 22:3110–3116
93. Liu S, Tang S, Zhou H, Fu C, Huang Z, Liu H, Kuang Y (2013) J Solid State Electrochem 17:1931–1938

94. Yin A, Guico RS, Xu J (2007) Nanotechnol 18:035304(1–6)
95. Li Y, Jin S, Qin Y, Hu X, Ling Z, Liu Q, Jin L (2015) J Electrochem Soc 162:E173–E178
96. Santos A, Formentín P, Pallarès J, Ferré-Borrull J, Marsal LF (2011) J Electroanal Chem 655:73–78
97. Li J, Li C, Chen C, Hao Q, Wang Z, Zhu J, Gao X (2012) ACS Appl Mater Interfaces 4:5678–5683
98. Law CS, Santos A, Nemati M, Losic D (2016) ACS Appl Mater Interfaces 8:13542–13554
99. Li SY, Wang J, Wang G, Wang JZ, Wang CW (2015) Mater Res Bull 68:42–48
100. Lee W, Ji R, Gösele U, Nielsch K (2006) Nature Mater 5:741–747
101. Lee W, Kim JC (2010) Nanotechnol 21:485304(1–8)
102. Lee W, Schwirn K, Steinhart M, Pippel E, Scholz R, Gösele U (2008) Nat Nanotechnol 3:234–239
103. Noormohammadi M, Moradi M, Almasi Kashi M, Ramazani A, Mayamai Y (2013) Surf Coat Tech 223:104–109
104. Noormohammadi M, Moradi M (2012) Mater Chem Phys 135:1089–1095
105. Losic D, Losic D Jr (2009) Langmuir 25:5426–5431
106. Resende PM, Sanz R, Ruiz-de Clavijo A, Caballero-Calero O, Martin-Gonzalez M (2016) Coatings 6:59(1–7)
107. Montero-Moreno JM, Sarret M, Müller C (2007) Surf Coat Tech 201:6352–6357
108. Bai A, Hu CC, Yang YF, Lin CC (2008) Electrochim Acta 53:2258–2264
109. Montero-Moreno JM, Sarret M, Müller C (2010) Micropor Mesopor Mater 136:68–74
110. Michalska-Domańska M, Norek M, Stępniowski WJ, Budner B (2013) Electrochim Acta 105:424–432
111. Zaraska L, Kurowska E, Sulka GD, Senyk I, Jaskula M (2014) J Solid State Electrochem 18:361–368
112. Veys-Renaux D, Chahboun N, Rocca E (2016) Electrochim Acta 211:1056–1065
113. Chen CC, Chen JH, Chao CG (2005) Jpn J Appl Phys 44:1529–1533
114. Kure-Chu SZ, Osaka K, Yashiro H, Segawa H, Wada K, Inoue S (2015) J Electrochem Soc 162:C24–C34
115. Kasi AK, Kasi JK, Afzulpurkar N, Hasan MM, Mahaisavariya B (2012) J Vac Sci Technol B 30:031805(1–7)
116. Chung CK, Zhou RX, Liu TY, Chang WT (2009) Nanotechnol 20:055301(1–5)
117. Chung CK, Chang WT, Liao MW, Chang HC, Lee CT (2011) Electrochim Acta 56:6489–6497
118. Chung CK, Chang WT, Liao MW, Chang HC (2012) Mater Lett 88:104–107
119. Chu SZ, Wada K, Inoue S, Isogai M, Yasumori A (2005) Adv Mater 17:2115–2119
120. Schwirn K, Lee W, Hillebrand R, Steinhart M, Nielsch K, Gösele U (2008) ACS Nano 2:302–310
121. Lee K, Tang Y, Ouyang M (2008) Nano Lett 8:4624–4629
122. Santos A, Formentín P, Ferré-Borrull J, Pallarès J, Marsal LF (2012) Mater Lett 67:296–299
123. Vega V, García J, Montero-Moreno JM, Hernando B, Bachmann J, Prida VM, Nielsch K (2015) ACS Appl Mater Interfaces 7:28682–28692
124. Lia J, Zhang Z, Lia Y, Ma Y, Chen L, Zhang Z, Suna R (2016) Electrochim Acta 213:14–20
125. Li D, Jiang C, Ren X, Long M, Jiang J (2008) Mater Lett 62:3228–3231
126. Han XY, Shen WZ (2011) J Electroanal Chem 655:56–64
127. Liu J, Liu S, Zhou H, Xie C, Huang Z, Fu C, Kuang Y (2014) Thin Solid Films 552:75–81
128. Norek M, Dopierała M, Stępniowski WJ (2015) J Electroanal Chem 750:79–88
129. Sulka GD, Brzózka A, Zaraska L, Jaskuła M (2010) Electrochim Acta 55:4368–4376
130. Ding GQ, Yang R, Ding JN, Yuan NY, Zhu YY (2010) Nanoscale Res Lett 5:1257–1263
131. Stępniowski WJ, Zasada D, Bojar Z (2011) Surf Coat Tech 206:1416–1422
132. Chung CK, Liao MW, Chang HC, Lee CT (2011) Thin Solid Films 520:1554–1558
133. Song Y, Jiang L, Qi W, Lu C, Zhu X, Jia H (2012) J Electroanal Chem 673:24–31
134. Cheng C, Ngan AHW (2013) Nanotechnol 24:215602(1–10)
135. Zaraska L, Stępniowski WJ, Ciepiela E, Sulka GD (2013) Thin Solid Films 534:155–161
136. Zaraska L, Stępniowski WJ, Jaskuła M, Sulka GD (2014) Appl Surf Sci 305:650–657

137. Stępniowski WJ, Nowak-Stępniowska A, Presz A, Czujko T, Varin RA (2014) Mater Charact 91:1–9
138. Akiya S, Kikuchi T, Natsui S, Sakaguchi N, Suzuki RO (2016) Electrochim Acta 190:471–479
139. Nishinaga O, Kikuchi T, Natsui S, Suzuki RO (2013) Sci Rep 3:2748(1–6)
140. Kikuchi T, Nishinaga O, Natsui S, Suzuki RO (2014) Electrochim Acta 137:728–735
141. Gordeeva EO, Roslyakov IV, Napolskii KS (2019) Electrochim Acta 307:13–19
142. Tsuchiya H, Berger S, Macak JM, Muñoz AG, Schmuki P (2007) Electrochem Commun 9:545–550
143. Kondo R, Kikuchi T, Natsui S, Suzuki RO (2016) Mater Lett 183:285–289
144. Kikuchi T, Nishinaga O, Natsui S, Suzuki RO (2014) ECS Electrochem Lett 3:C25–C28
145. Kikuchi T, Nakajima D, Kawashima J, Natsui S, Suzuki RO (2014) Appl Surf Sci 313:276–285
146. Kikuchi T, Nishinaga O, Natsui S, Suzuki RO (2015) Electrochim Acta 156:235–243
147. Takenaga A, Kikuchi T, Natsui S, Suzuki RO (2016) Electrochim Acta 211:515–523
148. Kikuchi T, Yamamoto T, Suzuki RO (2013) Appl Surf Sci 284:907–913
149. Chu SZ, Wada K, Inoue S, Isogai M, Katsuta Y, Yasumori A (2006) J Electrochem Soc 153:B384–B391
150. Lee W, Nielsch K, Gösele U (2007) Nanotechnol 18:475713(1–8)
151. Ren J, Zuo Y (2012) Appl Surf Sci 261:193–200
152. Ono S, Saito M, Asoh H (2005) Electrochim Acta 51:827–833
153. Sun B, Li J, Jin X, Zhou C, Hao Q, Gao X (2013) Electrochim Acta 112:327–332
154. Kikuchi T, Yamamoto T, Natsui S, Suzuki RO (2014) Electrochim Acta 123:14–22
155. Vrublevsky IA, Chernyakova KV, Ispas A, Bund A, Zavadski S (2014) Thin Solid Films 556:230–235
156. Almasi Kashi M, Ramazani A, Rahmandoust M, Noormohammadi M (2007) J Phys D Appl Phys 40:4625–4630
157. Almasi Kashi M, Ramazani A, Mayamai Y, Noormohammadi M (2010) Jpn J Appl Phys 49:015202(1–5)
158. Kao TT, Chang YC (2014) Appl Surf Sci 288:654–659
159. Qin X, Zhang J, Meng X, Deng C, Zhang L, Ding G, Zeng H, Xu X (2015) Appl Surf Sci 328:459–465
160. Stępniowski WJ, Moneta M, Norek M, Michalska-Domańska M, Scarpellini A, Salerno M (2016) Electrochim Acta 211:453–460
161. Azimi HAR, Zarei M, Rafati A, Noormohammadi M (2016) J Porous Mater 23:357–363
162. Li Y, Zheng M, Ma L, Shen W (2006) Nanotechnol 17:5101–5105
163. Zaraska L, Sulka GD, Jaskuła M (2010) Surf Coat Tech 204:1729–1737
164. Nguyen TN, Kim D, Jeong DY, Kim MW, Kim JU (2012) Electrochim Acta 83:288–293
165. Song Y, Wu H, Yang B, Wang J, Yang J, Xu C, Zhu X, Jia H (2012) J Electroanal Chem 682:110–115
166. Stępniowskia WJ, Forbota D, Norek M, Michalska-Domańska M, Król A (2014) Electrochim Acta 133:57–64
167. Manzano CV, Martín J, Martín-González MS (2014) Micropor Mesopor Mater 184:177–183
168. Qin X, Zhang J, Meng X, Wang L, Deng C, Ding G, Zeng H, Xu X (2014) Surf Coat Tech 254:398–401
169. Li Y, Qin Y, Jin S, Hu X, Ling Z, Liu Q, Liao J, Chen C, Shen Y, Jin L (2015) Electrochim Acta 178:11–17
170. Roy P, Berger S, Schmuki P (2011) Angew Chem Int Ed 50:2904–2939
171. Regonini D, Bowen CR, Jaroenworaluck A, Stevens R (2013) Mater Sci Eng R 74:377–406
172. Kowalski D, Kimb D, Schmuki P (2013) Nano Today 8:235–264
173. Zhou X, Nguyen NT, Özkan S, Schmuki P (2014) Electrochem Commun 46:157–162
174. Sopha H, Macak JM (2020) Recent advancements in the synthesis, properties, and applications of anodic self-organized TiO_2 nanotube layers. In: Sulka GD (ed) Nanostructured anodic metal oxides, synthesis and applications. Elsevier, Amsterdam
175. Gong D, Grimes CA, Varghese OK, Hu W, Singh RS, Chen Z, Dickey EC (2001) J Mater Res 16:3331–3334

176. Beranek R, Hildebrand H, Schmuki P (2003) Electrochem Solid-State Lett 6:B12–B14
177. Mor GK, Varghese OK, Paulose M, Mukherjee N, Grimes CA (2003) J Mater Res 18:2588–2593
178. Yu X, Li Y, Ge W, Yang Q, Zhu N, Kalantar-zadeh K (2006) Nanotechnol 17:808–814
179. Vega V, Prida VM, Hernández-Vélez M, Manova E, Aranda P, Ruiz-Hitzky E, Vázquez M (2007) Nanoscale Res Lett 2:355–363
180. Liu R, Yang WD, Qiang LS, Wu JF (2011) Thin Solid Films 519:6459–6466
181. Macak JM, Sirotna K, Schmuki P (2005) Electrochim Acta 50:3679–3684
182. Yu X, Li Y, Wlodarski W, Kandasamy S, Kalantar-zadeh K (2008) Sens Actuators B 130:25–31
183. Macak JM, Schmuki P (2006) Electrochim Acta 52:1258–1264
184. Regonini D, Satka A, Jaroenworaluck A, Allsopp DWE, Bowen CR, Stevens R (2012) Electrochim Acta 74:244–253
185. Yang DJ, Park H, Kim HG, Cho SJ, Choi WY (2009) J Electroceram 23:159–163
186. Berger S, Albu SP, Schmidt-Stein F, Hildebrand H, Schmuki P, Hammond JS, Paul DF, Reichlmaier S (2011) Surf Sci 605:L57–L60
187. Dronov A, Gavrilin I, Kirilenko E, Dronova D, Gavrilov S (2018) Appl Surf Sci 434:148–154
188. Macak JM, Albu SP, Schmuki P (2007) Phys Stat Sol RRL 1:181–183
189. Albu SP, Ghicov A, Aldabergenova S, Drechsel P, LeClere D, Thompson GE, Macak JM, Schmuki P (2008) Adv Mater 20:4135–4139
190. Albu SP, Schmuki P (2013) Electrochim Acta 91:90–95
191. Fang D, Luo Z, Huang K, Lagoudas DC (2011) Appl Surf Sci 257:6451–6461
192. Li H, Cao L, Liu W, Su G, Dong B (2012) Ceram Intern 38:5791–5797
193. So S, Hwang I, Riboni F, Yoo JE, Schmuki P (2016) Electrochem Commun 71:73–78
194. Wei W, Berger S, Hauser C, Meyer K, Yang M, Schmuki P (2010) Electrochem Commun 12:1184–1186
195. Galstyan V, Vomiero A, Comini E, Faglia G, Sberveglieri G (2011) RSC Adv 1:1038–1044
196. Choi J, Wehrspohn RB, Lee J, Gösele U (2004) Electrochim Acta 49:2645–2652
197. Chen B, Lu K, Geldmeier JA (2011) Chem Commun 47:10085–10087
198. Kondo T, Nagao S, Yanagishita T, Nguyen NT, Lee K, Schmuki P, Masuda H (2015) Electrochem Commun 50:73–76
199. Sopha H, Samoril T, Palesch E, Hromadko L, Zazpe R, Skoda D, Urbanek M, Ng S, Prikryl J, Macak JM (2017) Chem Open 6:480–483
200. Butail G, Ganesan PG, Teki R, Mahima R, Ravishankar N, Duquette DJ, Ramanath G (2011) Thin Solid Films 520:235–238
201. Ozkan S, Nguyen NT, Mazare A, Cerri I, Schmuki A (2016) Electrochem Commun 69:76–79
202. Ozkan S, Mazare A, Schmuki P (2018) Electrochim Acta 268:435–447
203. Tsuchiya H, Macak JM, Sieber I, Schmuki P (2005) Small 1:722–725
204. Berger S, Jakubka F, Schmuki P (2008) Electrochem Commun 10:1916–1919
205. Wang LN, Luo JL (2010) Electrochem Commun 12:1559–1562
206. Jiang W, He J, Zhong J, Lu J, Yuan S, Liang B (2014) Appl Surf Sci 307:407–413
207. Wierzbicka E, Syrek K, Sulka GD, Pisarek M, Janik-Czachor M (2016) Appl Surf Sci 388:799–804
208. Gomez Sanchez A, Katunar M, Ceré S (2020) Structural characteristics and barrier properties of anodic zirconium oxides for biomedical applications. In: Sulka GD (ed) Nanostructured anodic metal oxides, synthesis and applications. Elsevier, Amsterdam
209. Choi J, Lim JH, Lee SC, Chang JH, Kim KJ, Cho MA (2006) Electrochim Acta 51:5502–5507
210. Tzvetkov B, Bojinov M, Girginov A (2009) J Solid State Electrochem 13:1215–1226
211. Yoo JE, Choi J (2010) Electrochim Acta 55:5142–5147
212. Rani RA, Zoolfakar AS, Subbiah J, Ou JZ, Kalantar-zadeh K (2014) Electrochem Commun 40:20–23
213. Tsuchiya H, Schmuki P (2005) Electrochem Commun 7:49–52
214. Berger S, Jakubka F, Schmuki P (2009) Electrochem Solid State Lett 12:K45–K48
215. Wei W, Lee K, Shaw S, Schmuki P (2012) Chem Commun 48:4244–4246

216. Apolinário A, Sousa CT, Oliveira GNP, Lopes AML, Ventura J, Andrade L, Mendes A, Araújo JP (2020) Nanomater 10:382(1–15)
217. Wei W, Macak JM, Shrestha NK, Schmuki P (2009) J Electrochem Soc 156:K104–K109
218. Barton JE, Stender CL, Lia P, Odom TW (2009) J Mater Chem 19:4896–4898
219. Lee K, Schmuki P (2011) Electrochem Commun 13:542–545
220. Su Z, Grigorescu S, Wang L, Lee K, Schmuki P (2015) Electrochem Commun 50:15–19
221. Ni J, Sun M (2020) Anodic tantalum oxide: synthesis and energy-related applications. In: Sulka GD (ed) Nanostructured anodic metal oxides, synthesis and applications. Elsevier, Amsterdam
222. Tsuchiya H, Macak JM, Sieber I, Taveira L, Ghicov A, Sirotna K, Schmuki P (2005) Electrochem Commun 7:295–298
223. de Tacconi NR, Chenthamarakshan CR, Yogeeswaran G, Watcharenwong A, de Zoysa RS, Basit NA, Rajeshwar K (2006) J Phys Chem B 110:25347–25355
224. Watcharenwong A, Chanmanee W, de Tacconi NR, Chenthamarakshan CR, Kajitvichyanukul P, Rajeshwar K (2008) J Electroanal Chem 612:112–120
225. Ou JZ, Balendhran S, Field MR, McCulloch DG, Zoolfakar AS, Rani RA, Zhuiykov S, O'Mullane AP, Kalantar-zadeh K (2012) Nanoscale 4:5980–5988
226. Ou JZ, Rani RA, Balendhran S, Zoolfakar AS, Field MR, Zhuiykov S, O'Mullane AP, Kalantar-zadeh K (2013) Electrochem Commun 27:128–132
227. Kikuchi T, Kawashima J, Natsui S, Suzuki RO (2017) Appl Surf Sci 422:130–137
228. Prakasam HE, Varghese OK, Paulose M, Mor GK, Grimes CA (2006) Nanotechnol 17:4285–4291
229. Albu SP, Ghicov A, Schmuki P (2009) Phys Status Solidi RRL 3:64–66
230. Rangaraju RR, Raja KS, Panday A, Misra M (2010) Electrochim Acta 55:785–793
231. Habazaki H, Konno Z, Aoki Z, Skeldon P, Thompson GE (2010) J Phys Chem C 114:18853–18859
232. Jagminas A, Mažeika K, Bernotas N, Klimas V, Selskis A, Baltrūnas D (2011) Appl Surf Sci 257:3893–3897
233. Xie K, Li J, Lai Y, Lu W, Zhang Z, Liu Y, Zhou L, Huang H (2011) Electrochem Commun 13:657–660
234. Konno Y, Tsuji E, Skeldon P, Thompson GE, Habazaki H (2012) J Solid State Electrochem 16:3887–3896
235. Momeni MM, Ghayeb Y, Mohammadi F (2015) J Mater Sci Mater Electron 26:685–692
236. Yamaguchi A, Iimura T, Hotta K, Teramae N (2011) Thin Solid Films 519:2415–2420
237. Zaraska L, Czopik N, Bobruk M, Sulka GD, Mech J, Jaskuła M (2013) Electrochim Acta 104:549–557
238. Zaraska L, Syrek K, Hnida KE, Bobruk M, Krzysik A, Łojewski T, Jaskuła M, Sulka GD (2016) Electrochim Acta 205:273–280
239. Lu C, Wang J, Meng D, Wang A, Wang Y, Zhu Z (2016) J Alloy Compd 685:670–679
240. Wang M, Liu Y, Xue D, Zhang D, Yang H (2011) Electrochim Acta 56:8797–8801
241. Zaraska L, Gilek D, Gawlak K, Jaskuła M, Sulka GD (2016) Appl Surf Sci 390:31–37
242. Zaraska L, Gawlak K, Wiercigroch E, Malek K, Kozieł M, Andrzejczuk M, Marzec MM, Jarosz M, Brzózka A, Sulka GD (2019) Electrochim Acta 319:18–30
243. Tsuchiya H, Berger S, Macak JM, Ghicov A, Schmuki P (2007) Electrochem Commun 9:2397–2402
244. Berger S, Tsuchiya H, Schmuki P (2008) Chem Mater 20:3245–3247
245. Jha H, Hahn R, Schmuki P (2010) Electrochim Acta 55:8883–8887
246. Grigorescu S, Ungureanu C, Kirchgeorg R, Schmuki P, Demetrescu I (2013) Appl Surf Sci 270:190–196
247. Stoian AB, Vardaki M, Ionita D, Enachescu M, Prodana M, Brancoveanu O, Demetrescu I (2018) Ceram Intern 44:7026–7033
248. Wei W, Berger S, Shrestha N, Schmuki P (2010) J Electrochem Soc 157:C409–C413
249. Huang HH, Wu CP, Sun ZS, Lee TH (2013) Thin Solid Films 528:157–162
250. Sun YS, Liu JF, Wu CP, Huang HH (2015) J Alloy Compd 643:S124–S132

251. Li GZ, Zhao QM, Tang HP, Li G, Chi YD (2016) Comp Interf 23:223–230
252. Godoy Pérez DA, Junior AMJ, Asato GH, Lepretre JC, Roche V, Bolfarini C, Botta WJ (2019) J Alloy Compd 796:93–102
253. Hang R, Zong M, Bai L, Gao A, Liu Y, Zhang X, Huang X, Tang B, Chu PK (2016) Electrochem Commun 71:28–32
254. Hang R, Liu Y, Gao A, Zong M, Bai L, Zhang X, Huang X, Tang B, Chu PK (2017) Surf Coat Tech 321:136–145
255. Li T, Ding D, Li N (2019) Materials 12:1315(1–9)

Chapter 14
Nanostructures Obtained with Plasma Discharge Processes

14.1 Carbon Nanostructures: Nanotubes and Nanoonions

The discovery of the carbon nanotubes (CNTs) was related to the arc discharge process [1] formerly applied for the production of fullerenes (even though earlier and nearly forgotten observations of similar structures have also been documented [2, 3]). After an early expansion period of the field [4, 5], the arc discharge process was finely tuned for the production of nanotubes, too. Beside laser ablation and chemical vapour deposition, arc discharge remained one of the major production modes of CNTs due to its advantageous features.

Due to the popularity of the CNT research and the variety of the expected application field of the resulting carbon nanostructures, reviews focusing on various aspects of the field are available (general [6, 7], synthesis with any method [8, 9] or particularly with arc discharge [10–13], continuous production of nanotubes [14], growth mechanism of carbon nanotubes [15], solar cell applications [16]).

The basic structural properties of CNTs can be understood on the basis of the graphene sheet and the virtual roll-up of its segment cut with parallel edges to form single-wall carbon nanotubes (SWCNTs). This is shown in Fig. 14.1. Depending on the direction of the so-called chiral vector, one can obtain zig-zag, armchair and chiral nanotubes by folding a slice of the graphene sheet so that the carbon atoms at the initial and final points of the chiral vector overlap. While the structural character of a CNT depends on the direction of the chiral vector, its length determines the CNT diameter.

Multiwalled carbon nanotubes (MWCNTs, see Fig. 14.2) can be imagined as SWCNTs embedded into each other. For MWCNTs, the diameter difference between the neighbouring nanotubes is not arbitrary but it equals to that of the planar graphene sheets in graphite (0.34 nm), hence defining a regular spacing between the neighbouring nanotubes. The interaction between the curved graphene sheets in MWCNTs is an energetic stabilization factor of the resulting structure, even though the carbon atoms of a particular CNT in the MWCNT structure are not as regularly positioned

L. Péter, *Electrochemical Methods of Nanostructure Preparation*, Monographs in Electrochemistry, https://doi.org/10.1007/978-3-030-69117-2_14

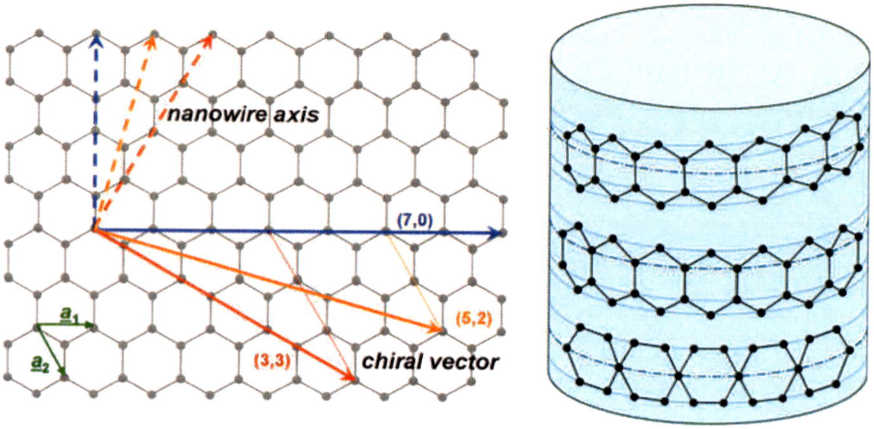

Fig. 14.1 Virtual formation of carbon nanotubes from slices of a graphene sheet. **Left**: A graphene sheet with the basal vectors of the lattice ($\underline{a}_1$, $\underline{a}_2$), chiral vectors (arrows with solid lines) and nanowire axes corresponding the chiral vectors (arrows with dashed line). Colours: zig-zag nanowire, chiral vector: $(m, 0)$—blue; armchair nanowire, chiral vector: (n, n)—red; chiral nanowire, chiral vector: (m, n), $m \neq n$ and $m, n \neq 0$—orange. **Right**: Hexagonal carbon ring arrangement on the virtual surface of a nanotobe. From top to down: chiral, zig-zag and armchair nanotube. Light grey dotted lines indicate the alignment of the carbon atoms in the neighbouring rings and thick dash-dot lines are parallel to the surface of the cylinder displayed (they coincide for the zig-zag and armchair configuration but are non-parallel for the chiral nanotube)

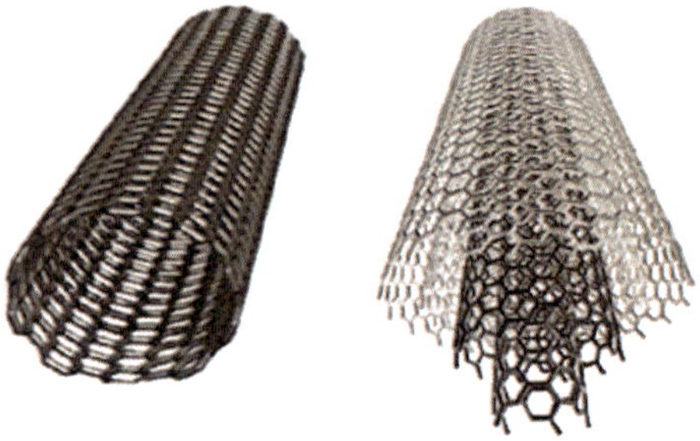

Fig. 14.2 Perspectivic image of a SWCNT of zig-zag structure (left) and a MWCNT with 3 tubes of zig-zag structure embedded into each other (right). Copyright (2011), Veena Choudhary and Anju Gupta. Originally published in [17] under Creative Common Licence 3.0. Available from: https://www.intechopen.com/books/carbon-nanotubes-polymer-nanocomposites/polymer-carbon-nanotube-nanocomposites

relative to the hexagonal carbon atom units in the neighbouring tube as in the graphite structure.

Carbon nanotubes can be open or closed. The end cap of the closed nanotubes is similar to a half fullerene molecule of the same diameter as that of the tube. Even if the nanotube itself is defect-free and composed of hexagonal rings only, the end curvature can be ensured with the incorporations of carbon pentagons. Therefore, closed SWCNTs can also be derived in another imaginary experiment as cutting a fullerene molecule into halves and packing hexagonal carbon atom units between these hemispherical caps.

Carbon nanoonions are multiwalled carbon nanostructured whose inner core is a fullerene-like nearly spherical molecule. Packing further carbon layers onto the core by sustaining the curvature with the required 0.34 nm spacing results in nanoonions. To obtain such objects, the proportion of the five-member rings should be decreased as the diameter gets larger.

Although the identification of various carbon nanostructures can be carried out on the basis of their physical (e.g., optical absorption) properties, the direct visualization of their structure is often decisive. This can be performed by TEM. A representative collection of TEM images of various carbon nanostructures obtained with arc discharge synthesis is shown in Fig. 14.3.

14.2 General Features of the Discharge Devices and Processes

The review of Aora and Sharma [12] yields a very comprehensive list of former publications on arc discharge-produced carbon nanostructures with detailed experimental conditions. As it can be seen from this collection, the common features of the synthesis devices and conditions do not provide a full transparency for this field, and minor differences in the experiments make the available information rather puzzled. Below the major routes of the synthesis methods are summarized only.

At high plasma current densities when the cathode is heated up by the impinging ions and hence, the electrons are thermally emitted from the cathode, we can speak about an *arc discharge*. Arc discharges are typically driven with d.c. excitation, even though a.c. arc plasmas can also be produced with frequencies not larger than a few tens of kHz. At 1 MHz and above, the electrons become the major carriers of the current in the plasma, and the motion of the ions can no longer follow the oscillation of the electric field *(radiofrequency* /RF/ plasma). Above the GHz frequency level (*microwave* /MW/ plasma), the heating mechanism depends on the density of the plasma, ranging from the surface-wave mode at high plasma density (and small penetration depth limited by the so-called skin effect) to resonator mode at small plasma density (and penetration depths much larger than the plasma dimension). Most RF and MW plasma sources are operated at specific frequencies assigned for industrial use in order to avoid interference with radio communication. Hence, the frequency used for RF and MW plasmas cannot be optimized for the specific goal of the research. The majority of the works to be cited below refers to arc discharges, and hence, this will be discussed below in detail.

Arc discharge devices work with electrodes positioned in close vicinity of each other. The electrodes can be momentarily contacted for the ignition of the discharge,

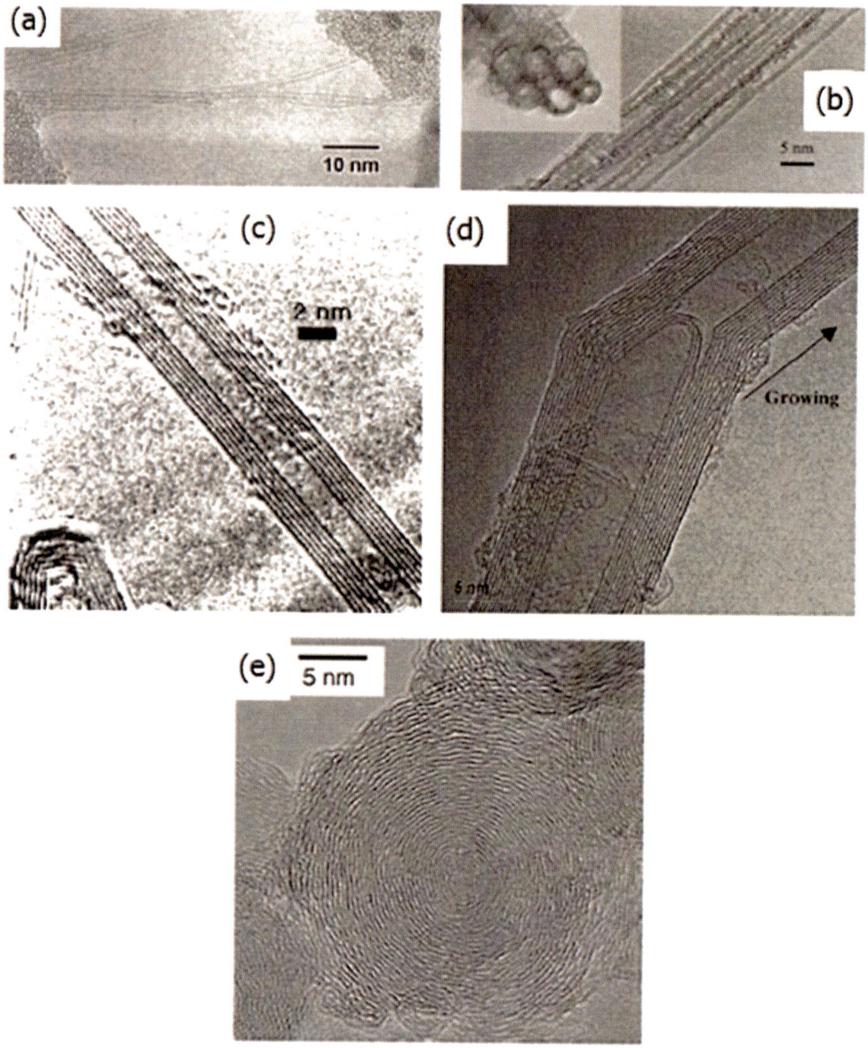

Fig. 14.3 TEM images of various arc discharge-produced carbon nanostructures. **a** Single-walled nanotubes. Reprinted from [18]. Copyright (2000), with permission from Elsevier. **b** Double-wall nanotubes with a cross-sectional image in the inset. Reprinted with permission from [19] Copyright (2003) American Chemical Society. **c** Multiwalled nanotubes with the typical end closure (left side, bottom of the image). Reprinted from [20]. Copyright (2000), with permission from Elsevier. **d** Multiwall nanotube with a partial closure of three inner tubes and an inclination at the defect site. Reprinted from [21]. Copyright (2004), with permission from Elsevier. **e** Nanoonion. Reprinted from [22]. Copyright (2006), with permission from Elsevier

and then they are kept separated typically with a gap of 0.25-3.0 mm between them. For the maintenance of the discharge process, a voltage of 18 to 30 V is necessary that leads to the flow of a very large current ranging from 30 to 200 A (the voltage tends to increase with the electrode distance and the current increases with the anode diameter under otherwise identical conditions) [4, 5, 18, 23–31]. The temperature in the core of the discharge is around 4000 K. In most cases, the discharge can be maintained for a few minutes. The arc discharge devices are equipped with a cooling system (water pipelines) and an electrode adjustment mechanism that precisely maintains a constant interelectrode gap which is the prerequisite for the discharge stability. Direct current discharges represent an overwhelming majority of the experiments. Although experiments with alternating current have also been reported [4, 32], this method could not lead to an improvement in either the yield or the product selectivity. Pulsed RF discharge is also mentioned occasionally [30, 33] where the maintenance of the discharge requires a voltage in the range of 3.5–5.5 kV.

The electrode materials are mostly graphite rods. The diameter of the anode is usually 6–8 mm, while the cathode is often larger and can be even a plate. When carbon is used as anode material, it is consumed due to the impact of the energetic electrons coming from the plasma, and the deposit forms at both the cathode and in various parts of the discharge chamber. A major difference as compared to the experimental setups making use of intercalation and upcoming graphene sheet detachment (discussed in Chap. 10.5) is that in the arc discharge experiment the carbon products with sp^2 hybridization form fully independently of the same bond structure in the anode since the high-temperature plasma leads to a nearly complete atomization. Therefore, the original graphene sheets are not conserved, and the final products do not form as a result of the roll-up of the original graphene sheets of the anode. This is particularly true for the experiments in which the carbon source is a gas, not the anode.

The electrodes are arranged as coaxially positioned rods in most of the setups reported (180° axis inclination). This is not the only possible electrode arrangement but configurations with an as low as 30° electrode inclination are also possible. This configuration was termed as arc jet method [34], referring to the fact that the plasma extends rather far from the interelectrode space. The inclined electrode configuration could be applied in both gas [35] phase and liquid-submerged discharges [36, 37].

In order to increase the stability of the discharge, a configuration with rotating one of the electrodes was developed. In accord with the concept of this arrangement [38], the rotation of the anode distributes the microdischarges uniformly along the anode surface, hence overwriting the impact of the lateral inhomogeneity of the electrodes, which leads to a stable plasma. Beside the improvement in the plasma stability, an increase of the product yield was also reported. Since the rotation could be increased up to 10^5 rpm, the primary location of the product deposition changed from the cathode to the cylindrical collector around the plasma.

Several attempts were published on the improvement of the arc discharge conditions by applying an external magnetic field [39–41], the maximum being as high as 10 T [42]. The impact of the external magnetic field manifests itself by the increase in the plasma density and the localization of the product deposition.

Although the electric field in the arc discharge is large, theoretical calculations show that it is not a major governing factor of the nanotube growth [43–45]. This is why nanotubes are often aligned parallel to the cathode surface. The majority of the nanotubes obtained with the arc discharge method is closed at the free end, and the preferred growth site is the anchored end of the nanotubes, which excludes the field-driven growth mechanism. This is particularly true when the nanotube is anchored at a catalyst particle.

The spatial distribution of the various carbon nanostructures is a persistent problem in the arc discharge experiments [1, 27, 33, 46–48]. On the cathode surface itself, the temperature distribution is a crucial factor. The nanotube formation zone exhibits a temperature >3000 K, and on parts of moderate temperature (<2500 K), the formation of non-tubular pyrolytic carbon is preferred [47]. While the major products are often collected from the deposit zone of the cathode surface, the carbon soot deposited in the peripheral parts of the discharge chamber exhibits a different product profile. Therefore, the regulation of the arc discharge process to obtain just one typical product has not been solved, although the application of various catalysts can lead to 80–90% purity in the sense of the preferred formation of single-walled or multiwalled nanotubes.

Regardless of the specific details of the carbon nanotube synthesis, purification is an indispensable step of the entire process. This is often an oxidative treatment in either oxygen-containing gas at moderate temperature (like 1000 K) or in hot oxidizing acid (like HNO_3). Under these conditions, the nanotubes are quite intact but other carbonaceous derivatives with a large number of terminal carbon atoms can be decomposed. A comparison of the SEM images before and after purification is shown in Fig. 14.4.

14.3 Arc Discharge Processes with Difference Gases

The first CNT synthesis was performed with argon [1]. Later, several works used rather helium gas of pressure ranging from 25 mbar to 1 bar [4, 18, 19, 23, 26–29, 39, 50, 51]. Helium proved to lead to faster nanotube growth, but the increase in He pressure up to 3 bar did not result is a yield increment as compared to atmospheric He pressure [4]. A theoretical calculation [46] underpins both the growth rate enhancement in He as compared to Ar by a factor of about three and the saturation of nanotube growth rate around 1 bar.

The application of helium and argon as noble gases sounds very logical to avoid the reaction with the carbon radicals that build up the nanotubes. Nevertheless, other gases have also been applied whose impact can be elucidated together with the other important experimental conditions such as the application of catalysts. One of these gases is hydrogen [24, 25]. When it is applied together with He for MWCNT production (no catalyst applied [52]), the larger H_2 pressure led to both higher anode consumption rate and larger deposit formation rate at the cathode. However, in the

Fig. 14.4 SEM images of low (**a**, **c**) and high (**b**, **d**) magnifications of as-received (**a**, **b**) and purified (**c**, **d**) carbon nanotubes. Reprinted from [49]. Copyright (2012), with permission from Elsevier

presence of a catalyst, H_2 can be associated with an increased preference to the formation of double-walled carbon nanotubes [53–56]. Nevertheless, it was not reported that hydrogen should cause a larger portion of hydrogenated nanotubes with sp^3-hybridized carbon atoms. This means that hydrogen can be considered rather a growth modifier than a reactant which is present in the final product. A similar experience was obtained for CO [57] whose impact was the increased selectivity for SWCNT formation.

Nitrogen gas (N_2) is a non-reactive gas under ambient conditions, but it is no longer true for the arc discharge processes. When the arc discharge process is conducted in the presence of nitrogen, its incorporation into the resulting structures is either assumed [21] or pronouncedly evidenced [58, 59]. Partly conflicting result was obtained for systems operated in virtually "open air" conditions [49, 60–62] in which the buffer gas was air of various pressure. In these studies, the presence of either nitrogen or oxygen in the product was not evidenced, nor could the oxidative atmosphere prevent the formation of the nanotubes. This indicates that the nanotube formation is much faster than the oxidation of carbon under the plasma conditions.

14.4 Plasma Discharge Setups Submerged into Various Liquids

The concept of liquid-submerged electrodes in the arc discharge process comes from the simplicity of the system not requiring a pressure- and atmosphere-regulated vacuum chamber. The possibility of continuous deposit preparation without the necessity of the chamber purification was also envisioned during the elaboration of this idea. This is supported by the sustainability of submerged discharges up to 1–2 h [37].

The first study of the liquid-submerged arc discharge electrode arrangement applied for CNT production applied liquid nitrogen [20]. Although the temperature gradient around the liquid-submerged discharge is much larger than in discharges applying buffer gases, the electrical parameters of the discharge were nearly identical to those applied in gas chambers [20, 31, 63, 64]. The similarity of the plasma conditions to the gas-based plasma is indicated by the fact that the product distribution, MWCNT vs. SWCNT, depends on the absence/presence of an appropriate catalyst, respectively. A beneficial aspect of this synthesis method was an enhanced surface purity of the nanotubes [20].

Arc discharge with electrodes submerged into water also became popular. Pure deionized water [22, 32, 65–70], hydrogen peroxide solution [71] as well as various metal salt solutions (LiCl [72], NaCl [73], iron-group metal sulphates [36]) proved to be suitable as discharge environment. Unlike in traditional electrochemical setups, the conductivity of the solutions applied has neither a role in the maintenance of the arc discharge, nor the ions of the electrolyte seem to influence the properties of the products.

Organic liquids have also been tested for media of liquid-submerged arc discharge experiments for producing carbon nanostructures. Examples range from ethanol [74, 75], ethylene glycol [76], various aromatic hydrocarbons [33, 77] to highly fluorinated ether derivatives [30]. By using organic liquids as the discharge medium, it is possible to achieve that the carbon source of the resulting nanostructure is not the anode but the organic liquid. A decisive experiment is when the arc discharge is ignited between non-carbonaceous electrodes [33, 75]. Another evidence for the incorporation of the fragments of the organic liquid used is that its constituents other than carbon (H, N) can also be detected in the products [33].

It was reported for various liquids that bubble formation around the discharge zone took place during the experiments. This means that the gas evolved plays a similar role in the maintenance of the discharge than the buffer gas in the gaseous environment. Due to the high density of the liquids as compared to the gases, no gas stream regulation is needed and the buffer gas is automatically refreshed around the liquid-submerged discharges, which contributes to the stability of the plasma formed. If the discharge is maintained between carbon electrodes under water [66], this gas is composed of hydrogen and carbon monoxide as a result of the reaction between carbon and water ($C + H_2O \rightleftarrows CO + H_2$. For hydrocarbon media, the buffer gas formed *in situ* was reported to be a mixture containing acetylene and hydrogen as the major components [77].

Liquid-submerged arrangements can be made suitable for the synthesis of both MWCNTs and SWCNTs, depending on the application of an appropriate catalyst (see the next subchapter). However, liquid discharge media proved to be particularly suitable for the production of nanoonions [22, 66] and other nanoparticles [30, 31]. This type of product can be collected from the surface of the liquid reservoir, unlike nanotubes that accumulate on the cathode.

14.5 Experimental Conditions Affecting the Type Distribution of Carbon Nanotubes

Carbon nanotubes obtained with arc discharge between carbon electrodes in the absence of any catalyst are typically MWCNTs. If an appropriate catalyst is applied, the preferred growth mode can be modified and single-wall or double-wall nanotubes can be grown. The widespread mode of the introduction of the catalysts is to drill a cavity in one of the carbon electrodes (mostly in the anode) that is filled up with the fine-grained catalyst mixed with carbon powder. During the anode erosion, the catalyst is also evaporated, and the nanotube growth proceeds at the condensed particles. Catalysts play a very important role in the regulation of the product profile of the arc discharge method since single-wall or double-wall nanotubes can be made dominant. Nevertheless, even catalysts can lead to a product of a preferred nanotube type but cannot significantly narrow the tube diameter distribution.

The first catalysts applied for the SWCNT synthesis were elemental 4d transition metals (Fe [5], Co [28, 78], Cu [79]). Later the Ni–Y-type mixed catalysts became nearly a standard [26, 29, 39, 51, 57, 80, 81] by using either a mixture of the two metals or their alloy. Molybdenum [82], tungsten [54], cerium [83] and holmium [84] were occasionally applied as a minor component in the catalyst, while calchogenide elements (S and Se) were identified as promoters increasing the nanotube yield [81]. While the Fe–Mo catalyst was first observed to promote the growth of double wall nanotubes [53], the catalysts applied for DWCNT growth are usually contain a number of elements (Ni, Co, Fe and S, either in elemental form or as metal sulphides) [19, 55, 56].

14.6 Arc Discharge Synthesis of Carbon-Encapsulated Nanostructures

Carbon nanostructures are prominent candidates to host various other nanostructures due to their cavity size of appropriate dimension. Although many *ex situ* methods are known to fill up carbon nanostructures (in particular, nanotubes [85]) with other materials, occasionally arc discharge synthesis can lead to nanostructures encapsulated in carbon. Such carbon-coated composite nanorods and nanoparticles containing a metal core are of great interest for their magnetic properties, electrochemical performance in electrodes of lithium ion batteries and controlled-release metal carriers. The relative intactness of the carbon-encapsulated metal originates from the low reactivity of the nearly perfect graphite layers covering them.

In some cases, the experimental setup for producing carbon-coated metal particles is nearly identical to those used for nanotube preparation with the exception that the metal content of the anode is higher than what is required to catalytic nanotube growth [25, 79, 86–88]. Since the discharge current is in the same range as for the catalytic preparation of the non-filled nanotubes or nanoonions, it is difficult to categorize the peculiar criteria that favour metal encapsulation instead of empty nanotube formation. In other cases, however, neither of the electrodes contained carbon, and the carbon content of the coating layer comes from the media surrounding the discharge, should it be either one of the components of the gas phase [89–91] or the liquid itself in which the discharge is generated [75]. The coating efficiency is generally higher for metals that do not form carbides. Representative TEM images of the metal-filled carbon nanostructures are presented in Fig. 14.5.

Both the diameter of the encapsulated nanoparticles and the thickness of the carbon coating layer vary in a wide range (and hence, so does their ratio). The particle diameter ranges from about 4 to 200 nm, while the lowest number of covering carbon layers is about three. The relationship between the experimental conditions and the characteristic length scales is not totally clear. The particle diameter seems to increase with the metal content of the anode [91]. The pressure of the carbon source gas has a dual impact on the particles since its increase leads to the enlargement of both the metal particles and their coating thickness [86, 89].

Ferromagnetic particles coated with carbon layers exhibit superparamagnetic behaviour since the carbon shield effectively hinders the exchange interaction between the electrons of the core metal particles [87, 90]. The carbon shield also prevents dipole-type particle interaction, which is important for applications. At the same time, the total magnetization of the nanoparticle assembly is consequently lower than that calculated merely on the basis of its metal content due the diamagnetic contribution of the carbon shield.

14.7 Non-carbonaceous Nanostructures Obtained by Plasma Synthesis

The plasma synthesis of non-carbonaceous nanomaterials became a wide and very diverse field for which it is difficult to offer a complete summary in the framework of the present book. A short highlight of the field is given below with examples ranging from doping of carbon through metal particle synthesis to ceramic particles.

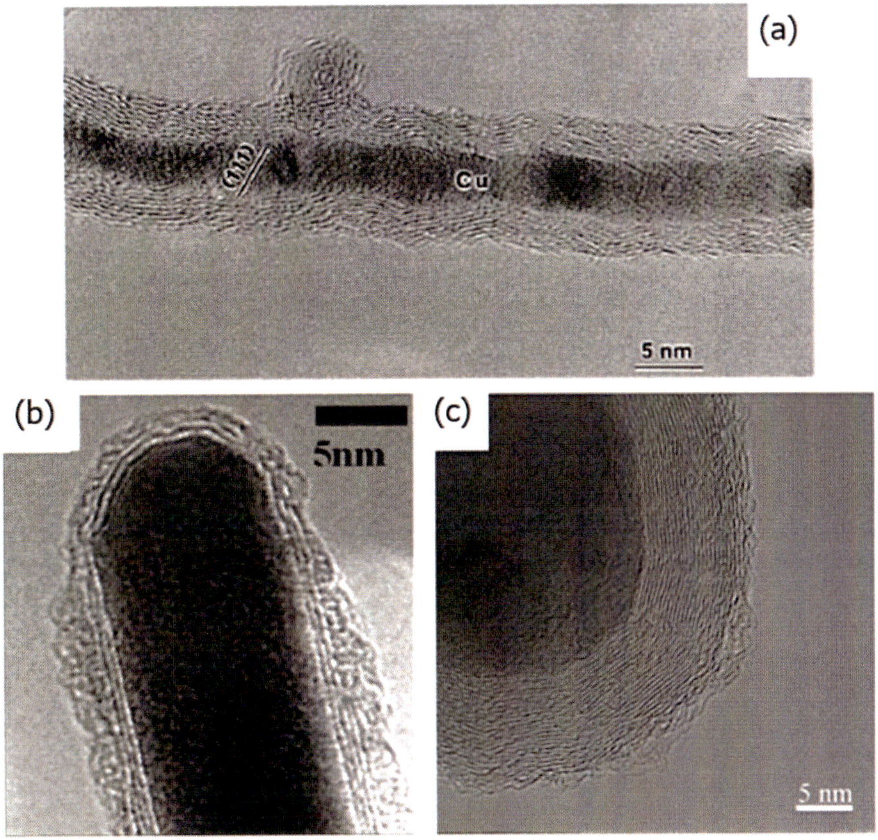

Fig. 14.5 Representative TEM images of various carbon-coated metal nanorods and nanoparticles. **a** Copper nanorod in which the carbon layers follow the surface undulation of the rod. Reprinted from [25], with the permission of AIP Publishing. **b** Nickel nanorod. Reprinted from [75]. Copyright (2014), with permission from Elsevier. **c** Iron nanoparticle. Reprinted from [86]. Copyright (2017), with permission from Elsevier. In Figs. (**b**) and (**c**), it is well visible that the highly conformal and regular inner carbon layers are followed by a less regularly graphitized outer layer

Doping of carbon nanotubes with boron and/or nitrogen was demonstrated soon after the discovery of the nanotube synthesis with arc discharge [58]. Since this first successful attempt, one principle of the B–C–N nanotube experiments remained unchanged; namely, the boron source material (either elemental boron or a boride compound) is added to the composite anode. The nitrogen source can be either N_2 as a buffer gas component or boron nitride as a part of the anode mixture. As it is shown in the literature summary given in Ref. [59], nanotubes with various B and N doping level could be achieved from ~ 1% of the doping element to BC_2N composition.

Similarly to doping processes, various stoichiometric carbide nanofibers and nanoparticles could by synthesized by plasma processes. For SiC nanorod formation, an anode with either Si [92] or SiC filling [50, 93, 94] proved to be appropriate. The

SiC nanofibers of 10–20 nm in diameter were highly crystalline and exhibiting an α- or β-SiC structure, occasionally covered with an SiO_2 shield. Niobium carbide nanoparticles [95] were obtained by using Nb electrodes and various carbohydrides as carbon source gas. The shape of the NbC nanoparticles of a few tens of nm in diameter could be regulated with the gas phase composition.

The preparation of silicon nanoparticles with arc discharge was successful by plasma etching of either the anode [96] or the cathode [97] under liquid. The discharge under nitrogen [96] resulted in Si nanoparticles, while the application of aqueous solution can lead to either Si particles covered with a thin oxide layer (when acidic solutions are applied) or SiO_2 (when the solution is alkaline) [97]. The crystallinity of the particles was the larger, the larger their size. The mean diameter of the Si nanoparticles obtained can be easily diminished with a chemical etching process in HF solution and particles with as low a 4 nm diameter can be obtained [96].

Etching of metal anodes leads to various nanoparticles. If the anode is a noble metal (Ag, Au, Pt), nanoparticles can be produced in discharges submerged into deionised water. The resulting nanoparticles were reported as being fully metallic [98–100] and they form a stable suspension that is often termed as nanofluid. This property of the particles gave rise to a speculation that the surface bonds may be influenced by the CO_2 dissolved in water and give the particle some charge, and ionic reaction products (like Ag^+) are often found in the solution without specifying the corresponding anion. A zeta potential sufficiently large for the suspension stabilization was also measured in some cases [98]. The mean particle diameter is between 6 and 25 nm.

Discharges in liquid result in noble metal nanoparticle formation if the precursor is a dissolved noble metal salt (Ag: $AgNO_3$, Au: $HAuCl_4$) and the electrodes used are non-reactive (Ti or W). Various plasma excitation mode (d.c. arc discharge [101], RF pulsed arc discharge [102] or MW plasma [103, 104]) are easily applicable for this goal since the effect in all cases is the fast evaporation of the solution in the near-plasma region. Although the mean diameter of the noble metal particles produced with any of the above-mentioned methods was below 20 nm, the lowest mean particle size (4–5 nm) was obtained by applying a reverse micelle solution of $HAuCl_4$ in which the micelles were reported to act as limited-volume nanoreactors [102]. This was evidenced by the experience that the increase in the water/surfactant ratio leads to an enhancement of the size of Au particles as the micelle size increases.

When the arc discharge method was used for the preparation of alloyed non-noble metallic particles, the precursor material was a master alloy comprised in the anode. Even though the experiments were conducted in a reducing atmosphere containing H_2, the resulting nanopowder exhibited a core-shell structure due to the oxidation of the particles upon removal from the discharge chamber. The post-synthesis oxidation is selective, and one of the alloy components is the preferred metal forming the outer oxide shell (Sn in Fe–Sn [105] and Cu in Co–Cu [106] particles).

The last group of materials can be classified as oxide nanorods and nanoparticles. The common feature of the synthesis method that the media was of oxidizing character, either because of the oxygen present in the gas phase [107] or water when the plasma was submerged into the liquid phase [97, 100, 103, 108]. The precursor material can be the elemental form of the metal to be oxidized (Ta [108], W [100],

Si [97], Zn and Mg [103]) or its compound than is catalytically oxidized (GaN to yield Ga_2O_3 [107]). In some cases, the setup used for oxide nanoparticle syntheses was fully analogous to those found to be appropriate to produce metallic nanostructures, only the metals more prone to be oxidized led to the formation of the corresponding oxide [97, 100, 103]. Gallium oxide produced during the catalytic oxidation of an oxygen-containing buffer gas was obtained as highly crystalline nanorods [107], while liquid-submerged setups were appropriate to yield oxide nanoparticles [97, 100, 103, 108].

References

1. Ijima S (1991) Nature 354:56–58
2. Radushkevich LV, Lukyanovich VM (1952) Zurn Fisic Chim 26:88–95
3. Oberlin A, Endo M, Koyama T (1976) J Cryst Growth 32:335–349
4. Ebbesen TW, Ajayan PM (1992) Nature 358:220–222
5. Ijima S, Ichihashi T (1993) Nature 363:603–605
6. Dresselhaus MS, Dresselhaus G, Avouris P (2001) Carbon nanotubes: Synthesis, structure, properties, and applications. Springer, Berlin
7. Dervishi E, Li Z, Xu Y, Saini V, Biris AR, Lupu D, Biris AS (2009) Particul Sci Technol 27:107–125
8. Baddour CE, Briens C (2005) Int J Chem React Eng 3:R3(1–22)
9. Szabó A, Perri C, Csató A, Giordano G, Vuono D, Nagy JB (2010) Materials 3:3092–3140
10. Ando Y, Zhao X (2006) New Diam Front C Tec 16:123–137
11. Ando Y (2010) J Nanosci Nanotechnol 10:3726–3738
12. Arora N, Sharma NN (2014) Diam Relat Mater 50:135–150
13. Su Y, Zhang Y (2015) Carbon 83:90–99
14. Ying LS, Salleh MAM, Yusoff HM, Rashid SBA, Razak JA (2011) J Ind Eng Chem 17:367–376
15. Harris PJF (2007) Carbon 45:229–239
16. Kumar U, Sikarwar S, Sonker RK, Yadav BC (2016) J Inorg Organomet Polym 26:1231–1242
17. Choudhary V, Gupta A (2011) Polymer/Carbon nanotube nanocomposites. In: Yellampalli S (ed) Carbon nanotubes—polymer nanocomposites. InTech Europe, Rijeka
18. Shi Z, Lian Y, Liao FH, Zhou X, Gu Z, Zhang Y, Iijima S, Li H, Yue KT, Zhang SL (2000) J Phys Chem Solid 61:1031–1036
19. Huang H, Kajiura H, Tsutsui S, Murakami Y, Ata M (2003) J Phys Chem B 107:8794–8798
20. Ishigami M, Cumings J, Zettl A, Chen S (2000) Chem Phys Lett 319:457–459
21. Cui S, Scharff P, Siegmund C, Schneider D, Risch K, Klötzer S, Spiess L, Romanus H, Schawohl J (2004) Carbon 42:931–939
22. Imasaka K, Kanatake Y, Ohshiro Y, Suehiro J, Hara M (2006) Thin Solid Films 506–507:250–254
23. Chang BH, Xie SS, Zhou WY, Qian LX, Pan ZW, Mao JM, Li WZ (1998) J Mat Sci Lett 17:1015–1017
24. Wang XK, Lin XW, Dravid VP, Ketterson JB, Chang RPH (1995) Appl Phys Lett 66:2430–2432
25. Setlur AA, Lauerhaas JM, Dai JY, Chang RPH (1996) Appl Phys Lett 69:345–347
26. Ando, Zhao X (2000) Mol Crys Liq Cryst Sci Technol A. 340:707–712
27. Tang DS, Zhou WY, Ci LJ, Yan XQ, Yuan HJ, Zhou ZP, Liang YX, Liu DF, Liu W (2002) Chinese Phys 11:496–501
28. Ajayan PM, Lambert JM, Bemier P, Barbedette L, Colliex C, Planeix JM (1993) Chem Phys Lett 215:509–517

29. Journet C, Maser WM, Bernier P, Loiseau A, Lamy de la Chapelle M, Lefrant S, Deniard P, Leek R, Fischer JE (1997) Nature 388:756–758
30. Hosoi K, Kuroda S, Kubota H (2008) Chem Lett 37:54–55
31. Charinpanitkul T, Kanjanaprapakul K, Leelaviwat N, Kurukitkoson N, Kim KS (2010) J Ind Eng Chem 16:912–917
32. Hsin YL, Hwang KC, Chen FR, Kai JJ (2001) Adv Mater 113:830–833
33. Leel H, Wada Y, Kaneko A, Li OL, Ishizaki T (2018) Jap J Appl Phys 57:0102BD(1–6)
34. Ando Y, Zhao X, Hirahara K, Suenaga K, Bandow S, Iijima S (2000) Chem Phys Lett 323:580–585
35. Choi SI, Nam JS, Lee CM, Choi SS, Kim JI, Park JM, Hong SH (2006) Curr Appl Phys 6:224–229
36. Zhu HW, Li XS, Jiang B, Xu CL, Zhu YF, Wu DH, Chen XH (2002) Chem Phys Lett 366:664–669
37. Horváth ZE, Kertész K, Pethő L, Koós AA, Tapasztó L, Vértesy Z, Osváth Z, Darabont A, Nemes-Incze P, Sárközi Z, Biró LP (2006) Curr Appl Phys 6:135–140
38. Lee SJ, Baik HK, Yoo JE, Han JH (2002) Diam Relat Mater 11:914–917
39. Keidar M, Levchenko I, Arbel T, Alexander M, Waas AM, Ostrikov K (2008) Appl Phys Lett 92:043129
40. Keidar M, Levchenko I, Arbel T, Alexander M, Waas AM, Ostrikov K (2008) J Appl Phys 103:094318 (1–8)
41. Li J, Volotskova O, Shashurin A, Keidar M (2011) J Nanosci Nanotechnol 11:10047–10052
42. Yokomichi H, Ichihara M, Kishimoto N (2014) Jap J App Phys 53:02BD05(1–4)
43. Maiti A, Brabec CJ, Roland C, Bernholc J (1995) Phys Rev B 52:14850–14860
44. Keidar M, Waas AM (2004) Nanotechnology 15:1571–1575
45. Keidar M, Waas AM, Raitses Y, Waldorff EI (2006) J Nanosci Nanotechnol 6:1309–1314
46. Hinkov I, Farhat S, Scott CD (2005) Carbon 43:2453–2462
47. Yeh YW, Raitses Y, Yao N (2016) Carbon 105:490–495
48. Fang X, Shashurin A, Teel G, Keidar M (2016) Carbon 107:273–280
49. Zhao J, Wei L, Yang Z, Zhang Y (2012) Physica E 44:1639–1643
50. Li YB, Xie SS, Zou XP, Tang DS, Liu ZQ, Zhou WY, Wang G (2001) J Crys Growth 223:125–128
51. Li H, Guan L, Shi Z, Gu Z (2004) J Phys Chem B 108:4573–4575
52. Durbach SH, Witcomb MJ, Coville NJ (2005) Fuller Nanotub Car N 13:155–169
53. Wang H, Li Z, Ghosh K, Maruyama T, Inoue S, Ando Y (2010) Carbon 48:2882–2889
54. Fang L, Sheng L, An K, Yu L, Ren W, Ando Y, Zhao X (2013) Physica E 50:116–121
55. Hutchison JL, Kiselev NA, Krinichnaya EP, Krestininc AV, Loutfy RO, Morawsky AP, Muradyan VE, Obraztsova ED, Sloan J, Terekhov SV, Zakharov DN (2001) Carbon 39:761–770
56. Saito Y, Nakahira T, Uemura S (2003) J Phys Chem B 107:931–934
57. Su Y, Yang Z, Wei H, Kong ESW, Zhang Y (2011) Appl Surf Sci 257:3123–3127
58. Stephan O, Ajayan PM, Colliex C, Redlich P, Lambert JM, Bernier P, Lefin P (1994) Science 266:1683–1685
59. Belgacem AB, Hinkov I, Yahia SB, Brinza O, Farhat S (2016) Mater Today Commun 8:183–195
60. Paladugu MC, Manesh K, Kesavan Nair P, Haridoss P (2005) J Nanosci Nanotechnol 5:747–752
61. Zhao J, Zhang J, Su Y, Yang Z, Wei L, Zhang Y (2012) J Mater Sci 47:6535–6541
62. Su Y, Zhou P, Zhao J, Yang Z, Zhang Y (2013) Mater Res Bull 48:3232–3235
63. Sano N, Nakano J, Kanki T (2004) Carbon 42:667–691
64. Scalese S, Scuderi V, Bagiante S, Gibilisco S, Faraci G, Piluso N, La Via F, Privitera V (2012) Physica E 44:1005–1008
65. Sano N, Wang H, Chhowalla M, Alexandrou I, Amaratunga GAJ (2001) Nature 414:506–507
66. Sano N, Wang H, Alexandrou I, Chhowalla M, Teo KBK, Amaratunga GAJ, Iimura K (2002) J Appl Phys 92:2783–2788

67. Yao Y, Wang R, Wei DZ, Du D, Liang J (2004) Nanotechnology 15:555–558
68. Bera D, Johnston G, Heinrich H, Seal S (2006) Nanotechnology 17:1722–1730
69. Yousef S, Khattab A, Osman TA, Zaki M (2013) J Nanomat 392126(1–9)
70. Lange H, Sioda M, Huczko A, Zhu YQ, Kroto HW, Walton DRM (2003) Carbon 41:1617–1623
71. Contini V, Mancini R, Marazzi R, Mirabile Gattia D, Vittori Antisari M (2007) Phil Mag 87:1123–1137
72. Dehghani Kiadehi A, Jahanshahi M, Mozdianfard MR, Vakili-Nezhaad GHR, Jabari Seresht R (2011) J Exp Nanosci 6:432–440
73. Wang SD, Chang MH, Lan KMD, Wu CC, Cheng JJ, Chang HK (2005) Carbon 43:1778–1814
74. Mao B, Kang Z, Wang E, Tian C, Wang C, Lan Y (2006) J Nanosci Nanotechnol 6:1392–1395
75. Sagara T, Kurumi S, Suzuki K (2014) Appl Surf Sci 292:39–43
76. Liu W, Miao Y, Meng Q (2012) Integr Ferroelectr 138:77–82
77. Ryzhkov VA (2002) Physica B 323:324–326
78. Bethune DS, Klang CH, de Vries MS, Gorman G, Savoy R, Vazquez J, Beyers R (1993) Nature 363:605–607
79. Lin X, Wang XK, Dravid VP, Chang RPH, Ketterson JB (1994) Appl Phys Lett 64:181–183
80. Wang M, Wang X, Li Z, Liu Z, He P (2006) Mater Chem Phys 97:243–246
81. Huang L, Wu B, Chen J, Xue Y, Liu Y, Kajiura H, Li Y (2011) Carbon 49:4792–4800
82. Jahanshahi M, Raoof J, Seresht RJ (2009) J Exp Nanosci 4:331–339
83. Zhu HW, Jiang B, Xu CL, Wu DH (2003) J Phys Chem B 107:6514–6518
84. Yao M, Liu B, Zou Y, Wang L, Li D, Cui T, Zou G, Sundqvist B (2005) Carbon 43:2894–2901
85. Poudel YR, Li W (2018) Mater Today Phys 7:7–34
86. Hua R, Furukawa T, Wang X, Nagatsu M (2017) Appl Surf Sci 416:731–741
87. Ang KH, Alexandrou I, Mathur ND, Amaratunga GAJ, Haq S (2004) Nanotechnology 15:520–524
88. Fang X, Cheng X, Zhang Y, Zhang LG, Keidar M (2018) J Coll Interf Sci 509:414–421
89. Shinde KP, Ranot M, Choi CJ, Kim HS, Chung KC (2017) AIP Adv 7:075013(1–7)
90. Sun X (2003) J Dispersion Sci Technol 24:557–567
91. Chiu CC, Lo JC, Teng MH (2012) Diam Relat Mater 24:179–183
92. Seeger T, Kohler-Redlich P, Rühle M (2000) Adv Mater 12:279–282
93. Zhu YQ, Kroto HW, Walton DRM, Lange H, Huczko A (2002) Chem Phys Lett 365:457–463
94. Li Y, Xie S, Wei B, Lian G, Zhou W, Tang D, Zou X, Liu Z, Wang G (2001) Solid State Commun 119:51–53
95. Grove DE, Gupta U, Castleman AW Jr (2010) Langmuir 26:16517–16521
96. Kobayashi M, Liu SM, Sato S, Yao H, Kimura K (2006) Jap J Appl Phys 45:6146–6152
97. Saito G, Sakaguchi N (2015) Nanotechnology 26:235602(1–8)
98. Lung JK, Huang JC, Tien DC, Liao CY, Tseng KH, Tsung TT, Kao WS, Tsai TH, Jwo CS, Lin HM, Stobinski L (2007) J Alloy Compd 434–435:655–658
99. Tien DC, Chen LC, Thai NV, Ashraf S (2010) J Nanomat 634757(1–9)
100. Hattori Y, Nomura S, Mukasa S, Toyota H, Inoue T, Usui T (2013) J Alloy Compd 578:148–152
101. Ashkarran AA (2010) Curr Appl Phys 10:1442–1447
102. Hieda J, Saito N, Takai O (2008) Surf Coat Technol 202:5343–5346
103. Hattoria Y, Mukasa S, Toyota H, Inoue T, Nomura S (2011) Mat Chem Phys 131:425–430
104. Sato S, Mori K, Ariyada O, Atsushi H, Yonezawa T (2011) Surf Coat Technol 206:955–958
105. Lei JP, Dong XL, Zhu XG, Lei MK, Huang H, Zhang XF, Lu B, Park WJ, Chung HS (2007) Intermetallics 15:1589–1594
106. Yang ZQ, You CY, He LL (2006) J Alloy Compd 423:128–131
107. Choi YC, Kim WS, Park YS, Lee SM, Bae DJ, Lee YH, Park GS, Choi WB, Lee NS, Kim JM (2000) Adv Mater 12:746–750
108. Delaportas D, Svarnas P, Alexandrou I (2010) J Electrochem Soc 157:K138–K143

Index

L. Péter, *Electrochemical Methods of Nanostructure Preparation*, Monographs in Electrochemistry, https://doi.org/10.1007/978-3-030-69117-2

D

S

T